THEORY OF
ORLICZ SPACES

MONOGRAPHS AND TEXTBOOKS IN
PURE AND APPLIED MATHEMATICS

1. *K. Yano,* Integral Formulas in Riemannian Geometry (1970)*(out of print)*
2. *S. Kobayashi,* Hyperbolic Manifolds and Holomorphic Mappings (1970) *(out of print)*
3. *V. S. Vladimirov,* Equations of Mathematical Physics (A. Jeffrey, editor A. Littlewood, translator) (1970) *(out of print)*
4. *B. N. Pshenichnyi,* Necessary Conditions for an Extremum (L. Neustadt, translation editor; K. Makowski, translator) (1971)
5. *L. Narici, E. Beckenstein, and G. Bachman,* Functional Analysis and Valuation Theory (1971)
6. *S. S. Passman,* Infinite Group Rings (1971)
7. *L. Dornhoff,* Group Representation Theory (in two parts). Part A: Ordinary Representation Theory. Part B: Modular Representation Theory (1971, 1972)
8. *W. Boothby and G. L. Weiss (eds.),* Symmetric Spaces: Short Courses Presented at Washington University (1972)
9. *Y. Matsushima,* Differentiable Manifolds (E. T. Kobayashi, translator) (1972)
10. *L. E. Ward, Jr.,* Topology: An Outline for a First Course (1972) *(out of print)*
11. *A. Babakhanian,* Cohomological Methods in Group Theory (1972)
12. *R. Gilmer,* Multiplicative Ideal Theory (1972)
13. *J. Yeh,* Stochastic Processes and the Wiener Integral (1973) *(out of print)*
14. *J. Barros-Neto,* Introduction to the Theory of Distributions (1973) *(out of print)*
15. *R. Larsen,* Functional Analysis: An Introduction (1973) *(out of print)*
16. *K. Yano and S. Ishihara,* Tangent and Cotangent Bundles: Differential Geometry (1973) *(out of print)*
17. *C. Procesi,* Rings with Polynomial Identities (1973)
18. *R. Hermann,* Geometry, Physics, and Systems (1973)
19. *N. R. Wallach,* Harmonic Analysis on Homogeneous Spaces (1973) *(out of print)*
20. *J. Dieudonné,* Introduction to the Theory of Formal Groups (1973)
21. *I. Vaisman,* Cohomology and Differential Forms (1973)
22. *B. -Y. Chen,* Geometry of Submanifolds (1973)
23. *M. Marcus,* Finite Dimensional Multilinear Algebra (in two parts) (1973, 1975)
24. *R. Larsen,* Banach Algebras: An Introduction (1973)
25. *R. O. Kujala and A. L. Vitter (eds.),* Value Distribution Theory: Part A; Part B: Deficit and Bezout Estimates by Wilhelm Stoll (1973)
26. *K. B. Stolarsky,* Algebraic Numbers and Diophantine Approximation (1974)
27. *A. R. Magid,* The Separable Galois Theory of Commutative Rings (1974)
28. *B. R. McDonald,* Finite Rings with Identity (1974)
29. *J. Satake,* Linear Algebra (S. Koh, T. A. Akiba, and S. Ihara, translators) (1975)

30. *J. S. Golan,* Localization of Noncommutative Rings (1975)
31. *G. Klambauer,* Mathematical Analysis (1975)
32. *M. K. Agoston,* Algebraic Topology: A First Course (1976)
33. *K. R. Goodearl,* Ring Theory: Nonsingular Rings and Modules (1976)
34. *L. E. Mansfield,* Linear Algebra with Geometric Applications: Selected Topics (1976)
35. *N. J. Pullman,* Matrix Theory and Its Applications (1976)
36. *B. R. McDonald,* Geometric Algebra Over Local Rings (1976)
37. *C. W. Groetsch,* Generalized Inverses of Linear Operators: Representation and Approximation (1977)
38. *J. E. Kuczkowski and J. L. Gersting,* Abstract Algebra: A First Look (1977)
39. *C. O. Christenson and W. L. Voxman,* Aspects of Topology (1977)
40. *M. Nagata,* Field Theory (1977)
41. *R. L. Long,* Algebraic Number Theory (1977)
42. *W. F. Pfeffer,* Integrals and Measures (1977)
43. *R. L. Wheeden and A. Zygmund,* Measure and Integral: An Introduction to Real Analysis (1977)
44. *J. H. Curtiss,* Introduction to Functions of a Complex Variable (1978)
45. *K. Hrbacek and T. Jech,* Introduction to Set Theory (1978)
46. *W. S. Massey,* Homology and Cohomology Theory (1978)
47. *M. Marcus,* Introduction to Modern Algebra (1978)
48. *E. C. Young,* Vector and Tensor Analysis (1978)
49. *S. B. Nadler, Jr.,* Hyperspaces of Sets (1978)
50. *S. K. Segal,* Topics in Group Rings (1978)
51. *A. C. M. van Rooij,* Non-Archimedean Functional Analysis (1978)
52. *L. Corwin and R. Szczarba,* Calculus in Vector Spaces (1979)
53. *C. Sadosky,* Interpolation of Operators and Singular Integrals: An Introduction to Harmonic Analysis (1979)
54. *J. Cronin,* Differential Equations: Introduction and Quantitative Theory (1980)
55. *C. W. Groetsch,* Elements of Applicable Functional Analysis (1980)
56. *I. Vaisman,* Foundations of Three-Dimensional Euclidean Geometry (1980)
57. *H. I. Freedman,* Deterministic Mathematical Models in Population Ecology (1980)
58. *S. B. Chae,* Lebesgue Integration (1980)
59. *C. S. Rees, S. M. Shah, and C. V. Stanojević,* Theory and Applications of Fourier Analysis (1981)
60. *L. Nachbin,* Introduction to Functional Analysis: Banach Spaces and Differential Calculus (R. M. Aron, translator) (1981)
61. *G. Orzech and M. Orzech,* Plane Algebraic Curves: An Introduction Via Valuations (1981)
62. *R. Johnsonbaugh and W. E. Pfaffenberger,* Foundations of Mathematical Analysis (1981)
63. *W. L. Voxman and R. H. Goetschel,* Advanced Calculus: An Introduction to Modern Analysis (1981)
64. *L. J. Corwin and R. H. Szcarba,* Multivariable Calculus (1982)
65. *V. I. Istrătescu,* Introduction to Linear Operator Theory (1981)
66. *R. D. Järvinen,* Finite and Infinite Dimensional Linear Spaces: A Comparative Study in Algebraic and Analytic Settings (1981)

67. *J. K. Beem and P. E. Ehrlich*, Global Lorentzian Geometry (1981)
68. *D. L. Armacost*, The Structure of Locally Compact Abelian Groups (1981)
69. *J. W. Brewer and M. K. Smith, eds.*, Emmy Noether: A Tribute to Her Life and Work (1981)
70. *K. H. Kim*, Boolean Matrix Theory and Applications (1982)
71. *T. W. Wieting*, The Mathematical Theory of Chromatic Plane Ornaments (1982)
72. *D. B. Gauld*, Differential Topology: An Introduction (1982)
73. *R. L. Faber*, Foundations of Euclidean and Non-Euclidean Geometry (1983)
74. *M. Carmeli*, Statistical Theory and Random Matrices (1983)
75. *J. H. Carruth, J. A. Hildebrant, and R. J. Koch*, The Theory of Topological Semigroups (1983)
76. *R. L. Faber*, Differential Geometry and Relativity Theory: An Introduction (1983)
77. *S. Barnett*, Polynomials and Linear Control Systems (1983)
78. *G. Karpilovsky*, Commutative Group Algebras (1983)
79. *F. Van Oystaeyen and A. Verschoren*, Relative Invariants of Rings: The Commutative Theory (1983)
80. *I. Vaisman*, A First Course in Differential Geometry (1984)
81. *G. W. Swan*, Applications of Optimal Control Theory in Biomedicine (1984)
82. *T. Petrie and J. D. Randall*, Transformation Groups on Manifolds (1984)
83. *K. Goebel and S. Reich*, Uniform Convexity, Hyperbolic Geometry, and Nonexpansive Mappings (1984)
84. *T. Albu and C. Năstăsescu*, Relative Finiteness in Module Theory (1984)
85. *K. Hrbacek and T. Jech*, Introduction to Set Theory, Second Edition, Revised and Expanded (1984)
86. *F. Van Oystaeyen and A. Verschoren*, Relative Invariants of Rings: The Noncommutative Theory (1984)
87. *B. R. McDonald*, Linear Algebra Over Commutative Rings (1984)
88. *M. Namba*, Geometry of Projective Algebraic Curves (1984)
89. *G. F. Webb*, Theory of Nonlinear Age-Dependent Population Dynamics (1985)
90. *M. R. Bremner, R. V. Moody, and J. Patera*, Tables of Dominant Weight Multiplicities for Representations of Simple Lie Algebras (1985)
91. *A. E. Fekete*, Real Linear Algebra (1985)
92. *S. B. Chae*, Holomorphy and Calculus in Normed Spaces (1985)
93. *A. J. Jerri*, Introduction to Integral Equations with Applications (1985)
94. *G. Karpilovsky*, Projective Representations of Finite Groups (1985)
95. *L. Narici and E. Beckenstein*, Topological Vector Spaces (1985)
96. *J. Weeks*, The Shape of Space: How to Visualize Surfaces and Three-Dimensional Manifolds (1985)
97. *P. R. Gribik and K. O. Kortanek*, Extremal Methods of Operations Research (1985)
98. *J.-A. Chao and W. A. Woyczynski, eds.*, Probability Theory and Harmonic Analysis (1986)
99. *G. D. Crown, M. H. Fenrick, and R. J. Valenza*, Abstract Algebra (1986)
100. *J. H. Carruth, J. A. Hildebrant, and R. J. Koch*, The Theory of Topological Semigroups, Volume 2 (1986)

101. *R. S. Doran and V. A. Belfi*, Characterizations of C*-Algebras: The Gelfand-Naimark Theorems (1986)

102. *M. W. Jeter*, Mathematical Programming: An Introduction to Optimization (1986)

103. *M. Altman*, A Unified Theory of Nonlinear Operator and Evolution Equations with Applications: A New Approach to Nonlinear Partial Differential Equations (1986)

104. *A. Verschoren*, Relative Invariants of Sheaves (1987)

105. *R. A. Usmani*, Applied Linear Algebra (1987)

106. *P. Blass and J. Lang*, Zariski Surfaces and Differential Equations in Characteristic p > 0 (1987)

107. *J. A. Reneke, R. E. Fennell, and R. B. Minton*. Structured Hereditary Systems (1987)

108. *H. Busemann and B. B. Phadke*, Spaces with Distinguished Geodesics (1987)

109. *R. Harte*, Invertibility and Singularity for Bounded Linear Operators (1988).

110. *G. S. Ladde, V. Lakshmikantham, and B. G. Zhang*, Oscillation Theory of Differential Equations with Deviating Arguments (1987)

111. *L. Dudkin, I. Rabinovich, and I. Vakhutinsky*, Iterative Aggregation Theory: Mathematical Methods of Coordinating Detailed and Aggregate Problems in Large Control Systems (1987)

112. *T. Okubo*, Differential Geometry (1987)

113. *D. L. Stancl and M. L. Stancl*, Real Analysis with Point-Set Topology (1987)

114. *T. C. Gard*, Introduction to Stochastic Differential Equations (1988)

115. *S. S. Abhyankar*, Enumerative Combinatorics of Young Tableaux (1988)

116. *H. Strade and R. Farnsteiner*, Modular Lie Algebras and Their Representations (1988)

117. *J. A. Huckaba*, Commutative Rings with Zero Divisors (1988)

118. *W. D. Wallis*, Combinatorial Designs (1988)

119. *W. Więsław*, Topological Fields (1988)

120. *G. Karpilovsky*, Field Theory: Classical Foundations and Multiplicative Groups (1988)

121. *S. Caenepeel and F. Van Oystaeyen*, Brauer Groups and the Cohomology of Graded Rings (1989)

122. *W. Kozlowski*, Modular Function Spaces (1988)

123. *E. Lowen-Colebunders*, Function Classes of Cauchy Continuous Maps (1989)

124. *M. Pavel*, Fundamentals of Pattern Recognition (1989)

125. *V. Lakshmikantham, S. Leela, and A. A. Martynyuk*, Stability Analysis of Nonlinear Systems (1989)

126. *R. Sivaramakrishnan*, The Classical Theory of Arithmetic Functions (1989)

127. *N. A. Watson*, Parabolic Equations on an Infinite Strip (1989)

128. *K. J. Hastings*, Introduction to the Mathematics of Operations Research (1989)

129. *B. Fine*, Algebraic Theory of the Bianchi Groups (1989)

130. *D. N. Dikranjan, I. R. Prodanov, and L. N. Stoyanov*, Topological Groups: Characters, Dualities, and Minimal Group Topologies (1989)

Other Volumes in Preparation

THEORY OF ORLICZ SPACES

M. M. Rao
University of California
Riverside, California

Z. D. Ren
Xiangtan University
Hunan, People's Republic of China

Marcel Dekker, Inc.   **New York • Basel • Hong Kong**

SEP/AE
mATH

ISBN 0-8247-8478-2

This book is printed on acid-free paper.

MARCEL DEKKER, INC.
270 Madison Avenue, New York, New York 10016

Current printing (last digit):
10 9 8 7 6 5 4 3 2 1

PRINTED IN THE UNITED STATES OF AMERICA

Preface

One of the real attractions of Orlicz spaces is that the subject is sufficiently concrete and yet the spaces have fine structure of importance for applications. Moreover the area is quite accessible for young people leading them to gain sophistication in mathematical analysis in a relatively short time during their graduate studies. It also helps widen the applicational potential for professional mathematicians. These features, among others, make the subject particularly interesting. This circumstance motivated us to undertake a preparation of the present monograph containing a detailed and general exposition of these function spaces. Let us explain the perspective and then the format employed together with the coverage of the work.

After the appearance of A. Zygmund's monograph (first in 1935, and more in the 1959 edition) containing the elements of Orlicz spaces, and then in A. C. Zaanen's volume (1953), mostly as adjuncts or illustrations of their main treatments, the first systematic account of these spaces was given in the treatise by M. A. Krasnosel'skii and Yu. B. Rutickii (1958). In their pioneering treatment, the latter authors restricted the spaces to be on a bounded Lebesgue space in $I\!R^n$, and also to nice Young (or N-) functions. Within this framework, they gave a penetrating analysis of these spaces and noted a number of open problems. Since then the theory has progressed in several directions and most of the above problems have been solved. Our treatment includes solutions of essentially all these questions, viewed from a more

iii

general point of view. We base the work on abstract measure spaces and Young functions without restrictions and specializations are included to indicate some possible refinements. Here is a brief description of our treatment.

The first seven chapters contain the fundamental theory of Orlicz spaces that everyone would like to know. This details in an updated and extended form most of the subject covered in the Krasnosel'skii-Rutickii volume, including solutions of the earlier noted open problems. The last three chapters, on the other hand, deal with the subject showing the many different directions along which the basic theory, motivated by current research needs, is progressing. To entice the newcomers to this area, we have included here at least one nontrivial result in each case and indicated many other possibilities.

In presenting the subject, there are two viewpoints. Since an Orlicz space is a Banach space, one can specialize various aspects of the general theory of the latter and obtain conditions for the particular case at hand. On the other hand, one may consider the structure of Orlicz spaces and discover their special properties that can lead to generalizations of the abstract theory. Although these are somewhat overlapping properties, we have generally leaned to the second point of view, while keeping the motivation and important applications in sight. However, we also employed some abstract results on occasion when it appeared to clarify and simplify the presentation.

Here we introduce the subject through the classical de la Vallée Poussin criterion of uniform integrability of functions as a motivating step and then classify various types of Young functions in the first two chapters. The third chapter is devoted to the fundamental properties of Orlicz spaces on general measure spaces, from the Lebesgue L^p-space point of view. Topological structures and a complete characterization of the adjoint spaces of Orlicz spaces together with their singular functionals are given in Chapter IV. We discuss compact sets, embedding theorems and products of Orlicz spaces in Chapter V. The next two chapters contain finer aspects of the spaces that are fundamental in applications. These include a substantial account of interpolation theory of (sub-) linear operators, linear integral operators, factorization theorems as well as applications to packing and fractional integration. Also (norm) differentiability and (uniform) convexity analysis is detailed.

Thus these seven chapters are devoted to the basic theory of Orlicz spaces, and the treatment is leisurely.

As noted before, the last three chapters contain further developments based on the preceding theory and a somewhat accelerated pace is employed to keep the book within reasonable bounds. Chapter VIII presents a novel treatment based on (uncountable) sets of measures. Here we include a special Radon-Nikodým theorem in the form of a conditional expectation for a vector measure of infinite variation, and the results are motivated by "sufficient σ-algebras," appearing in statistical inference theory. Then Chapter IX is devoted to an account of Hardy-Orlicz spaces including some interpolation theory for them complementing the earlier work in Chapter VI, as well as a brief treatment of Orlicz-Sobolev spaces of finite and (to a lesser extent) infinite order. We also discuss the Besicovitch-Orlicz spaces of almost periodic functions here. In the final chapter we indicated certain other directions, by including nonlocally convex Orlicz spaces (of the Fréchet type) motivated by the preceding analysis, and also considered Orlicz spaces of additive set functions and linear operators. This contrasts with our preoccupation of point functions in all the preceding work. A general formulation of Orlicz spaces which subsumes Lorentz spaces concludes this volume. These last three chapters show new directions indicating a vigorous growth of the subject. In these chapters we also presented some previously unpublished material for completeness. Further detail of the subject treated appears at the beginning of each chapter.

We included bibliographical notes at the end of the chapters detailing the various contributions of several workers influencing our treatment. An effort is made here to assign proper credits to the authors. In trying to present a unified account, we often had to rework the previous results since our underlying measure spaces are unrestricted and the Young functions need not be N-functions. We also have readjusted our own previously published work for unification, clarity and a better perspective. It is our hope that the present volume contains the core of the general subject, together with sufficient indications of promising areas for research. The book can be used as a text for a semester (or a two quarter) graduate course on Function Spaces, or for collateral reading in such a course.

The bibliography contains only the items related to the textual material that has been mentioned in the book. The results of various authors are referred to by name and by year of publication. Each chapter is divided into sections, and numbering starts afresh in each section. All definitions, propositions, theorems and lemmas are consecutively numbered. Thus an object is referred to as m.n.p with 'm' denoting the chapter, 'n' the section, and 'p' the particular item. In a given chapter, 'm' is omitted and in a section, 'n' is also dropped, following the current standard usage.

Our collaboration was made possible by partial support of the Office of Naval Research Contract No. N00014-84-K-0356; and a small UCR Academic Senate grant was used for the secretarial help. The difficult and hand-written manuscript, which often was amended as soon as it was typed, has been transformed into beautiful pages using the word processing technique, by Mrs. Joyce Kepler. We are deeply indebted to her for this preparation since she even caught a number of oversights in the manuscript copy, and brought the project to a happy conclusion. One of us (Ren) also wants to record his appreciation to the UCR Mathematics Department for providing him facilities to be on campus and participate in the completion of this monograph.

Finally, we feel that our labors will be amply rewarded if this book succeeds in stimulating interest in young scientists to enter mathematical research motivated by the concreteness and promise of this material.

M. M. Rao
Z. D. Ren

Contents

Contents

THEORY OF
ORLICZ SPACES

I
Introduction and Preliminaries

The uniform integrability concept through its equivalence with a condition discovered by de la Vallée Poussin in 1915 has given a powerful inducement for our study of Young's functions and the corresponding function spaces. This chapter motivates the development leading to Orlicz spaces and their extensions.

1.1. *Motivation and generalities*

Ever since Lebesgue's theory of integration has taken a center stage in concrete problems of analysis, the need for a more inclusive class of spaces than the L^p-families naturally arose. In retrospect, this could be seen as early as 1915 with the publications of de la Vallée Poussin, although it is the Banach space research of 1920's that formally gave birth to what are later called the Orlicz spaces, first proposed by Z. W. Birnbaum and W. Orlicz, and immediately followed and developed by Orlicz himself. However, the necessary tools, centering around the convex function theory were already well established in a study of Fourier series by W. H. Young, Jensen, and others early in the twentieth century. A systematic study of Orlicz spaces has taken place from the 1930's; and after the war years the study and applications have been vigorous in Poland, USSR and Japan, the latter under the lead of H. Nakano with the name "modulared spaces." The work of Zaanen, and especially his book on Linear Analysis in 1953 has enabled the western countries to develop the theory and applications in many directions.

1

Particularly, the monograph by M. A. Krasnoselskii and Ya. B. Rutickii, available by 1961 in English, devoted entirely to Orlicz spaces, has become a catalyst in having the theory used effectively in differential and integral equations, probability, statistics, as well as in harmonic analysis. It has also been instrumental in suggesting some new ideas in abstract theory of topological vector spaces.

Because of the above reasons, we shall present in this volume some central aspects of the Orlicz space theory concentrating on the developments since the appearance of the Krasnoselskii-Rutickii book. The latter confined itself to the spaces based on a bounded domain of $I\!R^n$, with the Lebesgue measure. In view of the manifold applications that we wish to cover, the treatment here will include abstract measure spaces and hence we plan to go from the general case to particular spaces to exhibit special aspects and fine structure when that is deemed necessary and illuminating.

To give a better idea of the scope, we start in the next section with the uniform integrability concept, prove a crucial theorem, and explain its implications in motivating the rest of the treatment. The reader is assumed to be familiar with the basic knowledge of the theory of measure and integration such as that given in the first six chapters of the recent book by Rao (1987).

1.2. Uniform integrability and Vallée Poussin's theorem

To start with the subject, let (Ω, Σ, μ) be a finite measure space which we shall later both specialize and generalize as the theory proceeds. Thus Ω is an abstract set, Σ is a σ-algebra of Ω, and μ is a σ-additive finite function on Σ, so that $\mu(\Omega) < \infty$. Let $\mathcal{F} = \{f_\alpha : \Omega \to I\!R, \alpha \in I\}$ be an arbitrary family of measurable real functions on Ω.

Definition 1. $\mathcal{F}$ is *uniformly integrable* if the following two conditions hold:

$$\text{(i)} \quad \sup_\alpha \int_\Omega |f_\alpha| d\mu = C < \infty, \quad \text{and} \quad \text{(ii)} \quad \lim_{\mu(A) \to 0} \int_A |f_\alpha| d\mu = 0, \quad (1)$$

uniformly in $\alpha \in I$. (Actually (i) follows from (ii) since $\mu(\Omega) < \infty$.)

It is desirable that an alternative and more effective characterization of this condition be found. The following theorem discovered by de la Vallée Poussin in 1915 satisfies this quest:

Theorem 2. *Let $\mathcal{F}$ be a family of scalar measurable functions on a finite measure space (Ω, Σ, μ). Then the following conditions are equivalent:*

(i) *$\mathcal{F} = \{f_\alpha, \alpha \in I\}$ is uniformly integrable.*

(ii)

$$\lim_{\lambda \to \infty} \int_{[|f_\alpha| > \lambda]} |f_\alpha| d\mu = 0, \quad \text{uniformly in } \alpha \in I. \tag{2}$$

(iii) *There exists a convex function $\Phi : \mathbb{R} \to \mathbb{R}^+$ such that $\Phi(0) = 0$, $\Phi(-x) = \Phi(x)$, and $(\Phi(x)/x) \nearrow \infty$ as $x \nearrow \infty$, in terms of which*

$$C_1 = \sup_\alpha \int_\Omega \Phi(f_\alpha) d\mu < \infty. \tag{3}$$

Proof: We first establish the result and then comment on some of its implications.

(i) $\Rightarrow$ (ii). Let $A_\alpha^\lambda = [|f_\alpha| > \lambda]$. Then by (1)

$$\mu(A_\alpha^\lambda) = \frac{1}{\lambda} \int_{A_\alpha^\lambda} \lambda d\mu \leq \frac{1}{\lambda} \int_\Omega |f_\alpha| d\mu \leq \frac{C}{\lambda} \to 0,$$

as $\lambda \to \infty$, uniformly in α. Let $\varepsilon > 0$ and choose $\delta_\varepsilon > 0$ such that by (1) $\mu(A) < \delta_\varepsilon \Rightarrow \int_A |f_\alpha| d\mu < \varepsilon$, uniformly in α. Hence if $\lambda = C/\delta_\varepsilon$ and $A = A_\alpha^\lambda$ here then $\mu(A_\alpha^\lambda) < \delta_\varepsilon$ and $\int_{A_\alpha^\lambda} |f_\alpha| d\mu < \varepsilon$ uniformly in α. Since $\varepsilon > 0$ is arbitrary, this gives (2) and hence (ii) follows.

(ii) $\Rightarrow$ (iii). This is the key part which gives the desired Φ. To construct it, by (ii) we can choose $0 < \lambda_n < \lambda_{n+1} \to \infty$, such that

$$\sup_\alpha \int_{[|f_\alpha| > \lambda_n]} |f_\alpha| d\mu < \beta_n, \quad n \geq 1, \tag{4}$$

where $\{\beta_n, n \geq 1\}$ is any sequence satisfying $\sum_n \beta_n < \infty$. Note that (2) shows that this is possible and λ_n depends only on $\mathcal{F}$ but not on the individual f_α. Let $a_0 = 0$ and for $n \geq 1$, let a_n be the number of λ_n's of (4) which lie in the interval $[n, n+1)$. Set $a_n = 0$ if there are no λ_n's. Let $\varphi(n) = \sum_{k=0}^n a_k$. Then $\varphi(n) \nearrow \infty$ as $n \to \infty$, and let $\varphi(t) = \varphi(n)$, $n \leq t < n+1$, and set

$$\Phi(x) = \int_0^{|x|} \varphi(t) dt, \quad x \in \mathbb{R}. \tag{5}$$

Since $\varphi(\cdot)$ is increasing, it follows that $\Phi(x) = \Phi(-x)$, $\frac{\Phi(x)}{x} \geq \varphi(k)\frac{x-k}{x}$ $\nearrow \infty$ as $k < x$ and $k \to \infty$. Moreover $\Phi(\cdot)$ is a convex function, i.e., $\Phi(\alpha x + \beta y) \leq \alpha\Phi(x) + \beta\Phi(y)$, $0 \leq \alpha, \beta \leq 1$, $\alpha + \beta = 1$. This follows from the fact that $\varphi(\cdot)$ is increasing and use a change of variables. To see that this Φ satisfies (iii), consider (3):

$$
\int_\Omega \Phi(f_\alpha)d\mu = \sum_{n=1}^\infty \int_{[n-1\leq|f_\alpha|<n]} \Phi(|f_\alpha|)d\mu
$$

$$
\leq \sum_{n=1}^\infty \Phi(n)\mu(n-1 \leq |f_\alpha| < n)
$$

$$
\leq \sum_{n=0}^\infty [\Phi(n+1) - \Phi(n)]\mu(|f_\alpha| \geq n)
$$

$$
\leq \sum_{n=0}^\infty \varphi(n)\mu(|f_\alpha| \geq n). \tag{6}
$$

Also

$$
\int_{[|f_\alpha|\geq\lambda_n]} |f_\alpha|d\mu = \sum_{r=\lambda_n}^\infty \int_{[r\leq|f_\alpha|<r+1]} |f_\alpha|d\mu \geq \sum_{r=\lambda_n}^\infty r\mu[r \leq |f_\alpha| < r+1]
$$

$$
\geq \sum_{r=\lambda_n}^\infty \mu[|f_\alpha| > r].
$$

It follows from this, on adding over n, that

$$
\sum_{n=1}^\infty \varphi(n)\mu[|f_\alpha| > n] = \sum_{n=1}^\infty \sum_{r=\lambda_n}^\infty \mu[|f_\alpha| > r]
$$

$$
\leq \sum_{n=1}^\infty \int_{[|f_\alpha|>\lambda_n]} |f_\alpha|d\mu
$$

$$
< \sum_{n=1}^\infty \beta_n < \infty. \tag{7}
$$

Thus (6) and (7) imply (3).

(iii) $\Rightarrow$ (ii). Given $\varepsilon > 0$, let $k_\varepsilon = C_1/\varepsilon$, and choose λ_n such that $x > \lambda_n$ implies $(\Phi(x)/x) > k_\varepsilon$. This can be done since $(\Phi(x)/x) \nearrow \infty$ as $x \nearrow \infty$. Hence

$$
\int_{[|f_\alpha|>\lambda_n]} |f_\alpha|d\mu \leq \frac{1}{k_\varepsilon} \int_{[|f_\alpha|\geq\lambda_n]} \Phi(f_\alpha)d\mu \leq \frac{C_1}{k_\varepsilon} = \varepsilon. \tag{8}
$$

Since this is uniform in α, and $\varepsilon > 0$ is arbitrary, this gives (ii).

(ii) $\Rightarrow$ (i). We have for $\lambda > 0$,

$$\int_\Omega |f_\alpha|d\mu = \int_{[|f_\alpha|\leq\lambda]} |f_\alpha|d\mu + \int_{[|f_\alpha|>\lambda]} |f_\alpha|d\mu \leq \lambda\mu(\Omega) + 1,$$

where λ is chosen such that the second integral is atmost 1 by (2), uniformly in α. Hence (i) of Definition 1 holds with $C = \lambda\mu(\Omega) + 1$. As for (ii) of that definition, we have

$$\int_A |f_\alpha|d\mu = \int_{A\cap[|f_\alpha|>\lambda]} |f_\alpha|d\mu + \int_{A\cap[|f_\alpha|\leq\lambda]} |f_\alpha|d\mu$$

$$\leq \int_{[|f_\alpha|>\lambda]} |f_\alpha|d\mu + \lambda\mu(A). \tag{9}$$

So given $\varepsilon > 0$, choose $\lambda_\varepsilon > 0$ such that by (2) the first integral on the right of (9) is $< \frac{\varepsilon}{2}$ and then for this $\lambda_\varepsilon > 0$, choose $\mu(A) < \delta_\varepsilon = \frac{\varepsilon}{2\lambda_\varepsilon}$. This shows that $\mu(A) < \delta_\varepsilon \Rightarrow$ the right side of (9) is $< \varepsilon$, uniformly in α. This gives (i), as desired. ∎

Several forms of this result appear in the literature. We shall have occasions to include some of these in later parts. Here some consequences are noted.

Let $L^1(\mu)$ be the usual Lebesgue space of integrable functions on (Ω, Σ, μ), and let $\tilde{L}^\Phi(\mu) = \{f : \Omega \to I\!\!R, \text{measurable} \mid \int_\Omega \Phi(|f|)d\mu < \infty\}$, where $\Phi : I\!\!R \to I\!\!R^+$ is a convex function such that $\Phi(x)/x \nearrow \infty$ as in (iii) of the theorem. For instance $\Phi_p(x) = |x|^p$, $p > 1$, gives $L^{\Phi_p}(\mu)$ as the usual Lebesgue space $L^p(\mu)$. The following interesting relation between these spaces is implied by the above theorem.

Corollary 3. *If (Ω, Σ, μ) is a finite measure space on which $L^1(\mu)$ is defined, then*

$$L^1(\mu) = \cup\{\tilde{L}^\Phi(\mu) : \Phi(\cdot) \text{ convex}, \frac{\Phi(x)}{x} \uparrow \text{ as defined above}\}. \tag{10}$$

Proof: Since Φ is convex, $\Phi(x) \geq ax + b$ for some constants a, b. [The curve is above the line which supports it.] Then for each $f \in \tilde{L}^\Phi(\mu)$,

$$\int_\Omega (a|f| + b)d\mu \leq \int_\Omega \Phi(|f|)d\mu < \infty.$$

But the left side is $(a \int_\Omega |f| d\mu + b\mu(\Omega))$ which is thus finite. Hence $f \in L^1(\mu)$. Since this holds for all such convex Φ, $\tilde{L}^\Phi(\mu) \subset L^1(\mu)$ so that the left side contains the right side of (10). On the other hand, if $f \in L^1(\mu)$, then it is trivially "uniformly" integrable. Hence by Theorem 2, there *exists* a convex function Φ of the above type such that $\int_\Omega \Phi(|f|) d\mu < \infty$ so that $f \in \tilde{L}^\Phi(\mu)$. Thus the left side of (10) is contained in the right side and hence equality holds.   ∎

This result raises the following questions. What are the sets $\tilde{L}^\Phi(\mu)$? Are they linear and do they have any properties analogous to $L^p(\mu)$ spaces? Since it is classical that there is strict inclusion in (10) if $\tilde{L}^\Phi(\mu)$ is replaced by $L^p(\mu)$, $p > 1$, it is essential that one admit more numerous convex functions in the spaces in (10) than $\Phi(x) = |x|^p$. These are nontrivial questions: We analyze them in the following chapters. It is first necessary to look at the structure of convex functions that are useful in this study and that are isolated by Theorem 2. This will be done in Chapter II. The study of $\tilde{L}^\Phi(\mu)$ and $L^\Phi(\mu)$ spaces and their applications start with Chapter III and will occupy the rest of this book. As a preliminary to the structural analysis, we make a few observations in the next section on the types of convex functions that play a key role in our work.

1.3. *Young's functions*

In his studies on Fourier series, W. H. Young has analyzed certain convex functions $\Phi : \mathbb{R} \to \bar{\mathbb{R}}^+$ which satisfy the conditions: $\Phi(-x) = \Phi(x)$, $\Phi(0) = 0$, and $\lim_{x \to \infty} \Phi(x) = +\infty$. With each such function Φ, one can associate another convex function $\Psi : \mathbb{R} \to \bar{\mathbb{R}}^+$ having similar properties, which is defined by

$$\Psi(y) = \sup\{x|y| - \Phi(x) : x \geq 0\}, \qquad y \in \mathbb{R}. \tag{1}$$

Then Φ is called a *Young function*, and Ψ the *complementary* function to Φ. It follows from the definition that $\Psi(0) = 0$, $\Psi(-y) = \Phi(y)$, and, what is important, $\Psi(\cdot)$ is a convex increasing function satisfying $\lim_{y \to +\infty} \Psi(y) = +\infty$. From (1) it is evident that the pair (Φ, Ψ) satisfies *Young's inequality:*

$$xy \leq \Phi(x) + \Psi(y), \qquad x, y \in \mathbb{R}. \tag{2}$$

This definition of the complementary function is direct and simple but not informatory. For an alternative definition, it is useful to present different types of examples of Young functions Φ which will motivate a constructive definition of Ψ.

Examples. (i) Let $\Phi(x) = |x|^p$, $p \geq 1$. Then Φ is a continuous Young function such that $\Phi(x) = 0$ iff $x = 0$, and $\lim_{x\to\infty} \Phi(x) = +\infty$ while $\Phi(x) < \infty$ for all $x \in \mathbb{R}$.

(ii) $\Phi(x) = 0$, $0 \leq x \leq a < \infty$; $= \Phi_1(x) > 0$, $a < x < b$, and $= +\infty$ for $x \geq b$, where Φ_1 is a continuous increasing convex function on (a, b). For instance $\Phi_1(x) = |x - a|^p$, $p \geq 1$, for $a \leq x < b$. Thus Φ is a Young function which is continuous on $(0, b)$, and jumps to $+\infty$ at $b > 0$, and $\Phi(-x) = \Phi(x)$.

(iii) $\Psi(x) = 0$ for $0 \leq x < a$; $= +\infty$ for $x > a$, and $\Psi(-x) = \Psi(x)$. This is also a Young function, but it jumps to $+\infty$ at $a > 0$. It can be verified that this Ψ is complementary to Φ given by $\Phi(x) = |x|$ with $a = 1$, implying that the complementary function of a continuous Young function Φ on $\mathbb{R}$ can be a jump function.

Since from (2) it follows that the complementary Ψ is the smallest convex (Young's) function that satisfies the Young inequality, it is clear that we have to analyze the structure of a Young function $\Phi : \mathbb{R} \to \mathbb{R}^+$. This will be done now, and then a constructive definition of Ψ, for which (2) holds, is presented.

We recall from advanced calculus that a convex function on an open interval is continuous. But it has much stronger properties, and in fact admits an integral representation such as (2.5). More precisely, we can establish the following:

Theorem 1. *Let $\Phi : (a, b) \to \mathbb{R}$ be a function. Then Φ is convex iff for each closed subinterval $[c, d] \subset (a, b)$ we have*

$$\Phi(x) = \Phi(c) + \int_c^x \varphi(t)dt, \qquad c \leq x \leq d, \tag{3}$$

where $\varphi : \mathbb{R} \to \mathbb{R}$ is a monotone nondecreasing and left continuous function. Also, Φ has a left and a right derivative at each point of (a, b) and they are equal except perhaps for at most a countable number of points.

Proof: Let $c \le x_i \le d$, $i = 1, 2$, and, to use the definition of convexity, let $c \le x < y < z \le d$. Setting $\alpha = (y - x)/(z - x)$ and $\beta = (z - y)/(z - x)$, so that $0 < \alpha, \beta < 1$ and $\alpha + \beta = 1$, we have with $x_1 = z$ and $x_2 = x$,

$$\Phi(y) = \Phi(\alpha x_1 + \beta x_2) \le \alpha \Phi(x_1) + \beta \Phi(x_2) = \alpha \Phi(z) + \beta \Phi(x). \quad (4)$$

Substituting for α, β and rearranging (4), we get with $z - x$ as $(z - y) + (y - x)$,

$$\frac{\Phi(y) - \Phi(x)}{y - x} \le \frac{\Phi(z) - \Phi(y)}{z - y}. \quad (5)$$

This implies that the difference quotient for Φ is nondecreasing in $[c, d]$. Hence if $c < c_1 \le x < y \le d_1 < d$, then (5) can be extended to get

$$\frac{\Phi(c_1) - \Phi(c)}{c_1 - c} \le \frac{\Phi(y) - \Phi(x)}{y - x} \le \frac{\Phi(d) - \Phi(c)}{d - c}. \quad (6)$$

From this we deduce immediately that

$$|\Phi(y) - \Phi(x)| \le K_1 |y - x| \quad (7)$$

where K_1 = maximum of the extreme terms in (6) in absolute value. This implies that $\Phi(\cdot)$ satisfies a *Lipschitz condition* in $[c, d]$, and hence, in particular is *absolutely continuous* in (a, b). Recall that the latter means: for each $\varepsilon > 0$, there is a $\delta_\varepsilon > 0$ such that for any disjoint intervals $[a_i, b_i) \subset (a, b)$ satisfying $\sum_{i=1}^{n} |\Phi(b_i) - \Phi(a_i)| < \varepsilon$. Hence by the classical Lebesgue-Vitali theorem we have

$$\Phi(x) = \Phi(a) + \int_a^x \Phi'(t) dt, \qquad a \le x \le b. \quad (8)$$

It remains to verify the properties of Φ'.

By (5), with $y = x + h$, $z = y + h'$, $h, h' \ge 0$, we get

$$(D^+ \Phi)(x) = \lim_{h \to 0+} \frac{\Phi(x + h) - \Phi(x)}{h} \le \frac{\Phi(d) - \Phi(c)}{d - c} < \infty$$

and

$$(D^- \Phi)(y) = \lim_{h \to 0+} \frac{\Phi(y) - \Phi(y - h)}{h} > -\infty.$$

Hence the right and left derivatives of Φ exist at each point of $[c, d]$ and for $x < y$,

$$(D^+\Phi)(x) \leq \frac{\Phi(y) - \Phi(x)}{y - x} \leq (D^-\Phi)(y). \qquad (9)$$

Since $(D^-\Phi)(x) \leq (D^+\Phi)(x)$ by (6), $(D^{\pm}\Phi)(\cdot)$ is increasing, and the set of discontinuities of these functions is at most countable. So $(D^+\Phi)(x) = (D^-\Phi)(x)$ at each continuity point of these functions, and this common value is Φ' of (8). This establishes (3) with $\varphi = \Phi'$.

For the converse, let the representation (3) hold for Φ. For $c < x < d$ consider the chord $L(x)$ joining $(c, \Phi(c))$ and $(d, \Phi(d))$ which is given by

$$L(x) = \Phi(c) + \frac{\Phi(c) - \Phi(d)}{c - d}(x - c).$$

To see that this is above the arc, i.e., $\geq \Phi(x)$, we have to show

$$\frac{\Phi(x) - \Phi(c)}{x - c} \leq \frac{\Phi(d) - \Phi(c)}{d - c}, \qquad c < x < d. \qquad (10)$$

Substituting (3) for $\Phi(\cdot)$, we verify it. But we always have

$$\frac{1}{x - c} \int_c^x \varphi(t)dt \leq \varphi(x) \leq \frac{1}{d - x} \int_x^d \varphi(u)du, \qquad (11)$$

since $\varphi(c) \leq \varphi(t) \leq \varphi(x) \leq \varphi(d)$, for $c < t < x < d$. The right side of (10) can be expressed as

$$\frac{\int_c^x \varphi(t)dt + \int_x^d \varphi(u)du}{(d - x) + (x - c)} \geq \min\left(\frac{1}{x - c}\int_c^x \varphi(t)dt, \frac{1}{d - x}\int_x^d \varphi(u)du\right)$$

$$= \frac{1}{x - c}\int_c^x \varphi(t)dt.$$

The last inequality is a consequence of the elementary relation

$$\min\left(\frac{a_1}{b_1}, \frac{a_2}{b_2}\right) \leq \left(\frac{a_1 + a_2}{b_1 + b_2}\right) \leq \max\left(\frac{a_1}{b_1}, \frac{a_2}{b_2}\right)$$

for any real numbers a_1, a_2 and positive numbers b_1, b_2. Thus (10) holds, and Φ given by (3) is convex in (a, b) as asserted. ∎

Recalling that a Young function $\Phi : \mathbb{R} \to \bar{\mathbb{R}}^+$ is convex and $\Phi(0) = 0$, but which may jump to $+\infty$ at a finite point, it is clear that if

$\Phi(a) = +\infty$ for some $a > 0$, then $\Phi(x) = +\infty$ for all $x > a$. Interpreting $\varphi(0) = 0$ and $\varphi(x) = \infty$ for $x \geq a$ in this case, (3) continues to hold. Hence we may state the representation (3) for any Young function Φ in the following form.

Corollary 2. *Let* $\Phi : \mathbb{R}^+ \to \bar{\mathbb{R}}^+$ *be a Young function. Then it can be represented as:*

$$\Phi(x) = \int_0^x \varphi(t)dt, \qquad x \in \mathbb{R}^+, \tag{12}$$

where $\varphi(0) = 0$, $\varphi : \mathbb{R}^+ \to \bar{\mathbb{R}}^+$ *is nondecreasing left continuous and if* $\varphi(x) = +\infty$ *for* $x \geq a$ *then* $\Phi(x) = +\infty$, $x \geq a \geq 0$.

The representation (12) and the complementary function Ψ in (1) motivates an alternative definition at least when the derivative φ of Φ is well behaved, i.e., continuous. In fact let $\psi(\cdot)$ be the generalized inverse of the monotone function φ of (12). To get left continuity we have to define it with some care in the general case. Thus we let

$$\psi(u) = \inf\{t : \varphi(t) > u\}, \qquad u \geq 0. \tag{13}$$

Then $\psi(0) = 0$, $\psi(\cdot)$ is increasing and uniquely defined. By the left continuity of φ, $\{t : \varphi(t) > u\}$ is the left open interval $(\psi(u), \infty]$ and $\psi(u) < t$ iff $\varphi(t) > u$. Clearly $\psi(\cdot)$ is a Borel function since $\varphi(\cdot)$ is. Now define

$$\Psi(y) = \int_0^y \psi(u)du, \qquad y \geq 0. \tag{14}$$

Then $\Psi(0) = 0$, $\Psi(\cdot)$ is convex by Theorem 1, and in fact Ψ is a Young function. It will follow that Ψ is complementary to Φ when we establish (2) for the pair (Φ, Ψ) and that equality can hold sometimes.

Theorem 3. *Let* $\Phi : \mathbb{R}^+ \to \bar{\mathbb{R}}^+$ *be a Young function and let* Ψ *be defined by* (13) *and* (14) *in terms of* Φ. *Then they satisfy the Young inequality,*

$$xy \leq \Phi(x) + \Psi(y), \qquad x \geq 0, y \geq 0, \tag{15}$$

with equality in (15) *when* $y = \varphi(x)$, *or* $x = \psi(y)$, *for* $x \geq 0, y \geq 0$.

Proof: If for some x_0, y_0, either $\Phi(x_0) = +\infty$ or $\Psi(y_0) = +\infty$, then (15) is true. So we only need to prove it when both $\Phi(x) < \infty$ and $\Psi(y) < \infty$ for all $0 \leq x < \infty$ and $0 \leq y < \infty$. Thus

$$0 \leq xy = \int_0^x \int_0^y du\,dv$$

$$= \iint_{\{u \leq x, v \leq y: 0 \leq u \leq \varphi(v), 0 \leq \psi(u) < v\}} du\,dv$$

$$+ \iint_{\{u \leq x, v \leq y: u > \varphi(v) \geq 0, 0 \leq v \leq \psi(u)\}} du\,dv$$

$$= \int_0^x du \int_0^{y \wedge \varphi(u)} dv + \int_0^y dv \int_0^{x \wedge \psi(v)} du, \quad \text{where } u \wedge v = \min(u, v),$$

and the Fubini-Tonelli theorem is used,

$$\leq \int_0^x \varphi(u)du + \int_0^y \psi(v)dv,$$

$$= \Phi(x) + \Psi(y), \tag{16}$$

where equality occurs in the penultimate step iff $y \geq \varphi(u)$ so that $\psi(v) = x$ by (13), or $y = \varphi(x)$ and $x \geq \psi(y)$. Thus (15) holds with equality under the stated conditions. ∎

The general Young complementary pair plays a key role in the study of Orlicz spaces subsuming the Lebesgue classes. Also it should be noted that the Ψ defined by (1) with Φ as an increasing (not necessarily convex) function, $\Phi(0) = 0$, is still a Young function and (2) is again valid. This generality, for which Theorem 3 need not hold, will be of interest in the last chapter, and thus we shall include this type of discussion from time to time. However, in the next chapter we restrict the classes of Young functions and carry out a finer analysis of them for applications.

Bibliographical Notes: As noted already, the importance of Young functions in various summability studies were well recognized since the work of W. H. Young (1912). However their key role in abstract analysis and function spaces emerged only with the fundamental researches of Z. W. Birnbaum and W. Orlicz (1931) and particularly the further papers of

W. Orlicz (1932 and later). The basic theory of convex functions was already known from the work of J. L. W. V. Jensen (1906) to whom Theorem 3.1 is due, and Theorem 2.2 was obtained by C. J. de la Vallée Poussin (1915). Results from Lebesgue's measure, and the abstract measures used here and in the following chapters are classical. For convenience, we shall often use them in the form presented in the recent book by M. M. Rao (1987), without further mention, although we also refer explicitly to other sources when a different method is needed. Theorem 3.3 was first considered in that way in A. C. Zaanen (1953), to include the "boundary cases" of L^1 and L^∞-spaces.

II
Some Classes of Young's Functions

The structure of continuous strictly increasing Young functions is considered in this chapter. These convex functions play an important role in many applications. They are compared and classified according to their growth properties. Also some generalized as well as normalized Young functions are discussed.

2.1. Relations between complementary pairs of Young functions

In Section 1.3 we have already introduced the complementary pairs of Young functions. Several interesting nontrivial properties and ordering relations can be analyzed if a Young function $\Phi : \mathbb{R} \to \mathbb{R}^+$ is continuous. A specially useful *nice Young function* Φ, termed an N-*function*, is a continuous Young function such that $\Phi(x) = 0$ iff $x = 0$ and $\lim_{x \to 0} \Phi(x)/x = 0$, $\lim_{x \to \infty} \Phi(x)/x = +\infty$ while $\Phi(\mathbb{R}) \subset \mathbb{R}^+$. It follows from Corollary I.3.2 that $\varphi(x) = \Phi'_-(x)$, the left derivative, which is nondecreasing, satisfies $0 < \varphi(x) < \infty$ for $0 < x < \infty$; $\varphi(0) = 0$ and $\lim_{x \to \infty} \varphi(x) = +\infty$, although the left continuous φ can have jump discontinuities.

The following properties also are immediate from Theorem I.3.1:

Proposition 1. *Let* (Φ, Ψ) *be a complementary pair of N-functions. Then* Φ, Ψ *are strictly increasing so that their inverses* Φ^{-1}, Ψ^{-1} *are uniquely defined, and*

(i) $\Phi(a) + \Phi(b) \leq \Phi(a+b)$; $\Phi^{-1}(a+b) \leq \Phi^{-1}(a) + \Phi^{-1}(b)$, $a, b \in \mathbb{R}^+$;

(ii) $a < \Phi^{-1}(a)\Psi^{-1}(a) \le 2a$, $a > 0$.

Proof: (i) follows from the integral representation

$$\Phi(a) + \Phi(b) = \int_0^a \varphi(t)dt + \int_0^b \varphi(u)du \le \int_0^a \varphi(t)dt + \int_a^{a+b} \varphi(u)du, \text{ since}$$

$$\varphi \text{ is increasing,}$$

$$= \int_0^{a+b} \varphi(t)dt = \Phi(a + b),$$

The second half follows from this by writing $x = \Phi^{-1}(y)$ for $x = a, b$ and $a + b$. Thus if $\alpha = \Phi(a), \beta = \Phi(b)$, then $\alpha + \beta \le \Phi(a + b)$ so that

$$\Phi^{-1}(\alpha + \beta) \le a + b = \Phi^{-1}(\alpha) + \Phi^{-1}(\beta), \quad \alpha \ge 0, \beta \ge 0.$$

(ii) Note that for any $a > 0$

$$\frac{\Phi(a)}{a} = \frac{1}{a}\int_0^a \varphi(t)dt = \varphi(t^*), \text{ for some } 0 < t^* < a,$$

by the mean value theorem for Lebesgue integrals. Hence

$$\Psi\left(\frac{\Phi(a)}{a}\right) = \int_0^{\Phi(a)/a} \psi(t)dt = \frac{\Phi(a)}{a}\psi(\tilde{t}), \quad 0 < \tilde{t} < \frac{\Phi(a)}{a} = \varphi(t^*)$$

$$< \frac{\Phi(a)}{a} \cdot \psi(\varphi(a)) \le \frac{\Phi(a)}{a} \cdot a = \Phi(a), \tag{1}$$

since φ and ψ are inverse to each other. Letting $\Phi(a) = \alpha$ we get from (1) that

$$\frac{\alpha}{\Phi^{-1}(\alpha)} < \Psi^{-1}(\alpha) \quad \text{or} \quad \alpha < \Phi^{-1}(\alpha)\Psi^{-1}(\alpha). \tag{2}$$

On the other hand, by the Young inequality if $\alpha = \Phi(a), \beta = \Psi(b)$,

$$\Phi^{-1}(\alpha)\Psi^{-1}(\beta) \le \alpha + \beta. \tag{3}$$

Taking $\alpha = \beta$ in (3) and combining it with (2), we get (ii). ∎

An explicit formula for the complementary function of a Young function is not always simple. The following examples illustrate this fact.

1. Let $\Phi(x) = |x|^p/p$, $p > 1$. Then $\Psi(y) = |y|^q/q$, with $p^{-1} + q^{-1} = 1$.
2. Let $\Phi(x) = e^{|x|} - |x| - 1$. Then $\Psi(y) = (1 + |y|)\log(1 + |y|) - |y|$.
3. Let $\Phi(x) = e^{|x|^\delta} - 1$ with $\delta > 1$. Then its derivative on $I\!R^+$ is increasing $\Phi(0) = 0$, so that Φ is a Young function. But its complementary function is not simple, admitting an explicit expression, although it is an N-function.

These already indicate the types of complementary pairs that appear in our analysis. We establish a simple inequality relation between them.

Proposition 2. *If (Φ_i, Ψ_i), $i = 1, 2$ are two complementary pairs of Young functions and $\Phi_1(x) \le \Phi_2(x)$, $x \ge x_0 \ge 0$, then $\Psi_2(y) \le \Psi_1(y)$ for all $y \ge y_0 \ge 0$ where $y_0 = \varphi_2(x_0)$.*

Proof: The argument typically depends on the equality conditions in the Young inequality (Theorem I.3.3), and we illustrate it here. Thus let us study Ψ_2 and the equality condition in its Young inequality:

$$\psi_2(y)y = \Phi_2(\psi_2(y)) + \Psi_2(y), \tag{4}$$

and then for the pair (Φ_1, Ψ_1)

$$\psi_2(y)y \le \Phi_1(\psi_2(y)) + \Psi_1(y), \tag{5}$$

Since $\Phi_1(x) \le \Phi_2(x)$, $x \ge x_0 \ge 0$, (4) and (5) imply the desired inequality at once. ∎

We shall introduce several partial orderings between pairs of Young functions and then study their corresponding complementary function relations systematically.

2.2. Comparisons between classes of N-functions

It is possible to consider different partial ordering relations between N-functions and some of these are introduced here. They will be useful in treating inclusion properties, and hence embeddings, of Orlicz spaces to be studied later. Here then are some of these relations.

Definition 1. Let Φ_1, Φ_2 be two Young functions. Then
(i) Φ_1 is *stronger* than Φ_2, $\Phi_1 \succ \Phi_2$ [or $\Phi_2 \prec \Phi_1$] in symbols, if

$$\Phi_2(x) \le \Phi_1(ax), \qquad x \ge x_0 \ge 0 \tag{1}$$

for some $a > 0$ and x_0 (depending on a).

(ii) Φ_1 is *essentially stronger* than Φ_2, denoted $\Phi_1 \succ\succ \Phi_2$, if (1) holds for each $a > 0$ with an $x_0(a)$. [This is also written $\Phi_2 \prec\prec \Phi_1$.]

(iii) Φ_1 is *completely stronger* than Φ_2, denoted $\Phi_1 \rightthreetimes \Phi_2$ [or $\Phi_2 \leftthreetimes \Phi_1$], if $\forall \varepsilon > 0$, $\exists K (= K_\varepsilon) > 0$, and $x_0 (= x_0(\varepsilon)) \geq 0$, such that

$$\Phi_2(\frac{x}{\varepsilon}) \leq K\Phi_1(x), \qquad x \geq x_0. \tag{2}$$

(iv) Φ_1 increases *more rapidly* than Φ_2, denoted $\Phi_1 \blacktriangleright \Phi_2$ [or $\Phi_2 \prec \Phi_1$], if $\forall \varepsilon > 0$, $\exists \delta (= \delta_\varepsilon) > 0$, and $x_0 (= x_0(\varepsilon)) \geq 0$, such that

$$\frac{1}{\delta}\Phi_2(\delta x) \leq \varepsilon\Phi_1(x), \qquad x \geq x_0. \tag{3}$$

As is clear from definitions, it is not necessary that a given pair of Young functions should satisfy any of these relations. The first two are classical while the last two ((2) and (3)) were introduced by T. Andô in the early 1960's. The consequences and implications of these orderings are now explored. In the above, if $x_0 = 0$, the orderings are *global* and if $x_0 > 0$, they are *local*. These distinctions will be noted when it is necessary, but are omitted in the general context.

Theorem 2. *Let* (Φ_i, Ψ_i), $i = 1, 2$, *be two complementary pairs of* N-*functions with (left continuous) densities* (φ_i, ψ_i) *respectively. Then*
(a) *the following are equivalent conditions:*

(i) $\Phi_1 \succ \Phi_2$;

(ii) $\exists b_1 > 0$, $k_1 > 0$, *and* $x_1 \geq 0$, *such that* $\varphi_2(x) \leq k_1\varphi_1(b_1 x)$, $x \geq x_1$;

(iii) $\exists b_2 > 0$, $k_2 > 0$, *and* $y_1 \geq 0$, *such that* $\psi_2(y) \geq k_2\psi_1(b_2 y)$, $y \geq y_1$;

(iv) $\Psi_1 \prec \Psi_2$.

(b) *the following are again equivalent conditions:*

(i) $\Phi_1 \succ\succ \Phi_2$;

(ii) $\forall \epsilon_1 > 0$, $\exists x_1 > 0$, *such that* $\varphi_2(x) \leq \epsilon_1\varphi_1(\epsilon_1 x)$, $x \geq x_1$;

(iii) $\forall k > 0$, $\exists y_1 > 0$, *such that* $\psi_2(y) \geq k\psi_1(ky)$, $y \geq y_1$;

(iv) $\Psi_1 \prec\prec \Psi_2$;

(v) $\forall \lambda > 0$, $\lim\limits_{x\to\infty} \frac{\Phi_2(\lambda x)}{\Phi_1(x)} = 0$;

(vi) $\lim\limits_{y\to\infty} \frac{\Phi_1^{-1}(y)}{\Phi_2^{-1}(y)} = 0$;

(vii) $\exists$ an N-function R such that $\Phi_1 \succ \Phi_2 \circ R$.

Proof: We first prove the equivalences (a(i)–(iv)), and then (b(i)–(vii)).

(a)(i) $\Rightarrow$ (ii) This follows from the inequalities

$$x\varphi_2(x) \leq \int_x^{2x} \varphi_2(t)dt \leq \int_0^{2x} \varphi_2(t)dt = \Phi_2(2x)$$

$$\leq \Phi_1(2bx), \quad \text{by hypothesis of (a)(i),}$$

$$= \int_0^{2bx} \varphi_1(t)dt \leq 2bx\varphi_1(2bx) = b_1 x\varphi_1(b_1 x) \tag{4}$$

where $b_1 = 2b > 0$, and letting $k_1 = b_1$, $x_1 = x_0/2 \geq 0$.

(a)(ii) $\Rightarrow$ (iii) Let $y_1 > 0$ be a number such that $\psi_2(y_1) \geq x_1$. Then by hypothesis

$$\psi_1\left(\frac{y}{2k_1}\right) \leq \psi_1\left(\frac{1}{2k_1}\varphi_2(\psi_2(y))\right)$$

$$\leq \psi_1\left(\frac{1}{2}\varphi_1(b_1\psi_2(y))\right)$$

$$\leq \psi_1(\varphi_1(b_1\psi_2(y)) - \varepsilon), \quad 0<\varepsilon<\frac{1}{2}\varphi_1(b_1\psi_2(y)),$$

$$\leq b_1\psi_2(y), \quad y \geq y_1. \tag{5}$$

If we set $b_2 = \frac{1}{2k_1}$, $k_2 = \frac{1}{b_1}$, then (a)(iii) follows.

(a)(iii) $\Rightarrow$ (iv) If (a)(iii) holds, by reducing b_2 if necessary, we may assume $0 < k_2 \leq b_2$, and then set $y_0 = k_2 y_1$, $c = 2/k_2$ so that

$$\Psi_2(cy) = \int_0^{2y/k_2} \psi_2(t)dt \geq \frac{y}{k_2}\psi_2\left(\frac{y}{k_2}\right)$$

$$\geq y\psi_1\left(\frac{b_2 y}{k_2}\right), \quad \text{by (a)(iii)}$$

$$\geq \frac{k_2}{b_2}\Psi_1\left(\frac{b_2 y}{k_2}\right) \geq \Psi_1(y), \quad y \geq y_0, \tag{6}$$

so $\Psi_2 \succ \Psi_1$.

(a)(iv) $\Rightarrow$ (i) Since $\Psi_1(y) \leq \Psi_2(cy)$, $y \geq y_0$, let $w = \Psi_2(cy)$, and $w_0 = \Psi_2(cy_0)$, and we have

$$\frac{\Psi_2^{-1}(w)}{c} \leq \Psi_1^{-1}(w), \quad w \geq w_0.$$

From this, using Proposition 1.1(ii), we get

$$\frac{w}{c\Phi_2^{-1}(w)} < \frac{\Psi_2^{-1}(w)}{c} \le \Psi_1^{-1}(w) \le \frac{2w}{\Phi_1^{-1}(w)},$$

so that

$$\Phi_1^{-1}(w) \le 2c\Phi_2^{-1}(w), \qquad w \ge w_0. \tag{7}$$

Letting $x_0 = \Phi_2^{-1}(w_0), x = \Phi_2^{-1}(w)$, in the above, one has $\Phi_2 \prec \Phi_1$, and this establishes all equivalences of (a). Let us turn to (b).

(b)(i) $\Rightarrow$ (ii)   Let $\varepsilon_1 > 0$ be given and if $\varepsilon = \frac{1}{2}\varepsilon_1$, then by (b)(i), there is $x_0 \ge 0$ such that $\Phi_2(x) \le \Phi_1(\varepsilon x), x \ge x_0$, and hence

$$x\varphi_2(x) \le \Phi_2(2x) \le \Phi_1\left(\frac{\varepsilon_1}{2} \cdot 2x\right) \le \varepsilon_1 x \varphi_1(\varepsilon_1 x),$$

$$\text{for } x \ge \frac{x_0}{2} = x_1 \ge 0, \tag{8}$$

and (ii) follows.

(b)(ii) $\Rightarrow$ (iii)   Let $y_1 > 0$ be such that $\psi_2(y_1) \ge x_1$, $\varepsilon_1 = \frac{1}{2k}$. We have

$$\psi_1(ky) \le \psi_1\big(k\varphi_2(\psi_2(y))\big) \le \psi_1\big[k\varepsilon_1\varphi_1(\varepsilon_1\psi_2(y))\big]$$
$$= \psi_1\left(\frac{1}{2}\varphi_1\left(\frac{1}{2k}\psi_2(y)\right)\right)$$
$$\le \psi_1\left(\varphi_1\left(\frac{1}{2k}\psi_2(y)\right) - \delta\right), \quad \text{with } 0<\delta<\frac{1}{2}\varphi_1\left(\frac{1}{2k}\psi_2(y)\right),$$
$$\le \frac{1}{2k}\psi_2(y) < \frac{1}{k}\psi_2(y), \qquad y \ge y_1 > 0. \tag{9}$$

(b)(iii) $\Rightarrow$ (iv)   For given $\delta > 0$, let $k = \frac{2}{\delta}$, and by hypothesis, there is $y_1 > 0$, such that $\psi_2(y) \ge k\psi_1(ky), y \ge y_1$ and hence

$$\Psi_1(w) \le w\psi_1(w) \le \frac{w}{k}\psi_2\left(\frac{w}{k}\right)$$
$$\le \Psi_2\left(\frac{2}{k}w\right) = \Psi_2(\delta w), \quad w \ge ky_1 = w_0. \tag{10}$$

(b)(iv) $\Rightarrow$ (v)   Given $\varepsilon > 0$, let $\delta = \frac{\varepsilon}{2}$, and by hypothesis $\Psi_2^{-1}(\Psi_1(w)) \le \delta w$, and so $\Psi_2^{-1}(x) \le \delta\Psi_1^{-1}(x), x \ge \Psi_1(w_0)$. Hence

$$\frac{x}{\Phi_2^{-1}(x)} < \Psi_2^{-1}(x) \le \delta\Psi_1^{-1}(x) \le \frac{2\delta x}{\Phi_1^{-1}(x)},$$

and $\Phi_1^{-1}(x) \leq \varepsilon \Phi_2^{-1}(x)$. Setting $x = \Phi_2(y)$, we get

$$\Phi_2(y) \leq \Phi_1(\varepsilon y), \qquad y \geq \Phi_2^{-1}(\Psi_1(w_0)) = x_0. \qquad (11)$$

So (b)(i) holds.

For given $\lambda > 0$, if we let $\varepsilon_n = (n\lambda)^{-1}$, then by (11), there is $x_n > 0$ such that $\Phi_2(y) \leq \Phi_1(\varepsilon_n y) = \Phi_1(\frac{1}{n}\frac{y}{\lambda})$, $y \geq x_n$. This gives (v) on setting $x = \frac{y}{\lambda}$, so that $\Phi_2(\lambda x) \leq \Phi_1(\frac{x}{n}) < \frac{1}{n}\Phi_1(x)$, $x \geq x_n/\lambda$.

(b)(v) $\Rightarrow$ (vi) If this implication is false, then there are $\varepsilon_0 > 0$ and a sequence $t_n \nearrow \infty$ such that

$$\frac{\Phi_1^{-1}(t_n)}{\Phi_2^{-1}(t_n)} \geq \varepsilon_0 > 0, \qquad n \geq 1. \qquad (12)$$

Taking $x_n = \Phi_1^{-1}(t_n)$, $\lambda_0 = 1/\varepsilon_0$ in the above, we see that (v) cannot hold.

(b)(vi) $\Rightarrow$ (vii) Considering the reciprocals, and setting $x = \Phi_1^{-1}(y)$, we get

$$\lim_{x \to \infty} \frac{\Phi_2^{-1}(\Phi_1(x))}{x} = +\infty.$$

Hence, defining

$$r(t) = \begin{cases} \inf[\Phi_2^{-1}(\Phi_1(x))/x : x > t], & t \geq 1, \\ t\, r(1), & t < 1, \end{cases}$$

we get $r(0) = 0$, $r(\cdot)$ is nondecreasing left continuous and tends to infinity. Now if R is its indefinite integral on $\mathbb{R}^+$, extended to $\mathbb{R}^-$ by setting $R(-x) = R(x)$, then R is an N-function and

$$R(x) \leq xr(x) \leq \Phi_2^{-1}(\Phi_1(x)), \qquad x \geq 1. \qquad (13)$$

Consequently $\Phi_1(x) \geq \Phi_2(R(x))$, $x \geq 1$ and (vii) holds.

(b)(vii) $\Rightarrow$ (i) By hypothesis, there are constants $b > 0, x_0 > 0$, satisfying $\Phi_2(R(\frac{x}{b})) \leq \Phi_1(x)$, $x \geq x_0$. For each $\varepsilon > 0$, $R(\varepsilon x/b)/x \to \infty$ as $x \to \infty$, so that there is $x_1 \geq x_0$ such that $x \leq R(\frac{\varepsilon x}{b})$, $x \geq x_1$. These two inequalities imply (i) since

$$\Phi_1(\varepsilon x) \geq \Phi_2(R(\frac{\varepsilon x}{b})) \geq \Phi_2(x), \text{ for } x \geq \frac{x_1}{\varepsilon}.$$

This implies all the equivalent relations of the theorem.   ∎

As a consequence of these conditions, we can define *equivalence of a pair* of N-functions: $\Phi_1 \sim \Phi_2$ iff $\Phi_1 \succ \Phi_2$ and $\Phi_2 \succ \Phi_1$. Another way of expressing $\Phi_1 \sim \Phi_2$ is therefore the following: There exist numbers $0 < a \leq b < \infty$, $x_0 \geq 0$ such that

$$\Phi_1(ax) \leq \Phi_2(x) \leq \Phi_1(bx), \qquad x \geq x_0. \tag{14}$$

Corollary 3. *Let* (Φ_i, Ψ_i), $i = 1,2$ *be two complementary pairs. Then each of the following is equivalent to the rest of the conditions:*

(i) $\Phi_1 \sim \Phi_2$;

(ii) $\exists\, a_i > 0$, $i = 1,2,3,4$ *and* $x_1 \geq 0$ *such that*

$$a_1\varphi_1(a_2 x) \leq \varphi_2(x) \leq a_3\varphi_1(a_4 x), \qquad x \geq x_1 \geq 0;$$

(iii) $\exists\, b_i > 0$, $i = 1,2,3,4$ *and* $y_1 \geq 0$ *such that*

$$b_1\psi_1(b_2 y) \leq \psi_2(y) \leq b_3\psi_1(b_4 y), \qquad y \geq y_1 \geq 0;$$

(iv) $\Psi_1 \sim \Psi_2$.

A simple illustration of this result is this: If Φ_2 is an N-function then

$$\Phi_1(x) = \int\limits_0^{|x|} \frac{\Phi_2(t)}{t}\, dt,$$

which is a Young function, satisfies $\Phi_1 \sim \Phi_2$ (also a consequence of the mean value theorem for integrals).

Remark. It should be emphasized that '$\prec$' is only a *partial* ordering and not a linear ordering. Thus one can construct a pair of N-functions Φ_1, Φ_2 for which neither $\Phi_1 \prec \Phi_2$ nor $\Phi_2 \prec \Phi_1$ need hold. Regarding the ordering '$\succ$' we have the following

Proposition 4. *Let* (Φ_i, Ψ_i), $i = 1,2$, *be two pairs of complementary N-functions. Then the following are equivalent conditions:*

(i) $\Phi_1 \succ\!\!\prec \Phi_2$;

(ii) $\exists$ *another N-function* Φ_3 *such that* $\Phi_2(xy) \leq \Phi_1(x)\Phi_3(y)$, $x,y \geq x_0 \geq 0$;

(iii) $\Psi_2 \succ \Psi_1$.

Proof: (i) $\Rightarrow$ (ii) Define Φ_3 by the expression

$$\Phi_3(x) = \sup\{\Phi_2(xy)/\Phi_1(y) : y \geq 0\}.$$

Then $\Phi_3 : I\!\!R^+ \to I\!\!R^+$ is an N-function and (ii) holds for it with $x_0 = 1$.

(ii) $\Rightarrow$ (i) For given $\varepsilon > 0$, choose $y_0 \geq \max(x_0, \frac{1}{\varepsilon})$. Let $K = \Phi_3(y_0)$. Then

$$\Phi_2(\frac{x}{\varepsilon}) \leq \Phi_2(xy_0) \leq \Phi_3(y_0)\Phi_1(x) = K\Phi_1(x), \quad x \geq x_0.$$

So (i) holds.

(i) $\Rightarrow$ (iii) By definition of the order here, for each $\varepsilon > 0$, there is a $k > 0$, and an $x_0 \geq 0$

$$\frac{1}{k}\Phi_2(\frac{x}{\varepsilon}) \leq \Phi_1(x), \quad x \geq x_0. \tag{15}$$

But if $\tilde{\Phi}_2(x)$ is the left side of (15), then it is a Young function and its complementary function $\tilde{\Psi}_2$ is given by $x \mapsto \frac{1}{k}\Psi_2(\varepsilon kx)$, since $(\tilde{\Phi}_2)'(x) = \frac{1}{k\varepsilon}\Phi_2'(\frac{x}{\varepsilon})$, and hence its inverse is

$$(\tilde{\Psi}_2)'(y) = \sup\{t : \tilde{\Phi}_2'(t) < y\} = \sup\{\frac{\varepsilon t}{\varepsilon} : \frac{1}{k\varepsilon}\Phi_2'(\frac{t}{\varepsilon}) < y\}$$
$$= \varepsilon \sup\{\tau : \Phi_2'(\tau) < k\varepsilon y\} = \varepsilon\Psi_2'(k\varepsilon y),$$

and

$$\tilde{\Psi}_2(y) = \int_0^{|y|} \varepsilon\Psi_2'(k\varepsilon t)dt = \frac{1}{k}\Psi_2(k\varepsilon y). \tag{16}$$

Thus there is $y_0 \geq 0$ such that (15) and Proposition 1.2 imply

$$\frac{1}{k}\Psi_2(k\varepsilon y) \geq \Psi_1(y), \quad y \geq y_0.$$

So using this, with $\frac{1}{k\varepsilon} = \delta$, $w = \frac{y}{\delta}$ here we have

$$\frac{1}{\delta}\Psi_1(\delta w) \leq \varepsilon\Psi_2(w), \quad w \geq \frac{y_0}{\delta} \geq 0, \tag{17}$$

which is (iii).

(iii) $\Rightarrow$ (i) This follows from the above by reversing the procedure. This gives all the equivalences. ∎

The preceding results can be stated schematically as follows:
$$(\Phi_1 \succ\succ \Phi_2) \Rightarrow (\Phi_1 \bowtie \Phi_2 \text{ and } \Phi_1 \succ \Phi_2), \text{ and}$$
$$(\Phi_1 \bowtie \Phi_2) \Rightarrow (\Phi_1 \succ \Phi_2) \Leftarrow (\Phi_1 \succ \Phi_2).$$
As an *example*, let $\Phi_i(u) = |u|^{\alpha_i}$, $i = 1, 2$. Then $\Phi_1 \succ\succ \Phi_2$ iff $\alpha_1 > \alpha_2$ (> 1), and $\Phi_1 \succ \Phi_2$ iff $\alpha_1 \geq \alpha_2$. Similarly, if $\Phi_3(u) = |u|^{\alpha_2}(1 + |\log|u||)$, and $\alpha_1 > \alpha_2$ (> 1), then $\Phi_1 \succ\succ \Phi_3 \succ\succ \Phi_2$. (These relations hold for large enough $u > 0$.)

2.3. Classification by growth properties

The comparisons of N-functions given in the preceding section become more useful in the theory when a corresponding classification based on the rapidity of their growth is added. In fact the latter plays a central role in the structure theory of Orlicz spaces. We thus introduce these conditions, and use them systematically.

Definition 1. A Young function $\Phi : I\!R \to I\!R^+$ is said to satisfy the Δ_2-condition (globally), denoted $\Phi \in \Delta_2$ ($\Phi \in \Delta_2$ (globally)) if

$$\Phi(2x) \leq K\Phi(x), \qquad x \geq x_0 \geq 0 \ (\, x_0 = 0 \,) \qquad (1)$$

for some absolute constant $K > 0$.

Considering $\Phi(x) = |x|^p$, $p \geq 1$, we see that $K \geq 2$. Also in (1), 2 can be replaced by $\alpha > 1$, and one gets a condition equivalent to (1). Further note that all $\Phi : x \mapsto a|x|^p$, $p \geq 1, a > 0$, belong to Δ_2. On the other hand, if $\Phi_0 : x \mapsto e^{|x|} - 1$, then this $\Phi_0 \notin \Delta_2$.

In the opposite direction to (1), one can introduce the following condition studied by Andô (1960).

Definition 2. A Young function $\Phi : I\!R \to I\!R^+$ is said to satisfy the ∇_2-condition (globally), denoted $\Phi \in \nabla_2$ ($\Phi \in \nabla_2$ (globally)) if

$$\Phi(x) \leq \frac{1}{2\ell}\Phi(\ell x), \qquad x \geq x_0 > 0 \ (\, x_0 = 0 \,) \qquad (2)$$

for some $\ell > 1$.

For instance, if $\Phi(x) = (1 + |x|)\log(1 + |x|) - |x|$, then its complementary function Ψ is given by $\Psi(y) = e^{|y|} - |y| - 1$. It is quickly verified that $\Phi \in \Delta_2$ (but not ∇_2) and $\Psi \in \nabla_2$ (but not Δ_2). We may present a characterization of these conditions:

Theorem 3. *Let Φ be an N-function with Ψ as its complementary function. If φ, ψ are the (left) derivatives of Φ, Ψ as usual, then the* *following equivalent conditions on Δ_2 and ∇_2 obtain in 1. and 2., respectively.*

1. (i) $\Phi \in \Delta_2$;
   (ii) $\exists 1 < \alpha < \infty$ and $x_0 \geq 0$ such that $\frac{x\varphi(x)}{\Phi(x)} < \alpha$, $x \geq x_0$;
   (iii) $\exists 1 < \beta < \infty$ and $y_0 \geq 0$ such that $\frac{y\psi(y)}{\Psi(y)} > \beta$, $y \geq y_0$;
   (iv) $\Psi \in \nabla_2$;
   (v) $\exists \delta > 0$, $x_0 \geq 0$ such that $\Phi((1+\delta)x) \leq 2\Phi(x)$, $x \geq x_0$;
   (vi) $\limsup\limits_{x \to \infty} \frac{\Phi^{-1}(x)}{\Phi^{-1}(2x)} < 1$.

2. (i) $\Phi \in \nabla_2$;
   (ii) $\exists \delta > 0$, $x_0 \geq 0$ such that $\Phi(2x) \geq (2+\delta)\Phi(x)$, $x \geq x_0$;
   (iii) $\exists 0 < \eta < 1$, $x_1 \geq 0$ such that $\Phi((2-\eta)x) \geq 2\Phi(x)$, $x \geq x_1$;
   (iv) $\liminf\limits_{x \to \infty} \frac{\Phi^{-1}(x)}{\Phi^{-1}(2x)} > \frac{1}{2}$.

Proof: We first establish the equivalences of 1., and then prove 2.

1(i) $\Rightarrow$ (ii) Using (1) we have

$$K\Phi(x) \geq \Phi(2x) = \int\limits_0^{2x} \varphi(t)dt \geq \int\limits_x^{2x} \varphi(t)dt \geq x\varphi(x),$$

since $\varphi(\cdot)$ is increasing. Since $K > 1$, (ii) follows with $\alpha > K$.

1(ii) $\Rightarrow$ (iii) Let $y_0 \geq 1$ be chosen to satisfy $\psi(y_0) > x_0$, and since by the natural relations between φ and ψ (see (I.3.13)) $\varphi(\psi(y)) > y$, we have on letting $x = \psi(y)$,

$$\frac{y\psi(y)}{\Psi(y)} \geq \frac{\varphi(\psi(y)) \cdot \psi(y)}{\Psi(\varphi(\psi(y)))} = \frac{x\varphi(x)}{\Psi(\varphi(x))}, \quad \text{since } \frac{\Psi(y)}{y} \nearrow \text{ as } y \nearrow,$$

$$= \frac{x\varphi(x)}{x\varphi(x) - \Phi(x)}, \quad \text{using the equality condition in Young's}$$

$$\text{inequality,}$$

$$= \frac{x\varphi(x)/\Phi(x)}{[x\varphi(x)/\Phi(x)] - 1} > \frac{\alpha}{\alpha - 1} = \beta > 1, \quad y \geq y_0 \geq 0, \qquad (3)$$

by (ii) and the fact that $u/(u-1)$ is decreasing for $u > 1$. So (iii) holds.

1(iii) $\Rightarrow$ (iv) For $\ell > 1$, we have

$$\log\frac{\Psi(\ell y)}{\Psi(y)} = \int\limits_y^{\ell y} \frac{\psi(t)}{\Psi(t)}dt \geq \int\limits_y^{\ell y} \frac{\beta}{t}dt = \beta \log \ell, \text{ by (iii)}.$$

Taking ℓ to satisfy $\ell^{\beta-1} \geq 2$ here, we get (iv).

1(iv) $\Rightarrow$ (i) Let $\tilde{\Psi}(y) = \frac{1}{2\ell}\Psi(\ell y)$ so that $\tilde{\Phi}(x) = \frac{1}{2\ell}\Phi(2x)$, as seen in (2.16). But by (iv) we also have $\tilde{\Psi}(y) \geq \Psi(y)$ for $y \geq y_0 \geq 0$, so that $\tilde{\Phi}(x) \leq \Phi(x)$ for $x \geq x_0$ (cf. Theorem 2.2a). This gives $\Phi(2x) \leq k\Phi(x)$ for $x \geq x_0$ with $k = 2\ell > 2$. So (i) holds.

1(i) $\Rightarrow$ (v) If $k \geq 2$ is the constant of the Δ_2-condition for Φ, then $0 < \delta = \frac{1}{k-1} \leq 1$, and since Φ is convex we get

$$
\begin{aligned}
\Phi((1+\delta)x) &= \Phi((1-\delta)x + 2\delta x) \\
&\leq (1-\delta)\Phi(x) + \delta\Phi(2x) \\
&\leq (1-\delta)\Phi(x) + \delta k\Phi(x) = 2\Phi(x), \qquad x \geq x_0, \quad (4)
\end{aligned}
$$

giving (v).

1(v) $\Rightarrow$ (vi) This is immediate since (4) gives, on letting $x = \Phi^{-1}(y)$,

$$(1+\delta)\Phi^{-1}(x) \leq \Phi^{-1}(2x),$$

so that $\limsup_{x\to\infty}(\Phi^{-1}(x)/\Phi^{-1}(2x)) \leq (1+\delta)^{-1} < 1$.

1(vi) $\Rightarrow$ (i) If $\Phi \notin \Delta_2$ then we can find $x_n \nearrow \infty$ such that $\Phi((1+\frac{1}{n})x_n) > 2\Phi(x_n)$. Hence, with $y_n = \Phi(x_n)$, one has

$$1 > \limsup_n \Big[\frac{\Phi^{-1}(y_n)}{\Phi^{-1}(2y_n)}\Big] \geq \liminf_n \Big[\frac{\Phi^{-1}(y_n)}{\Phi^{-1}(2y_n)}\Big] \geq \lim_n\Big(\frac{n}{1+n}\Big) = 1.$$

This contradiction shows that $\Phi \in \Delta_2$ if (vi) holds. Thus 1. is established.

We now consider the equivalences of 2., which are similar to the last three of 1.

2(i) $\Rightarrow$ (ii) By 1(iii) $\Leftrightarrow$ 1(iv) established in 1., there is a $\beta > 1$ such that $\frac{x\varphi(x)}{\Phi(x)} > \beta$ for $x \geq x_0 \geq 0$. Then we have

$$\log\frac{\Phi(2x)}{\Phi(x)} = \int_x^{2x}\frac{\varphi(t)}{\Phi(t)}dt \geq \beta\int_x^{2x}\frac{dt}{t} = \log 2^\beta, \qquad x \geq x_0 \geq 0.$$

This gives 2(ii), with $\delta = 2^\beta - 2 > 0$.

2(ii) $\Rightarrow$ (i) This follows on choosing n_0 such that $(1 + \frac{\delta}{2})^{n_0} \geq 2$ and letting $\ell = 2^{n_0}$. Thus by 2(ii),

$$\Phi(x) \leq \frac{1}{2+\delta}\Phi(2x) \leq \cdots \leq \frac{1}{(2+\delta)^{n_0}}\Phi(2^{n_0}x) \leq \frac{1}{2\ell}\Phi(\ell x), \quad x \geq x_0.$$

So (i) holds. The other equivalences are also similar.

2(i) $\Rightarrow$ (iii) By 1(i) $\Leftrightarrow$ 1(iv), $\Phi \in \nabla_2$ now. Hence by 1(v) $\exists \delta > 0$, and $x_0 \geq 0$ such that $\Psi((1 + \delta)x) \leq 2\Psi(x)$, $x \geq x_0$, one has (by Proposition 1.2)

$$2\Phi(\frac{y}{2}) \leq \Phi(\frac{y}{1+\delta}), \quad y \geq y_0.$$

Letting $w = \frac{y}{2}$ and $\eta = \frac{2\delta}{1+\delta}$, this gives $2\Phi(w) \leq \Phi((2 - \eta)w)$, $w \geq y_0/2 = x_1$, as asserted.

2(iii) $\Rightarrow$ (iv) Let $x = \Phi^{-1}(y)$ in (iii). Then

$$\frac{\Phi^{-1}(y)}{\Phi^{-1}(2y)} \geq \frac{1}{2-\eta} > \frac{1}{2}, \quad y \geq y_1 = \Phi(x_1). \tag{5}$$

This gives 2(iv). Finally,

2(iv) $\Rightarrow$ (i) If (i) is false, then (ii) is false; there exist $x_n \nearrow \infty$ such that

$$\Phi(2x_n) < (2 + \frac{1}{n})\Phi(x_n).$$

By convexity of Φ, we always have $\Phi(2x_n) \geq 2\Phi(x_n)$, so that

$$\lim_{n\to\infty} \frac{\Phi(x_n)}{\Phi(2x_n)} = \frac{1}{2}. \tag{6}$$

We claim that for the x_n sequence above, on letting $y_n = \frac{1}{2}\Phi(2x_n)$,

$$\lim_{n\to\infty} \frac{\Phi^{-1}(y_n)}{\Phi^{-1}(2y_n)} = \frac{1}{2}. \tag{7}$$

Indeed, if (7) is not true, since $2\Phi^{-1}(y) \geq \Phi^{-1}(2y)$, there exist a subsequence $\{y_{n_i}, i \geq 1\}$ and an $\varepsilon_0 > 0$, such that

$$\frac{\Phi^{-1}(y_{n_i})}{\Phi^{-1}(2y_{n_i})} > \frac{1}{2-\varepsilon_0}, \quad i \geq 1. \tag{8}$$

Hence, substituting for y_{n_i} in terms of x_{n_i}, we get from (8)

$$(2 - \varepsilon_0)\Phi^{-1}(\tfrac{1}{2}\Phi(2x_{n_i})) > \Phi^{-1}(\Phi(2x_{n_i})) = 2x_{n_i}.$$

Consequently

$$\Phi(x_{n_i}) < \Phi\left(\frac{2 - \varepsilon_0}{2}\Phi^{-1}\left(\frac{1}{2}\Phi(2x_{n_i})\right)\right)$$
$$< \frac{2 - \varepsilon_0}{2}\left(\frac{1}{2}\Phi(2x_{n_i})\right) = \frac{2 - \varepsilon_0}{4}\Phi(2x_{n_i}), \qquad i \geq 1.$$

But this contradicts (6). So our claim is proved. This implies that if 2(i) is false, then so is 2(iv). This completes the proof of the equivalences.
∎

The following consequence of the above result may be recorded. If (Φ, Ψ) is a complementary pair of N-functions, let

$$a_\Phi = \liminf_{x \to \infty} \frac{x\varphi(x)}{\Phi(x)}, \qquad b_\Phi = \limsup_{x \to \infty} \frac{x\varphi(x)}{\Phi(x)}, \qquad (9)$$

and similarly a_Ψ, b_Ψ be defined. Then we have:

Corollary 4. *For a pair of complementary N-functions Φ, Ψ, the following four relations are mutually equivalent.*

   (i) $\Phi \in \Delta_2$; (ii) $\Psi \in \nabla_2$; (iii) $b_\Phi < \infty$; (iv) $a_\Psi > 1$.
Moreover, $\Phi \in \Delta_2 \cap \nabla_2$ iff $1 < a_\Phi \leq b_\Phi < \infty$.

As an illustration, let $\Phi(x) = (1 + |x|)\log(1 + |x|) - |x|$ so that $\Psi(y) = e^{|y|} - |y| - 1$. Then $a_\Phi = b_\Phi = 1$, $a_\Psi = b_\Psi = +\infty$ so that $\Phi \notin \nabla_2$ (but $\Phi \in \Delta_2$) and $\Psi \notin \Delta_2$ (but $\Psi \in \nabla_2$). On the other hand, if $\Phi(x) = |x|^\alpha(1 + |\log|x||)$, $\alpha > 1$, then $\lim \frac{x\varphi(x)}{\Phi(x)} = \alpha$, $a_\Phi = b_\Phi = \alpha$ so that $\Phi \in \Delta_2 \cap \nabla_2$.

Another consequence is given by:

Corollary 5. *If $\Phi \in \Delta_2$ is an N-function, then $\Phi(x) \leq C|x|^\alpha$, $x \geq x_0$, for some $C > 0$ and $\alpha > 1$, and its complementary function Ψ (which is in ∇_2 by Corollary 4) satisfies $\Psi(y) \geq D|y|^\beta$, $y \geq y_0 > 0$ for some $D > 0$ and $\beta > 1$.*

Proof: Both assertions follow from Theorem 3(1.ii and 1.iii), since, for instance,

$$\log\frac{\Psi(y)}{\Psi(y_0)} = \int_{y_0}^{y} \frac{\psi(t)}{\Psi(t)}dt \geq \beta\int_{y_0}^{y} \frac{dt}{t} = \log(\frac{y}{y_0})^\beta, \qquad y \geq y_0 > 0, \beta > 1.$$

Set $D = \Psi(y_0)/y_0^\beta$. The other one is similar. ∎

The second part of Theorem 3 and Corollary 4 give the following:

Corollary 6. *If for an N-function Φ, a_Φ and b_Φ are the associated numbers given by (9), and Ψ is complementary to Φ, then $(a_\Psi)^{-1} + (b_\Phi)^{-1} = 1$ whenever $\Phi \in \Delta_2$ (or $\Psi \in \nabla_2$).*

Proof: By definition of a_Ψ, b_Φ in (9), for each $\varepsilon > 0$, we can find $x_0 > 0$ such that $\frac{x\varphi(x)}{\Phi(x)} < b_\Phi + \varepsilon$, $x \geq x_0$, and then by Theorem 3(part 1), there is $y_0 > 0$ such that

$$\frac{y\psi(y)}{\Psi(y)} > \frac{(b_\Phi + \varepsilon)}{(b_\Phi + \varepsilon) - 1}, \qquad y \geq y_0 \text{ (cf. (3))}.$$

This implies $a_\Psi > (b_\Phi + \varepsilon)/[b_\Phi + \varepsilon - 1]$ so that $a_\Psi b_\Phi \geq a_\Psi + b_\Phi$.

For the opposite inequality, $\Phi \in \Delta_2 \Rightarrow \Psi \in \nabla_2$ so that $a_\Psi > 1$. Hence if $0 < \delta < a_\Psi - 1$, then for some $y_0 > 0$, one has $\frac{y\psi(y)}{\Psi(y)} > a_\Psi - \delta$, $y \geq y_0$. But then by Theorem 3, we can find $x_1 > 0$, such that

$$\frac{x\varphi(x)}{\Phi(x)} < \frac{(a_\Psi - \delta)}{(a_\Psi - \delta) - 1}, \qquad x \geq x_1,$$

and then $b_\Phi < (a_\Psi - \delta)/(a_\Psi - \delta - 1)$. This gives $b_\Phi a_\Psi \leq a_\Psi + b_\Phi$ since $\delta > 0$ is allowed to go to zero. Thus $a_\Psi b_\Phi = a_\Psi + b_\Phi$ as desired. ∎

As an illustration, if $\Phi(x) = |x|^p/p$, $p > 1$, so that $\Psi(x) = |x|^q/q$, $p^{-1} + q^{-1} = 1$, one has $b_\Phi = p$, $a_\Psi = q$.

Remark. In this connection, one should note that, in general, for a complementary pair (Φ, Ψ) of N-functions, neither Φ nor Ψ need to satisfy a Δ_2 (hence ∇_2) condition, even if one of them has growth property of at most a polynomial (as in Corollary 5). The following simple example of Krasnosel'skii and Rutickii illustrates this point.

Example. Let $\varphi : \mathbb{R}^+ \to \mathbb{R}^+$ be given by $\varphi(t) = t$, if $0 \leq t < 1$, and $= n!$ if $(n-1)! \leq t < n!$, $n \geq 2$. Its inverse function then is given by $\psi(t) = t$, if $0 \leq t < 1$, and $= (n-1)!$ if $(n-1)! \leq t < n!$, $n \geq 2$. If Φ, Ψ are their indefinite integrals, then they are complementary N-functions such that $\Psi(t) \leq \frac{t^2}{2}$ for $t \geq 0$, and neither Φ nor Ψ satisfies a Δ_2-condition. In fact it is immediately seen that $\Phi(2 \cdot n!) > n\Phi(n!)$ and $\Psi(2 \cdot n!) > n\Psi(n!)$, $n \geq 1$.

We shall now consider another pair of growth conditions.

Definition 7. Let Φ be an N-function. Then Φ is said to satisfy a $\Delta' - (\nabla'-)$ condition, in symbols $\Phi \in \Delta'$ ($\Phi \in \nabla'$) if $\exists\, c > 0$ (and $b > 0$) such that

$$\Phi(xy) \le c\Phi(x)\Phi(y), \quad x, y \ge x_0 \ge 0 \qquad (10)$$

(and

$$\Phi(x)\Phi(y) \le \Phi(bxy), \quad x, y \ge y_0 \,). \qquad (11)$$

If $x_0 = 0$ ($y_0 = 0$) here, then these conditions are said to hold *globally*.

Let us start with an alternative form of the Δ' condition showing also that it is a subclass of Δ_2 of the N-functions.

Lemma 8. *An N-function $\Phi \in \Delta'$ iff there are $a > 0, x_1 \ge 0$ such that*

$$\Phi(axy) \le \Phi(x)\Phi(y), \quad x, y \ge x_1. \qquad (12)$$

Moreover Δ'-functions form a proper subclass of Δ_2-functions.

Proof: If $\Phi \in \Delta'$, let $a = \min(1, \frac{1}{c})$ where $0 < c < \infty$ appearing in (10). Then

$$\Phi(axy) \le a\Phi(xy) \le a \cdot c\Phi(x)\Phi(y), \quad x, y \ge x_0 \ge 0, \qquad (13)$$

by (10). Since $a \cdot c \le 1$, (12) follows.

For the converse, first observe that (12) implies that $\Phi \in \Delta_2$. Indeed, given $0 < \varepsilon < 1$, choose $y_1 \ge \max(x_1, \frac{1}{\varepsilon a})$. Then

$$\Phi(\frac{1}{\varepsilon}x) \le \Phi(ay_1 x) \le \Phi(y_1)\Phi(x), \quad x \ge x_1 \ge 0. \qquad (14)$$

This shows $\Phi \in \Delta_2$ with $K = \Phi(y_1) > 0$ as the desired constant. But then we have

$$\Phi(\frac{x}{\sqrt{a}}) \le K_1\Phi(x), \quad x \ge x_2 \ge 0 \qquad (15)$$

for any $a > 0$. Letting $c = K_1^2$, $x_0 = \max(\sqrt{a}x_1, x_2)$, we get

$$\Phi(xy) = \Phi(a\frac{x}{\sqrt{a}} \cdot \frac{y}{\sqrt{a}}) \le \Phi(\frac{x}{\sqrt{a}})\Phi(\frac{y}{\sqrt{a}}) \le K_1^2\Phi(x)\Phi(y), \quad x, y \ge x_0,$$

by (15). So (10) holds and $\Phi \in \Delta'$. But the direct part and (14) imply that each element of Δ'-class is also an element of Δ_2-class. That the latter is a properly bigger set of N-functions is shown by the following classical example:

Let $\Phi(x) = x^2/\log(e + |x|)$. Then one checks easily that Φ' exists and increases faster than x on $I\!\!R^+$ so that it is an N-function. Moreover, it is in Δ_2 since $\lim\limits_{x\to\infty} [\Phi(2x)/\Phi(x)] < \infty$. But it is not in Δ' since

$$\lim_{x\to\infty} [\Phi(x^2)/\Phi(x)^2] = +\infty,$$

contradicting (10), so that Δ_2 functions from a bigger class than the Δ'-ones. ∎

Let us now present various other equivalent conditions for the Δ'-class, in terms of complementary functions.

Proposition 9. *Let (Φ_i, Ψ_i), $i = 1, 2, 3$ be three complementary N-function pairs. Then the following conditions are mutually equivalent:*

(i) *$\exists\, a > 0, x_0 \geq 0$, such that $\Phi_1(axy) \leq \Phi_2(x)\Phi_3(y)$, $x, y \geq x_0$;*

(ii) *$\exists\, b > 0, y_0 \geq 0$, such that $\Psi_2(x)\Psi_3(y) \leq \Psi_1(bxy)$, $x, y \geq y_0$.*

Proof: (i) $\Rightarrow$ (ii). Let $u = \Phi_2(x)$, $v = \Phi_3(y)$. Then letting $u_0 = \max(\Phi_2(x_0), \Phi_3(x_0))$, we get

$$\frac{uv}{\Psi_2^{-1}(u)\Psi_3^{-1}(v)} < \Phi_2^{-1}(u)\Phi_3^{-1}(v) \leq \tfrac{1}{a}\Phi_1^{-1}(uv),\ u, v \geq u_0,\ \text{by}$$

$$\text{Prop. 1.1(ii) and (i) above,}$$

$$\leq \tfrac{1}{a}\frac{2uv}{\Psi_1^{-1}(uv)},\ \text{by Prop. 1.1(ii) again.}$$

From this chain of inequalities we get

$$\Psi_1^{-1}(uv) \leq \frac{2}{a}\Psi_2^{-1}(u)\Psi_3^{-1}(v), \qquad u, v \geq u_0. \tag{16}$$

Let $x' = \Psi_2^{-1}(u)$, $y' = \Psi_3^{-1}(v)$, $b = \frac{2}{a}$ and $y_0 = \max(\Psi_2^{-1}(u_0), \Psi_3^{-1}(u_0))$ in (16). It reduces to (ii) after applying Ψ_1 to both sides:

$$\Psi_2(x')\Psi_3(y') \leq \Psi_1(bx'y'), \qquad x', y' \geq y_0.$$

(ii) $\Rightarrow$ (i). The argument is analogous. Thus by Proposition 1.1(ii), and (ii) here

$$\frac{uv}{\Phi_1^{-1}(uv)} < \Psi_1^{-1}(uv) \le b\Psi_2^{-1}(u)\Psi_3^{-1}(v)$$

$$\le \frac{2buv}{\Phi_2^{-1}(u)\Phi_3^{-1}(v)}, \quad \text{for large } u, v.$$

Thus $\Phi_2^{-1}(u)\Phi_3^{-1}(v) \le 2b\Phi_1^{-1}(uv)$. Letting $a = \frac{1}{2b}$ and applying $\Phi_1(\cdot)$ to both sides, one gets (i). ∎

As a consequence of this result we have the following;

Corollary 10. *Let* (Φ_i, Ψ_i), $i = 1, 2$ *be a complementary pair of N-functions. Then* $\Psi_1 \succ \Psi_2$ *iff there exist* (Φ_3, Ψ_3), *another such pair, satisfying*

$$\Psi_3(x)\Psi_2(y) \le \Psi_1(xy), \qquad x, y \ge y_0 \ge 0.$$

This is a statement about the relation between complementary pairs (which are also N-functions!) and it follows from the above proposition and Proposition 2.4. The detail can be omitted.

A characterization of elements of Δ', especially in terms of complementary N-functions, was raised in the Krasnosel'skii and Rutickii monograph, and the following result due to Andô (1960) contains a solution.

Theorem 11. *Let* (Φ, Ψ) *be a complementary pair of N-functions with* φ, ψ *as their (left) derivatives. Then the following conditions are equivalent.*

  (i) $\Phi \in \Delta'$;
  (ii) $\exists$ *constants* $D > 0, x_1 \ge 0$ *such that* $\varphi(xy) \le D\varphi(x)\varphi(y)$, $x, y \ge x_1$;
  (iii) $\exists$ *constants* $c > 0, y_1 \ge 0$ *such that* $\psi(x)\psi(y) \le \psi(cxy)$, $x, y \ge y_1$;
  (iv) $\Psi \in \nabla'$.

Proof: (i) $\Rightarrow$ (ii). Since $\Phi \in \Delta'$ and hence $\Phi \in \Delta_2$ (cf. Lemma 8), we have for $x, y \ge x_1 \ge 0$,

$$xy\varphi(xy) \le \Phi(2xy) \le k\Phi(xy) \le k \cdot c \cdot \Phi(x)\Phi(y) \le kc \cdot xy\varphi(x)\varphi(y).$$

Setting $D = ck$, this gives (ii).

(ii) $\Rightarrow$ (iii). If $x_2 = \max(2, x_1)$, then by (ii)

$$\varphi(2y) \leq \varphi(x_2 y) \leq D\varphi(x_2)\varphi(y) = k_0 \varphi(y), \quad y \geq x_1, \qquad (17)$$

where $k_0 = D\varphi(x_2) > 0$. Choose $y_1 > 0$ such that $\psi(y_1) > 2x_1$, and if $0 < \varepsilon < \frac{1}{2}\varphi(y_1)$ then we get from (ii) and (17)

$$\begin{aligned}
\psi(x)\psi(x) &\leq \psi\big(\varphi(\psi(x) \cdot \psi(y))\big) \\
&\leq \psi\big(D\varphi(\psi(x))\varphi(\psi(y))\big) \leq \psi\big(Dk_0^2\varphi(\tfrac{1}{2}\psi(x))\varphi(\tfrac{1}{2}\psi(y))\big), \\
&\leq \psi\big(Dk_0^2\varphi(\psi(x) - \varepsilon)\varphi(\psi(y) - \varepsilon)\big) \\
&\leq \psi(Dk_0^2 xy), \qquad x, y \geq y_1.
\end{aligned}$$

This gives (iii) if we let $c = Dk_0^2$.

(iii) $\Rightarrow$ (iv). Since ψ is increasing, we may take $c > 1$ in (iii). Then setting $a = 2c$ and $y_0 = y_1$, we get, using (iii),

$$\Psi(x)\Psi(y) \leq cx\psi(x)y\psi(y) \leq cxy\psi(cxy) \leq \Psi(2cxy) = \Psi(axy),$$

$x, y \geq y_0$. So (iv) holds.

(iv) $\Rightarrow$ (i). This is just Proposition 9 (cf. also Lemma 8). $\blacksquare$

In Definition 7 we also introduced a condition ∇' which implies that the N-functions should not grow slower than a polynomial. Thus the following statement is natural and not unexpected.

Proposition 12. *If $\Phi \in \Delta' \cap \nabla'$, then $\Phi \sim \Phi_p$ where $\Phi_p(x) = |x|^p$, $p > 1$, and conversely. Thus Φ common to these classes is equivalent to Φ_p for some $p > 1$.*

Proof: We only need to establish the direct part since the converse is obvious. Thus let $\Phi \in \Delta' \cap \nabla'$. Then for some $a_1 > 0, a_2 > 0$ and $x_0 \geq 0$,

$$\Phi(a_1 xy) \leq \Phi(x)\Phi(y) \text{ and } \Phi(x)\Phi(y) \leq \Phi(a_2 xy), \quad x, y \geq x_0. \qquad (18)$$

Since Φ is convex, we may take $a_2 \geq a_1$ here. Then letting $f_i(u) = \log \Phi(\frac{1}{a_i}e^u)$, and $u_0 = \max(1, \log a_2 x_0)$ we have (18) expressible as

$$f_1(u+v) \leq f_1(u) + f_1(v), \quad f_2(u) + f_2(v) \leq f_2(u+v), \quad u, v \geq u_0. \qquad (19)$$

It may be assumed that $u \geq v$ so that $nv \leq u < (n+1)v$ for some $n \geq 1$. So the first inequality of (19), after substitution, gives

$$f_1(u) \leq f_1((n+1)v) \leq (n+1)f_1(v) \leq \frac{u+v}{v}f_1(v).$$

Hence

$$\limsup_{u \to \infty} \frac{f_1(u)}{u} \leq \limsup_{u \to \infty} \left(\frac{u+v}{u} \cdot \frac{f_1(v)}{v}\right) \leq \frac{f_1(v)}{v}. \qquad (20)$$

It follows from this that

$$\limsup_{u \to \infty} \frac{f_1(u)}{u} \leq \liminf_{v \to \infty} \frac{f_1(v)}{v} < \infty. \qquad (21)$$

Thus there is equality in (21), and let $p_1 = \lim\limits_{u \to \infty} (f_1(u)/u) < \infty$. It is clear from (20) that

$$0 < p_1 \leq [\log \Phi(\tfrac{x}{a_1})/\log x], \qquad x \geq x_1 = e^{u_0},$$

so that

$$(a_1 x)^{p_1} \leq \Phi(x), \qquad \text{for } x \geq (x_1/a_1).$$

Next using the second part of (19), we get by a similar argument,

$$\Phi(x) \leq (a_2 x)^{p_2}, \qquad x \geq (x_1/a_2),$$

for some $0 < p_2 < \infty$. But since $\lim\limits_{x \to \infty} (\Phi(x)/x) = \infty$, $p_i > 1$. Moreover,

$$p_1 = \lim_{x \to \infty} \left(\frac{\log \Phi(x)}{\log a_1 x}\right) = \lim_{x \to \infty} \left(\frac{\log \Phi(x)}{\log x}\right) = \lim_{x \to \infty} \left(\frac{\log \Phi(x)}{\log a_2 x}\right) = p_2$$

$$(= p, \text{ say}). \qquad (22)$$

Hence $(a_1 x)^p \leq \Phi(x) \leq (a_2 x)^p$, $x \geq (x_1/a_1)$ so $\Phi \sim \Phi_p$ as desired. ∎

The situation for the Δ_2-case is somewhat different and is given by

Proposition 13. *An N-function Φ is in $\Delta_2 \cap \nabla_2$ iff there exists an N-function $\tilde{\Phi}$ such that*

$$\Phi\left(\frac{x}{\sqrt{2}}\right)\tilde{\Phi}\left(\frac{y}{\sqrt{2}}\right) \leq \Phi(xy) \leq \Phi(x)\tilde{\Psi}(y), \qquad x, y \geq y_0, \qquad (23)$$

where $\tilde{\Psi}$ is complementary to $\tilde{\Phi}$.

Proof: Suppose that $\Phi \in \Delta_2 \cap \nabla_2$. Then $\Phi \rightarrowtail \Phi$ and by Proposition 2.4(ii) there exists another N-function Ψ_1 such that the second half of (23) holds with Ψ_1 for $\tilde{\Phi}$. Next considering the complementary function Ψ of Φ, since $\Phi \in \nabla_2$ also, $\Psi \in \Delta_2$ (by Theorem 3), and hence there is an N-function Ψ_2 such that the last part of (23) holds with Ψ and Ψ_2 in place of Φ and $\tilde{\Phi}$. Let $\tilde{\Psi} = \max(\Psi_1, \Psi_2)$, $w_0 = \max(x_0, x_0')$, the latter going with Ψ_1, Ψ_2. Then $\tilde{\Psi}$ is an N-function and

$$\Phi(xy) \leq \Phi(x)\tilde{\Psi}(y), \qquad \Psi(xy) \leq \Psi(x)\tilde{\Psi}(y), \qquad x, y \geq w_0. \qquad (24)$$

But by Proposition 9, this implies (with identification of this Ψ as Φ_1 there etc.) that

$$\Phi(x)\tilde{\Phi}(y) \leq \Phi(bxy), \qquad x, y \geq x_0 \qquad (25)$$

where $\tilde{\Phi}$ is complementary to $\tilde{\Psi}$. The proof there shows that $b = \frac{2}{a} = 2$ since $a = 1$ here. Hence (23) holds with $x, y \geq \max(w_0, \sqrt{2}x_0)$.

For the converse, suppose (23) holds. Then by Corollary 10 and Proposition 2.4(iii), we get $\Phi \succ \Phi$ and $\Phi \rightarrowtail \Phi$ so that $\Phi \in \nabla_2$ and Δ_2 at the same time, giving the assertion. ∎

It should be observed that both Δ' and Δ_2 are conditions that make the N-functions grow *moderately*. Thus they have no more than a polynomial growth. Moreover, if $\Phi \in \Delta'$ or $\Phi \in \Delta_2$ then any N-function $\tilde{\Phi}$ equivalent to Φ (i.e., $\Phi \prec \tilde{\Phi}$ and $\tilde{\Phi} \prec \Phi$), satisfies $\tilde{\Phi} \in \Delta'$ or $\tilde{\Phi} \in \Delta_2$, respectively. This follows from the preceding work without much difficulty. In contrast to Δ_2 which uses inequalities based on a single variable, Δ' needs the corresponding condition based on two variables. Consequently, the following natural question was raised by Krasnosel'skii and Rutickii in their monograph: Does there exist in each equivalence class of N-functions from Δ', a function which satisfies Δ' globally? A negative solution to this problem is furnished by the following example (see Andô (1960)).

Example. Let $\Phi(x) = (1 + |x|) \log(1 + |x|) - |x|$, $x \in \mathbb{R}$. We assert that $\Phi \in \Delta'$, but not globally, and that there is no $\tilde{\Phi}$ belonging to Δ' globally, with $\Phi \sim \tilde{\Phi}$.

Since $\varphi(x) = \log(1+x)$, $x > 0$, it follows from Theorem 11(ii) that $\Phi \in \Delta'$ because

$$\varphi(xy) = \log(1 + xy) \leq \log(1 + x) + \log(1 + y) \leq \varphi(x) \cdot \varphi(y),$$

$x, y > x_0 = e^2 - 1$. But if this $\Phi \in \Delta'$ globally, then (10) holds for all $x, y \geq 0$ and for a fixed $c > 0$. Hence $f(x, y) = (\Phi(x)\Phi(y)/\Phi(xy)) \geq \frac{1}{c} > 0$ for all $x, y \geq 0$. But $f(n, \frac{1}{n}) \to 0$ as $n \to \infty$, for this Φ, and not bounded below by $c^{-1} > 0$. So it is not in Δ' globally.

If possible let there be an N-function $\tilde{\Phi} \sim \Phi$ such that $\tilde{\Phi}$ is in Δ' globally. Then $\tilde{\Phi}(xy) \leq c\tilde{\Phi}(x)\tilde{\Phi}(y)$, $c > 0$, all $x, y \geq 0$, and for some $0 < a \leq b < \infty$, $\Phi(ax) \leq \tilde{\Phi}(x) \leq \Phi(bx)$, $x \geq x_0 > 0$. To see that these two inequalities cannot hold, let $0 < x < 1, y \geq x_0/x$. Then

$$\frac{\tilde{\Phi}(xy)}{\tilde{\Phi}(y)} \geq \frac{\Phi(axy)}{\Phi(by)} = \frac{(1 + axy)\log(1 + axy) - axy}{(1 + by)\log(1 + by) - by},$$

and hence,

$$\liminf_{y \to \infty} \frac{\tilde{\Phi}(xy)}{\tilde{\Phi}(y)} \geq \frac{ax}{b}, \quad \text{and} \quad \limsup_{y \to \infty} \frac{\tilde{\Phi}(xy)}{\tilde{\Phi}(y)} \leq c\tilde{\Phi}(x).$$

Thus $\frac{a}{cb} \leq \frac{\tilde{\Phi}(x)}{x}$, for all $0 < x < 1$. But this is impossible since for an N-function $\lim_{x \to 0} \frac{\tilde{\Phi}(x)}{x} = 0$. So $\tilde{\Phi}$ cannot be in Δ' globally.

Note that there are elements of ∇' whose complementary functions are in Δ'. Here is an example. Let $\Phi : I\!\!R^+ \to I\!\!R^+$ be defined by

$$\Phi(x) = (1 + x)^{\sqrt{\log(1+x)}} - 1, \qquad \Phi(-x) = \Phi(x).$$

Then $\Phi \in \nabla'$ and its complementary function $\Psi \in \Delta'$. In fact, one can verify that $\Phi \in \Delta' \cap \nabla'$ iff there exist $a > 0, b > 0$ $x_0 \geq 0$, such that

$$a \leq \frac{\Phi^{-1}(xy)}{\Phi^{-1}(x)\Phi^{-1}(y)} \leq b \qquad \text{for } x, y \geq x_0.$$

2.4. Generalized and normalized Young functions

In both the general study of function spaces and their applications it will be very convenient to consider normalized as well as generalized Young functions. The former are especially useful in comparing the results with the Lebesgue space theory. Hence we present these in this brief section.

Recall that if $\Phi(x) = |x|^p/p$, $p \geq 1$, then its complementary function Ψ is given by $\Psi(y) = |y|^q/q$, where $p^{-1} + q^{-1} = 1$. In fact, using the equality condition in Young's inequality we see that $\Phi(1) + \Psi(1) = 1$, since for $p > 1$, $(p-1)(q-1) = 1$ and this identity still holds as $p \searrow 1$ so that $q \nearrow \infty$ in the last relation. This situation can be transferred to the general case of Young functions, which was first noted by M. Billik (1957) and effectively used by A. Zygmund (1959).

Thus let Φ be a continuous Young function on $I\!\!R^+ \to I\!\!R^+$. Suppose that $(\Phi(x)/x) \nearrow \infty$ as $x \to \infty$, and let Ψ be the complementary function of Φ. Then, as we have seen in Section 1.3, Ψ has the similar properties. Let φ, ψ be the (left) derivatives of Φ, Ψ. But $\varphi : k \mapsto \varphi(k)$ is an increasing function and hence the area of the rectangle with vertices $(0,0)$, $(0,k)$, $(k,0)$ and $(k, \varphi(k))$ has area $k\varphi(k)$ which is continuous in k. Choose k_0 such that $\varphi(k_0) = k_0^{-1}$. Then the area of this rectangle at $(k_0, 0)$ will be unity, and $\tilde{\Phi} : x \mapsto \Phi(k_0 x)$ is a Young function, and if $\tilde{\Psi}$ is its complementary function, we have $\tilde{\Phi}'(x) = k_0 \varphi(k_0 x)$ for which $k_0 \varphi(k_0) = 1$ holds. Hence there is equality in Young's inequality and $\tilde{\Psi}(y) = \Psi(\frac{y}{k_0})$ so that

$$\tilde{\Phi}(1) + \tilde{\Psi}(1) = 1. \tag{1}$$

But observe that $\tilde{\Phi}(x) = \Phi(k_0 x)$ implies $\Phi \sim \tilde{\Phi}$, i.e., the functions are (globally) equivalent. This shows that every continuous Young function $\Phi : I\!\!R^+ \to I\!\!R^+$ such that $(\Phi(x)/x) \nearrow \infty$, can be replaced by another Young function of a similar nature for which (1) always holds. If, however, $(\Phi(x)/x) \nearrow \alpha < \infty$ as $x \nearrow \infty$, so that $\varphi(x) \le \alpha$ for all $x > 0$, we can still use the same procedure except that the inverse ψ of φ is finite on the interval $[0, \alpha)$ and $= +\infty$ on its complement. For such a pair (Φ, Ψ) one can define an equivalent pair $(\tilde{\Phi}, \tilde{\Psi})$ satisfying (1) again, only the (essentially) trivial two valued (i.e., 0 and $+\infty$) Φ may not be used for this definition. In the latter case we consider instead that $\tilde{\Phi}(x) = 0$ for, $0 \le x \le 1$, and $\tilde{\Psi}(y) = 0$ if $y = 0$, and $+\infty$ if $y > 0$. With such a definition we can thus replace each non-two-valued complementary pair (Φ, Ψ) with an equivalent pair $(\tilde{\Phi}, \tilde{\Psi})$ for which (1) holds. When this is done, we say that $(\tilde{\Phi}, \tilde{\Psi})$ is a *normalized complementary Young pair*. Then we drop the "tildes" and explicitly state it as a normalized pair. Thus $\Phi(x) = |x|^p/p$, $p \ge 1$ and it complementary function $\Psi(\cdot) : y \mapsto |y|^q/q$ ($= 0$ on $(0,1)$ if $p = 1$) together form a classical normalized Young pair (Φ, Ψ).

It may be noted that we may summarize the above as: Given a complementary pair (Φ, Ψ) of Young functions at least one of which is (continuous, finite, and) positive on an open interval of $I\!\!R^+$, then it can be replaced by an equivalent normalized complementary Young pair. All the preceding classifications based on the growth and order properties evidently remain valid if these N-functions are replaced by the corresponding normalized ones. We shall do this according to the convenience of a given problem.

For some function spaces, especially those considered in Chapter X and in certain applications, it will be necessary to consider Φ which are monotone increasing but not necessarily convex. Since they serve a similar purpose as in the case that Φ is also convex, we shall say that a function $\Phi_1 : I\!R^+ \to I\!R^+$ such that $\Phi_1(0) = 0$, $\Phi_1(x) \nearrow \infty$ as $x \nearrow \infty$ will be termed a *generalized Young function*, and its complementary function Ψ_1 is defined as in Section 1.3, namely,

$$\Psi_1(y) = \sup\{xy - \Phi_1(x) : x \geq 0\}, \qquad y \geq 0. \tag{2}$$

Evidently $\Psi_1(0) = 0$, and if $0 < \alpha < 1$, then

$$\Psi_1(\alpha y) = \sup\{\alpha xy - \Phi_1(x) : x \geq 0\}$$
$$\leq \sup\{\alpha(xy - \Phi_1(x)) : x \geq 0\} = \alpha \Psi_1(y),$$

and similarly $\Psi_1(\alpha y_1 + \beta y_2) \leq \alpha \Psi_1(y_1) + \beta \Psi_1(y_2)$, $0 < \alpha < 1, \beta = 1 - \alpha$, $y_i \geq 0$ so that Ψ_1 is a Young function. We extend it as $\Phi_1(-x) = \Phi_1(x)$, $\Psi_1(-y) = \Psi_1(y)$.

An example of the generalized function is $\Phi_1(x) = |x|^p$, $0 < p < \infty$. We even have $\Phi_1(2x) \leq C\Phi_1(x)$, $x \geq x_0$ as a meaningful growth condition for the generalized case. However the complementary function of Ψ_1 is not necessarily Φ_1 since it need not be convex. Thus many of the relations of the preceding two sections do not necessarily hold for the generalized case.

Other increasing functions $\Phi_2 : I\!R^+ \to I\!R^+$ that are of interest in some applications of Chapter IX are given by $\tilde{\Phi} : x \mapsto \Phi_2(|x|^s)$, $0 < s \leq 1$ with Φ_2 a Young function. Such a $\tilde{\Phi}$ is called an *s-convex* function. Thus for x, y in $I\!R^+$,

$$\tilde{\Phi}(\alpha x + \beta y) \leq \alpha^s \tilde{\Phi}(x) + \beta^s \tilde{\Phi}(y), \qquad \alpha \geq 0, \beta \geq 0, \alpha^s + \beta^s = 1, \tag{3}$$

and it becomes a Young function for $s = 1$. These are useful in generalizing the Hardy classes; and for an example, let $\tilde{\Phi}(x) = |x|^{ps}$, $0 < s \leq 1$ and $p \geq 1$ so that $0 < ps < \infty$ is possible. We shall discuss more about these functions at appropriate places.

Thus far Φ was always defined in $I\!R$. If the latter is replaced by $I\!R^n$, $n \geq 2$, we cannot use the definition of Ψ as in (I.3.13) and (I.3.14). However, the direct method of (I.3.1) applies even when $I\!R^n$ is replaced by an infinite dimensional space. We discuss this in Chapter X.

2.5. Supplementary results on N-functions

Although we considered several types of N-functions in the preceding sections, we have not analyzed in detail those that grow essentially more rapidly than a polynomial. We present some of these and related results in this section.

Definition 1. Let $\Phi : I\!R^+ \to I\!R^+$ be an N-function. Then it is said to satisfy: (i) a Δ_3-condition, denoted $\Phi \in \Delta_3$, if for some $b > 0$,

$$x\Phi(x) \leq \Phi(bx), \qquad x \geq x_0 \geq 0. \tag{1}$$

and (ii) a ∇_3-condition, denoted $\Phi \in \nabla_3$, if for some $k > 0$,

$$\Phi^{-1}(x)\Phi(x) \leq kx^2, \qquad x \geq x_0 \geq 0. \tag{2}$$

Since (1) can be expressed as

$$\Phi(x) \leq \frac{1}{x}\Phi(bx) \leq \frac{1}{2b}\Phi(bx), \quad x \geq \max(x_0, 2b) > 0,$$

it follows that $\Phi \in \Delta_3 \Rightarrow \Phi \in \nabla_2$. Also Δ_3 is a class condition in that $\Phi \sim \tilde{\Phi}$ and $\Phi \in \Delta_3 \Rightarrow \tilde{\Phi} \in \Delta_3$. If $x_0 = 0$ in (1) and (2), then they are *global* conditions. The functions satisfying the Δ_3-condition grow at the following rate:

Proposition 2. *If $\Phi \in \Delta_3$, then for some $\alpha > 0$ and $x_1 > 0$, we have*

$$\Phi(x) \geq x^{\alpha \log x}, \qquad x \geq x_1. \tag{3}$$

Proof: Since Φ is increasing and satisfies a Δ_3-condition we may take $b > 1$ and $x_0 > 1$ in (1) and $\Phi(x_0) > 1$, so that with $y = \log x$, $y_0 = \log x_0$, (1) becomes

$$\exp(y + \log \Phi(e^y)) \leq \exp(\log \Phi(e^{y+\log b})), \quad y \geq y_0.$$

Letting $f(y) = \log \Phi(e^y)$, this gives

$$f(y + \log b) \geq y + f(y), \quad y \geq y_0.$$

Since $f(\cdot) > 0$ and increasing, for each $y > y_0$ we can choose an integer $n \geq 1$ such that $y_0 + (n-1)\log b \leq y < y_0 + n\log b$ so that

$$f(y) \geq f(y_0 + (n-1)\log b) \geq y_0 + (n-2)\log b + f(y_0 + (n-2)\log b)$$
$$\geq \cdots \geq y_0 + (y_0 + \log b) + \cdots + (y_0 + (n-2)\log b) + f(y_0)$$
$$= (n-1)y_0 + \frac{(n-1)(n-2)}{2}\log b + f(v_0) \geq \frac{1}{2}(n-1)(n-2)\log b$$
$$\geq \frac{\log b}{2}\left(\frac{y - y_0}{\log b} - 1\right)\left(\frac{y - y_0}{\log b} - 2\right). \tag{4}$$

Choose $y_1 \geq y_0$ such that for $y \geq y_1$ we have $y - y_0 - 2\log b \geq \frac{1}{2}y$. Then we get from (4) immediately

$$f(y) \geq y^2/(8\log b), \quad y \geq y_1.$$

Consequently,

$$\log \Phi(x) \geq \frac{1}{8\log b}(\log x)^2, \quad x \geq e^{y_1}. \tag{5}$$

Set $\alpha = (8\log b)^{-1}$, $x_1 = e^{y_1}$, in (5). It reduces to (3). ■

As examples of Φ in Δ_3, we have $\Phi(x) = e^{|x|} - |x| - 1$ or $e^{x^2} - 1$, $x^{\log x}$ for $x > e$ and $= \frac{x}{e}$ for $0 \leq x \leq e$, but $\Phi(x) = [(1 + |x|)^{\sqrt{\log(1+|x|)}} - 1]$, is not in Δ_3.

We shall now present equivalent conditions for Δ_3 (and ∇_3) in:

Theorem 3. *Let (Φ, Ψ) be a complementary pair of N-functions. Then the following are equivalent conditions:*

(i) $\Phi \in \Delta_3$;

(ii) $\exists$ *constants* $k_1 > 0, x_1 \geq 0$, *such that* $x\varphi(x) \leq \varphi(k_1 x)$, $x \geq x_1$;

(iii) $\exists$ *constants* $k_2 > 0, y_1 \geq 0$, *such that* $\psi(y\psi(y)) \leq k_2\psi(y)$, $y \geq y_1$;

(iv) $\Psi \in \nabla_3$; *[As usual φ, ψ are the (left) derivatives of Φ, Ψ in (ii) and (iii) above.]*

(v) $\exists$ *constants* $k > 0, y_0 \geq 0$, *such that* $\Psi(y) \leq ky\Phi^{-1}(y)$, $y \geq y_0$;

(vi) $\Phi \succ \Psi \circ \Phi$, *so that* $\Psi(\Phi(x)) \leq \Phi(cx)$ *for some* $c > 0$, $x \geq \tilde{x} \geq 0$.

Proof: (i) $\Rightarrow$ (ii) Using the conditions of (i) we get for some $b > 0$

$$x\varphi(x) \le \Phi(2x) \le \frac{1}{2x}\Phi(2bx), \qquad x \ge x_0 > 0,$$

$$\le \frac{1}{2x} \cdot 2bx\varphi(2bx) \le 2b\varphi(2bx),$$

so that replacing x by $2bx$ the extreme terms give

$$2bx\varphi(2bx) \le 2b\varphi(4b^2x), \qquad x \ge x_0/2b.$$

Letting $k_1 = 4b^2$, $x_1 = \max(x_0/2b, 2b)$ here we get (ii) since

$$x\varphi(x) \le 2b\varphi(2bx) \le x\varphi(2bx) \le \varphi(4b^2x) = \varphi(k_1x), \qquad x \ge x_1.$$

(ii) $\Rightarrow$ (iii) Because of the monotonicity of φ, $k_1 \ge 1$ may be assumed. If we choose $y_1 > 0$ such that $\psi(y_1) \ge \max(x_1, 2)$, and set $k_2 = k_1^2$, then (ii) gives

$$\psi(y\psi(y)) \le \psi[\varphi(\psi(y)) \cdot \psi(y)] \le \psi(\varphi(k_1\psi(y)))$$

$$\le \psi[\frac{1}{k_1\psi(y)}\varphi(k_1^2\psi(y))]$$

$$\le \psi(\frac{1}{2}\varphi(k_2\psi(y))) \le \psi(\varphi(k_2\psi(y)) - \varepsilon) \le k_2\psi(y), \qquad y \ge y_1,$$

if $0 < \varepsilon < \frac{1}{2}\varphi(k_2\psi(y))$ is taken. Thus (iii) holds.

(iii) $\Rightarrow$ (iv) Let $w_0 \ge 2y_1$ to satisfy $\psi(\frac{w_0}{2}) \ge 2$. Let $k = 4k_2^3$. Then for $w \ge w_0$,

$$w\Psi(\Psi(w)) \le w\Psi(w\psi(w)) \le w^2\psi(w) \cdot \psi(w \cdot \psi(w))$$

$$\le w^2k_2\psi^2(w) \le 4k_2\left(\frac{w}{2}\right)^2\psi^2\left[\frac{w}{2} \cdot \psi(\frac{w}{2})\right]$$

$$\le 4k_2^3\left(\frac{w}{2}\right)^2\psi^2\left(\frac{w}{2}\right) \le k\Psi^2(w).$$

Let $y = \Psi(w)$, $y_0 = \Psi(w_0)$. Then this gives $\Psi^{-1}(y)\Psi(y) \le ky^2$, $y \ge y_0$, so $\Psi \in \nabla_3$.

(iv) $\Rightarrow$ (v) This is immediate, since

$$\Psi(y) \le \frac{ky^2}{\Psi^{-1}(y)} \le ky\Phi^{-1}(y), \qquad y \ge y_0, \text{ by Prop. 1.1(ii).}$$

(v) $\Rightarrow$ (vi) Let $c = 2k$, $\tilde{x} = \Phi^{-1}(y_0)$, so that (v) becomes

$$2\Psi(\Phi(x)) \le 2kx\Phi(x) \le \Phi(2kx) + \Psi(\Phi(x)), \qquad x \ge \tilde{x} \ge 0.$$

Hence $(\Psi \circ \Phi)(x) \le \Phi(cx)$, $x \ge \tilde{x}$, and (vi) holds.

(vi) $\Rightarrow$ (i) We may take $c \geq 1$, so that by Young's inequality again

$$x\Phi(x) \leq \Phi(x) + \Psi(\Phi(x)) \leq \Phi(x) + \Phi(cx), \quad \text{by (vi)}$$
$$\leq 2\Phi(cx) \leq \Phi(2cx), \quad \text{by convexity, and } x \geq \tilde{x}.$$

This implies that $\Phi \in \Delta_3$, and (i) holds. ■

We have the following useful consequence.

Corollary 4. *If an N-function* $\Phi \in \nabla_3$, *then* $\Phi \in \Delta_2$.

Proof: Let Ψ be the complementary Young function to Φ. Then by the above theorem (part (iv)), $\Psi \in \Delta_3$ iff $\Phi \in \nabla_3$. But as noted after Definition 1, $\Psi \in \Delta_3 \Rightarrow \Psi \in \nabla_2$. However by Theorem 3.3(1), the last holds iff $\Phi \in \Delta_2$. This is precisely the assertion. ■

Definition 5. An N-function Φ satisfies a Δ^2-condition, denoted $\Phi \in \Delta^2$, if $\Phi \sim \Phi^2$ so that there exist $b > 0$ and $x_0 \geq 0$ such that

$$\Phi^2(x) \leq \Phi(bx), \qquad x \geq x_0. \tag{6}$$

Examples of $\Phi \in \Delta^2$ are the Young functions of exponential growth, i.e., $\Phi(x) = e^{|x|^\delta} - 1$, $\delta \geq 1$. Also note that $\Phi \in \Delta^2 \Rightarrow \Phi \in \Delta_3$, since by (6), $\Phi^2(x) \leq \Phi(kx)$, $x \geq x_0$ and $k > 1$ and $\Phi(x) > x$ for $x > x_1$ so that for $x \geq \max(x_0, x_1)$ we get $x\Phi(x) \leq \Phi^2(x) \leq \Phi(kx)$, which is Δ_3. The following example gives a better insight. Consider

$$\Phi_r(x) = \exp[(\log x)^{r+1}], \text{ if } x \geq e; = x^{r+1}/e^r, \text{ if } 0 \leq x < e,$$

where $r > 0$. Then Φ_r is an N-function and one may verify that for $r \geq 1$, $\Phi_r \in \Delta_3$ but $\Phi_r \notin \Delta^2$, and if $0 < r < 1$, then $\Phi_r \notin \Delta_3$. A computation (use Prop. 2 for the second part) shows the truth of both assertions, and we omit the detail.

To see that the Φ satisfying Δ^2-condition have exponential growth, we establish the following.

Proposition 6. *If* $\Phi \in \Delta^2$, *then for some* $\alpha > 0$ *and* $x_1 \geq 0$, *we have*

$$\Phi(x) \geq \exp(x^\alpha), \qquad x \geq x_1. \tag{7}$$

Proof: Since Φ is convex ($\Phi \in \Delta^2$), in (6) we may take $b \geq 2, x_0 \geq 1$ so that

$$2 \log \Phi(x) \leq \log \Phi(bx), \qquad x \geq x_0.$$

Hence, given $x \geq x_0$, $\exists$ an integer $n \geq 1$, such that $b^n x_0 \leq x < b^{n+1} x_0$. From the above inequality we have for each such n,

$$
\begin{aligned}
\log \Phi(x) &\geq \log \Phi(b^n x_0) \geq 2^n \log \Phi(x_0) \\
&\geq \left(\frac{x}{bx_0}\right)^{\log_b 2} \log \Phi(x_0) \\
&= \frac{\log \Phi(x_0)}{(bx_0)^{\log_b 2}} x^{\log_b 2},
\end{aligned}
\tag{8}
$$

where $\log_b 2$ is the logarithm of 2 to the base b. Let $0 < \alpha < \log_b 2$ and $x_1 \geq x_0$, such that

$$
x^{\log_b 2 - \alpha} \geq \frac{(bx_0)^{\log_b 2}}{\log \Phi(x_0)}, \qquad x \geq x_1.
\tag{9}
$$

From (8) and (9) we get $\log \Phi(x) \geq x^\alpha$, $x \geq x_1$, giving (7). ∎

We now introduce a condition "opposite" to Δ^2 to characterize the latter.

Definition 7. An N-function Φ satisfies the ∇^2-condition, denoted $\Phi \in \nabla^2$, if for some $c > 0$ and $x_0 \geq 0$, we have

$$
\frac{\Phi(x)}{x} \leq c \frac{\Phi(\sqrt{x})}{\sqrt{x}}, \qquad x > x_0.
\tag{10}
$$

Both Δ^2 and ∇^2 are class conditions in that equivalent N-functions either do or do not satisfy these simultaneously. We now prove a characterization of Δ^2 (and ∇^2) in

Theorem 8. *Let (Φ, Ψ) be a complementary pair of N-functions. Then the following are mutually equivalent statements:*
  (i) $\Phi \in \Delta^2$;
  (ii) $\exists k_1 > 0$, $x_1 \geq 0$, *such that* $\varphi^2(x) \leq \varphi(k_1 x)$, $x \geq x_1$;
  (iii) $\exists k_2 > 0$, $y_1 \geq 0$, *such that* $\psi(y^2) \leq k_2 \psi(y)$, $y \geq y_1$;
  (iv) $\Psi \in \nabla^2$; *[As usual φ, ψ are the (left) derivatives of Φ, Ψ in the above.]*

Proof: (i) $\Rightarrow$ (ii) Since $\Phi \in \Delta^2$, as we noted preceding Proposition 6, $\Phi \in \Delta_3$, and so by Theorem 3(ii) there are $b > 1$, $x_0 \geq 1$ such that

$$
x\varphi(x) \leq \varphi(bx), \quad \text{and also } \Phi^2(x) \leq \Phi(bx), \ x \geq x_0.
\tag{11}
$$

Hence using $\Phi(2x) \geq x\varphi(x)$, $x \geq 0$, and setting $k_1 = 2b^2$, we get

$$\varphi^2(x) \leq \frac{1}{x^2}\Phi^2(2x) \leq \frac{1}{x^2}\Phi(2bx)$$

$$\leq \frac{1}{x^2} \cdot 2bx\varphi(2bx) \leq 2bx\varphi(2bx), \quad x \geq x_0 \geq 1,$$

$$\leq \varphi(2b^2 x) = \varphi(k_1 x), \quad \text{by the first part of (11).}$$

Taking $x_1 = x_0 \geq 1$, proves (ii).

(ii) $\Rightarrow$ (iii) When (ii) holds, if we let $y = \varphi(x)$, and $k_2 = 4k_1$, then

$$\psi(y^2) \leq \psi(\varphi(k_1 x)) \leq 2k_1 x \leq 4k_1\psi(y) = k_2\psi(y), \quad y \geq y_1 = \varphi(x_1).$$

Thus (iii) holds.

(iii) $\Rightarrow$ (iv) If $y \geq y_0 = \max(2y_1, 4)$, then by (iii) we have

$$\psi(y) = \psi(2 \cdot \frac{y}{2}) \leq \psi\left(\left(\frac{y}{2}\right)^2\right) \leq k_2\psi(\frac{y}{2}). \tag{12}$$

Let $c = 2k_2^2$ in (12). Hence (iii) implies

$$\Psi(y^2) \leq y^2\psi(y^2) \leq k_2 y^2 \psi(y) \leq 2k_2^2 y \cdot \frac{y}{2}\psi(\frac{y}{2}) \leq cy\Psi(y), \quad y \geq y_0.$$

Thus (iv) holds.

(iv) $\Rightarrow$ (i) Since Ψ is an N-function, if we define $\xi(y) = \frac{\Psi(y)}{y}$, we get
$\xi(y) \searrow 0$ as $y \searrow 0$ and $\xi(y) \nearrow \infty$ as $y \nearrow \infty$, so that $\Psi_1(y) = \int_0^{|y|} \xi(t)dt$
is an N-function. Also $\Psi_1(y) \leq \Psi(y)$ and $\Psi_1(2y) \geq \Psi(y)$. Thus
$\Psi \sim \Psi_1$ and since ∇^2 is a class condition we have $\Psi_1 \in \nabla^2$, with
derivative ξ. But $\Psi \in \nabla^2 \Rightarrow \Psi$ satisfies (10) so that $\xi(y) < k\xi(\sqrt{y})$ or
letting $w = \xi(y)$ we get $\frac{w}{k} < \xi(\sqrt{\xi^{-1}(w)})$. Hence $\xi^{-1}(\frac{w}{k}) < [\xi^{-1}(w)]^{\frac{1}{2}}$.
Let $\varphi_1(x) = \xi^{-1}(x)$. Then the last inequality can be expressed as

$$\varphi_1^2(x) \leq \varphi_1(kx), \quad \text{for } x \geq x_0. \tag{13}$$

We now iterate the procedure as in the proof of Proposition 6 (see
equation (8)) to deduce that $\varphi_1(x) > \exp(x^\alpha)$ for some $\alpha > 0$ and
$x \geq x_1 > 0$. Hence there is an $x_2 > 0$ such that $\varphi_1(x) \geq x$ for $x \geq x_2$.
If $x_2 = \max(x_0, x_1)$, then we get from (13), with the last inequality, for
the N-function Φ_1 determined by φ_1 the following:

$$\Phi_1^2(x) \leq x^2\varphi_1^2(x) \leq x^2\varphi_1(kx) \leq x\varphi_1(x) \cdot \varphi_1(kx)$$

$$\leq x\varphi_1^2(kx) \leq x\varphi_1(k^2 x) \leq k^2 x\varphi_1(k^2 x)$$

$$\leq \Phi_1(2k^2 x). \tag{14}$$

This implies that $\Phi_1 \in \Delta^2$. Since $\Phi \sim \Phi_1$ and Δ^2 is a class condition, we conclude that $\Phi \in \Delta^2$. This is (i). ∎

Remark. Another argument for the last part can be given, basing it on Proposition 1.1 which implies with the hypothesis (iv), that $\Phi^2(x) \leq \frac{x}{k}\Phi(2kx)$ for $x \geq x_1 > 0$. This replaces (13) and then the rest of the argument leading to (14) applies and gives *directly* that $\Phi \in \Delta^2$ without going through the equivalent function Φ_1. We omit the details.

Let us note the following consequence of the theorem.

Corollary 9. *An N-function $\Phi \in \nabla^2$ implies that $\Phi \in \nabla_3$.*

Proof: By the theorem $\Phi \in \nabla^2 \Leftrightarrow \Psi \in \Delta^2 \Rightarrow \Psi \in \Delta_3 \Leftrightarrow \Phi \in \nabla_3$, as desired. ∎

There is also a relation between Δ^2 and ∇' as given by

Proposition 10. *If an N-function $\Phi \in \Delta^2$, then $\Phi \in \nabla'$.*

Proof: By definition of $\Phi \in \Delta^2$, there are $b > 0$, $x_0 \geq 1$, such that $\Phi(bx) \geq \Phi^2(x)$, $x \geq x_0$. If $x, y \geq x_0$ we have by the monotonicity of Φ,

$$\Phi(bxy) \geq \Phi(bx) \geq \Phi^2(x) \geq \Phi(x)\Phi(y), \quad \text{if } x \geq y \geq x_0$$

and similarly, by interchanging x, y for the other possibility

$$\Phi(bxy) \geq \Phi(by) \geq \Phi^2(y) \geq \Phi(x)\Phi(y), \qquad y \geq x \geq x_0.$$

Hence $\Phi(bxy) \geq \Phi(x) \cdot \Phi(y)$, $x, y \geq x_0$, so that $\Phi \in \nabla'$. ∎

Thus far we have introduced four sets of (Δ, ∇)-conditions. Their interrelations, proved above, can be schematically expressed as follows.

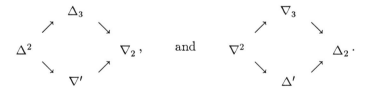

There are no comparisons between Δ_3 and ∇', or ∇_3 and Δ' and none of the arrows can be reversed, as counterexamples have been constructed by Andô (1960).

Finally we make several remarks on other possible relations that one can introduce now. These are motivated by the Δ^2-condition. If

Δ^{α} is defined as a condition for the N-function Φ as $\Phi^{\alpha} \sim \Phi$ with $\alpha > 1$, then one can verify that this implies that $\Phi^2 \sim \Phi$ and, conversely if $\Phi^2 \sim \Phi$, then it is easily seen that $\Phi^{\alpha} \sim \Phi$ for all $\alpha > 1$. Thus Δ^{α} is the same as Δ^2 for $\alpha > 1$. But if Φ_1, Φ_2 are two N-functions since obviously $\Phi_1 \circ \Phi_2$ can never be equivalent to Φ_1, one can see the other possibility which, with $\Phi_1(x) = x^2$, is denoted above as Δ^2. In this way a new condition is Δ_{Φ}, obtained if we define $\Phi_1 \in \Delta_{\Phi}$ iff $\Phi \circ \Phi_1 \sim \Phi_1$. But this also gives, if $\tilde{\Phi} = \Phi \circ \cdots \circ \Phi$ then $\Phi_1 \in \Delta_{\Phi} \Rightarrow \Phi_1 \in \Delta_{\tilde{\Phi}}$. Proceeding further, Y. W. Wang (1981) introduced a class $\cup \Delta_{\Phi}$, where the union is for all N-functions Φ, so that $\Phi_1 \in \cup \Delta_{\Phi}$ iff $\Phi_1 \in \Delta_{\Phi_0}$ for some Φ_0. Several other conditions can be considered and analyzed for the N-functions. But that will lead to the function theory of N-functions. We shall leave this line of investigation here, since that seems to have very few immediate applications in functional analysis, and turn to the definition and analysis of Orlicz spaces.

Bibliographical Notes: Most of the analysis of this chapter is both a completion and extension of the theory given in the monograph by M. A. Krasnosel'skii and Ya. B. Rutickii, and in fact contains solutions of the many problems raised there. Several of these are due to Andô (1960, 1962). Further additions and simplifications are taken from Ren (1981, 1985, 1987), Zhou (1986) among others. Some earlier work of Salehov (1968) should also be noted. We have presented a unified account and streamlined results of several authors in this account, but restricted to those aspects that are of interest in our treatment of function spaces to follow.

III
Orlicz Function Spaces

The basic structural analysis of function spaces on arbitrary measure spaces with general Young functions is contained in this chapter. These are the Orlicz spaces, and it is shown that they are Banach spaces relative to the equivalent Orlicz and gauge norms. Their separability properties and subspaces having simple functions dense are studied. The work of this chapter will be fundamental for the rest of the treatment in the book.

3.1. *Orlicz classes on general measure spaces*

As indicated in Chapter I, we shall first introduce and study the structure of Orlicz spaces on arbitrary measure spaces, and then specialize to obtain particular results. This will eliminate the extraneous restrictions on the underlying measures in the general theory of these function spaces.

Thus we start with an abstract measure space (Ω, Σ, μ) where Ω is some point set and Σ is a σ-algebra of its subsets on which a σ-additive function $\mu : \Sigma \to \bar{I\!\!R}^+$ is given. Let $\Phi : I\!\!R \to \bar{I\!\!R}^+$ be a Young function so that $\Phi(x) = \Phi(-x)$, $\Phi(0) = 0$. $\Phi(x) \nearrow \infty$ as $x \nearrow \infty$, but $\Phi(x_0) = +\infty$ for some $x_0 \in I\!\!R$ is permitted. If $f : \Omega \to \bar{I\!\!R}$ is a measurable function, then $\Phi(f) : \Omega \to \bar{I\!\!R}^+$ is measurable (for Σ), since a Young function by definition is an extended real Borel function.

Definition 1. Let $\tilde{\mathcal{L}}^\Phi(\mu)$ be the set of all $f : \Omega \to \bar{I\!\!R}$, measurable for Σ, such that $\int_\Omega \Phi(|f|)d\mu < \infty$.

Although we are considering only real functions in $\tilde{\mathcal{L}}^\Phi(\mu)$, the basic structure theory extends to the complex valued case also by considering the real and imaginary parts separately. If an additional (nontrivial) idea is needed for the complex case, it will be pointed out. Mostly the real case is considered without mention. Also to avoid some annoying repetition, we *assume that our measures μ have the finite subset property*, i.e., $E \in \Sigma$, $\mu(E) > 0 \Rightarrow \exists F \in \Sigma$, $F \subset E$ and $0 < \mu(F) < \infty$. This only eliminates the measures such as $\mu(A) = 0$ if $A = \emptyset$, $= +\infty$ if $A \neq \emptyset$. Otherwise it does not restrict the generality of our theory.

Recall that a set $A \in \Sigma$ is an *atom* for μ (or a μ-*atom*) if $\mu(A) > 0$ and for each $B \subset A, B \in \Sigma$, either $\mu(B) = 0$ or $\mu(A - B) = 0$. A set $D \in \Sigma$ is *diffuse for* μ if it does not contain any μ-atoms. This implies that for $0 \leq \lambda \leq \mu(D)$, we can find a set $D_1 \subset D, D_1 \in \Sigma$ such that $\mu(D_1) = \lambda$. Such a D is sometimes called "nonatomic" for μ.

We now discuss the structure of $\tilde{\mathcal{L}}^\Phi(\mu)$. It will follow from the next result that this space is *not* generally linear, and for this reason we put a wave (or a tilde) on top of the symbol. The precise result is given by:

Theorem 2. (i) *The space $\tilde{\mathcal{L}}^\Phi(\mu)$ introduced above is* absolutely convex, *i.e., if $f, g \in \tilde{\mathcal{L}}^\Phi(\mu)$ and α, β are scalars such that $|\alpha| + |\beta| \leq 1$, then $\alpha f + \beta g \in \tilde{\mathcal{L}}^\Phi(\mu)$. Also $h \in \tilde{\mathcal{L}}^\Phi(\mu)$, $|f| \leq |h|$, f measurable $\Rightarrow f \in \tilde{\mathcal{L}}^\Phi(\mu)$.*

(ii) *The space is linear (i.e., a vector space) if $\Phi \in \Delta_2$ globally when $\mu(\Omega) = \infty$, and locally if $\mu(\Omega) < \infty$. Conversely, this Δ_2-condition (in these two cases) is necessary if μ is diffuse on a set of positive measure.*

Remark. Hereafter we say that a Δ- (or ∇-) condition is *regular* if it holds for Φ locally when the measure in $\tilde{\mathcal{L}}^\Phi(\mu)$ is finite and globally when the measure there is infinite. This will save annoying repetition hereafter.

Proof: (i) Let $f, g \in \tilde{\mathcal{L}}^\Phi(\mu)$. Then by the monotonicity and convexity of Φ, we get with $0 < \gamma = |\alpha| + |\beta| \leq 1$,

$$\Phi(|\alpha f + \beta g|) \leq \Phi(|\alpha||f| + |\beta||g|) \leq \gamma\Phi\left(\tfrac{|\alpha|}{\gamma}|f| + \tfrac{|\beta|}{\gamma}|g|\right)$$
$$\leq |\alpha|\Phi(|f|) + |\beta|\Phi(|g|), \tag{1}$$

and the right side is integrable. Hence $\alpha f + \beta g \in \tilde{\mathcal{L}}^\Phi(\mu)$. The second statement is clear since $\Phi(|f|) \leq \Phi(|h|)$.

(ii) For linearity, it is sufficient to verify that with each $f \in \tilde{\mathcal{L}}^{\Phi}(\mu)$ $2f \in \tilde{\mathcal{L}}^{\Phi}(\mu)$, since then $nf \in \tilde{\mathcal{L}}^{\Phi}(\mu)$ for any integer n and hence also for each $\alpha > 0, \alpha f \in \tilde{\mathcal{L}}^{\Phi}(\mu)$, by the last part of (i). This then yields $af_1 + bf_2 = \gamma(\frac{a}{\gamma}f_1 + \frac{b}{\gamma}f_2) \in \tilde{\mathcal{L}}^{\Phi}(\mu)$, $\gamma = |a| + |b| > 0$, for any $f_i \in \tilde{\mathcal{L}}^{\Phi}(\mu)$, $i = 1, 2$, by (i). Thus we need to show only that $2f \in \tilde{\mathcal{L}}^{\Phi}(\mu)$ with each f in $\tilde{\mathcal{L}}^{\Phi}(\mu)$. If Φ is Δ_2-regular, then we have when $\mu(\Omega) = +\infty$, $\Phi(2|f|) \leq K\Phi(|f|)$, $K > 0$, and hence $2f$ is in $\tilde{\mathcal{L}}^{\Phi}(\mu)$ with f. In case $\mu(\Omega) < \infty$, then $\Phi(2x) \leq K\Phi(x)$, for $x \geq x_0 \geq 0$. Now let $f_1 = f$ if $|f| \leq x_0$ and $= 0$ otherwise. Set $f_2 = f - f_1$ so that $f = f_1 + f_2$ and

$$\Phi(2|f|) = \Phi(2|f_1|) + \Phi(2|f_2|) \leq \Phi(2|f_1|) + K\Phi(|f_2|).$$

Hence

$$\int_{\Omega} \Phi(2|f|)d\mu \leq \Phi(2x_0)\mu(\Omega) + K\int_{\Omega} \Phi(|f|)d\mu < \infty,$$

and $2f \in \tilde{\mathcal{L}}^{\Phi}(\mu)$. By the initial reduction, this shows that $\tilde{\mathcal{L}}^{\Phi}(\mu)$ is linear when Φ is Δ_2-regular.

For the converse, let $E \in \Sigma$ be a set of positive measure on which μ is diffuse and that Φ is not Δ_2-regular. We now construct an f in $\tilde{\mathcal{L}}^{\Phi}(\mu)$ such that $2f \notin \tilde{\mathcal{L}}^{\Phi}(\mu)$ which then proves the necessity. If $0 < \alpha < \mu(E) \leq \infty$, then by hypothesis on μ, there is an $F \subset E, F \in \Sigma$ with $\mu(F) = \alpha < \infty$. We thus construct a function supported by F to satisfy our assertions. We also assume, for nontriviality, that $\Phi(\mathbb{R}) \subset \mathbb{R}^+$.

Since $\Phi \notin \Delta_2$, there exists a sequence $x_n \geq n$ such that $\Phi(2x_n) > n\Phi(x_n)$, $n \geq 1$. Let n_0 be an integer such that

$$\sum_{n \geq n_0} \frac{1}{n^2} < \alpha, \quad \text{and} \quad \Phi(x_n) \geq 1 \text{ for all } n \geq n_0. \tag{2}$$

This is clearly possible. By diffuseness of μ on F, there is a measurable $F_0 \subset F$ such that $\mu(F_0) = \sum_{n \geq n_0} \frac{1}{n^2} < \alpha$. Similarly we can find a set $D_1 \in \Sigma, D_1 \subset F_0$ such that $\mu(D_1) = n_0^{-2}$. Since $\mu(F_0 - D_1) > 0$, we can again find $D_2 \in \Sigma, D_2 \subset F_0 - D_1$ such that $\mu(D_2) = (n_0+1)^{-2}$. Repeating the process, we find disjoint sets $D_n \in \Sigma, \mu(D_n) = (n_0+n-1)^{-2}$, $n \geq 1$. Now let $F_k \subset D_k$, $F_k \in \Sigma$, be chosen such that $\mu(F_k) =$

$\mu(D_k)/\Phi(x_n)$. Let $f = \sum\limits_{n=1}^{\infty} x_n \chi_{F_n}$. Then f is measurable and

$$\int_{\Omega} \Phi(f) d\mu = \sum_{n=1}^{\infty} \Phi(x_n) \mu(F_n) = \sum_{n \geq n_0} n^{-2} < \infty.$$

So $f \in \tilde{\mathcal{L}}^{\Phi}(\mu)$. However,

$$\int_{\Omega} \Phi(2f) d\mu = \sum_{n=1}^{\infty} \Phi(2x_n) \mu(F_n) \geq \sum_{n \geq n_0} n\Phi(x_n) \mu(F_n) = \sum_{n \geq n_0} \frac{1}{n} = +\infty.$$

Hence $2f \notin \tilde{\mathcal{L}}^{\Phi}(\mu)$, so that the latter is not a linear space. ∎

In the future we use this type of technique for converse statements, and the reader should note the basic idea. It should also be observed that in the complex case the α, β of part (i) of the theorem are complex numbers satisfying the same conditions. Any space of functions with the property of (i) is termed a *circled* and *solid* space, i.e., $f \in \tilde{\mathcal{L}}^{\Phi}(\mu) \Rightarrow h \cdot f \in \tilde{\mathcal{L}}^{\Phi}(\mu)$ for any scalar h of absolute value 1 (e.g. $h(\theta) = e^{i\theta}$), and $|g| \leq |f|, g$ measurable $\Rightarrow g \in \tilde{\mathcal{L}}^{\Phi}(\mu)$. We state this property for reference as:

Corollary 3. *The set $\tilde{\mathcal{L}}^{\Phi}(\mu)$ is a circled solid class of scalar functions and it is linear iff it is closed under multiplication by positive numbers.*

Let us note that for the necessity of Theorem 2(ii), it is essential that μ be diffuse on some set of positive measure. The following simple example illustrates this remark.

Example 4. Let $\Omega = \{1, 2, \ldots\}$, $\Sigma = $ power set of Ω, and μ be the counting measure, i.e., $\mu(\{i\}) = 1$, $i \geq 1$. Let $\Phi(x) = e^{x^2} - 1$. Then Φ is an N-function and $\Phi \notin \Delta_2$. We assert that $\tilde{\mathcal{L}}^{\Phi}(\mu)$ is a linear space. In fact, if $f \in \tilde{\mathcal{L}}^{\Phi}(\mu)$ then

$$\int_{\Omega} \Phi(f) d\mu = \sum_{n=1}^{\infty} \left(\exp(|f(n)|^2) - 1 \right) < \infty. \tag{3}$$

So the terms on the right of (3) are bounded. Let $K > 0$ be the bound so that $\exp(|f(n)|^2) \le K + 1$, $n \ge 1$. Then

$$
\begin{aligned}
\int_\Omega \Phi(2f)d\mu &= \sum_{n=1}^\infty \left(\exp(4|f(n)|^2) - 1 \right) \\
&\le \sum_{n=1}^\infty \left(\exp(|f(n)|^2) - 1 \right)(K+2)((K+1)^2 + 1) \\
&= (K+2)((K+1)^2 + 1) \int_\Omega \Phi(f)d\mu < \infty.
\end{aligned}
$$

Hence $2f \in \tilde{\mathcal{L}}^\Phi(\mu)$, and the space is linear.

Motivated by Corollary 3, we can introduce the desired space in:

Definition 5. Let $\tilde{\mathcal{L}}^\Phi(\mu)$ be the set, introduced in Definition 1, on an arbitrary measure space (Ω, Σ, μ). Then the space $\mathcal{L}^\Phi(\mu)$ of all measurable $f : \Omega \to \bar{I\!R}$, such that $\alpha f \in \tilde{\mathcal{L}}^\Phi(\mu)$ for *some* $\alpha > 0$, is called an *Orlicz space*. Thus

$$
\mathcal{L}^\Phi(\mu) = \{f : \Omega \to \bar{I\!R}, \text{ measurable } | \int_\Omega \Phi(\alpha f)d\mu < \infty
$$

$$
\text{for } some \ \alpha > 0\}. \tag{4}
$$

The following simple property of this space is noted for later use:

Proposition 6. *The set* $\mathcal{L}^\Phi(\mu)$ *of* (4) *is a vector space. Moreover for each* $f \in \mathcal{L}^\Phi(\mu)$, *there is an* $\alpha > 0$ *such that*

$$
\alpha f \in B_\Phi = \{g \in \tilde{\mathcal{L}}^\Phi(\mu) : \int_\Omega \Phi(g)d\mu \le 1\}, \tag{5}
$$

and B_Φ *is a circled solid subset of* $\tilde{\mathcal{L}}^\Phi(\mu)$.

Proof: Let $f_i \in \mathcal{L}^\Phi(\mu)$, $i = 1, 2$. Then there exist $\alpha_i > 0$ such that $\alpha_i f_i \in \tilde{\mathcal{L}}^\Phi(\mu)$, by definition. Let $\alpha = \min(\alpha_1, \alpha_2)$. Then $\alpha > 0$ and

$$
\int_\Omega \Phi(\tfrac{\alpha}{2}(f_1 + f_2))d\mu \le \tfrac{1}{2}[\int_\Omega \Phi(\alpha_1 f_1)d\mu + \int_\Omega \Phi(\alpha_2 f_2)d\mu] < \infty,
$$

where we used the convexity and monotonicity of Φ. Since $\frac{\alpha}{2} > 0$, it follows from (4) that $f_1 + f_2 \in \mathcal{L}^\Phi(\mu)$. In particular, with each f in $\mathcal{L}^\Phi(\mu)$, $2f$ and then $nf \in \mathcal{L}^\Phi(\mu)$ for all integers $n > 1$ so that

$\beta f \in \mathcal{L}^{\Phi}(\mu)$ for any scalar β. Hence $\mathcal{L}^{\Phi}(\mu)$ is a vector space. It is also obviously solid and circled. The later property is again shared by B_{Φ}.

To see that (5) holds, let $f \in \mathcal{L}^{\Phi}(\mu)$ so that $\alpha f \in \tilde{\mathcal{L}}^{\Phi}(\mu)$ for some $\alpha > 0$. Let $a_n \searrow 0$ be arbitrary and set $\alpha_n = \min(\alpha, a_n)$. Then $\alpha_n \searrow 0$ and $\Phi(\alpha_n f) \leq \Phi(\alpha f)$ and $\Phi(\alpha_n f) \to 0$ when Φ is a continuous Young function. Hence by the dominated convergence then $\int_{\Omega} \Phi(\alpha_n f) d\mu \to 0$ so that for some n_0, $\int_{\Omega} \Phi(\alpha_{n_0} f) d\mu \leq 1$. Thus $\alpha_{n_0} f \in B_{\Phi}$. If $\Phi(x) = +\infty$ for $x > x_0 > 0$, then all f's must be bounded and the result trivially holds in this case and (5) follows. ∎

We will now include two more properties related to the classical Lebesgue spaces $\mathcal{L}^1(\mu)$ and $\mathcal{L}^{\infty}(\mu)$, depending crucially on the Orlicz space theory.

Proposition 7. (a) *Let (Ω, Σ, μ) be a finite measure space. Then*
(i) $\mathcal{L}^1(\mu) = \cup\{\tilde{\mathcal{L}}^{\Phi}(\mu) : \Phi$ *ranges over all N-functions*$\}$,
(ii) $\mathcal{L}^{\infty}(\mu) = \cap\{\tilde{\mathcal{L}}^{\Phi}(\mu) : \Phi$ *ranges over all N-functions*$\}$.

(b) *If (Ω, Σ, μ) is any measure space, (Φ, Ψ) is a complementary pair of N-functions, then $\mathcal{L}^1(\mu) = \tilde{\mathcal{L}}^{\Phi}(\mu) \cdot \tilde{\mathcal{L}}^{\Psi}(\mu)$ where $\tilde{\mathcal{L}}^{\Phi}(\mu) \cdot \tilde{\mathcal{L}}^{\Psi}(\mu) = \{f \cdot g : f \in \tilde{\mathcal{L}}^{\Phi}(\mu), g \in \tilde{\mathcal{L}}^{\Psi}(\mu)\}$.*

Proof: (a)(i) We have already established this fact in Corollary I.2.3.

(ii) It is clear that $\mathcal{L}^{\infty}(\mu) \subset \tilde{\mathcal{L}}^{\Phi}(\mu)$. Indeed $f \in \mathcal{L}^{\infty}(\mu) \Rightarrow |f(\omega)| \leq k < \infty$ for all $\omega \in \Omega - \Omega_0, \mu(\Omega_0) = 0$, so that for any N-function Φ we have

$$\int_{\Omega} \Phi(f) d\mu \leq \int_{\Omega} \Phi(k) d\mu = \Phi(k) \mu(\Omega) < \infty.$$

Thus $\mathcal{L}^{\infty}(\mu) \subset \cap\{\tilde{\mathcal{L}}^{\Phi}(\mu) : \Phi$ ranges over all N-functions$\}$.

Suppose that the inclusion is strict. Then there exists $f \in \tilde{\mathcal{L}}^{\Phi}(\mu)$ for all such Φ but is not essentially bounded. We now show that there is an N-function Φ_0 such that $f \notin \tilde{\mathcal{L}}^{\Phi_0}(\mu)$ to derive a contradiction.

Let $n_1 < n_2 < \cdots$ be a sequence of integers ($n_k \to \infty$) such that $\Omega_k = \{\omega : n_k \leq |f(\omega)| < n_{k+1}\}$ satisfies $\mu(\Omega_k) \geq \mu(\Omega_{k+1}) > 0$. This is possible since f is unbounded. Also $\sum_{k=1}^{\infty} \mu(\Omega_k) \leq \mu(\Omega) < \infty$ since Ω_k are disjoint, $\Omega_k \in \Sigma$. So $\mu(\Omega_k) \to 0$ as $k \to \infty$. Define $\varphi_0(t) = 2t[n_1 \mu(\Omega_{n_1})]^{-1}$, $0 \leq t < \frac{n_1}{2}$, and in general $\varphi_0(t) = (\mu(\Omega_k))^{-1}$, if $\frac{n_k}{2} \leq t < \frac{n_{k+1}}{2}$, $k \geq 1$. Then φ_0 is nondecreasing $\varphi_0(0) = 0$ and

$\varphi_0(t) \nearrow \infty$ as $t \nearrow \infty$. Hence $\Phi_0(x) = \int\limits_0^{|x|} \varphi_0(t)dt$ defines an N-function, and

$$\int\limits_\Omega \Phi_0(f)d\mu \geq \sum_{k=1}^\infty \int\limits_{\Omega_k} \Phi_0(f)d\mu \geq \sum_{k=1}^\infty \Phi_0(n_k) \cdot \mu(\Omega_k)$$

$$\geq \sum_{k=1}^\infty \frac{n_k}{2} \varphi_0(\tfrac{n_k}{2})\mu(\Omega_k) = \frac{1}{2} \sum_{k=1}^\infty n_k = \infty.$$

So $f \notin \tilde{\mathcal{L}}^{\Phi_0}(\mu)$. Thus (ii) holds as stated.

(b) Let $0 \leq h \in L^1(\mu)$, and (Φ, Ψ) a complementary pair of N-functions. The general case of h can be reduced to this since $|h| = h\,\mathrm{sgn}\,h$. Thus define f, g by:

$$f(\omega) = \Phi^{-1}(h(\omega)), \omega \in A_1 = \{\omega : 0 < h(\omega) < \infty\};$$
$$= 0, \omega \in A_2 = \{\omega : h(\omega) = 0\};$$
$$\text{and } = +\infty \text{ for } \omega \in (A_1 \cup A_2)^c.$$
$$g(\omega) = h(\omega)/\Phi^{-1}(h(\omega)), \omega \in A_1; \quad = 0, \omega \in A_2;$$
$$= +\infty \text{ for } \omega \in (A_1 \cup A_2)^c.$$

Note that $h = f \cdot g$ and $\mu((A_1 \cup A_2)^c) = 0$ since h is integrable. Also

$$\int\limits_\Omega \Phi(f)d\mu = \int\limits_{A_1} \Phi(f)d\mu = \int\limits_\Omega h\,d\mu < \infty, \text{ since } \Phi \text{ is strictly increasing,}$$

$$\int\limits_\Omega \Psi(g)d\mu = \int\limits_{A_1} \Psi\left(\frac{h}{\Phi^{-1}(h)}\right)d\mu$$

$$\leq \int\limits_{A_1} \Psi(\Psi^{-1}(h))d\mu, \text{ by Prop. II.1.1(ii)}$$

$$= \int\limits_\Omega h\,d\mu < \infty.$$

Thus $f \in \tilde{\mathcal{L}}^\Phi(\mu)$ and $g \in \tilde{\mathcal{L}}^\Psi(\mu)$. So $L^1(\mu) \subset \tilde{\mathcal{L}}^\Phi(\mu) \cdot \tilde{\mathcal{L}}^\Psi(\mu)$. The opposite inclusion is an immediate consequence of the Young inequality, and hence equality holds. ∎

Remark. The classical $\mathcal{L}^p$-theory indicates that part (a) of the above proposition no longer holds, if the Young functions Φ are restricted to $\Phi(x) = |x|^p$, $1 < p < \infty$.

3.2. *Orlicz spaces: basic structure*

Although the $\mathcal{L}^\Phi(\mu)$ are motivated by (and are a generalization
of) the $\mathcal{L}^p(\mu)$-spaces of Lebesgue, we should note some important differences before introducing and studying certain topologies on these
vector spaces. The similarity is that $\mathcal{L}^\Phi(\mu) = \mathcal{L}^p(\mu)$ if $\Phi(x) = |x|^p$, $p \geq$
1, and moreover, if $\Phi_0(x) = 0$ for $0 \leq x < 1$, $= +\infty$ for $x > 1$
($\Phi_0(-x) = \Phi_0(x)$), then $\mathcal{L}^{\Phi_0}(\mu) = \mathcal{L}^\infty(\mu)$. Thus all the $\mathcal{L}^p(\mu)$-spaces
$1 \leq p \leq +\infty$, are included in the $\mathcal{L}^\Phi(\mu)$ collection with Φ as a Young
function. The differences will become clear from what follows.

Definition 1. For each $f \in \mathcal{L}^\Phi(\mu)$, let $I_\Phi^f(\tilde{\mathcal{L}}^\Phi(\mu)) = I_\Phi^f = \{k \geq 0 :$
$kf \in \tilde{\mathcal{L}}^\Phi(\mu)\}$, and set $k_f = \sup I_\Phi^f$ ($\in \bar{\mathbb{R}}^+$), and $\ell_f : k \mapsto \int_\Omega \Phi(kf)d\mu$,
$k \in I_\Phi^f$.

The set I_Φ^f and the functional ℓ_f reveal some new and interesting
aspects of Orlicz spaces that are not present for the $\mathcal{L}^p$-spaces.

Proposition 2. *Let $0 \neq f \in \mathcal{L}^\Phi(\mu)$ and I_Φ^f, ℓ_f be as in Definition
1. Then $\ell_f : I_\Phi^f \to \bar{\mathbb{R}}^+$ is increasing and continuous on $[0, k_f)$ and
$\lim\limits_{k \to k_f} \ell_f(k) \leq +\infty$. If μ is diffuse on a set $E \in \Sigma$ of positive μ-measure,
and if $0 < \alpha, \beta < \infty$ are arbitrarily given, then we can find a Young
function Φ, and $f \in \mathcal{L}^\Phi(\mu)$ such that $k_f = \alpha$, $\ell_f(k_f) = \beta$. Further one
can also find a $g \in \mathcal{L}^\Phi(\mu)$ such that $k_g = \alpha$ and $\ell_g(k_g) = +\infty$.*

Proof: From definitions of Φ and ℓ_f one has

$$\ell_f(k) = \int_\Omega \Phi(kf)d\mu, \qquad f \in \mathcal{L}^\Phi(\mu), \qquad (1)$$

so that $\ell_f(\cdot)$ is increasing on $[0, k_f)$ since Φ is. If $k_n \to k_0 \in [0, k_f]$,
then for $k_0 < k_f$, (1) implies that $\ell_f(k_n) \nearrow \ell_f(k_0)$ by the monotone convergence theorem. If $k_0 = k_f$, then $\ell_f(k_n) \leq \ell_f(k_0)$ so that
$\limsup\limits_{n \to \infty} \ell_f(k_n) \leq \ell_f(k_0)$. Since $\Phi(k_n f) \geq 0$, and $\Phi(k_n f) \to \Phi(k_0 f)$ a.e.,
we have by Fatou's lemma $\liminf\limits_{n \to \infty} \ell_f(k_n) \geq \ell_f(k_0)$ so that $\ell_f(k_n) \to$
$\ell_f(k_0)$ in all cases, and hence $\ell_f(\cdot)$ is continuous. Note that if Φ is
strictly increasing and continuous, then so is ℓ_f, since $f \neq 0$ on a set
of positive measure.

For the last part, let $0 < \alpha, \beta < \infty$ be given. Let Φ be the
Young function: $\Phi(x) = e^{|x|} - 1$ so that $\Phi \notin \Delta_2$. Since by hypothesis
there is an $E \in \Sigma$ with $\mu(E) > 0$ and μ is diffuse on E, there exists

$F \subset E$, $0 < \mu(F) < \infty$, and μ is diffuse on F. Hence by a standard result in measure theory (cf., e.g., Rao (1987), p. 507, the first half of Exercise 2, or Royden (1988), p. 399ff), $\mathcal{L}^\Phi(F, \Sigma(F), \mu)$ is "isometrically isomorphic" (i.e., identifiable) with $\mathcal{L}^\Phi(I, \mathcal{B}, \lambda)$ where $I = [0, \mu(F))$, $\Sigma(F)$ being the trace or restriction σ-algebra of Σ to F, and $(I, \mathcal{B}, \lambda)$ the Lebesgue interval. Hence for the present construction we can assume (Ω, Σ, μ) is $(J, \mathcal{B}, \lambda)$ with $J = [0, \mu(F))$ and denote the corresponding space $\mathcal{L}^\Phi(\lambda)$.

Choose $0 \le x_1 < x_2 < \cdots < x_n < \cdots$, from J such that $x_{n+1} - x_n = \beta/n(n+1)(2^{n+1}-1)$, $n \ge 1$. Then $x_n \to \tilde{x}_\beta < \infty$. Define

$$f = \frac{1}{\alpha} \sum_{n=1}^{\infty} (\log 2^{n+1}) \chi_{[x_n, x_{n+1})}. \tag{2}$$

From this one has

$$\int_J \Phi(\alpha f) d\lambda = \sum_{n=1}^{\infty} \int_{x_n}^{x+1} [\exp(\log 2^{n+1}) - 1] d\lambda$$

$$= \sum_{n=1}^{\infty} \left[(2^{n+1} - 1) \frac{\beta}{n(n+1)(2^{n+1} - 1)} \right] = \beta. \tag{3}$$

On the other hand for each $0 < \varepsilon < 1$, the same computation yields

$$\int_J \Phi((1 + \varepsilon)\alpha f) d\lambda = \beta \sum_{n=1}^{\infty} \left[(2^{(1+\varepsilon)(n+1)} - 1)/n(n+1)(2^{n+1} - 1) \right]$$

$$> \beta \sum_{n=1}^{\infty} \left[2^{\varepsilon(n+1)}/n(n+1) \right] = +\infty. \tag{4}$$

Here $\alpha = k_f$ and $\beta = \ell_f(k_f)$.

With the same f as in (2), where $x_{n+1} - x_n = 1/n(2^{n+1} - 1)$, $n \ge 1$ (call this new f by g) and Φ as given there, we have

$$\int_J \Phi(\alpha g) d\lambda = \sum_{n=1}^{\infty} \int_{x_n}^{x_n} [\exp(\log 2^{n+1}) - 1] d\lambda$$

$$= \sum_{n=1}^{\infty} \frac{1}{n} = +\infty.$$

But for $0 < \varepsilon < 1$, we have

$$\int_J \Phi((1-\varepsilon)\alpha g)d\lambda = \sum_{n=1}^{\infty} \left[(2^{(1-\varepsilon)(n+1)} - 1)/n(2^{n+1} - 1) \right]$$

$$< \sum_{n=1}^{\infty} \left(2^{-(n+1)\varepsilon}/n \right) < \infty. \qquad (5)$$

Hence $k_g = \alpha$ and $\ell_g(k_g) = +\infty$. In view of the isomorphism noted earlier, this establishes the last part also. ∎

Remark. Unlike in the Lebesgue space case, the values of k_f and $\ell_f(k_f)$ can be independently prescribed. As is evident from the examples given in (2)–(5), all the difficulties and new behavior appear when the Young function is not of a power (or a polynomial) growth. More surprizes appear below as the theory progresses, and this engenders considerable interest for Orlicz spaces and their applications.

We are now ready to introduce a norm functional on $\mathcal{L}^{\Phi}(\mu)$. Consider B_{Φ} of Proposition 1.6. It was noted there that it is a circled solid subset of $\mathcal{L}^{\Phi}(\mu)$. Define the functional

$$N_{\Phi}(f) = \inf\{k > 0 : \tfrac{1}{k}f \in B_{\Phi}\} = \inf\{k > 0 : \int_{\Omega} \Phi(\tfrac{f}{k})d\mu \leq 1\}. \qquad (6)$$

We shall now establish some important facts about $N_{\Phi}(\cdot)$:

Theorem 3. $(\mathcal{L}^{\Phi}(\mu), N_{\Phi}(\cdot))$ *is a normed linear space when equivalent functions are identified. Moreover,* $N_{\Phi}(f) \leq 1$ *iff* $\int_{\Omega} \Phi(f)d\mu \leq 1$.

Proof: To see that $N_{\Phi}(\cdot)$ is a norm, we have to verify, by the definition of norm, the following conditions: (i) $N_{\Phi}(f) = 0$ iff $f = 0$ a.e., (ii) $N_{\Phi}(\alpha f) = |\alpha| N_{\Phi}(f)$, $\alpha \in \mathbb{R}$ (or $\mathbb{C}$), and (iii) $N_{\Phi}(f + g) \leq N_{\Phi}(f) + N_{\Phi}(g)$. Thus, if $f = 0$, a.e., then evidently $N_{\Phi}(f) = 0$. Conversely if $N_{\Phi}(f) = 0$, let $|f| > 0$ on a set of positive measure, if possible. Then there exists a number $\delta > 0$ such that $A = \{\omega : |f(\omega)| \geq \delta\}$ satisfies $\mu(A) > 0$. But by (6), we must have $f/k \in B_{\Phi}$ for all $k > 0$, so that $nf \in B_{\Phi}$ for all $n \geq 1$. This implies in particular

$$\Phi(n\delta)\mu(A) = \int_A \Phi(n\delta)d\mu \leq \int_A \Phi(nf)d\mu \leq \int_{\Omega} \Phi(nf)d\mu \leq 1, \quad n \geq 1.$$

Since $\mu(A) > 0$ and $\Phi(n\delta) \nearrow \infty$ as $n \to \infty$, this is impossible so that $\mu(A) = 0$. Thus $f = 0$ a.e. and (i) holds.

For (ii) consider the nontrivial case that $\alpha \neq 0$,

$$N_\Phi(\alpha f) = \inf\{k > 0 : \int_\Omega \Phi(\frac{\alpha f}{k})d\mu \leq 1\}$$

$$= |\alpha| \inf\{\frac{k}{|\alpha|} > 0 : \int_\Omega \Phi(\frac{f}{(k/|\alpha|)})d\mu \leq 1\}$$

$$= |\alpha| \inf\{\beta > 0 : \int_\Omega \Phi(\frac{f}{\beta})d\mu \leq 1\} = |\alpha| N_\Phi(f).$$

Finally for the triangle inequality (iii), let $a_i > N_\Phi(f_i)$, $i = 1, 2$. Then $0 < a_i < \infty$, and let $b = a_1 + a_2 > 0$. Since $f_1 + f_2 \in \mathcal{L}^\Phi(\mu)$, $N_\Phi(f_1 + f_2) < \infty$. Consider

$$\int_\Omega \Phi\left(\frac{f_1 + f_2}{b}\right) d\mu = \int_\Omega \Phi\left[\frac{f_1}{a_1} \cdot \frac{a_1}{b} + \frac{f_2}{a_2} \cdot \frac{a_2}{b}\right] d\mu$$

$$\leq \frac{a_1}{b} \int_\Omega \Phi\left(\frac{f_1}{a_1}\right) d\mu + \frac{a_2}{b} \int_\Omega \Phi\left(\frac{f_2}{a_2}\right) d\mu, \text{ by the}$$

$$\text{convexity of } \Phi,$$

$$\leq \frac{a_1}{b} + \frac{a_2}{b} = 1, \text{ since } \frac{f_i}{a_i} \in B_\Phi, \quad i = 1, 2.$$

Hence $\frac{1}{b}(f_1 + f_2) \in B_\Phi$ so that $N_\Phi(f_1 + f_2) \leq b = a_1 + a_2$. Letting $a_i \to N_\Phi(f_i)$, (iii) follows. This shows $(\mathcal{L}^\Phi(\mu), N_\Phi(\cdot))$ is a normed linear space, since $\mathcal{L}^\Phi(\mu)$ was already seen in Proposition 1.6 to be linear.

For the last part, let $a = N_\Phi(f)$, $f \in \mathcal{L}^\Phi(\mu)$ and $a > 0$, since $a = 0$ is the trivial case. Then by definition, $\frac{1}{a}f \in B_\Phi$. If $a \leq 1$, then

$$\int_\Omega \Phi(f)d\mu \leq \int_\Omega \Phi(\frac{f}{a})d\mu \leq 1 \qquad (7)$$

so that $N_\Phi(f) \leq 1$ implies the left side of (7) is bounded by 1. If on the other hand, $f \in B_\Phi$ then by definition of (6), $N_\Phi(f) \leq 1$ holds. Note that if $a > 1$, then by Proposition 2, $\int_\Omega \Phi(\frac{f}{a})d\mu \leq 1$ but $\int_\Omega \Phi(f)d\mu = +\infty$ is possible. Thus only $0 \leq a \leq 1$ is relevant here. ∎

The preceding result, together with Proposition 1.6 can be restated as follows: B_Φ is a circled, solid, and convex subset of $\mathcal{L}^\Phi(\mu)$. It is classical (going back at least to the 1930's) that a nonnegative real functional $p(\cdot)$ on such a set B_Φ defined as

$$p_{B_\Phi}(f) = \inf\{k > 0 : \frac{f}{k} \in B_\Phi\} \tag{8}$$

is called the *gauge, or the Minkowski functional*. By (6) $p_{B_\Phi}(\cdot)$ is just $N_\Phi(\cdot)$ and thus it is a specialization of that notion to the Orlicz spaces. Since by Theorem 3 it is a norm when a.e. equal spaces are identified, we hereafter call $N_\Phi(\cdot)$ the *gauge norm* of the Orlicz space $\mathcal{L}^\Phi(\mu)$.

We observe that if a Young pair (Φ, Ψ) is normalized (cf. Section 2.4), then we replace (6) by the following modified formula:

$$N_\Phi(f) = \inf\{k > 0 : \int_\Omega \Phi(\tfrac{f}{k})d\mu \le \Phi(1)\}. \tag{9}$$

The difference here and before is that $0 < \Phi(1) \le 1$ replaces 1 to reflect the normalization of equation (2.4.1). Note that $k_0 = N_\Phi(f) < \infty \Rightarrow \exists k_n \to k_0$ and then $\Phi(\tfrac{f}{k_n}) \to \Phi(\tfrac{f}{k_0}) \ge 0$, and so by Fatou's lemma $\int_\Omega \Phi(\tfrac{f}{k_0})d\mu \le 1$ or $\Phi(1)$ in (6) or (9). We use this fact below without further mention. Also when $f \ne 0$, $\int_\Omega \Phi(\tfrac{f}{N_\Phi(f)})d\mu < 1$ (or $\Phi(1)$) iff $k_f < \infty$ and $\int_\Omega \Phi(k_f f)d\mu < 1$ (cf. Definition 1), and hence $k_f N_\Phi(f) = 1$ must hold. Another useful property of $N_\Phi(\cdot)$ is given by:

Proposition 4. *Let $f_n \in \mathcal{L}^\Phi(\mu)$, $n \ge 1$, be a sequence such that $f_n \to f$ a.e., with $f \in \mathcal{L}^\Phi(\mu)$, and $\Phi(x) = 0$ iff $x = 0$. Then $N_\Phi(f) \le \liminf_{n \to \infty} N_\Phi(f_n)$, i.e., the gauge norm is lower semi-continuous on $\mathcal{L}^\Phi(\mu)$.*

Proof: We may assume that $f \ne 0$, a.e. since the result is trivial in the other case. Thus $N_\Phi(f) > 0$ so that for all large enough n, $N_\Phi(f_n) > 0$. If $k_0 = \liminf_{n \to \infty} N_\Phi(f_n) = +\infty$, the result is again true. Suppose that $0 \le k_0 < \infty$. But $k_0 = 0$ implies the existence of a subsequence f_{n_i} such that $\lim_{i \to \infty} N_\Phi(f_{n_i}) = 0$. Consequently from some i_0, $N_\Phi(f_{n_i}) \le 1$, $i \ge i_0$ and

$$\frac{1}{N_\Phi(f_{n_i})} \int_\Omega \Phi(|f_{n_i}|)d\mu \le \int_\Omega \Phi\left(\frac{|f_{n_i}|}{N_\Phi(f_{n_i})}\right)d\mu \le 1,$$

using the convexity of Φ. Hence

$$\int_\Omega \Phi(|f_{n_i}|)d\mu \leq N_\Phi(f_{n_i}) \to 0, \tag{10}$$

as $i \to \infty$. Since $|f_{n_i}| \to |f|$ a.e., we have by Fatou's lemma and the fact that $\Phi(x) > 0$ for $|x| > 0$,

$$0 < \int_\Omega \Phi(|f|)d\mu = \int_\Omega \lim_{i \to \infty} \Phi(|f_{n_i}|)d\mu \leq \liminf_{i \to \infty} \int_\Omega \Phi(|f_{n_i}|)d\mu = 0,$$

by (10). This contradiction shows that $0 < k_0 < \infty$.

Finally let $0 < k_0 < t$ be arbitrary. Then $k_0 < k_i < t$ for some i so that

$$\int_\Omega \Phi\left(\frac{f}{t}\right)d\mu = \int_\Omega \lim_{n \to \infty} \Phi\left(\frac{f_{n_i}}{t}\right)d\mu$$

$$\leq \liminf_{i \to \infty} \int_\Omega \Phi\left(\frac{f_{n_i}}{t}\right)d\mu$$

$$\leq \liminf_{i \to \infty} \int_\Omega \Phi\left(\frac{f_{n_i}}{k_i}\right)d\mu \leq 1. \tag{11}$$

But then (6) and (11) imply that $N_\Phi(f) \leq t$. Since $t > k_0$ is arbitrary, we get $N_\Phi(f) \leq k_0 = \liminf_{n \to \infty} N_\Phi(f_n)$ so that $N_\Phi(\cdot)$ is lower semi-continuous. ∎

It is easily seen that $N_\Phi(\cdot)$ has the following further properties.

(i) $|f_1| \leq |f_2|$ a.e. $\Rightarrow N_\Phi(f_1) \leq N_\Phi(f_2)$ (f_i measurable),

(ii) $0 \leq f_n \uparrow f$ a.e. $\Rightarrow N_\Phi(f_n) \nearrow N_\Phi(f) \leq \infty$ (f_i measurable).

The first one is monotonicity, and the second one is called the *strong Fatou property* of the (function) norm $N_\Phi(\cdot)$ which is related to the preceding proposition. For a deeper analysis of the $\mathcal{L}^\Phi(\mu)$, it will be useful to introduce another norm which is equivalent to $N_\Phi(\cdot)$. This is studied in the following section.

3.3. *Orlicz and gauge norms*

To motivate Orlicz's norm and for further analysis, we first derive an extension of Hölder's inequality. Thus let (Φ, Ψ) be a pair of *normalized* Young functions (cf. Section 2.4). So $\Phi(1) + \Psi(1) = 1$, and let

$\mathcal{L}^\Phi(\mu), \mathcal{L}^\Psi(\mu)$ be the corresponding Orlicz spaces with the gauge norms $N_\Phi(\cdot)$ and $N_\Psi(\cdot)$ respectively. Then we have:

Proposition 1. *If $f \in \mathcal{L}^\Phi(\mu)$ and $g \in \mathcal{L}^\Psi(\mu)$, with (Φ, Ψ) as a normalized complementary Young pair, then one has*

$$\int_\Omega |fg| d\mu \leq N_\Phi(f) N_\Psi(g). \tag{1}$$

Proof: If either $N_\Phi(f) = 0$ or $N_\Psi(g) = 0$ (so $f = 0$ or $g = 0$, a.e.), then (1) is true and trivial. So let $N_\Phi(f) > 0$ and $N_\Psi(g) > 0$. Then by Young's inequality (cf. Theorem I.3.3),

$$\frac{|fg|(\omega)}{N_\Phi(f) N_\Psi(g)} \leq \Phi\left(\frac{|f|}{N_\Phi(f)}\right)(\omega) + \Psi\left(\frac{|g|}{N_\Psi(g)}\right)(\omega), \quad \omega \in \Omega. \tag{2}$$

Integration of (2), and (2.9) imply (1) since

$$\int_\Omega \left[\frac{|fg|}{N_\Phi(f) N_\Psi(g)}\right] d\mu \leq \Phi(1) + \Psi(1) = 1. \quad \blacksquare \tag{3}$$

Remark. If the (Φ, Ψ) are not normalized, then on the right side of (3) $\Phi(1)$ and $\Psi(1)$ are replaced by 1 each time. In this case one gets

$$\int_\Omega |fg| d\mu \leq 2 N_\Phi(f) N_\Psi(g) \tag{4}$$

instead of (1). This corresponds, in the Lebesgue case, to $\Psi(x) = |x|^p$, $p \geq 1$, so that $\Psi(y) = \dfrac{y^q}{qp^{\frac{1}{p-1}}}$ if $p > 1$; and if $p = 1$, then $\Psi(y) = 1$ or $= +\infty$ accordingly as $0 \leq y < 1$ or $y \geq 1$. Thus the inequality (4) is not the best possible.

An improvement of (4) can be obtained by using Orlicz's norm to be introduced below.

Definition 2. If (Ω, Σ, μ) is a measure space (with the finite subset property), $f : \Omega \to \mathbb{R}$ is measurable for Σ, (Φ, Ψ) is a Young complementary pair, then we define the *Orlicz norm* $\|\cdot\|_\Phi : f \mapsto \|f\|_\Phi$ as:

$$\|f\|_\Phi = \sup\{\int_\Omega |fg| d\mu : \int_\Omega \Psi(|g|) d\mu \leq 1\}. \tag{5}$$

To call the functional $\|\cdot\|_\Phi$ a norm, we have to verify that it is a positively homogeneous subadditive functional, i.e., the three conditions recalled in the beginning of the proof of Theorem 2.3. The facts that $\|cf\|_\Phi = |c|\|f\|_\Phi$, $c \in \mathbb{R}$, and $\|f_1 + f_2\|_\Phi \leq \|f_1\|_\Phi + \|f_2\|_\Phi$ are easily seen from (5). Also $f = 0$ a.e. $\Rightarrow \|f\|_\Phi = 0$. Conversely let $\|f\|_\Phi = 0$. Then if $f \neq 0$ a.e., let $E = \{\omega : |f(\omega)| > 0\}$ and $\mu(E) > 0$. By the finite subset property, there is $F \subset E$, $F \in \Sigma$, $0 < \mu(F) < \infty$. Let $g = k\chi_F$ where $k > 0$ is chosen so that $\int_\Omega \Psi(|g|)d\mu \leq 1$. Hence by (5)

$$0 = \|f\|_\Phi \geq \int_\Omega |fg|d\mu \geq \int_F k|f|d\mu, \tag{6}$$

so that $f = 0$ a.e. on F contradicting the choice of F. Thus $f = 0$, a.e., so that $\|\cdot\|_\Phi$ is a norm. If μ does not have the finite subset property however, we cannot conclude that $\|f\|_\Phi = 0 \Rightarrow f = 0$ a.e., and in that generality one has to say that $\|\cdot\|_\Phi$ is a seminorm. Since this is not restrictive (cf. e.g. Rao (1987), pp. 68–69 for a discussion on this point), *we assume the finite subset property without further mention*. All σ-finite and the somewhat more general "localizable" measures (cf. the last reference on this point also) have this property. Since not even this (simple) restriction is required of $N_\Phi(\cdot)$-functionals, the latter are a trite more general.

It should also be noted that in (5) $\int_\Omega |fg|d\mu$ may be replaced by $|\int_\Omega fgd\mu|$ as g varies over B_Ψ and one gets the same $\|f\|_\Phi$ on the left. This follows from the fact that

$$\left|\int_\Omega fgd\mu\right| \leq \int_\Omega |f||g|d\mu = \int_\Omega f\tilde{g}d\mu \tag{7}$$

where $\tilde{g} = |g|\operatorname{sgn} f$. Then $\tilde{g} \in B_\Psi$ since $\int_\Omega \Psi(|\tilde{g}|)d\mu = \int_\Omega \Psi(|g|)d\mu$. So the extreme members of (7) have the same supremum as g ranges over B_Ψ and $\|f\|_\Phi$ is obtained by either side, i.e.,

$$\|f\|_\Phi = \sup\{\left|\int_\Omega fgd\mu\right| : g \in B_\Psi\} = \sup\{\int_\Omega |fg|d\mu : g \in B_\Psi\}. \tag{8}$$

To show that the two norms $\|\cdot\|_\Phi$ and $N_\Phi(\cdot)$ are *equivalent*, we first need to establish the following fact.

Proposition 3. *If $0 \neq f \in \mathcal{L}^{\Phi}(\mu)$, then we have*

$$\int_{\Omega} \Phi\left(\frac{f}{\|f\|_{\Phi}}\right) d\mu \leq 1. \tag{9}$$

Proof: Let $\rho_{\Psi}(g) = \int_{\Omega} \Psi(|g|) d\mu$, and denote by $\rho_{\Psi}'(g) = \max(1, \rho_{\Psi}(g))$. Then

$$\int_{\Omega} |fg| d\mu \leq \rho_{\Psi}'(g) \|f\|_{\Phi}, \qquad g \in \mathcal{L}^{\Phi}(\mu), \tag{10}$$

which follows from (5) if $\rho_{\Psi}(g) \leq 1$, and replacing g by $\frac{g}{\rho_{\Psi}(g)}$ in case $\rho_{\Psi}(g) > 1$. Also note that from (4) we always have $\|f\|_{\Phi} \leq 2N_{\Phi}(f)$ so that $f \in \mathcal{L}^{\Phi}(\mu) \Rightarrow \|f\|_{\Phi} < \infty$. From this it follows that $\nu_f(\cdot) : E \mapsto \int_E \Phi(\frac{|f|}{\|f\|_{\Phi}}) d\mu$ is a measure with the finite subset property. In fact, the standard measure properties imply that $\nu_f(\cdot) : \Sigma \to \bar{I\!R}^+$ is σ-additive, and if $\nu_f(E) > 0$ then $\mu(E) > 0$ so that $\int_E \Phi(\frac{|f|}{k\|f\|_{\Phi}}) d\mu \leq 1$ for a $k > 0$. Hence $\nu_{kf}(E) < \infty$. By Proposition 2.2, the functional is continuous on the interval $(\frac{1}{k_0}, \infty)$ where $k_0 = \inf\{k > 0 : \frac{f}{k} \in \tilde{\mathcal{L}}^{\Phi}(\mu)\}$. So we find some $k_0 < k < \infty$ such that $F = \{\omega \in E : \Phi(\frac{|f(\omega)|}{k\|f\|_{\Phi}}) > 0\}$, $0 < \nu(F) < \infty$. We claim that (since $\mu(F) > 0$, and using the finite subset property of μ), (9) holds for all sets $A \in \Sigma$ with $\mu(A) < \infty$, i.e., replacing f by $f\chi_A$, (9) holds.

The idea of proof is to use the equality in Young's inequality. Thus let f be a simple function, so that $g = \varphi(\frac{f}{\|f\|_{\Phi}})$ is also a simple function and is in $\mathcal{L}^{\Phi}(\mu)$. Moreover, by Theorem I.3.3, with $\Omega = A$,

$$\left| \int_{\Omega} \left(\frac{f}{\|f\|_{\Phi}}\right) g d\mu \right| = \int_{\Omega} \Phi\left(\frac{f}{\|f\|_{\Phi}}\right) d\mu + \rho_{\Psi}(g), \tag{11}$$

and the left side of (11) is at most $\rho_{\Psi}'(g)$ by (10) which is $\rho_{\Psi}(g)$ when $\rho_{\Psi}(g) > 1$, so that $\int_{\Omega} \Phi(\frac{f}{\|f\|_{\Phi}}) d\mu = 0$, and $\rho_{\Psi}'(g) = 1$, if $\rho_{\Psi}(g) \leq 1$. Thus in all cases (9) holds if f is a simple function. In the general case of f, there exist simple $0 \leq f_n \uparrow |f|$ pointwise, so that $\|f_n\|_{\Phi} \leq \|f\|_{\Phi}$ and we have

$$\int_{\Omega} \Phi\left(\frac{f_n}{\|f\|_{\Phi}}\right) d\mu \leq \int_{\Omega} \Phi\left(\frac{f_n}{\|f_n\|_{\Phi}}\right) d\mu \leq 1, \tag{12}$$

by the special case. Letting $n \to \infty$, by the monotone convergence theorem, the left side of (12) becomes (9) so that the result holds if $\mu(\Omega) < \infty$.

We now extend the result if $\mu(\Omega) = +\infty$, and μ is unrestricted. Let $\Sigma_1 = \{A \in \Sigma : \mu(A) < \infty\}$. Then by the preceding paragraph $\sup\{\nu_f(A) : A \in \Sigma_1\} = \alpha \leq 1$ where ν_f is the measure introduced above. Since Σ_1 is a ring (in fact a δ-ring) there exists a sequence $A_n \subset A_{n+1}$ from Σ_1 such that $\lim_{n\to\infty} \nu_f(A_n) = \alpha$. If $B = \bigcup_{n=1}^{\infty} A_n$, then $\nu_f(B) = \alpha$. We now claim that $\nu_f(B^c) = 0$. Let E_0 be the set where $\Phi(\frac{f}{\|f\|_\Phi}) > 0$. Then we need to show that $\nu_f(E_0 \cap B^c) = 0$. If not, there is a set $F \subset E_0 \cap B^c$, $F \in \Sigma_1$, $\mu(F) > 0$. Consequently,

$$\alpha < \nu_f(B) + \nu_f(F) = \nu_f(B \cup F) = \lim_{n\to\infty} \nu_f(A_n \cup F)$$
$$\leq \sup\{\nu_f(D) : D \in \Sigma_1\} = \alpha. \qquad (13)$$

This contradiction shows that $\nu_f(B^c \cap E_0) = 0$, $E_0 \in \Sigma_1$ and then $\nu_f(B^c) = 0$. So

$$\nu_f(\Omega) = \nu_f(B) = \alpha \leq 1,$$

and hence (9) holds in all cases. ∎

The preceding result gives a useful relation between the Orlicz and gauge norms.

Proposition 4. *For any $f \in \mathcal{L}^\Phi(\mu)$, Φ being a Young function, we have*

$$N_\Phi(f) \leq \|f\|_\Phi \leq 2N_\Phi(f). \qquad (14)$$

In case (Φ, Ψ) is normalized, so (2.9) is used for $N_\Phi(\cdot)$, the relation then becomes

$$\Phi(1)N_\Phi(f) \leq \|f\|_\Phi \leq 2N_\Phi(f). \qquad (15)$$

In all cases, both $\|\cdot\|_\Phi$ and $N_\Phi(\cdot)$ are monotone in that for $0 \leq f_1 \leq f_2$, with $f_i \in \mathcal{L}^\Phi(\mu)$, one has $\|f_1\| \leq \|f_2\|$ where $\|\cdot\|$ denotes $\|\cdot\|_\Phi$ or $N_\Phi(\cdot)$.

Proof: The relation (14) is an immediate consequence of (4) and (9), and the definition of $N_\Phi(\cdot)$ given by (2.6). As for (15), if we set $\alpha = \Phi(1)$ so that $0 < \alpha \leq 1$, then (9) implies

$$\int_\Omega \Phi\left(\frac{\alpha f}{\|f\|_\Phi}\right) d\mu \leq \alpha \int_\Omega \Phi\left(\frac{f}{\|f\|_\Phi}\right) d\mu \leq \alpha \leq 1.$$

Hence $N_\Phi(f) \le \|f\|_\Phi/\alpha$, giving the left side. Regarding the right side of (15), note that by Young's inequality with $\rho_\Psi(g) \le 1$,

$$\int_\Omega |fg|d\mu = \beta \int_\Omega \tfrac{|f|}{\beta}|g|d\mu \le \beta \Big(\int_\Omega \Phi(\tfrac{|f|}{\beta})d\mu + \int_\Omega \Psi(g)d\mu \Big)$$

$$\le \beta(\Phi(1)+1) \le 2\beta$$

with $\beta = N_\Phi(f)$. The last statement being obvious, we have the result. ∎

Remarks. 1. As a consequence of (5) and (14) the following improvement of the Hölder inequality (4) is possible: If $f \in \mathcal{L}^\Phi(\mu), g \in \mathcal{L}^\Psi(\mu)$, then

$$\int_\Omega |fg|d\mu = N_\Psi(g) \int_\Omega |f| \tfrac{g}{N_\Psi(g)}|d\mu \le N_\Psi(g)\|f\|_\Phi, \qquad (16)$$

and (Φ, Ψ) can be interchanged here. Hence

$$\int_\Omega |fg|d\mu \le \min \big(N_\Phi(f)\|g\|_\Psi, N_\Psi(g)\|f\|_\Phi \big). \qquad (17)$$

For the normalized (Φ, Ψ), (17) takes the form (1).

2. If the Young pair (Φ, Ψ) is normalized, then one may define $\|f\|_\Phi$ in (5) as $\|f\|_\Phi' = \sup\{|\int_\Omega fg d\mu| : N_\Psi(g) \le 1\}$ so that by (1) $\|f\|_\Phi' \le N_\Phi(f)$. But (9) is not generally improvable. (See Proposition 17 below.) Hence (15) becomes for a $C > 0$ (by the Closed Graph Theorem)

$$CN_\Phi(f) \le \|f\|_\Phi' \le N_\Phi(f), \qquad f \in \mathcal{L}^\Phi(\mu). \qquad (18)$$

Thus for simplicity, we use Definition 2 with (5) for $\|f\|_\Phi$ in this book.

We now present an interesting supplement to the Hölder inequality which extends (4). For this we need two preliminary results of some independent interest.

Proposition 5 (Jensen's inequality). *Let (Ω, Σ, μ) be a probability space, i.e., $\mu(\Omega) = 1$. If $\Phi : \mathbb{R} \to \mathbb{R}$ is a convex function, $f : \Omega \to \mathbb{R}$ is μ-measurable, and $\int_\Omega fd\mu$ and $\int_\Omega \Phi(f)d\mu$ exist, then*

$$\Phi\Big(\int_\Omega fd\mu \Big) \le \int_\Omega \Phi(f)d\mu. \qquad (19)$$

Proof: By Theorem I.3.1, we have

$$\Phi(x) = \Phi(a) + \int_a^x \varphi(t)dt \geq \Phi(a) + \varphi(a)(x - a) \qquad (20)$$

since $\varphi(\cdot)$ is increasing. This is often called the *support line property* of a convex function. Let $x = f(\omega)$, $a = \int_\Omega f d\mu$ in (20), and integrate. Thus we get

$$\int_\Omega \Phi(f)d\mu - \Phi(\int_\Omega f d\mu) \geq \varphi(a)(\int_\Omega f d\mu - \int_\Omega f d\mu) = 0.$$

This gives (19) as desired.   ∎

Although we use this for Young functions Φ, for (19), Φ need not be positive. The next result is related to Proposition II.2.9 but here the Φ's need not be convex.

Lemma 6. *Let $\Phi_i : I\!\!R^+ \to \bar{I\!\!R}$, $i = 1, 2, 3$ be monotone nondecreasing, left continuous functions such that*

$$\Phi_1^{-1}(x)\Phi_2^{-1}(x) \leq \Phi_3^{-1}(x), \qquad x \geq 0 \qquad (21)$$

where $\Phi_i^{-1}(x) = \inf\{y : \Phi_i(y) > x\}$ with $\inf(\emptyset) = +\infty$ as usual. Then

$$\Phi_3(xy) \leq \Phi_1(x) + \Phi_2(y), \qquad x \geq 0, y \geq 0. \qquad (22)$$

Proof: The definition of the inverse above implies $\Phi_i(\Phi_i^{-1}(x)) \leq x \leq \Phi_i^{-1}(\Phi_i(x))$. Let $x, y \ (\in I\!\!R^+)$ be arbitrarily fixed. Then $\Phi_1(x) \leq \Phi_2(y)$ or its opposite holds. In the first case, we have

$$xy \leq \Phi_1^{-1}(\Phi_1(x))\Phi_2^{-1}(\Phi_2(y)) \leq \Phi_1^{-1}(\Phi_2(y))\Phi_2^{-1}(\Phi_2(y))$$
$$\leq \Phi_3^{-1}(\Phi_2(y)), \qquad \text{by (21)}.$$

Hence $\Phi_3(xy) \leq \Phi_3(\Phi_3^{-1}(\Phi_2(y))) \leq \Phi_2(y)$. If the opposite case is true, then $\Phi_1(x) > \Phi_2(y)$ and interchanging 1 and 2, we get $\Phi_3(xy) \leq \Phi_1(x)$, so that

$$\Phi_3(xy) \leq \max(\Phi_1(x), \Phi_2(y)) \leq \Phi_1(x) + \Phi_2(y),$$

giving (22); and it generalizes Young's inequality.   ∎

We can now present a generalization of Hölder's inequality, due to R. O'Neil (1965), which is of interest in applications:

Theorem 7. *Let* Φ_i, $i = 1, 2, 3$ *be Young's functions for which* (21) *holds. If* $f_i \in \mathcal{L}^{\Phi_i}(\mu)$, $i = 1, 2$, *where* (Ω, Σ, μ) *is any measure space, then* $f_1 \cdot f_2 \in \mathcal{L}^{\Phi_3}(\mu)$ *and we have*

$$N_{\Phi_3}(f_1 f_2) \leq 2 N_{\Phi_1}(f_1) N_{\Phi_2}(f_2). \tag{23}$$

Proof: We may assume $N_{\Phi_i}(f_i) = 1$, $i = 1, 2$, for simplicity. Then by the convexity of Φ_i (≥ 0), and (22) we have

$$\int_\Omega \Phi_3 \left(\frac{|f_1 f_2|}{2} \right) d\mu \leq \frac{1}{2} \int_\Omega \Phi_3(|f_1||f_2|) d\mu$$

$$\leq \frac{1}{2} \left[\int_\Omega \Phi_1(|f_1|) d\mu + \int_\Omega \Phi_2(|f_2|) d\mu \right]$$

$$\leq \frac{1}{2}(1 + 1) = 1, \quad \text{by Theorem 2.3.}$$

Hence $N_{\Phi_3}(f_1 f_2) \leq 2$, and (23) follows. ∎

Taking $\Phi_1(x) = |x|^p/p$, $\Phi_2(x) = |x|^q/q$, $p^{-1} + q^{-1} = r^{-1}$, $p, q, r \geq 1$, so that $\Phi_3(x) = |x|^r/r$, one gets a less sharp form of a classical inequality.

Corollary 8. *If* $f_1 \in \mathcal{L}^p(\mu)$, $f_2 \in \mathcal{L}^q(\mu)$, *then* $f_1 f_2 \in \mathcal{L}^r(\mu)$ *and*

$$\|f_1 f_2\|_r \leq 2 \|f_1\|_p \|f_2\|_q. \tag{24}$$

(In the classical case 2 can be replaced by 1 on the right side of (24)*.)*

If there is group structure in the underlying measure space, then a related inequality may be presented which supplements Theorem 7. But now the pointwise multiplication of $f_1 f_2$ will be replaced by the convolution operation. Here is the corresponding result, also due to O'Neil (1965). In it Ω can be $I\!\!R^n$ or the n-torus $\mathbf{T}^n$ with μ as the Lebesgue measure t^{-1} as $-t$. But the statement is a little more general.

Theorem 9. *Let* Φ_i, $i = 1, 2, 3$ *be Young functions satisfying*

$$\Phi_1^{-1}(x) \Phi_2^{-1}(x) \leq x \Phi_3^{-1}(x), \qquad x \geq 0. \tag{25}$$

Let (Ω, Σ, μ) *be a measure space where* Ω *is a locally compact unimodular group with* μ *as its Haar measure and* Σ *is the Borel* σ-*algebra of*

Ω. *Then for any* $f_i \in \mathcal{L}^{\Phi_i}(\mu)$, $i = 1, 2$, *the convolution* $f_1 * f_2 \in \mathcal{L}^{\Phi_3}(\mu)$ *and moreover*

$$N_{\Phi_3}(f_1 * f_2) \leq 2N_{\Phi_1}(f_1)N_{\Phi_2}(f). \tag{26}$$

Proof: We first record that (25) implies an inequality similar to that of (22) and then use it together with Proposition 5 to establish (26).

As before, for given $x \geq 0, y \geq 0$, either $\Phi_1(x) \leq \Phi_2(y)$ or its opposite holds. In the first case, since $x \leq \Phi_i^{-1}(\Phi_i(x))$,

$$x\Phi_2^{-1}(\Phi_1(x)) \leq \Phi_1^{-1}(\Phi_1(x))\Phi_2^{-1}(\Phi_1(x)) \leq \Phi_1(x)\Phi_3^{-1}(\Phi_1(x)), \text{ by (25)}.$$

Dividing by $\Phi_2^{-1}(\Phi_1(x))$ and noting that $\frac{x}{\Phi_i^{-1}(x)}$ is increasing for Young functions, we get (with $\Phi_1(x) \leq \Phi_2(y)$)

$$x \leq \frac{\Phi_1(x)}{\Phi_2^{-1}(\Phi_1(x))}\Phi_3^{-1}(\Phi_1(x)), \quad x \geq 0, \tag{27a}$$

and

$$\frac{\Phi_1(x)}{\Phi_2^{-1}(\Phi_1(x))} \leq \frac{\Phi_2(y)}{\Phi_2^{-1}(\Phi_2(y))} \leq \frac{\Phi_2(y)}{y}. \tag{27b}$$

Multiplying (27a) and (27b) we find

$$xy \leq \Phi_2(y)\Phi_3^{-1}(\Phi_1(x)). \tag{28}$$

If $\Phi_1(x) > \Phi_2(y)$, we interchange $\Phi_2(y)$ and $\Phi_1(x)$ in the above to get

$$xy \leq \Phi_1(x)\Phi_3^{-1}(\Phi_2(y)). \tag{29}$$

Hence xy is bounded by the maximum of (28) and (29) so that

$$xy \leq \Phi_1(x)\Phi_3^{-1}(\Phi_2(y)) + \Phi_2(y)\Phi_3^{-1}(\Phi_1(x)), \qquad x, y \geq 0 \tag{30}$$

To establish (26), we again may take $N_{\Phi_i}(f_i) = 1$, $i = 1, 2$, for simplicity. Recall that $f(x) = (f_1 * f_2)(x) = \int_\Omega f_1(t)f_2(t^{-1}x)dt$. To show that f exists, it suffices to establish (26) directly. Thus consider

$$\int_\Omega \Phi_3\left(\frac{|f|}{2}\right)d\mu \leq \int_\Omega \Phi_3\left(\frac{1}{2}\int_\Omega |f_1(t)| \cdot |f_2(t^{-1}x)|d\mu(t)\right)d\mu(x)$$

$$\leq \int_\Omega \Phi_3\left[\frac{1}{2}\int_\Omega \Phi_1(|f_1(t)|)\Phi_3^{-1}\left(\Phi_2(|f_2(t^{-1}x)|)\right)d\mu(t)\right.$$

$$\left. + \frac{1}{2}\int_\Omega \Phi_2(|f_2(t^{-1}x)|)\Phi_3^{-1}(\Phi_1(|f_1(t)|))d\mu(t)\right]d\mu(x), \text{ by (30)},$$

$$\leq \frac{1}{2}\int_\Omega \Phi_3\Big[\int_\Omega \Phi_1(|f_1(t)|)\Phi_3^{-1}\big(\Phi_2(|f_2(t^{-1}x)|)\big)d\mu(t)\Big]d\mu(x)$$

$$+\frac{1}{2}\int_\Omega \Phi_3\Big[\int_\Omega \Phi_2(|f_2(t^{-1}x)|)\Phi_3^{-1}\big(\Phi_1(|f_1(t)|)\big)d\mu(t)\Big]d\mu(x)$$

$$=\frac{(I_1+I_2)}{2} \qquad \text{(say)}. \tag{31}$$

We shall now show that $I_j \leq 1$, $j = 1,2$. By symmetry it suffices to consider one of them, say I_1. Thus letting $\alpha = \int_\Omega \Phi_1(|f_1(t)|)d\mu(t)$, $0 < \alpha \leq 1$ we have

$$I_1 = \int_\Omega \Phi_3\Big[\int_\Omega \Phi_1(|f_1(t)|)\Phi_3^{-1}\big(\Phi_2(|f_2(t^{-1}x)|)\big)d\mu(t)\Big]d\mu(x)$$

$$\leq \alpha \int_\Omega \Phi_3\Big[\int_\Omega \Phi_3^{-1}\big(\Phi_2(|f_2(t^{-1}x)|)\big)d\nu(t)\Big]d\mu(x), \quad \text{where } \nu(A) =$$

$$\frac{1}{\alpha}\int_A \Phi_1(|f_1|)d\mu(t) \text{ is a probability measure,}$$

$$\leq \alpha \int_\Omega \Big[\int_\Omega \Phi_3\Big(\Phi_3^{-1}\big(\Phi_2(|f_2(t^{-1}x)|)\big)d\nu(t)\Big)\Big]d\mu(x), \text{ by Prop. 5,}$$

$$\leq \alpha \cdot \int_\Omega\int_\Omega \Phi_2(|f_2(t^{-1}x)|)d\nu(t)d\mu(x)$$

$$= \alpha \int_\Omega\Big[\int_\Omega \Phi_2(|f_2(t^{-1}x)|)d\mu(x)\Big]d\nu(t)$$

$$\leq \alpha \int_\Omega 1\cdot d\nu(t) = \alpha \leq 1, \qquad \text{by Thm. 2.3.}$$

Similarly $I_2 \leq 1$, so that (31) implies $\int_\Omega \Phi_3(\frac{|f|}{2})d\mu \leq 1$. This shows that $N_{\Phi_3}(f) \leq 2$ implying (26), and the result follows. ∎

The hypotheses for both Theorems 7 and 9 have a difficulty; namely, the key inequalities (21) and (25) on Φ_i, involving their inverses, are not really easy to verify. Moreover in Fourier analysis one employs analogous inequalities with pairs of complementary Young functions. However, they need a different set of arguments, and also use some results on interpolation theory to be considered in Chapter VI. So we omit further discussion here.

Maintaining the general point of view, we have the following theorem.

Theorem 10. *The normed linear space* $(\mathcal{L}^\Phi(\mu), N_\Phi(\cdot))$, *considered in Theorem 2.3, is complete so that it is a Banach space if a.e. equal functions are identified. The basic measure space* (Ω, Σ, μ) *is unrestricted.*

Proof: If $\{f_n, n \geq 1\} \subset \mathcal{L}^\Phi(\mu)$ is a Cauchy sequence, so that $N_\Phi(f_n - f_m) \to 0$ as $n, m \to \infty$, we need to construct an $f \in \mathcal{L}^\Phi(\mu)$ satisfying $N_\Phi(f_n - f) \to 0$ as $n \to \infty$. Since Φ is merely a Young function we have to consider two cases. Let $x_0 = \sup\{x \in \mathbb{R}^+ : \Phi(x) = 0\}$. Then $0 \leq x_0 < \infty$ since the set in braces $\{ \}$ is relatively compact by definition of Φ. Now by hypothesis there are numbers $k_{mn} \geq 0$ ($k_{mn}^{-1} \leq N_\Phi(f_n - f_m)$) such that

$$\int_\Omega \Phi(k_{mn}|f_n - f_m|)d\mu \leq 1. \tag{32}$$

First note that $A_{mn} = \{\omega : k_{mn}|f_n - f_m|(\omega) > x_0\} \in \Sigma$ is at most σ-finite for μ. Indeed, if $B_k = B_k^{mn} = \{\omega : k_{mn}|f_n - f_m|(\omega) > x_0 + k^{-1}\}$, then $A_{mn} = \overset{\infty}{\underset{k=1}{\cup}} B_k$, and $\mu(B_k) < \infty$ for each k. This is because

$$\mu(B_k) \leq \frac{1}{\Phi(x_0 + k^{-1})} \int_{B_k} \Phi(k_{mn}|f_n - f_m|)d\mu \leq 1, \qquad \text{by (32).}$$

Therefore each A_{mn} is σ-finite and so is $A = \underset{m,n \geq 1}{\cup} A_{mn}$. On A^c, $k_{mn}|f_n - f_m|(\omega) \leq x_0$ so that for $\omega \in A^c$, $|f_n(\omega) - f_m(\omega)| \to 0$ uniformly. Hence there is a measurable g_0 on A^c such that $f_n(\omega) \to g_0(\omega)$, and $|g_0| \leq x_0$, $\omega \in A^c$.

Let us write Ω for A temporarily. Then $\{f_n, n \geq 1\}$ is Cauchy on $\mathcal{L}^\Phi(\Omega, \mu)$ and hence for each $B \in \Sigma$, $\mu(B) < \infty$ we have

$$\mu(B \cap \{|f_m - f_n| \geq \varepsilon\}) = \mu(B \cap \{\Phi(k_{mn}|f_n - f_m|) \geq \Phi(k_{mn}\varepsilon)\})$$
$$\leq \frac{1}{\Phi(k_{mn}\varepsilon)} \int_B \Phi(k_{mn}|f_n - f_m|)d\mu$$
$$\leq [\Phi(k_{mn}\varepsilon)]^{-1}, \qquad \text{by (32).}$$

Since $k_{mn} \to \infty$, and $\varepsilon > 0$ is fixed, this shows that $\{f_n, n \geq 1\}$ is Cauchy in μ-measure on each such B. By the σ-finiteness this implies

$\{f_n, n \geq 1\}$ is Cauchy in measure. But then by a standard measure theory result (cf. e.g., Rao (1987), p. 117) $f_n \to \tilde{f}$ in measure. Then there is a subsequence $\{f_{n_i}, i \geq 1\}$ such that $f_{n_i} \to \tilde{f}$, a.e. Let $f = \tilde{f} \chi_A + g_0 \, \chi_{A^c}$. Hence $f_{n_i} \to f$ a.e. But for $\{f_n, n \geq 1\}$, which is a Cauchy sequence in $N_\Phi(\cdot)$, we get $N_\Phi(f_n) \to \rho$ and hence $N_\Phi(f_{n_i}) \to \rho$ also. Now by Fatou's inequality

$$\int_\Omega \Phi\left(\frac{|f|}{\rho}\right) d\mu \leq \liminf_{i \to \infty} \int_\Omega \Phi\left(\frac{f_{n_i}}{N_\Phi(f_{n_i})}\right) d\mu \leq 1.$$

Thus $f \in \mathcal{L}^\Phi(\Omega, \mu)$. Also, if m is fixed and $k \geq 0$ is given, then $\Phi(|f_{n_i} - f_{n_j}|k) \to \Phi(|f - f_{n_j}|k)$ as $i \to \infty$, a.e. If $n_0 \geq 1$ is chosen such that $n_i, n_j \geq n_0$ implies $k_{n_i n_j} \geq k$, then

$$\int_\Omega \Phi(k|f_{n_i} - f_{n_j}|) d\mu \leq \int_\Omega \Phi(k_{n_i n_j}|f_{n_i} - f_{n_j}|) d\mu \leq 1. \qquad (33)$$

Letting $n_i \to \infty$, we get by Fatou's lemma that $N_\Phi(f - f_{n_j}) \leq \frac{1}{k}$. Since $k > 0$ is arbitrary $N_\Phi(f_{n_j} - f) \to 0$. If f_{n_j} is any other subsequence with limit f', then $\{f_{n'_j}, f_{n_i}, i \geq 1, j \geq 1\} \subset \{f_n, n \geq 1\}$ so that $f = f'$ a.e. because $f_n \to f$ in measure. So for every convergent subsequence and hence for the whole sequence, $N_\Phi(f_n - f) \to 0$. This shows that every Cauchy sequence of $(\mathcal{L}^\Phi(\mu), N_\Phi(\cdot))$ converges to an element in the space. ∎

A similar result holds for the Orlicz norm $\|\cdot\|_\Phi$. However, we need to assume that μ has the finite subset property as noted before since otherwise $\|\cdot\|_\Phi$ is only a seminorm. But with the latter (nonrestrictive) condition and Proposition 4 above we have the following immediate consequence of the preceding theorem.

Proposition 11. *The set $(\mathcal{L}^\Phi(\mu), \|\cdot\|_\Phi)$ is a Banach space when μ-equivalent functions are identified and μ has the finite subset property.*

Hereafter we set $\mathcal{N}_\Phi = \{f \in \mathcal{L}^\Phi(\mu) : N_\Phi(f) = 0\}$ which is the same set if $N_\Phi(\cdot)$ is replaced by $\|\cdot\|_\Phi$ when μ is assumed to have the finite subset property. In these cases therefore $f \in \mathcal{N}_\Phi$ iff $f = 0$ a.e. Hence the factor space $L^\Phi(\mu) = \mathcal{L}^\Phi(\mu)/\mathcal{N}_\Phi$ with $\bar{N}_\Phi([f]) = N_\Phi(f)$ is well defined where $[f] = \{g \in \mathcal{L}^\Phi(\mu) : g - f \in \mathcal{N}_\Phi\}$. Then the above two results imply the following.

Corollary 12. $(\mathcal{L}^{\Phi}(\mu), \bar{N}_{\Phi})$ $((L^{\Phi}(\mu), \| \cdot \|_{\bar{\Phi}}))$ *is a Banach space where* $\| \cdot \|_{\bar{\Phi}}$ *is similarly defined.*

Following custom we shall conveniently (but somewhat imprecisely) refer to the elements of $L^{\Phi}(\mu)$ as functions, instead of equivalence classes. *Unless a distinction is essential in a particular situation, we shall not distinguish $\mathcal{L}^{\Phi}(\mu)$ and $L^{\Phi}(\mu)$ hereafter. The symbols $N_{\Phi}(\cdot)$ and $\| \cdot \|_{\Phi}$ will be used for $\bar{N}_{\Phi}(\cdot)$ and $\| \cdot \|_{\bar{\Phi}}$ also.*

Before leaving this section, we present two results on these norms. The first one relates to calculation of $\| \cdot \|_{\Phi}$ which is generally more involved since it always needs a knowledge of the complementary function Ψ of Φ. The second one shows that in general, equality does not obtain between the two norms. These will give us all the information needed on the use of either norm for our work in this book.

Theorem 13. *Let (Φ, Ψ) be a complementary pair of N-functions, and $f \in L^{\Phi}(\mu)$, on a measure space (Ω, Σ, μ). Then the Orlicz norm $\|f\|_{\Phi}$ is given in terms of Φ alone by the formula*

$$\|f\|_{\Phi} = \inf\{\tfrac{1}{k}(1 + \int_{\Omega} \Phi(kf)d\mu) : k > 0\}. \tag{34}$$

Moreover, there is a number k^* ($= k_f^*$) attaining the infimum in (34), so that

$$\|f\|_{\Phi} = \tfrac{1}{k^*}(1 + \int_{\Omega} \Phi(k^*f)d\mu), \tag{35}$$

iff $k_1 \leq k^* \leq k_2$ where

$$k_1 = \inf\{k > 0 : \rho_{\Psi}(\varphi(k|f|)) \geq 1\}, \tag{36}$$

$$k_2 = \sup\{k > 0 : \rho_{\Psi}(\varphi(k|f|)) \leq 1\}, \tag{37}$$

and $\rho_{\Psi}(g) = \int_{\Omega} \Psi(g)d\mu$, φ being the left derivative of Φ as usual.

The main part, namely (34), is due to Krasnoselskii and Rutickii, and is found in their monograph if $\Omega \subset \mathbb{R}^n$ is a bounded set with μ as the Lebesgue measure. But their proof extends to the general case. Also the second part (35) is an extension of their work and is obtained by C. X. Wu (1978). We now present the details of proof as propositions for completeness.

Proposition 14. *Let* (Φ, Ψ) *be a complementary pair of N-functions such that for* $f \in L^\Phi(\mu)$ *there is a* k_0 *(= k(f) > 0) satisfying* $\rho_\Psi(\varphi(k_0|f|)) = 1$. *Then*

$$\|f\|_\Phi = \int_\Omega \varphi(k_0|f|)|f|d\mu. \qquad (38)$$

Proof: This formula is an exact analog of the classical result in the Lebesgue case with $\Phi(x) = \frac{1}{p}|x|^p$, $1 < p < \infty$. It follows from (5) that the right side of (35) *never* exceeds $\|f\|_\Phi$. For the opposite inequality one uses Young's inequality and equality conditions in the latter. Thus

$$\|f\|_\Phi = \sup\left\{ \int_\Omega |fg|d\mu : \rho_\Psi(g) \le 1 \right\} \le \frac{1}{k_0}\left(\int_\Omega \Phi(k_0 f)d\mu + 1 \right) \quad (39)$$

$$= \frac{1}{k_0}\left(\int_\Omega \Phi(k_0 f)d\mu + \int_\Omega \Psi(\varphi(k_0|f|))d\mu \right), \quad \text{by hypothesis,}$$

$$= \frac{1}{k_0}\int_\Omega |k_0 f\varphi(k_0|f|)|d\mu = \int_\Omega \varphi(k_0|f|)|f|d\mu, \qquad (40)$$

using the equality in Young's inequality. This is (38) as asserted. ∎

Note that in the Lebesgue case $k_0^{-1} = \|f\|_p^{p-1}$ with essentially the same proof (cf. e.g., Rao (1987), p. 205). In the general case k_0 is harder to compute.

Using this result we have

Theorem 15. *Under the hypothesis of Theorem 13, formula (34) holds.*

Proof: Since Φ and Ψ are N-functions so that $\Phi(x) = 0$ [$\Psi(y) = 0$] iff $x = 0$ [$y = 0$], it follows that the support of each simple function $h \in L^\Phi(\mu)$ [$\tilde{h} \in L^\Psi(\mu)$] is of finite measure and each $f \in L^\Phi(\mu)$ [$\tilde{f} \in L^\Psi(\mu)$] has σ-finite support. Hence $f \in L^\Phi(\mu)$ implies the existence of a sequence of simple functions $0 \le f_n \uparrow |f|$ and hence $f_n \in L^\Phi(\mu)$. We first note that if φ is continuous, then (38) holds with some $k_n > 0$ for the f_n. Indeed by Proposition 2.2, $\ell_n(k) = \int_\Omega \Phi(kf_n)d\mu$, is continuous and increasing in $[0, \infty)$ with $\ell_n(0) = 0$, and $\lim_{k \to \infty} \ell_n(k) = +\infty$, and the same is true of $\tilde{\ell}_n(k) = \int_\Omega \Psi(\varphi(k|f_n|))d\mu$. Hence by the mean value theorem there is a $0 < k_n < \infty$ such that $\ell_n(k_n) = 1$. Moreover, since $f_n \uparrow$ the last equation implies that $k_n \geq k_{n+1}$. Let $k_n \to k^* > 0$. [$k^* = 0$ is impossible since $f \neq 0$.] Then we get, by Fatou's inequality and the monotone property of $\| \cdot \|_\Phi$, the following: By (39) and (40) we have for any $k > 0$,

$$\|f\|_\Phi \leq \frac{1}{k}\left(\int_\Omega \Phi(kf)d\mu + 1 \right), \tag{41}$$

and for the simple $f_n \leq |f|$, with the particular $k_n > 0$, there is equality. Hence

$$\|f_n\|_\Phi = \inf \left\{ \frac{1}{k}\left(1 + \int_\Omega \Phi(kf_n)d\mu\right) : k > 0 \right\}, \tag{41'}$$

and

$$\|f_n\|_\Phi = \frac{1}{k_n}\left(1 + \int_\Omega \Phi(k_n f_n)d\mu\right). \tag{42}$$

Letting $n \to \infty$ in (42), one has on using $\|f\|_\Phi = \lim_{n\to\infty} \|f_n\|_\Phi$

$$k^*\|f\|_\Phi = 1 + \lim_{n\to\infty} \int_\Omega \Phi(k_n f_n)d\mu \geq 1 + \int_\Omega \Phi(k^* f)d\mu. \tag{43}$$

Thus (41) with $k = k^*$ and (43) imply via (41'), the truth of (34) when φ is continuous.

In the general case that φ is only left continuous, we approximate φ with a continuous φ_1 arbitrarily closely so that for small $\varepsilon > 0$, Φ and Φ_1 satisfy the following inequalities

$$\Phi((1 - \varepsilon)x) < \Phi_1(x) < \Phi(x), \qquad x > 0. \tag{44}$$

This can be done in many ways by joining the discontinuities of φ with straight line segments to obtain a continuous φ_1 for which (44) holds. (Another is to note that φ is lower semicontinuous and apply known lower approximations.) Also $L^\Phi(\mu) = L^{\Phi_1}(\mu)$ as sets and one sees immediately that

$$(1 - \varepsilon)\|f\|_\Phi \leq \|f\|_{\Phi_1} \leq \|f\|_\Phi. \tag{45}$$

But (34) is shown to be true for Φ_1 so that for the Φ-case

$$\inf\left\{\frac{1}{k}\Big(1+\int_\Omega \Phi(kf)d\mu\Big) : k>0\right\}$$

$$\geq \inf\left\{\frac{1}{k}\Big(1+\int_\Omega \Phi_1(kf)d\mu\Big) : k>0\right\}$$

$$= \|f\|_{\Phi_1}$$

$$\geq \inf\left\{\frac{1}{k}\Big(\frac{1-\varepsilon}{1-\varepsilon}+\frac{1}{1-\varepsilon}\int_\Omega \Phi(kf)d\mu\Big) : k>0\right\}$$

$$\geq (1-\varepsilon)\inf\left\{\frac{1}{k}\Big(1+\int_\Omega \Phi(kf)d\mu\Big) : k>0\right\}$$

$$\geq (1-\varepsilon)\|f\|_\Phi, \quad \text{by (45)}.$$

Since $\varepsilon > 0$ is arbitrary and recalling (39), we find that (34) holds in the general case as well. ∎

We finally establish

Proposition 16. *Under the conditions of Theorem 13, (35) holds.*

Proof: From definitions of k_1 and k_2, it follows that $0 < k_1 \leq k_2 < \infty$. But by Proposition 2.2, the mapping $k \mapsto \int_\Omega \Phi(kf)d\mu$ is increasing and continuous on $[0, k_f)$ where k_f is the supremum of the numbers k for which the preceding integral is finite. Hence the function $J_f : k \mapsto \frac{1}{k}(1 + \int_\Omega \Phi(kf)d\mu)$, $0 < k < k_f$ is continuous. We also note that k_2 of (37) satisfies $k_2 \leq k_f$. This follows from the computation:

$$J_f(k) = \frac{1}{k}\Big(1+\int_\Omega \Phi(kf)d\mu\Big)$$

$$= \frac{1}{k}\Big(1+\int_\Omega k|f|\varphi(k|f|)d\mu - \rho_\Psi\big(\varphi(k|f|)\big)\Big), \quad \text{using the equality}$$

in Young's inequality,

$$\leq \frac{1}{k}+\int_\Omega |f|\varphi(k|f|)d\mu \leq \frac{1}{k}+\|f\|_\Phi, \quad \text{by (5)}.$$

Hence $k_2 < k_f$. But Proposition 2.2 also implies that $\lim_{k\to 0} J_f(k) = +\infty$ and $k_2 \leq k_f$, and for $0 < k_1 \leq k \leq k_2$, $J_f(k) < \infty$, the continuous function $J_f(\cdot)$ must attain its minimum on $[k_1, k_2]$ since $k_2 < \infty$ from (37). Hence any $k^* \in [k_1, k_2]$ satisfies (35), and this proves the proposition, and with it Theorem 13 is also established. ∎

To finish the comparison between the gauge and Orlicz norms we present the following.

Proposition 17. *If* (Φ, Ψ) *is a complementary pair of N-functions and* $f \in L^\Phi(\mu)$*, then* $\|f\|_\Phi = N_\Phi(f)$ *iff* $f = 0$ *a.e.*

Proof: Since the "if" is clear, consider the "only if" part. Assuming that $\|f\|_\Phi = N_\Phi(f)$, we get from (35) for $f \neq 0$,

$$1 = \left\| \frac{f}{N_\Phi(f)} \right\|_\Phi = \frac{1}{k^*}\left(1 + \int_\Omega \Phi\left(\frac{k^* f}{N_\Phi(f)}\right) d\mu\right) > \frac{1}{k^*} > 0.$$

But then for any $0 < \varepsilon < 1 - (\frac{1}{k^*})$, so that $(1 - \varepsilon)k^* > 1$, we have

$$1 = \frac{1}{k^*}\left(1 + \int_\Omega \Phi\left(\frac{k^*(1-\varepsilon)f}{(1-\varepsilon)N_\Phi(f)}\right) d\mu\right)$$

$$\geq \frac{1}{k^*}\left(1 + k^*(1-\varepsilon)\int_\Omega \Phi\left(\frac{f}{(1-\varepsilon)N_\Phi(f)}\right) d\mu\right), \text{ by the convexity}$$

$$\text{of } \Phi,$$

$$> \frac{1}{k^*}(1 + k^*(1-\varepsilon)), \text{ by definition of } N_\Phi(f),$$

$$= \frac{1}{k^*} + (1-\varepsilon)$$

Since k^* is independent of ε and $\varepsilon > 0$ is arbitrary the above inequality becomes $1 \geq 1 + \frac{1}{k^*}$ which is impossible. Thus $f = 0$ a.e. must hold. ∎

Although $\|\cdot\|_\Phi$ and $N_\Phi(\cdot)$ are not equal, they define the same topology and are equivalent by Proposition 5. This allows us to study the linear topological (and more precisely the Banach space) aspects of $L^\Phi(\mu)$ spaces on general measure spaces (Ω, Σ, μ). We thus turn to this aspect of our study.

3.4. The spaces L^Φ and M^Φ

Recall that L^Φ or $L^\Phi(\mu)$ is the Orlicz space $\{\mathcal{L}^\Phi(\mu), N_\Phi(\cdot)\}$ in which a.e. equivalent functions are identified. It is thus a Banach space and is based on an arbitrary measure space (Ω, Σ, μ). Similarly $\tilde{L}^\Phi(\mu)$ is the space $\{\tilde{\mathcal{L}}^\Phi(\mu), N_\Phi(\cdot)\}$ with a.e. equal elements are identified. As seen in Section 1 above, $\tilde{L}^\Phi(\mu)$ is a solid circled class which need not be linear unlike $L^\Phi(\mu)$, which is always linear. However, as Theorem 1.2

shows, if Φ satisfies a Δ_2-regular condition, then $\tilde{L}^\Phi(\mu) = L^\Phi(\mu)$. But as Example 1.4 amply illustrates, this equality can hold for the Young functions which do not satisfy a Δ_2-condition if the measure space is discrete. *Thus for Orlicz spaces $L^\Phi(\mu)$ more than the classical Lebesgue spaces $L^p(\mu)$, the topological vector space properties depend both on the Young functions Φ and the measures μ.* This point is emphasized here, since failure to recognize it leads to erroneous claims and assertions, as the existing literature amply illustrates. In this book the measure space is unrestricted unless the contrary is stated.

For a large part of the analysis, especially when $\mu(\Omega) < \infty$, a key role is played by the Young functions Φ for large values of the argument. Consequently, one says that a convex function $Q : I\!\!R^+ \to I\!\!R^+$ is the *principal part* of a Young function Φ if $Q = \Phi$ on $[x_0, \infty)$ for some large enough $x_0 > 1$. Thus we can state:

Lemma 1. *A convex function $Q : I\!\!R^+ \to I\!\!R^+$ satisfying $Q(x) \nearrow \infty$ as $x \to \infty$ is the principal part of some Young function $\Phi : I\!\!R^+ \to I\!\!R^+$, and Φ will be an N-function if $\frac{Q(x)}{x} \nearrow \infty$ as $x \nearrow \infty$ in addition.*

Proof: Since Q is convex, we have by Theorem I.3.1, on (x_0, ∞),

$$Q(x) = Q(x_0) + \int_{x_0}^x Q'(t)dt, \qquad x > x_0, \tag{1}$$

where $Q'(\cdot)$ is the left derivative of Q and is nondecreasing and left continuous. Let $\delta = \lim_{x \to \infty} Q'(t) \ (\leq \infty)$. Select $x_1 > x_0 + 1$ such that $x_1 Q'(x_1) > Q'(x_0) + Q(x_0)$. This is possible since $Q(x) \nearrow \infty$ as $x \nearrow \infty$, and $Q'(x_1) \geq Q'(x_0)$. Then

$$Q(x_1) = \left(\int_{x_0}^{x_0+1} + \int_{x_0+1}^{x_1} \right) Q'(t)dt + Q(x_0)$$
$$\leq Q'(x_0 + 1) + Q'(x_1)(x_1 - x_0 - 1) + Q(x_0)$$
$$< Q'(x_1)(x_1 - x_0) < Q'(x_1)x_1.$$

If we set $\alpha = x_1 \frac{Q'(x_1)}{Q(x_1)}$, then $\alpha > 1$ and if we define Φ by the equation

$$\Phi(x) = \begin{cases} \frac{Q(x_1)}{x_1^\alpha}|x|^\alpha, & \text{for } x_1 > |x| \\ Q(x), & \text{for } |x| \geq x_1, \end{cases}$$

then Φ is a Young function and an N-function in the case that $\frac{Q(x)}{x} \to \infty$, so that Φ has Q as its principal part. ∎

The point of this lemma is that two Young functions Φ_1, Φ_2 with the same principal parts are (locally) equivalent, but, by the device used in defining Φ above can also be modified to be (globally) equivalent. Such an alteration is clearly nonunique, but for most of the analysis on $L^\Phi(\mu)$-spaces this change will cause no problems.

To analyze the linear structure of $L^\Phi(\mu)$, we introduce:

Definition 2. Let $M^\Phi = \{f \in L^\Phi(\mu) : kf \in \tilde{L}^\Phi(\mu), \text{ for all } k > 0\}$. Also let $\mathcal{M}^\Phi$ be the closed linear span of all step functions from $L^\Phi(\mu)$. [A *step* function takes finitely many finite values on measurable sets of Ω, and a *simple* function is a step function which vanishes outside a set of finite μ-measure. (The M^Φ space was introduced by Morse and Transue (1950).)]

Interrelations between these spaces is given by:

Proposition 3. *In the above notation the linear spaces satisfy $M^\Phi \subseteq \mathcal{M}^\Phi \subseteq L^\Phi(\mu)$. There is equality between the first two if Φ is a continuous Young function with $\Phi(x) > 0$ for $x > 0$, and there is equality between the last two if Φ is discontinuous and $\mu(\Omega) < \infty$, or Φ is of the form $\Phi(x) = 0$ for $0 < x \leq x_0$, $= \Phi_1(x)$ for $x_0 \leq x < x_1$, $= \infty$ for $x \geq x_1$ (and $\mu(\Omega) \leq \infty$), where Φ_1 is a continuous convex function such that $\Phi_1(x) > 0$, for $x > x_0$.*

Proof: The inclusion relations between the spaces are immediate from the definition. Note that if Φ is discontinuous, then $M^\Phi = \{0\} \underset{\neq}{\subset} \mathcal{M}^\Phi$. We assert that $M^\Phi = \mathcal{M}^\Phi$ if Φ is continuous and $\Phi(x) > 0$ for $|x| > 0$.

Indeed, since every step function in M^Φ is simple, each such element of M^Φ is also in $\mathcal{M}^\Phi$. If $f \in M^\Phi$ is arbitrary, then $\exists 0 \leq f_n \uparrow |f|$ pointwise, f_n simple, so that for each $k > 0$, $kf_n \in M^\Phi$ and

$$\int_\Omega \Phi(k(|f| - f_n))d\mu \to 0, \qquad \text{as } n \to \infty \qquad (2)$$

by the dominated convergence theorem since $\Phi(kf) \geq \Phi(k(|f| - f_n))$. Hence for any $\varepsilon > 0$, choose $k > \frac{2}{\varepsilon}$ and then n_0 such that $n \geq n_0 \Rightarrow$

the integral in (2) for this k is at most 1. Then

$$\||f| - f_n\|_\Phi = \tfrac{1}{k}\sup\{\int_\Omega k(|f| - f_n)|g|d\mu : \rho_\Psi(g) \le 1\}$$

$$\le \tfrac{1}{k}\{\int_\Omega \Phi(k(|f| - f_n))d\mu + 1\} \le \tfrac{2}{k} < \varepsilon.$$

Consequently $|f| \in \mathcal{M}^\Phi$. But since $f \in M^\Phi$ iff $f^\pm$ and hence $|f|$ are in M^Φ, this shows $f \in \mathcal{M}^\Phi$ so that $M^\Phi \subset \mathcal{M}^\Phi$ and both have all μ-simple functions from $L^\Phi(\mu)$. On the other hand, if $f \in \mathcal{M}^\Phi$, and $k > 0$ is arbitrarily fixed, and $0 < \varepsilon \le 1$ is given, choose $f_\varepsilon \in \mathcal{M}^\Phi$, simple, such that $N_\Phi(f - f_\varepsilon) < \frac{\varepsilon}{2k}$. But $2kf_\varepsilon \in M^\Phi$ and

$$\int_\Omega \Phi(2k(f - f_\varepsilon))d\mu = \int_\Omega \Phi\left(\frac{2k(f - f_\varepsilon)N_\Phi(2k(f - f_\varepsilon))}{N_\Phi(2k(f - f_\varepsilon))}\right) d\mu$$

$$\le N_\Phi(2k(f - f_\varepsilon)) \le 1. \tag{3}$$

Hence $2k(f - f_\varepsilon) \in M^\Phi$, and by the convexity of this space, we get

$$kf = \tfrac{1}{2}(2k(f - f_\varepsilon)) + \tfrac{1}{2}(2kf_\varepsilon) \in M^\Phi.$$

Thus $M^\Phi = \mathcal{M}^\Phi$ and at the same time we have shown that $(M^\Phi, N_\Phi(\cdot))$ $((M^\Phi, \|\cdot\|_\Phi))$ is a Banach space (since $\mathcal{M}^\Phi$ is) having simple functions dense in it. This proves the first half of the proposition with more information.

Regarding the second part, let Φ be discontinuous as given. Then (cf., Equation (2.6)) every element of $L^\Phi(\mu)$ is essentially bounded so that $L^\Phi(\mu) \subset L^\infty(\mu)$. If $\Phi(x) = 0$, $0 \le x < x_0$, then $f \in L^\infty(\mu) \Rightarrow$ $\frac{|f|}{k} < x_0$ for some $k > 0$, so that $f \in L^\Phi(\mu)$ and hence $L^\Phi(\mu) = L^\infty(\mu)$. Next, let $x_0 = 0$ and $\Phi = \Phi_1$ on $(0, x_1)$ and $\Phi(x) = \infty$ for $x > x_1$ with Φ_1 continuous and $\mu(\Omega) < \infty$. Then each element of $L^\Phi(\mu)$ is essentially bounded, and if $f \in L^\infty(\mu)$ then for some $k_1 > 0$, $\frac{|f|}{k_1} < x_1 - \varepsilon$ for an $\varepsilon > 0$, so that

$$\int_\Omega \Phi(\tfrac{f}{k_1})d\mu \le \Phi(x_1 - \varepsilon)\mu(\Omega) < \infty. \tag{4}$$

Thus $f \in L^\Phi(\mu)$ and $L^\Phi(\mu) = L^\infty(\mu)$. Note that (4) implies that $\|f\|_\infty \le x_1 \cdot N_\Phi(f)$ and from this and the classical Banach closed graph theorem it follows that both $\|\cdot\|_\infty$ and $N_\Phi(\cdot)$ are equivalent. However, $L^\infty(\mu)$ has step functions dense in it. Indeed for each $f \in L^\infty(\mu)$, there is a μ-null set Ω_0 and given $\varepsilon > 0$, there are disjoint Borel sets $A_1, \ldots, A_n$ such that $f(\Omega - \Omega_0) \subset \bigcup_{i=1}^{n} A_i$ where the diameter of each A_i is at most ε. If $f_\varepsilon = \sum_{i=1}^{n} a_i \chi_{f^{-1}(A_i)}$, $a_i \in A_i$, then $\|f - f_\varepsilon\|_\infty < \varepsilon$. This implies that $f_\varepsilon \in M^\Phi$ and by the equivalence of norms, $f \in M^\Phi = L^\infty(\mu) = L^\Phi(\mu)$. However, if $\mu(\Omega) = +\infty$, and $\Phi(x) > 0$ for $0 < x < x_1$, then $L^\Phi(\mu) \underset{\neq}{\subset} L^\infty(\mu)$ obtains since (3) no longer holds and counterexamples may be constructed, as desired. ∎

The following result proved for (3) is of interest when the principal part of Φ grows exponentially.

Corollary 4. *If Φ is a continuous Young function, $\Phi(x) > 0$ for $x > 0$, then $M^\Phi = \mathcal{M}^\Phi \subset \tilde{L}^\Phi(\mu)$ with equality in the last inclusion iff $\tilde{L}^\Phi(\mu)$ is linear. In fact $\tilde{L}^\Phi(\mu)$ is linear iff $\tilde{L}^\Phi(\mu) = M^\Phi$ and hence also $= \mathcal{M}^\Phi = L^\Phi(\mu)$.*

We analyze the case of Φ in this corollary in more detail. As a consequence of Theorem 1.2 and the above result we have the useful

Corollary 5. *Let Φ be a Young function and $L^\Phi(\mu)$ be an Orlicz space on a measure space (Ω, Σ, μ). Then $M^\Phi = \mathcal{M}^\Phi = L^\Phi(\mu)$ if Φ satisfies a Δ_2-regular condition. On the other hand, this condition is also necessary if μ is diffuse on a set of positive measure. Thus if Φ is Δ_2-regular, then simple functions are dense in $L^\Phi(\mu)$.*

The result of this corollary gives the following.

Proposition 6. *If $L^\Phi(\mu)$ is an Orlicz space on (Ω, Σ, μ), Φ satisfies a Δ_2-regular condition then $0 \neq f \in L^\Phi(\mu)$ implies*

$$\int_\Omega \Phi\left(\frac{f}{N_\Phi(f)}\right) d\mu = 1. \tag{5}$$

Conversely, if Φ is an N-function and μ is diffuse on a set of positive measure then the Δ_2-regularity of Φ is also necessary for (5) to hold (so Φ is continuous and $\Phi(x) > 0$ for $x > 0$).

Proof: If Φ is Δ_2-regular, then $\tilde{L}^\Phi(\mu) = L^\Phi(\mu)$ and that the mapping $\ell_f : k \mapsto \int_\Omega \Phi(kf) d\mu$ is continuous and monotone increasing on

$I\!\!R^+ \to I\!\!R^+$, for each $f \in L^\Phi(\mu)$. Hence, there is a k_0 such that $\ell_f(k_0) = 1$ and from the definition of $N_\Phi(f)$, $k_0 = [N_\Phi(f)]^{-1}$ and (5) follows.

The converse need not hold for arbitrary μ, as seen in Example 1.4 for discrete measures. However, when μ is diffuse on a set of positive measure we show that Φ must be Δ_2-regular. Using the same idea as in the proof of Theorem 1.2, it is enough to prove the result when (Ω, Σ, μ) is a diffuse measure space with Ω as a set such that $\mu(\Omega) < \infty$. We assume this and suppose, if possible, that $\Phi \notin \Delta_2$ and derive a contradiction exactly as in that proof.

By the fact that $\Phi(x) \nearrow \infty$, we can find an increasing sequence $x_n \in I\!\!R^+$ such that $\Phi((1 + \frac{1}{n})x_n) > 2^n \Phi(x_n)$, $n \geq 1$ and by the diffuseness of μ, a disjoint sequence $\Omega_n \in \Sigma$ such that for $\alpha > \mu(\Omega) > 0$,

$$\mu(\Omega_n) = \frac{\mu(\Omega)}{2^n \alpha \Phi(x_n)}, \qquad n \geq 1.$$

If $f = \sum_{x=1}^{\infty} x_n \chi_{\Omega_n}$, then $f \in L^\Phi(\mu)$ and $N_\Phi(f) = 1$. Indeed,

$$\int_\Omega \Phi(f) d\mu = \sum_{n=1}^{\infty} \Phi(x_n) \mu(\Omega_n) = \frac{\mu(\Omega)}{\alpha} < 1. \qquad (6)$$

Moreover, for any $\beta \geq 1$, the same computation shows that $\int_\Omega \Phi(\frac{f}{\beta}) d\mu$ < 1. But if $\beta < 1$ so that $\frac{1}{\beta} \geq 1 + \frac{1}{n_0}$ for some $n_0 \geq 1$, we have

$$\int_\Omega \Phi(\frac{f}{\beta}) d\mu = \sum_{n=1}^{\infty} \Phi(\frac{x_n}{\beta}) \mu(\Omega_n)$$

$$> \sum_{n=n_0}^{\infty} \Phi((1 + \frac{1}{n})x_n) \mu(\Omega_n) \geq \sum_{n \geq n_0} \frac{\mu(\Omega)}{\alpha} = +\infty.$$

Hence $N_\Phi(f) = 1$, and then (6) shows that (5) is not true for this f. ∎

Since (5) always holds for simple functions f when Φ is continuous, and $\Phi(x) > 0$ for $x > 0$, we have the following consequence.

Corollary 7. *If (Ω, Σ, μ) is a measure space, $F \in \Sigma$ with $\mu(F) < \infty$, and Φ is a continuous Young function with $\Phi(x) = 0$ iff $x = 0$, then*

$$N_\Phi(\chi_F) = [\Phi^{-1}(\frac{1}{\mu(F)})]^{-1}. \qquad (7)$$

and if Φ is an N-function, then $\|\chi_F\|_\Phi = \mu(F)\Psi^{-1}(\frac{1}{\mu(F)})$ can be established easily.

Another property of the Δ_2-condition is given by the following.

Proposition 8. *Let $L^\Phi(\mu)$ be an Orlicz space on (Ω, Σ, μ). Consider (i) Φ is Δ_2-regular and (ii) for the complementary function Ψ of Φ we have $\sup\{\int_\Omega \Psi(\varphi(|f|))d\mu : f\in L^\Phi(\mu), N_\Phi(f) \le 1\} < \infty$: Then (i) $\Rightarrow$ (ii), and if μ is diffuse on a set of positive measure then (i) and (ii) are equivalent.*

Proof: (i) $\Rightarrow$ (ii). The equality in Young's inequality and the Δ_2-regularity of Φ imply for the pair (Φ, Ψ), and for a $k > 1$,

$$\Phi(x) + \Psi(\varphi(x)) = x\varphi(x) \le \Phi(2x) \le k\Phi(x), \qquad x \ge 0.$$

Hence for $f \in L^\Phi(\mu)$ with $N_\Phi(f) \le 1$ we have

$$\int_\Omega \Psi(\varphi(|f|))d\mu \le \int_\Omega \Phi(2f)d\mu - \int_\Omega \Phi(f)d\mu \le (k-1)\int_\Omega \Phi(f)d\mu < \infty.$$

Taking supremum on f with $N_\Phi(f) \le 1$ we get (ii).

Conversely, under the additional hypothesis, we proceed with the by-now-familiar argument to construct a function f_0 for which $N_\Phi(f_0) \le 1$ and $\int_\Omega \Psi(\varphi(|f_0|))d\mu = \infty$ when Φ does not satisfy the Δ_2-condition. Thus we assume that μ is finite and diffuse on Ω. Since $\Phi \notin \Delta_2$, there exist $1 < x_n \nearrow \infty$, and $x_n\varphi(x_n) > (2^n + 1)\Phi(x_n)$, and disjoint sets $\Omega_n \in \Sigma$, such that $\mu(\Omega_n) = (2^n\Phi(x_n))^{-1}$. If $f_0 = \sum_{n=1}^\infty x_n\chi_{\Omega_n}$, then $N_\Phi(f_0) = 1$ but $\varphi(|f_0|) \notin \tilde{L}^\Psi(\mu)$. In fact

$$\int_\Omega \Psi(\varphi(|f_0|))d\mu = \sum_{n=1}^\infty \Psi(\varphi(x_n))\mu(\Omega_n)$$

$$= \sum_{n=1}^\infty [x_n\varphi(x_n) - \Phi(x_n)]\mu(\Omega_n) \ge \sum_{n=1}^\infty 2^n\Phi(x_n)\mu(\Omega_n)$$

$$= \infty.$$

Hence (ii) does not hold.   ■

It is also of interest at this point, to discuss the equality conditions for the Hölder inequalities given by equations (3.1), and (3.16) [or (3.17)].

Proposition 9. *Let* (Φ, Ψ) *be a continuous Young pair and* $L^\Phi(\mu)$, $L^\Psi(\mu)$ *be the corresponding Orlicz spaces. Let* $0 \neq f \in L^\Phi(\mu), 0 \neq g \in L^\Psi(\mu)$. *Then*

(a) *there is equality in (3.16), i.e.,* $\int_\Omega |fg| d\mu = N_\Phi(f)\|g\|_\Psi$, *iff*

   (i) $\int_\Omega \Phi\left(\frac{f}{N_\Phi(f)}\right) d\mu = 1$, *and*

   (ii) *there is a constant* $0 < k^* < \infty$ *such that*
   $$\left(\frac{|f|}{N_\Phi(f)}\right)\left(\frac{k^*|g|}{\|g\|_\Psi}\right) = \Phi\left(\frac{f}{N_\Phi(f)}\right) + \Psi\left(\frac{k^*g}{\|g\|_\Psi}\right),\ \text{a.e.};$$

(b) *when* (Φ, Ψ) *are normalized, equality obtains in (3.1) iff*

   (i) $\int_\Omega \Phi\left(\frac{f}{N_\Phi(f)}\right) d\mu = \Phi(1)$, $\int_\Omega \Psi\left(\frac{g}{N_\Psi(g)}\right) d\mu = \Psi(1)$, *and*

   (ii) $\left(\frac{|f(\omega)|}{N_\Phi(f)}, \frac{|g(\omega)|}{N_\Psi(g)}\right)$ *is, for a.a.* (ω), *a point on the curve* $y = \varphi(x)$ *by adjoining the discontinuities of* φ *with vertical segments.*

Proof: We shall prove (a). The proof of (b) is similar and slightly simpler (cf. Zygmund (1959), Vol. I, p. 175).

For sufficiency, let (a)(i) and (a)(ii) be satisfied. Then

$$1 = \left\|\frac{g}{\|g\|_\Psi}\right\|_\Psi \leq \frac{1}{k^*}\left(1 + \int_\Omega \Psi\left(\frac{k^*g}{\|g\|_\Psi}\right) d\mu\right),\ \text{by (3.34),}$$

$$= \frac{1}{k^*}\left(\int_\Omega \Phi\left(\frac{f}{N_\Phi(f)}\right) d\mu + \int_\Omega \Psi\left(\frac{k^*g}{\|g\|_\Psi}\right) d\mu\right),\ \text{by (i),}$$

$$= \frac{1}{k^*}\int_\Omega \frac{|f|}{N_\Phi(f)}\frac{k^*|g|}{\|g\|_\Psi} d\mu,\ \text{by (ii).}$$

Since the opposite inequality always holds, we have equality.

For necessity, suppose there is equality in the Hölder inequality as given. By (3.35) there is a $k^* > 0$ such that

$$1 = \left\|\frac{g}{\|g\|_\Psi}\right\|_\Psi = \frac{1}{k^*}\left(1 + \int_\Omega \Psi\left(\frac{k^*g}{\|g\|_\Psi}\right) d\mu\right). \tag{8}$$

On the other hand, the equality condition gives

$$1 = \int_\Omega \frac{|fg|}{N_\Phi(f)\|g\|_\Psi} d\mu$$

$$\leq \frac{1}{k^*}\left[\int_\Omega \Phi\left(\frac{f}{N_\Phi(f)}\right) d\mu + \int_\Omega \Psi\left(\frac{k^*g}{\|g\|_\Psi}\right) d\mu\right],\ \text{by Young's inequality,}$$

$$\leq \frac{1}{k^*}\left[1 + \int_\Omega \Psi\left(\frac{k^*g}{\|g\|_\Psi}\right) d\mu\right] = 1,\ \text{by (8).} \tag{9}$$

Thus there is equality throughout, so that Young's inequality must be an equality, and (9) implies both (a)(i) and (a)(ii). ∎

The subspaces M^Φ and $\mathcal{M}^\Phi$ introduced in Definition 2 play an important role in the analysis on Orlicz spaces. Corollaries 4 and 5 give useful information, but the following result complements them in that the $\mathcal{M}^\Phi$ is a norm determining set for $L^\Psi(\mu)$. More precisely, we have

Proposition 10. *Let* (Φ, Ψ) *be a Young complementary pair and let* $\mathcal{M}^\Phi, \mathcal{M}^\Psi$ *be the subspaces of* $L^\Phi(\mu), L^\Psi(\mu)$, *introduced in Definition 2. Then each determines the norm in* $L^\Psi(\mu)$ *and* $L^\Phi(\mu)$ *respectively, in the sense that*

$$N_\Phi(f) = \sup\{|\int_\Omega fg d\mu| : \|g\|_\Psi \le 1, g \in \mathcal{M}^\Psi\}, \tag{10}$$

and

$$\|f\|_\Phi = \sup\{|\int_\Omega fg d\mu| : N_\Psi(g) \le 1, g \in \mathcal{M}^\Psi\}. \tag{11}$$

(Here Φ *and* Ψ *can be interchanged to get valid formulas for* $N_\Psi(\cdot)$ *and* $\|\cdot\|_\Psi$ *.) If* Φ, Ψ *are N-functions then* $\mathcal{M}^\Phi, \mathcal{M}^\Psi$ *become* M^Φ *and* M^Ψ.

Proof: In view of Proposition 3, it suffices to consider the case that Φ is continuous on $[0, x_0)$ where $x_0 \le \infty$. Also by the Hölder inequality (cf. equation (3.16)) in both (10) and (11) we only need to establish that the left sides are bounded by the right side suprema. So consider (10).

Let $N_\Phi(f) = 1$ for convenience, and $f \ge 0$. Observe that by Theorem 2.3, for any $\varepsilon > 0$, since $N_\Phi((1+\varepsilon)f) > 1$, $\int_\Omega \Phi((1+\varepsilon)f) > 1$ also. In fact this is an immediate consequence of the convexity and the definition of $N_\Phi(f)$: for any $\tilde{f}$ with $N_\Phi(\tilde{f}) > 1$, we have, for small enough $\eta > 0$ such that $N_\Phi(\tilde{f}) - \eta > 1$,

$$1 < \int_\Omega \Phi\Big(\frac{\tilde{f}}{N_\Phi(\tilde{f}) - \eta}\Big) d\mu \le \frac{1}{N_\Phi(\tilde{f}) - \eta} \int_\Omega \Phi\Big(\frac{(N_\Phi(\tilde{f}) - \eta)\tilde{f}}{N_\Phi(\tilde{f}) - \eta}\Big) d\mu, \tag{12}$$

so that $N_\Phi(\tilde{f}) - \eta < \int_\Omega \Phi(\tilde{f}) d\mu$, giving the assertion by letting $\eta \to 0$. Now in our case, there exist simple $0 \le f_n \uparrow f$ pointwise. So $\rho_\Phi((1+\varepsilon)f) = \int_\Omega \Phi((1+\varepsilon)f) d\mu = \lim_{n\to\infty} \int_\Omega \Phi((1+\varepsilon)f_n) d\mu$. Hence for some n_0, $n \ge n_0$

$\Rightarrow \rho_\Phi((1+\varepsilon)f_n) \geq 1 + \frac{\varepsilon}{2}$. Also $\varphi((1+\varepsilon)f_n) \in \tilde{L}^\Psi(\mu)$, f_n being simple.
Define a simple g_n as

$$g_n = \frac{\varphi((1+\varepsilon)f_n)}{\rho_\Psi(\varphi((1+\varepsilon)f_n)) + 1}, \qquad n \geq 1. \tag{13}$$

Then $g_n \in \mathcal{M}^\Psi$, and with Young's inequality

$$\|g_n\|_\Psi = \sup\left\{ \int_\Omega \frac{\varphi((1+\varepsilon)f_n)|h|}{\rho_\Psi(\varphi((1+\varepsilon)f_n)) + 1} d\mu : \rho_\Phi(h) \leq 1 \right\} \leq 1. \tag{14}$$

Consequently

$$\sup\left\{ |\int_\Omega fg d\mu| : \|g\|_\Psi \leq 1, g \in \mathcal{M}^\Phi \right\}$$

$$\geq \sup\left\{ |\int_\Omega f_n g_n d\mu| : n \geq n_0, g_n \text{ of } (13) \right\}$$

$$= \frac{1}{1+\varepsilon} \sup\left\{ \frac{\rho_\Phi((1+\varepsilon)f_n) + \rho_\Psi(\varphi((1+\varepsilon)f_n))}{\rho_\Psi(\varphi((1+\varepsilon)f_n)) + 1} : n \geq n_0 \right\}$$

$$> \frac{1}{1+\varepsilon}.$$

Since $\varepsilon > 0$ is arbitrary, this shows that the left side is at least $N_\Phi(f) = 1$, and (10) is proved. The proof of (11) is easy and the last statement is evident. ∎

The functional $\rho_\Phi : f \mapsto \int_\Omega \Phi(f) d\mu$ is called a *modular* by H. Nakano (1950) who developed a theory for them. In the Orlicz space context it is clear that $\rho_\Phi(\tilde{L}^\Phi(\mu)) \subset \mathbb{R}^+$ and $\rho_\Phi(f) = +\infty$ for $f \in L^\Phi(\mu) - \tilde{L}^\Phi(\mu)$. But by Corollary 4, $\mathcal{M}^\Phi \subset \tilde{L}^\Phi(\mu)$. Since $\mathcal{M}^\Phi$ is linear, ρ_Φ is finite on $\mathcal{M}^\Phi$, and hence also on $L^\Phi(\mu)$ when $\tilde{L}^\Phi(\mu) = L^\Phi(\mu)$ (and only then), or when Φ is Δ_2-regular. This has another interesting property given by:

Definition 11. A sequence $\{f_n, n \geq 1\} \subset L^\Phi(\mu)$, is said to be *mean* (or *modular*) *convergent* to $f \in L^\Phi(\mu)$, if $\rho_\Phi(f_n - f) \to 0$ as $n \to \infty$. It is *strongly or norm convergent* to f if $N_\Phi(f_n - f) \to 0$ as $n \to \infty$ (or equivalently $\|f_n - f\|_\Phi \to 0$).

The relation between these two concepts is clarified by the following result, the second part being the Orlicz space extension of a classical result due to F. Riesz for the L^p-spaces.

Theorem 12. *Let $\{f_n, n \geq 1\}$ be a sequence from $L^\Phi(\mu)$ and $f \in \tilde{L}^\Phi(\mu)$. Then the following assertions hold:*

(a) *The modular (or mean) convergence is implied by the norm (or strong) convergence of f_n to f. The converse holds if $\{f_n, f, n \geq 1\} \subset L^\Phi(\mu)$ and Φ is Δ_2-regular.*

(b) *If Φ is a Δ_2-regular Young function, or if Φ is continuous and concave $\Phi(0) = 0$, $\Phi \nearrow$ as well, $\rho_\Phi(f_n) \to \rho_\Phi(f)$ as $n \to \infty$ and $f_n \to f$ a.e., or in μ-measure, then $f_n \to f$ in norm.*

Proof: (a) Replacing f_n by $(f_n - f)$, we may assume that $f = 0$ here. Then let $N_\Phi(f_n) \to 0$ as $n \to \infty$. Hence for each $k > 0$, there is an n_0, $n \geq n_0 \Rightarrow k N_\Phi(f_n) \leq 1$. Then

$$\int_\Omega \Phi(k f_n) d\mu \leq k N_\Phi(f_n) \int_\Omega \Phi\left(\frac{f_n}{N_\Phi(f_n)}\right) d\mu \leq k N_\Phi(f_n) \to 0$$

as $n \to \infty$. Since $k f_n \in M^\Phi$ for each k, $f_n \to 0$ in mean.

Conversely let $f_n \to 0$ in mean, as in the hypothesis. Then for any $\varepsilon > 0$, $\rho_\Phi(\frac{f}{\varepsilon}) = \int_\Omega \Phi(\frac{f}{\varepsilon}) d\mu \leq C_\varepsilon \int_\Omega \Phi(f_n) d\mu$, by the Δ_2-regularity. So, since $\rho_\Phi(f_n) \to 0$, and $C_\varepsilon > 0$ is some fixed constant that goes with the Δ_2-condition, there is an $n_0 \geq 1$ such that $n \geq n_0 \Rightarrow C_\varepsilon \rho_\Phi(f_n) \leq 1$. Hence $\rho_\Phi(\frac{f_n}{\varepsilon}) \leq 1$ so that $N_\Phi(f_n) \leq \varepsilon$. This shows $N_\Phi(f_n) \to 0$.

(b) If Φ is concave increasing, then for $x, y \geq 0$,

$$\Phi(x + y) = \int_0^{x+y} \varphi(t) dt = \Phi(x) + \int_x^{x+y} \varphi(t) dt,$$

φ being the left derivative of Φ, and φ nonincreasing,

$$= \Phi(x) + \int_0^y \varphi(x + t) dt \leq \Phi(x) + \int_0^y \varphi(t) dt = \Phi(x) + \Phi(y).$$

On the other hand, if Φ is a convex Δ_2-regular Young function, then

$$\Phi(x + y) \leq \tfrac{1}{2}[\Phi(2x) + \Phi(2y)] \leq \tfrac{C}{2}[\Phi(x) + \Phi(y)], \quad x, y \geq x_0 \geq 0.$$

Let $\tilde{C} = \max(1, \frac{C}{2})$. Then for $x, y \geq x_0 \geq 0$,

$$\Phi(x + y) \leq \tilde{C}[\Phi(x) + \Phi(y)]. \qquad\qquad (15)$$

Now let $f_n \to f$ a.e. and $\Omega_1 = \{\omega : |f_n(\omega)| \geq x_0, |f(\omega)| \geq x_0, n \geq 1\}$. Then letting $x = |f_n(\omega)|$, $y = |f(\omega)|$ in (15), we get

$$0 \leq \widetilde{C}[\Phi(f_n) + \Phi(f)](\omega) - \Phi(|f_n - f|)(\omega), \quad \omega \in \Omega_1, \qquad (16)$$

since Φ is also increasing. But $\Phi(f_n) \to \Phi(f)$ a.e. so that by the Fatou inequality when $\mu(\Omega) = +\infty$ so that $x_0 = 0$ and $\Omega_1 = \Omega$, one has

$$2\widetilde{C}\int_\Omega \Phi(f)d\mu \leq \liminf_{n\to\infty} \int_\Omega \{\widetilde{C}[\Phi(f_n) + \Phi(f)] - \Phi(f_n - f)\}d\mu$$

$$= \lim_{n\to\infty} \widetilde{C}\int_\Omega [\Phi(f_n) + \Phi(f)]d\mu - \limsup_{n\to\infty} \int_\Omega \Phi(f_n - f)d\mu$$

$$= 2\widetilde{C}\int_\Omega \Phi(f)d\mu - \limsup_{n\to\infty} \int_\Omega \Phi(f_n - f)d\mu, \qquad (17)$$

where we used the hypothesis that $\rho_\Phi(f_n) \to \rho_\Phi(f)$. But (17) implies $\int_\Omega \Phi(f_n - f)d\mu \to 0$. Since Φ is Δ_2-regular, by (a) this implies $f_n \to f$ in norm. If the statement is false for $f_n \to f$ in measure, then there is a subsequence $f_{n_i} \to f$ a.e. for which the result must be false, contradicting the preceding result for a.e. convergence.

Finally, if $\mu(\Omega) < \infty$ and $x_0 > 0$ in (16), then the above argument applies if Ω is replaced by Ω_1. But on Ω_1^c, the sequence $\{f_n, n \geq 1, f\}$ is bounded by x_0. Since $f_n \to f$ a.e. on Ω_1^c and Φ is continuous, $\Phi(f_n - f) \to 0$ a.e. and boundedly on this set. Hence by the bounded convergence theorem we get $\int_{\Omega_1^c} \Phi(f_n - f)d\mu \to 0$. This together with (17) for Ω_1 implies $\int_\Omega \Phi(f_n - f)d\mu \to 0$, and hence the last part of the preceding paragraph applies, and the result holds in all cases. ∎

To understand another property of the space M^Φ, not necessarily shared by $\mathcal{M}^\Phi$ or $L^\Phi(\mu)$, we introduce the following.

Definition 13. An element $f \in L^\Phi(\mu)$ is said to have an *absolutely continuous norm* if $N_\Phi(f\chi_{A_n}) \to 0$ for each sequence of measurable sets $A_n \downarrow \emptyset$. The space $L^\Phi(\mu)$ has an absolutely continuous norm if every element of $L^\Phi(\mu)$ has such a property.

An alternative form of this concept and its importance in the M^Φ spaces can be obtained from the following in which Φ is continuous.

Theorem 14. *For each* $f \in L^\Phi(\mu)$ *the following are mutually equivalent statements:*

(i) *f has an absolutely continuous norm;*

(ii) *for each measurable $f_n, |f| \geq f_n \geq \cdots \to 0$ pointwise, we have $N_\Phi(f_n) \searrow 0$ as $n \to \infty$;*

(iii) *for each measurable f_n such that $f_n \to \tilde{f}$ a.e. and $|f_n| \leq |f|$, a.e., we have $N_\Phi(f_n - \tilde{f}) \to 0$ as $n \to \infty$.*

If further we have $\Phi(x) = 0$ only for $x = 0$, then M^Φ has an absolutely continuous norm and hence $L^\Phi(\mu)$ has also this property iff $M^\Phi = \tilde{L}^\Phi(\mu)$.

Proof: (i) $\Rightarrow$ (ii). Let $|f| \geq f_n \searrow 0$; and let $A_n \searrow \emptyset$ be a sequence such that $A_n \in \Sigma$. Then by (i), given $\varepsilon > 0$, $\exists n_0$ such that $n \geq n_0 \Rightarrow$

$$N_\Phi(f_m \chi_{A_n}) \leq N_\Phi(f \chi_{A_n}) < \tfrac{\varepsilon}{2}, \quad m \geq 1. \tag{18}$$

Also by (i), $\exists \eta > 0$ such that for each $B \in \Sigma$ with $\mu(B) < \eta$, one has $N_\Phi(f \chi_B) < \tfrac{\varepsilon}{4}$. Let $C_{n_0} \subset A_{n_0}^c$ be any set such that $C_{n_0} \in \Sigma$ and $0 < \mu(C_{n_0}) < \infty$. Such sets exist because $\mu(A_{n_0}^c) > 0$ and μ has the finite subset property. If $\alpha = \frac{\varepsilon}{4N_\Phi(\chi_{C_{n_0}})}$, then $0 < \alpha < \infty$ and the sets $B_m = \{\omega \in C_{n_0} : f_m(\omega) > \alpha\}$ satisfy $B_m \searrow \emptyset$, $B_m \in \Sigma$, and $\mu(B_m) < \eta$. Hence on $C_{n_0} - B_m$, $f_m \leq \alpha$ and $\mu(B_m) < \eta$ for $m \geq m_0$ since $B_n \searrow \emptyset$, so that

$$\begin{aligned}
N_\Phi(f_m \chi_{C_{n_0}}) &\leq N_\Phi(f_m \chi_{B_m}) + N_\Phi(f_m \chi_{C_{n_0} - B_m}) \\
&\leq N_\Phi(f \chi_{B_m}) + \alpha N_\Phi(\chi_{C_{n_0}}) \\
&< \tfrac{\varepsilon}{4} + \tfrac{\varepsilon}{4} = \tfrac{\varepsilon}{2}, \quad \text{if } m \geq m_0.
\end{aligned}$$

But by the Fatou property of the norm (i.e., for $0 \leq g_n \in L^\Phi(\mu)$, $g_n \nearrow g \Rightarrow N_\Phi(g_n) \nearrow N_\Phi(g) \leq \infty$ and the same is true of $\|\cdot\|_\Phi$ also), we get

$$\sup_{C_{n_0} \subset A_{n_0}^c} N_\Phi(f_m \chi_{C_{n_0}}) < \tfrac{\varepsilon}{2} \Rightarrow N_\Phi(f_m \chi_{A_{n_0}^c}) \leq \tfrac{\varepsilon}{2}.$$

Hence by (18) $N_\Phi(f_m \chi_{A_{n_0}}) < \varepsilon$, $m \geq 1$, so that $N_\Phi(f_m) < \varepsilon$ if m is large enough, i.e., $N_\Phi(f_m) \searrow 0$ and (ii) holds.

(ii) $\Rightarrow$ (iii). Let $|f_n| \leq |f|$ and $f_n \to \tilde{f}$ a.e. as in (iii). If we set

$$g_n(\omega) = \sup\{|f_{n+m} - \tilde{f}|(\omega) : m \geq 0\},$$

then $g_n \searrow 0$ a.e. and $g_n \leq 2|f|$. By (ii), $N_\Phi(g_n) \searrow 0$. Since $|f_n - \tilde{f}| \leq g_n$, by definition we get $N_\Phi(f_n - \tilde{f}) \leq N_\Phi(g_n) \to 0$, and hence (iii) holds.

(iii) $\Rightarrow$ (i). If $f \in L^{\Phi}(\mu)$, and $f_n = |f| \, \chi_{A_n}$ for any $A_n \in \Sigma, A_n \searrow$
$\emptyset$, then $f_n \to 0$ a.e., $f_n \leq |f|$ a.e. So $N_{\Phi}(f_n) \to 0$ and (i) holds for f.

Finally, if $f \in M^{\Phi}$ and Φ is as given, then $\int_{\Omega} \Phi(f \chi_{A_n}) d\mu \to 0$ for
any such $A_n \searrow \emptyset$. This implies (i) by Theorem 12(a). So M^{Φ} has an
absolutely continuous norm. Also if $M^{\Phi} = \tilde{L}^{\Phi}(\mu)$, then $M^{\Phi} = L^{\Phi}(\mu)$
so that $L^{\Phi}(\mu)$ has absolutely continuous norm. On the other hand,
if $L^{\Phi}(\mu)$ has this property, then for each $0 \leq f \in L^{\Phi}(\mu)$, there is a
sequence of simple functions $0 \leq f_n \uparrow f$ (since $\Phi(x) = 0$ iff $x = 0$
implies f has σ-finite support) and so $0 \leq f - f_n \leq f$ and $f - f_n \downarrow 0$.
Hence by (ii) above $N_{\Phi}(f - f_n) \to 0$. This implies that simple functions
are dense in $L^{\Phi}(\mu)$ so that $M^{\Phi} = L^{\Phi}(\mu)$ and then $M^{\Phi} = \tilde{L}^{\Phi}(\mu)$. This
proves the last part also. ∎

The following consequence of the theorem has applications.

Corollary 15. *If the Young function Φ is Δ_2-regular and $f, g_n \in$
$L^{\Phi}(\mu)$, then $\rho_{\Phi}(g_n) \to 0 \Rightarrow \rho_{\Phi}(f + g_n) \to \rho_{\Phi}(f)$. Moreover, a set S is
modular bounded (i.e., for each $f \in S \Rightarrow \rho_{\Phi}(f) \leq K_0$ for a constant
$K_0 > 0$) iff it is norm bounded.*

Proof: Since Φ is Δ_2-regular, Theorem 12(a) implies $\rho_{\Phi}(g_n) \to 0$ iff
$N_{\Phi}(g_n) \to 0$, and hence by the triangle inequality

$$|N_{\Phi}(f + g_n) - N_{\Phi}(f)| \leq N_{\Phi}(f + g_n - f) = N_{\Phi}(g_n) \to 0.$$

By the converse part of Theorem 12(a), this implies $\rho_{\Phi}(f + g_n) \to$
$\rho_{\Phi}(f)$, since clearly $f + g_n \to f$ in measure.

For the second part, we may take $K_0 \geq 1$. Then for any $f \in S$,

$$1 \geq \frac{1}{K_0} \int_{\Omega} \Phi(f) d\mu \geq \int_{\Omega} \Phi\left(\frac{f}{K_0}\right) d\mu \Rightarrow N_{\Phi}(f) \leq K_0. \qquad (19)$$

Thus S is bounded in norm $N_{\Phi}(\cdot)$ by K_0 also. Conversely if this holds,
and S is not modular bounded, then $\exists f_n \in S, \rho_{\Phi}(f_n) \geq n, n \geq 1$. So
(19) becomes, for $n \geq n_0 \geq \max(1, K_0)$,

$$\rho_{\Phi}(f_n) = \int_{\Omega} \Phi\left(\frac{n_0 f_n}{n_0}\right) d\mu \leq C_{n_0} \int_{\Omega} \Phi\left(\frac{f_n}{n_0}\right) d\mu \leq C_{n_0},$$

by the Δ_2 condition.

But this is a contradiction.   ∎

Remark. One can show, using the arguments of Proposition 8 and others before it, that if μ is diffuse on a set of positive measure and if a set $S \subset L^\Phi(\mu)$ is modular bounded, then Φ must be Δ_2-regular. In an analogous manner it is possible to translate the mean and strong convergences from one to the other in an $L^\Phi(\mu)$-space when the Φ is Δ_2-regular. In these statements either norm may be used because of their equivalence.

The important roles of the Δ_2-condition and the subspace M^Φ in the study of $L^\Phi(\mu)$ spaces are thus demonstrated. The converse part of Theorem 12(a) shows that the work on M^Φ is indeed more general than on $L^\Phi(\mu)$ with Φ being Δ_2-regular. This was already remarked by Morse and Transue ((1950), p. 614). They also showed that $\rho_\Phi(rf_n) \to 0$, all $r \geq 0$, $\Rightarrow N_\Phi(f_n) \to 0$, which is slightly weaker than assuming the Δ_2.

3.5. Separability, quotients, and related results

Recall that a topological space is *separable* if it contains a dense denumerable subset. A topological vector space is similarly called separable if it has a countable set of elements whose closed linear span is the whole space. Since it is classical that the $L^p(\mu)$ for $0 < p < \infty$ is separable iff its measure space (Ω, Σ, μ) is separable (cf., e.g., Rao (1987), p. 207, Example 5), we need to consider the $L^\Phi(\mu)$-space in a similar manner which however is more involved. First note that the triple (Ω, Σ, μ) is called a *separable measure* space if the set (Σ, d) is a separable metric space where

$$d(A, B) = f(\mu(A \triangle B)), \qquad A, B \in \Sigma, \tag{1}$$

with $f : \bar{R}^+ \to R^+$ is a bounded increasing continuous concave function such that $f(x) = 0$ iff $x = 0$. For instance $f(x) = x(1+x)^{-1}$ or $= \arctan x$ can be used in (1). It is a standard result that (Σ, d) is a complete (semi-) metric space when (Ω, Σ, μ) is a complete measure space (cf., e.g., the above reference, p. 81). In particular, the Lebesgue triple $(R^n, \mathcal{B}, \mu)$ with μ as Lebesgue measure is a separable measure space.

We now establish the following result for Orlicz spaces:

Theorem 1. *Let Φ be a continuous Young function with $\Phi(x) = 0$ iff $x = 0$. Then (a) the space M^Φ is separable iff the measure triple*

(Ω, Σ, μ) *is separable, and* (b) $L^{\Phi}(\mu)$ *is separable iff* (i) $L^{\Phi}(\mu) = M^{\Phi}$
and (ii) (Ω, Σ, μ) *is separable. In particular,* (c) *if Φ is Δ_2-regular,*
then $L^{\Phi}(\mu)$ is separable iff (Ω, Σ, μ) is separable.

Proof: (a) Suppose M^{Φ} is separable. Then there exists $\{f_n, n \geq 1\} = \mathcal{F}$, a countable set, which is dense in M^{Φ}. Let $\mathcal{A}$ be a countable collection of subsets generating the Borel σ-algebra of $I\!R$. For instance $\mathcal{A}$ can be taken as the set of all intervals with rational end points. Then $\mathcal{D} = \overset{\infty}{\underset{n=1}{\cup}} f_n^{-1}(\mathcal{A}) \subset \Sigma$ and $\mathcal{D}$ is countable. If $\widetilde{\Sigma} = \sigma(\mathcal{D})$, then $(\Omega, \widetilde{\Sigma}, \mu)$ is separable and since for each $f \in M^{\Phi}$, and $\varepsilon > 0$, $\exists f_{\varepsilon} \in \mathcal{F}$ such that $N_{\Phi}(f - f_{\varepsilon}) < \varepsilon$, f is $\widetilde{\Sigma}$-measurable. We claim that each element of Σ is μ-equivalent to some element element of $\widetilde{\Sigma}$ so that ($\widetilde{\Sigma} \subset \Sigma$) the triple (Ω, Σ, μ) is separable. But $\tilde{f} \in L^{\Phi}(\mu) \Rightarrow \exists$ simple g_n such that $g_n \to \tilde{f}$ pointwise and $|g_n| \nearrow |\tilde{f}|$ by the structure theorem and then $[\Phi(x) = 0$ iff $x = 0$ and Φ continuous $\Rightarrow M^{\Phi} = \mathcal{M}^{\Phi}$ by Proposition 4.3]$g_n \in M^{\Phi}$, it follows that $\tilde{f}$ is μ-equivalent to a measurable function relative to $\widetilde{\Sigma}$. This implies at once the desired conclusion that (Ω, Σ, μ) is separable.

Conversely, suppose (Ω, Σ, μ) is separable so that the metric space (Σ, ρ) is separable where $\rho(A, B) = \arctan \mu(A \triangle B)$. Let $\{A_n, n \geq 1\} \subset \Sigma$ be a countable set that is dense in Σ for ρ. Since every element f of M^{Φ} vanishes outside a set A_f of σ-finite measure, and since every set $A \in \Sigma$ of finite measure can be approximated in ρ by one from $\{A_n, n \geq 1\}$ of finite measure, it follows from a standard measure theory argument that every simple function from M^{Φ} can be approximated in N_{Φ}-norm by a simple function composed of linear combinations of χ_{A_k}, $A_k \in \{A_n, n \geq 1\}$, $\mu(A_k) < \infty$, with rational coefficients. But the set of simple functions of the latter type is countable and lies in M^{Φ}. Hence M^{Φ} is separable, because simple functions are dense in M^{Φ} by Proposition 4.3.

From (a) and Corollary 4.5, it follows immediately that (c) is true as given and hence only (b) remains to be established.

(b) If (i) and (ii) are true, then by (a), $L^{\Phi}(\mu)$ is separable. So the really new part is the converse which is also a difficult result. The idea of proof is to show that if $L^{\Phi}(\mu)$ is separable but does not have an absolutely continuous norm, then a contradiction results. But then Theorem 4.14 and part (a) will imply the desired conclusion. Thus it suffices to show that a separable $L^{\Phi}(\mu)$ has an absolutely continuous norm.

Suppose then that $L^{\Phi}(\mu)$ is separable so that there is a dense denumerable set $\{f_n, n \geq 1\}$ in $L^{\Phi}(\mu)$. If $\{g_n, n \geq 1\}$ is any sequence from $L^{\Psi}(\mu)$ such that $N_{\Psi}(g_n) \leq 1$, then we *assert* that there is a subsequence g_{n_i} of $\{g_n, n \geq 1\}$ such that $\int_{\Omega} f(g_{n_i} - \tilde{g})d\mu \to 0$ for each $f \in L^{\Phi}(\mu)$ and some $\tilde{g} \in L^{\Psi}(\mu)$, $N_{\Psi}(\tilde{g}) \leq 1$, i.e., $g_{n_i} \to \tilde{g}$ "weakly." This is a special case of results that we consider in the next chapter (see Sec. IV.4), but we shall present the details here. Using this fact, the desired contradiction is obtained. The argument is only a simple modification of a classical one going back to G. Köthe in the late 1930's, and is employed by several authors since. Unfortunately we could find no easier method than this one.

To establish the desired convergence, consider $\{\int_{\Omega} f_1 g_n d\mu, n \geq 1\}$. By the Hölder inequality this set of numbers is bounded (by $\|f_1\|_{\Phi}$). Hence it contains a convergent subsequence (by the Bolzano-Weierstrass theorem). Say this is $\{g_{1n}, n \geq 1\} \subset \{g_n, n \geq 1\}$. Next the set $\{\int_{\Omega} f_2 g_{1n} d\mu, n \geq 1\}$ is similarly bounded and let $\{g_{2n}, n \geq 1\} \subset \{g_{1n}, n \geq 1\}$ be such that $\int_{\Omega} f_2 g_{2n} d\mu \to$ a limit. Repeating this process with each f_k, we find $\{g_{kn}, n \geq 1\}$ so that $\int_{\Omega} f_k g_{kn} d\mu \to$ a limit. If $\{g_{nn}, n \geq 1\}$ is the diagonal sequence, then $\int_{\Omega} f_k g_{nn} d\mu \to$ a limit for each $k = 1, 2, \ldots$. Using the density of $\{f_n, n \geq 1\}$ at this point, we find for any $f \in L^{\Phi}(\mu)$, and $\varepsilon > 0$, an f_{n_ε} such that $\|f - f_{n_\varepsilon}\|_{\Phi} < \varepsilon$. Hence

$$\left| \int_{\Omega} f g_{nn} d\mu - \int_{\Omega} f_{n_\varepsilon} g_{nn} d\mu \right| \leq \int_{\Omega} |(f - f_{n_\varepsilon}) g_{nn}| d\mu$$

$$\leq \|f - f_{n_\varepsilon}\|_{\Phi} < \varepsilon. \tag{2}$$

Since $\int_{\Omega} f_{n_\varepsilon} g_{nn} d\mu \to$ a limit, this implies that $\int_{\Omega} f g_{nn} d\mu \to a(f)$, the limit depending on $f \in L^{\Phi}(\mu)$, i.e., $\{g_{nn}, n \geq 1\}$ is a "weak* Cauchy" sequence. We now show that there is a $\tilde{g} \in L^{\Psi}(\mu)$ such that $a(f) = \int_{\Omega} f \tilde{g} d\mu$, $f \in L^{\Phi}(\mu)$. It is clear that $|a(f)| \leq \liminf_{n \to \infty} |\int_{\Omega} f g_{nn} d\mu| \leq \|f\|_{\Phi}$ and $a(\alpha f + \beta \tilde{f}) = \alpha a(f) + \beta a(\tilde{f})$ so that $a(\cdot)$ is a bounded linear functional on $L^{\Phi}(\mu)$. Note that if $a(f) = 0$ for all $f \in M^{\Phi}$, then $a \equiv 0$. In fact, since simple functions are dense in M^{Φ}, and h is a simple function, then

$$\int_{\Omega} h g_{nn} d\mu \to a(h) = 0 \tag{3}$$

implies $\left| \int_{\Omega} h g_{nn} d\mu \right| < \varepsilon$ if n is large enough for any $\varepsilon > 0$. But then $\int_{A} g_{nn} d\mu \to a(\chi_A) = 0$, for each $A \in \Sigma$, $\mu(A) < \infty$, and since each $f \in L^{\Phi}(\mu)$ has σ-finite support ($\Phi(x) = 0$ iff $x = 0$ is used again), assuming for convenience that $N_{\Phi}(f) = 1$ so that $\Phi(f) \in L^1(\mu)$, there is a set $A_{\varepsilon} \in \Sigma$, $\mu(A_{\varepsilon}) < \infty$ and $\int_{A_{\varepsilon}^c} \Phi(f) d\mu < \varepsilon$ and hence $f\chi_{A_{\varepsilon}} \in \tilde{L}^{\Phi}(\mu) \cap L^1(\mu)$ by Proposition 1.7. Consequently (3) gives, since $\|f\chi_{A_{\varepsilon}} - h_{\varepsilon}\|_1 < \varepsilon$ for a simple h_{ε},

$$\int_{\Omega} f\chi_{A_{\varepsilon}} g_{nn} d\mu \to a(f\chi_{A_{\varepsilon}}) = 0. \tag{4}$$

From this, and replacing $\chi_{A_{\varepsilon}}$ by $\chi_{A_{\varepsilon}^c}$ in (4), one gets $a(f\chi_{A_{\varepsilon}^c}) = 0$ so that $a(M^{\Phi}) = 0 \Rightarrow a(L^{\Phi}(\mu)) = 0$ for the limit functional. But we have seen that $a(\cdot)$ is a bounded linear functional, and the above analysis implies that it *vanishes identically if it does so on* M^{Φ}. Such a functional has the representation, if the measure space is σ-finite, as

$$a(f) = \int_{\Omega} f\tilde{g} d\mu, \qquad f \in M^{\Phi} \tag{5}$$

for a unique $\tilde{g} \in L^{\Psi}(\mu)$. This important fact is a special case of the main characterization of the adjoint space of M^{Φ}, proved in the next chapter (cf. Theorem IV.1.6). Since the fixed set $\{f_n, n \geq 1\}$ vanishes outside a σ-finite set, the σ-finiteness is also satisfied. It follows *that each countable set* $\{g_n, n \geq 1\} \subset L^{\Psi}(\mu)$ *with* $\|g_n\|_{\Psi} \leq 1$, *has a subsequence* g_{n_i} *such that* $\int_{\Omega} f(g_{n_i} - \tilde{g}) d\mu \to 0$ *for a* $\tilde{g} \in L^{\Psi}(\mu)$, $\forall f \in L^{\Phi}(\mu)$. With this fact we can conclude the desired result.

If the norm of $L^{\Phi}(\mu)$ is not absolutely continuous, then there exists an element $f_0 \in L^{\Phi}(\mu) - M^{\Phi}$, and a $\delta > 0$, such that for each sequence $A_n \in \Sigma$, $A_n \searrow \emptyset$, $N_{\Psi}(f_0 \chi_{A_n}) \geq \delta > 0$, $n \geq 1$, by Theorem 3.14. Since the support of f_0 is σ-finite, we may assume that this holds if $\mu(A_n) < \infty$ from some n onwards. [Otherwise we may work with

$f_0\chi_A$ for some A with $\mu(A) < \infty$, which has the same property, as
easily seen.] Thus $\mu(A_n) \searrow 0$ also. By Proposition 3.10, there ex-
ists a sequence $g_n \in L^\Psi(\mu)$ [even in $\mathcal{M}^\Psi(\mu)$], $\|g_n\|_\Psi \le 1$ such that
$N_\Phi(f_0\chi_{A_k}) \ge \delta > 0 \Rightarrow |\int_\Omega f_0\chi_{A_k}g_nd\mu| \ge \frac{\delta}{2}, \; k \ge 1$. Replacing g_n by
$g_n\chi_{A_k}\mathrm{sgn}(f_0)$ here, we may remove the absolute values and assume g_n
to vanish outside A_k, for large n ($\ge n_0$, say). Since $\mu(A_k) \searrow 0$, we
may consider a subsequence, denoted by the same symbols, so that
$\sum_{k=1}^\infty \mu(A_k) < \infty$. Hence $\limsup_{n\to\infty} A_n = \tilde{A}$ satisfies $\mu(\tilde{A}) = 0$. But by the
preceding paragraph, there is a subsequence $g_{n_i} \to \tilde{g}$ in the sense of
(2) and hence $\int_\Omega f_0\chi_{A_k}\tilde{g}d\mu \ge \frac{\delta}{2}$ also, $k \ge 1$. If $B \in \Sigma, B \cap \bigcup_{n\ge n_0} A_n$,
$\mu(B) < \infty$, then $\int_\Omega \chi_B g_{n_i}d\mu \to \int_\Omega \chi_B\tilde{g}d\mu = 0$. This implies that $\tilde{g} = 0$
a.e., off $\bigcup_{n\ge n_0} A_n$ for all large enough n_0 so that it is zero off $\tilde{A}$. But
$\mu(\tilde{A}) = 0$ and hence $\tilde{g} = 0$ a.e. This implies the following contradiction:

$$0 = \int_\Omega f_0\tilde{g}d\mu = \lim_{i\to\infty}\int_\Omega f_0g_{n_i}d\mu \ge \frac{\delta}{2}. \tag{6}$$

Thus the norm of $L^\Phi(\mu)$ must be absolutely continuous. By the initial
reduction, this proves the converse part of (b) completely. ∎

Remarks. If (Ω, Σ, μ) is a finite diffuse measure space and Φ is an N-
function which does not satisfy the Δ_2-condition, then it was shown in
the Krasnoselskii and Rutickii monograph by an explicit construction
that there exist continuum many functions $\{f_\alpha, \alpha \in I\}$ in $L^\Phi(\mu)$ such
that $\|f_\alpha - f_\beta\|_\Phi \ge \delta > 0$ for $\alpha \ne \beta$, so that $L^\Phi(\mu)$ is not separable.
Theorem 1 does not assume that Φ is an N-function nor is the mea-
sure space restricted. Even in the restricted case the proof is involved.
However, the general result is functional analytic and also motivates
one to proceed for a study of weak topologies in the next chapter.

We present the following special case of the preceding theorem for
comparison and applications.

Proposition 2. *Let (Ω, Σ, μ) be a measure space, Φ be a Δ_2-
regular Young function. Then $L^\Phi(\mu)$ is separable iff the measure space
is separable. In particular, let $\Omega \subset \mathbb{R}^n$, Σ be the Borel σ-algebra, and
μ be a finite diffuse measure. Then $L^\Phi(\mu)$ is separable iff $\Phi \in \Delta_2$.*

We only need to observe, for the last part, that (Ω, Σ, μ) can be
put equivalent to a Lebesgue measure on a bounded interval (a result

originally due to V. A. Rohlin (1949), used also before), and hence is separable. Thus the proposition follows from the first part which is a consequence of the last theorem.

In view of these results, it will be useful to analyze $L^\Phi(\mu) - M^\Phi$. However, this is better studied by considering the *quotient* or *factor* spaces $L^\Phi(\mu)/M^\Phi$ [or $L^\Phi(\mu)/\mathcal{M}^\Phi$] which reduce to the trivial space $\{0\}$, iff step functions are dense in $L^\Phi(\mu)$. Let N^Φ [or $\mathcal{N}^\Phi$] denote the corresponding factor space. Observe that if $f_i \in L^\Phi(\mu)$, $i = 1, 2$, then f_1 and f_2 belong to the same equivalence (or factor) class, also termed a coset, iff $f_1 = f_2 + h$ for some $h \in M^\Phi$ [or $\mathcal{M}^\Phi$]. We write $\hat{f} = f + M^\Phi$, $f \in L^\Phi(\mu)$, and define $d_\Phi : N^\Phi \to I\!\!R^+$ by

$$d_\Phi(\hat{f}) = \inf\{N_\Phi(f + g) : g \in M^\Phi\}, \qquad \hat{f} \in N^\Phi, \qquad (7)$$

and replacing N^Φ, M^Φ by $\mathcal{N}^\Phi, \mathcal{M}^\Phi$ we get another (related) functional. It is not hard to verify that $d_\Phi(\cdot)$ is a norm functional on the factor space and in fact if X is any Banach space and M a closed subspace, then X/M with the corresponding norm $d_M(\cdot)$ given by (7) is also a Banach space. However, the distinguished subspace M^Φ of absolutely continuous norm, in $L^\Phi(\mu)$, makes the particular quotient space N^Φ very special and the following properties are thus pertinent to it.

Proposition 3. *Let Φ be a continuous Young function and $\Phi(x) = 0$ iff $x = 0$. If $f \in L^\Phi(\mu)$ and $\hat{f} = f + M^\Phi \in N^\Phi$, then (7) becomes*

$$d_\Phi(\hat{f}) = \lim_{n \to \infty} N_\Phi(f - f_n), \qquad (8)$$

for a sequence $\{f_n, n \geq 1\} \subset M^\Phi$, such that $f_n \to f$ a.e. and $|f_n| \uparrow |f|$. Moreover, in (7) and (8) one may replace $N_\Phi(\cdot)$ by $\|\cdot\|_\Phi$, and the $d_\Phi(\cdot)$ remains unchanged so that the quotient norm in N^Φ is the same if one uses the equivalent (but unequal) norms $N_\Phi(\cdot)$ or $\|\cdot\|_\Phi$.

Proof: First we let $d(\cdot)$ denote $d_\Phi(\cdot)$ and $\|\cdot\|$ denote $N_\Phi(\cdot)$ or $\|\cdot\|_\Phi$ since most of the computations are common to both norms. Let $h \in M^\Phi$ be arbitrary and choose $f \in L^\Phi(\mu)$ such that $\|f - h\| < 1$. Hence there exists $0 < \varepsilon < 1$ such that $\|f - h\| < 1 - \varepsilon$. Then $\frac{1}{\varepsilon} h \in M^\Phi \subset \tilde{L}^\Phi(\mu)$ by Corollary 4.4 and the linearity of M^Φ. Because of the definition of $N_\Phi(\cdot)$ and Proposition 3.3, we deduce that $\frac{f-h}{1-\varepsilon} \in \tilde{L}^\Phi(\mu)$ so that

$$f = \varepsilon\left(\frac{h}{\varepsilon}\right) + (1 - \varepsilon)\frac{f - h}{1 - \varepsilon} \in \tilde{L}^\Phi(\mu), \qquad h \in M^\Phi. \qquad (9)$$

Now to prove (8) let $f \in L^{\Phi}(\mu)$, satisfying $d(\hat{f}) > 0$. Consider for any $\eta > 0$, an α such that

$$\frac{1}{d(\hat{f}) + 2\eta} < \alpha < \frac{1}{d(\hat{f}) + \eta}. \tag{10}$$

But from the definition of d (the factor norm), we get ($\alpha\hat{f} = \widehat{\alpha f}$)

$$d(\alpha\hat{f}) = \alpha d(\hat{f}) < \frac{d(\hat{f})}{d(\hat{f}) + \eta} < 1, \qquad \text{by (10)}.$$

This means by (7), there is an $h \in M^{\Phi}$ such that $\|\alpha f - h\| < 1$, and then by (9) we see that $\alpha f \in \tilde{L}^{\Phi}(\mu)$. Let f_n be a sequence of step functions (by the structure theorem) such that $f_n \to f$ pointwise and $|f_n| \uparrow |f|$ so that $f - f_n \to 0$, pointwise, and $\Phi(\alpha(f - f_n)) \to 0$ pointwise since Φ is continuous. But the f_n vanish outside the support of f (this can be assumed because of the fact that $|f_n| \uparrow |f|$). Since $\Phi(\alpha(f - f_n)) \leq \Phi(\alpha f)$ and the latter is integrable, we get, by the dominated convergence,

$$\int_{\Omega} \Phi(\alpha(f - f_n))d\mu \to 0, \qquad \text{as } n \to \infty, \tag{11}$$

and f_n is a simple function so that $f_n \in M^{\Phi}$, $n \geq 1$. Let $\| \cdot \| = N_{\Phi}(\cdot)$ and $d(\cdot) = d_{\Phi}(\cdot)$. Then there is an n_0 such that $n \geq n_0 \Rightarrow$ the integral in (11) is ≤ 1. Thus $N_{\Phi}(\alpha(f - f_n)) \leq 1$ or $N_{\Phi}(f - f_n) \leq \frac{1}{\alpha} \leq d(\hat{f}) + 2\eta$ by (10). This shows that since $f - f_n \to 0$ and $\eta > 0$ is arbitrary, $\lim_{n\to\infty} N_{\Phi}(f - f_n) \leq d(\hat{f})$. On the other hand by (7), the inequality $N_{\Phi}(f - h) \geq d(\hat{f})$ holds for all $h \in M^{\Phi}$, and in particular for $h = f_n$. Thus $d_{\Phi}(\hat{f}) = \lim_{n\to\infty} N_{\Phi}(f - f_n)$ and (8) is proved in this case.

Suppose now we use the Orlicz norm, $\| \cdot \| = \| \cdot \|_{\Phi}$ and let $\bar{d}(\cdot)$ be the corresponding $d(\cdot)$. In (1) choose n_0' such that $n \geq n_0' \Rightarrow$ the integral is at most $\alpha\eta$. Consequently from the Young inequality

$$\|\alpha(f - f_n)\|_{\Phi} \leq \int_{\Omega} \Phi(\alpha(f - f_n))d\mu + 1 \leq 1 + \alpha\eta.$$

Thus

$$\|(f - f_n)\|_{\Phi} \leq \frac{1}{\alpha} + \eta \leq \bar{d}(\hat{f}) + 3\eta, \quad \text{by (10)}.$$

This gives, from the arbitrariness of $\eta > 0$ and the fact that $f_n \in M^\Phi$, that $\lim\limits_{n\to\infty} \|f - f_n\|_\Phi = \bar{d}(\hat{f})$, as before. This gives (8) for both norms. It remains to show that $d_\Phi(\hat{f}) = \bar{d}(\hat{f})$.

In the above it is clear that $\hat{f} = \tilde{0}$ iff $f \in M^\Phi$ so that $d_\Phi(\hat{f}) = 0 = \bar{d}(\hat{f})$. Thus let $f \in L^\Phi(\mu) - M^\Phi$. Define a functional $p(\cdot) : L^\Phi(\mu) \to I\!\!R^+$ by the formula

$$p(f) = \inf\{\alpha > 0 : \int_\Omega \Phi(\tfrac{f}{\alpha})d\mu < \infty\}. \tag{12}$$

Note that $p(f) = \frac{1}{k_f}$ where $k_f = \sup I_\Phi(f)$ as given by Definition 2.1. Thus $p(f) = 0$ iff $k_f = +\infty$ (cf. Proposition 2.2 also) and hence iff $\int_\Omega \Phi(kf)d\mu < \infty$ for all $k > 0$, i.e., iff $f \in M^\Phi$. Now if f is as above, then $k_f < \infty$ so that $p(f) > 0$. Thus $\int_\Omega \Phi((k_f + \varepsilon)f)d\mu = \infty$ for each $\varepsilon > 0$, or $\int_\Omega \Phi(\frac{f}{p(f)-\varepsilon})d\mu = +\infty$, for any $0 < \varepsilon < p(f)$. Let $f_n = f_n^+ - f_n^-$, $f_n^\pm = \sum\limits_{k=0}^{n2^n-1} \frac{k}{2^n}\chi_{A_{kn}^\pm}$, $A_{kn}^\pm = [\frac{k}{2^n} \le f^\pm < \frac{k+1}{2^n}]$, then $f_n \to f$ a.e. and $f_n \in M^\Phi$ as in (11). If Ω_n is the support of f_n, then for each $\varepsilon > 0$,

$$\infty = \int_\Omega \Phi\left(\frac{f}{p(f) - \varepsilon}\right)d\mu = \int_{\Omega_n^c} \Phi\left(\frac{f - f_n}{p(f) - \varepsilon}\right)d\mu$$

$$+ \int_{\Omega_n} \Phi\left(\frac{f_n}{p(f) - \varepsilon}\right)d\mu. \tag{13}$$

But the last integral of (13) is finite since $f_n \in M^\Phi$ and is zero off Ω_n. Hence we may replace Ω_n^c by Ω in the first integral on the right of (13) and conclude that $N_\Phi(f - f_n) \ge p(f) - \varepsilon$. This shows, on using the inequality between these norms $N_\Phi(\cdot)$ and $\|\cdot\|_\Phi$ (cf. (3.14)) and the arbitrariness of $\varepsilon > 0$, that

$$\lim_{n\to\infty} \|f - f_n\|_\Phi \ge \lim_{n\to\infty} N_\Phi(f - f_n) \ge p(f). \tag{14}$$

However, we also have by (12) and (13) a result similar to (11):

$$\lim_{n\to\infty} \int_\Omega \Phi\left(\frac{f - f_n}{p(f) + \varepsilon}\right)d\mu = 0. \tag{15}$$

Thus using the Young inequality one has

$$\|f - f_n\|_\Phi \le (p(f) + \varepsilon)\left(1 + \int_\Omega \Phi\left(\frac{f - f_n}{p(f) + \varepsilon}\right)d\mu\right) \to p(f) + \varepsilon, \text{ by } (15).$$

Hence $\lim_{n\to\infty} N_\Phi(f - f_n) \le \lim_{n\to\infty} \|f - f_n\|_\Phi \le p(f)$, since $\varepsilon > 0$ is arbitrary. This inequality, (14), and the first part imply $d_\Phi(\hat{f}) = \bar{d}(\hat{f}) = p(f)$. Thus in the formula (8), we may use either norm, as asserted. ∎

The argument leading to (9) has information regarding the position of $\tilde{L}^\Phi(\mu)$ in N^Φ. To make this precise, we denote the *quotient mapping* by $\lambda : L^\Phi(\mu) \to N^\Phi = L^\Phi(\mu)/M^\Phi$, also called the *canonical mapping*. It is an open mapping of the normed vector group $L^\Phi(\mu)$ onto the quotient group N^Φ. The question then is to find $\lambda(\tilde{L}^\Phi(\mu))$ in N^Φ. This is answered by:

Proposition 4. *The open unit ball of N^Φ is contained in $\lambda(\tilde{L}^\Phi(\mu))$ which in turn is contained in the closed unit ball of N^Φ. The containments can be proper.*

Proof: We have already shown in (9) above that if $f \in L^\Phi(\mu)$ and $d(\hat{f}) < 1$ where $\hat{f} = f + M^\Phi$, then $f \in \tilde{L}^\Phi(\mu)$ or equivalently $\hat{f} \in \lambda(\tilde{L}^\Phi(\mu))$. Since f with $d(\hat{f}) < 1$ is arbitrary, it follows that the open unit ball of N^Φ is contained in $\lambda(\tilde{L}^\Phi(\mu))$.

For the next statement, let $f \in L^\Phi(\mu)$ be such that $d(\hat{f}) \le 1$. Then for each $\varepsilon > 0$ we can find a simple function $f_\varepsilon \in M^\Phi$, $|f_\varepsilon| \le |f|$, a.e., satisfying

$$\int_\Omega \Phi(f - f_\varepsilon)d\mu < \varepsilon$$

exactly as in (11) or (13). The Young inequality then implies that

$$\|f - f_\varepsilon\|_\Phi \le \int_\Omega \Phi(f - f_\varepsilon)d\mu + 1 < 1 + \varepsilon.$$

Hence using Proposition 3, we get $d(\hat{f}) = \inf\{\|f - h\|_\Phi : h \in M^\Phi\} < 1 + \varepsilon$ so that $d(\hat{f}) \le 1$, since we can use either norm in the definition of $d_\Phi(\cdot)$ proved above.

The fact that the inclusions can be proper is constructed, when μ is diffuse on a set of positive measure, with the arguments similar to those of Proposition 2.2 and will be omitted here. ∎

An interesting supplement to the preceding results of this section is to consider the "size" of the quotient space N^Φ when $L^\Phi(\mu)$ does not have an absolutely continuous norm, or equivalently, is not separable. It may be answered effectively as follows.

Theorem 5. *Let Φ be a continuous Young function such that $\Phi(x) = 0$ iff $x = 0$. Suppose that $L^\Phi(\mu)$ does not have an absolutely continuous norm. Then the quotient space $N^\Phi = L^\Phi(\mu)/M^\Phi$ is infinite dimensional if either μ is σ-finite or if μ is diffuse on a set of positive measure. In the latter case, $(N^\Phi, d_\Phi(\cdot))$ is even nonseparable.*

Sketch of Proof: Suppose first that (Ω, Σ, μ) is separable. Then by Theorem 1 M^Φ is separable. Since $L^\Phi(\mu)$ does not have an absolutely continuous norm, $L^\Phi(\mu)$ is not separable by Theorem 1 again. If the quotient space is finite dimensional, then there exist linearly independent elements $\hat{f}_1, \ldots, \hat{f}_n$, which form a basis for N^Φ. Since $\hat{f}_i = f_i + M^\Phi$, $f_i \in L^\Phi(\mu)$, it can be verified that $(f_1, \ldots, f_n)$ must be linearly independent in $L^\Phi(\mu)$, since otherwise some linear combination of them will be in $M^\Phi(\mu)$, and then $\hat{f}_i$ cannot be linearly independent. In this case, a dense set $\{h_n, n \geq 1\}$ of the separable M^Φ together with $(f_1, \ldots, f_n)$ will form a dense denumerable set in $L^\Phi(\mu)$, contradicting the fact that the latter is nonseparable. Hence N^Φ must be infinite dimensional. In the general case one has to consider the possibilities that μ has no set of positive measure on which it is diffuse, and the opposite that there is such a set. The latter possibility takes us to the second part to be treated. If there is no such set, then (Ω, Σ, μ) is purely atomic. If μ is σ-finite, then the atomic case can be shown to lead to the separability of the measure case and hence reduces to the problem already treated. [If μ is not σ-finite, then M^Φ may be nonseparable and we have no solution at this time. The σ-finite case has been considered in detail by E. de Jonge (1973).]

Regarding the second part, with the technique used in Proposition 2.2, we may consider the problem when (Ω, Σ, μ) is the Lebesgue unit interval. Then again $L^\Phi(\mu)$ is nonseparable, and one may adapt an argument from Krasnoselskii and Rutickii ((1961), pp. 108–110) to construct a family $\{f_\alpha, \alpha \in I\} \subset L^\Phi(\mu)$ such that its cardinality is that of the continuum, and each $f_\alpha \notin M^\Phi$, $\|f_\alpha - f_\beta\|_\Phi \geq \frac{1}{2}$ for $\alpha \neq \beta$, so that $d(\hat{f}_\alpha - \hat{f}_\beta) \geq \frac{1}{2}$ also. This will then imply that N^Φ itself will be nonseparable, completing our sketch. ∎

We omit the detail, since this result is only auxiliary for us. Also the work of Section IV.2 will show that N^{Φ} is always infinite dimensional, if it is not $\{0\}$.

If Φ is not continuous, one has to consider $\mathcal{N}^{\Phi}$, the factor space $L^{\Phi}(\mu)/\mathcal{M}^{\Phi}$ in lieu of N^{Φ}. The situation needs a consideration of several types of the structure of Φ. In that case it is possible that $\mathcal{M}^{\Phi} = L^{\infty}(\mu)$ even if (Ω, Σ, μ) is separable and $\mathcal{M}^{\Phi}$ is nonseparable and $\mathcal{N}^{\Phi}$ has to be analyzed differently since $\mathcal{M}^{\Phi}$ need not have an absolutely continuous norm. The problem has not been satisfactorily treated in the literature, and we shall omit further discussion here. However, the work of the next chapter implies that the structure of $\mathcal{N}^{\Phi}$ is in some ways analogous to that of the N^{Φ} space, especially as regards the adjoint space.

It should be noted that, by considering the atomic and diffuse parts of (Ω, Σ, μ) and the behavior of μ on these parts, one can give more detailed conditions in order that $L^{\Phi}(\mu)$ has an absolutely continuous norm or other properties. Since one can produce specialized results with further assumptions, we shall not proceed in that direction but wish to consider the general theory of Orlicz spaces with occasional visits into applications to show how this type of analysis can be carried out. Thus we move on to a study of the adjoint space and various topologies that come into play in $L^{\Phi}(\mu)$ spaces in the next chapter.

Bibliographical Notes: One of the earliest and important papers on the structure theory of Orlicz spaces is that of Morse and Transue (1950). This paper seems to have been overlooked by most authors, including the influential monograph authors Krasnoselskii and Rutickii (1961). The introduction and use of the gauge norm as well as most of the work of Section 2 follows the paper of Morse and Transue. The use of this norm is also independently seen in Nakano (1950). The fact that this norm is definable on $L^{\Phi}(\mu)$ spaces without any restrictions on the measures or Φ was emphasized by G. Weiss (1956). The same norm was used by Luxemburg (1955) for the σ-finite measures. Since the $N_{\Phi}(\cdot)$ is an example of the classical Minkowski functional, we use the short name "gauge norm" without adding the half-a-dozen or so names to indicate the proper credit to all the early users. Theorem 1.2 and Proposition 1.6 follow Weiss's paper. The very useful normalization of a Young pair and the resulting improved Hölder inequality (Proposition 3.1) are due to M. Billik (1957), and this unpublished work was made known and streamlined in Zygmund (1959). The generalized Hölder inequalities given in Theorems 3.7 and 3.9 are adapted from R. O'Neil (1965). The main part of Theorem 3.13 is in Krasnoselskii and Rutickii (1961), and the last part is due to C. X. Wu (1978). In many of these works the measure space and/or the Φ are often restricted. As far as possible, we have removed these restrictions in our treatment. Proposition 3.17 and the first part of Section 4 follow Rao (1968). Part of Proposition 4.9(a) and the present proof of Proposition 4.10 are due to Ren. Theorem 4.12(b) is classical (due to F. Riesz) in the L^p-case, and it extends to the Orlicz spaces (cf. e.g., Rao (1981), p. 25). Theorem 4.14 is due to Luxemburg and Zaanen (1963), with (Ω, Σ, μ) being σ-finite.

Theorem 5.1 evolved from general considerations from the classical results of Banach, and Köthe-Toeplitz in the 1930's, and in that context was given by Lorentz and Wertheim (1953), their measure space being the Lebesgue unit interval. This was extended to σ-finite measures in the context of Banach function spaces and Orlicz spaces by Luxemburg (1955). Because of its usefulness, we have spelled out all the details here for the general case. Propositions 5.2 and 5.3 are essentially in Krasnoselskii and Rutickii (1961) although they have not stated them in this sharp form, which was used by Ando (1960), Rao (1968), and others. Part of Theorem 5.5 is from de Jonge (1973) as already indicated in the text.

We have attempted here to unify and give a general formulation of the results, freeing them from unnecessary, albeit simplifying, restrictions throughout. The work of this chapter plays a basic role in the ensuing analysis on Orlicz spaces, and may be specialized easily for appropriate applications. We intend to show this in the following chapters.

IV
Linear Functionals and Weak Topologies

Starting with a characterization of the adjoint space of the simpler subspace $\mathcal{M}^{\Phi}$, the general structure of the adjoint space of $L^{\Phi}(\mu)$ with an unrestricted measure and/or Φ is given. We study the function type and the singular linear functionals and then consider the weak topologies on Orlicz spaces. Further we analyze the weakly compact subsets of $L^{\Phi}(\mu)$, its reflexivity and related results.

4.1. *The adjoint space of* $\mathcal{M}^{\Phi}$

It was already seen in the proof of Theorem III.5.1 that we need to know the representation of continuous linear functionals on $\mathcal{M}^{\Phi}$. A characterization of such functionals will be given in this section. To understand the structure of the (possibly) larger $L^{\Phi}(\mu)$-spaces, one actually needs to find a solution of the same problem on $L^{\Phi}(\mu)$ itself. But this is somewhat more involved and will be treated in the following two sections. However, the $(L^{\Phi}(\mu))^*$ characterization uses the work here. There is also a similarity and a subtle difference between M^{Φ}- and $\mathcal{M}^{\Phi}$-spaces and both are considered here as it will illuminate the subject.

The space of all continuous linear functions on a normed linear space $\mathcal{X}$ is denoted by $\mathcal{X}^*$, called the *adjoint* or *dual* space of $\mathcal{X}$. Recall that a linear mapping $\ell : \mathcal{X} \to \mathbb{R}$ (or $\mathbb{C}$) is continuous iff for each $x_n \in \mathcal{X}$, $\|x_n\| \to 0 \Rightarrow |\ell(x_n)| \to 0$, as $n \to \infty$ and this is equivalent to the seemingly stronger condition that $|\ell(x)| \leq K_0 \|x\|$ for all $x \in \mathcal{X}$ ($\| \cdot \|$ being the norm in $\mathcal{X}$) and the smallest constant K_0 here is denoted by $\|\ell\|^*$. Thus $\sup\{|\ell(x)| : \|x\| \leq 1\} = K_0 = \|\ell\|^* < \infty$. We assume the standard properties of these functionals usually given in Real Analysis. For instance, $(\mathcal{X}^*, \|\cdot\|^*)$ is always a complete normed linear (or Banach) space even if $(\mathcal{X}, \|\cdot\|)$ is only a normed linear space. We usually write $\mathcal{X}$ for $(\mathcal{X}, \|\cdot\|)$ and $\mathcal{X}^*$ for $(\mathcal{X}^*, \|\cdot\|^*)$ and even the same norm symbol $\|\cdot\|$ is used in both spaces since the context makes the situation clear and prevents an ambiguity. Another fact is this: If $\{\ell_\alpha : \mathcal{X} \to I\!\!R \text{ (or } \mathbb{C}), \alpha \in I\}$ is a family of continuous linear functionals such that $\sup_\alpha |\ell_\alpha(x)| \leq K_x < \infty$ for each $x \in \mathcal{X}$, then there is a fixed constant $0 < K < \infty$ such that $\sup_\alpha \|\ell_\alpha\| \leq K < \infty$. (This is a form of the *uniform boundedness* principle.)

We start with a result which in a sense is converse to Proposition III.3.1, and call it the *inverse Hölder inequality* for Orlicz spaces.

Proposition 1. *Let (Φ, Ψ) be a complementary pair of Young functions and $(L^\Phi(\mu), L^\Psi(\mu))$ be the corresponding Orlicz spaces on a measure space (Ω, Σ, μ). If $f : \Omega \to I\!\!R$ (or $\mathbb{C}$) is a measurable function such that fg is integrable for each $g \in L^\Psi(\mu)$, then $f \in L^\Phi(\mu)$.*

Proof: Let $\Sigma_1 = \{A \in \Sigma : \mu(A) < \infty\}$, and set $f_A = f\chi_A, A \in \Sigma_1$. Then there is a sequence of functions $0 \leq f_{n,A} \uparrow |f_A|$, by the structure theorem, such that

$$\ell_{n,A}(g) = \int_\Omega f_{n,A} g d\mu, \qquad g \in L^\Psi(\mu), n \geq 1, A \in \Sigma_1, \qquad (1)$$

defines $\ell_{n,A}$, a continuous linear functional, on $L^\Psi(\mu)$, since

$$|\ell_{n,A}(g)| \leq \int_\Omega f_{n,A} |g| d\mu \leq \int_\Omega |f_A g| d\mu \leq \int_\Omega |fg| d\mu < \infty. \qquad (2)$$

Thus $\{\ell_{n,A}(g), n \geq 1, A \in \Sigma_1\}$ is uniformly bounded for each $g \in L^\Psi(\mu)$. Hence, by the uniform boundedness principle, $\sup_{n,A} \|\ell_{n,A}\| = K_0 < \infty$. Consequently,

$$\|f_A\|_\Phi = \sup\left\{|\int_\Omega f_A g d\mu| : N_\Psi(g) \leq 1\right\}$$

$$\leq \sup\left\{\lim_n |\ell_{n,A}(g)| : N_\Psi(g) \leq 1\right\}$$

$$\leq \sup\{K_0 N_\Psi(g) : N_\Psi(g) \leq 1\} = K_0 < \infty. \qquad (3)$$

Thus $f_A \in L^\Phi(\mu)$ for each $A \in \Sigma_1$.

To get rid of A, fix a $g \in L^\Psi(\mu)$ with $N_\Psi(g) \le 1$. Since for $A, B \in \Sigma_1$, on $A \cap B$ we have $f_A = f_B = f_{A \cap B}$, define $\nu_g : \Sigma_1 \to \mathbb{R}^+$ (or $\mathbb{C}$) as:

$$\nu_g(A) = \int_\Omega |f_A g| d\mu = \int_\Omega |fg| d\mu, \qquad A \in \Sigma_1.$$

Then ν_g is a measure on the δ-ring Σ_1, and $\sup\{\nu_g(A) : A \in \Sigma_1\} \le K_0$ by (3). Hence there exists a sequence $A_n \in \Sigma_1, A_n \subset A_{n+1}$ (since ν_g is σ-additive on the ring Σ_1) such that if $B = \overset{\infty}{\underset{n=1}{\cup}} A_n$, then $\nu_g(B) = \lim_{n \to \infty} \nu_g(A_n) \le K_0 < \infty$. If $E \subset B^c$, $E \in \Sigma$, and if $\mu(E) > 0$ then $\exists F \in \Sigma_1, \mu(F) > 0$ by the finite subset property of μ, and one has, if $\alpha = \nu_g(B)$,

$$\alpha < \nu_g(B) + \nu_g(F) = \nu_g(B \cup F) = \lim_{n \to \infty} \nu_g(A_n \cup B)$$

$$\le \sup\{\nu_g(D) : D \in \Sigma_1\} = \alpha.$$

This contradiction shows that $\mu(F) = 0$, so that $\nu_g(F) = 0$ and this implies $\nu_g(B^c) = 0$. Hence

$$\|f\|_\Phi = \sup\left\{\int_\Omega |fg| d\mu : N_\Psi(g) \le 1\right\} = \sup\{\nu_g(\Omega) : N_\Psi(g) \le 1\}$$

$$= \sup\{\nu_g(B) : N_\Psi(g) \le 1\} \le K_0 < \infty.$$

This shows that $f \in L^\Phi(\mu)$, as asserted. $\blacksquare$

We need to present some more preliminary concepts and results before the characterization of the adjoint space is considered.

Definition 2. Let Φ be a Young function and $A_\Phi(\mu)$ denote the set of all scalar additive functions G on Σ vanishing on μ-null sets of (Ω, Σ, μ) such that (i)(a) if the complementary function Ψ of Φ is not *purely discontinuous* (i.e., it is of the form $\Phi(x) = 0, 0 < x \le x_0$; $\Phi(x) > 0$ for $x_0 \le x \le x_1$; and $= +\infty$ for $x > x_1$), then $G(A) \equiv 0$ for all $A \in \Sigma_1 = \{B \in \Sigma : \mu(B) < \infty\}$ and (b) if Ψ is purely discontinuous, then G is defined on all of Σ, and (ii) $N_\Phi(G) < \infty$ where

$$N_\Phi(G) = \inf\{k > 0 : I_\Phi(\tfrac{G}{k}) \le 1\}, \tag{4}$$

and

$$I_\Phi(G, E) =$$
$$\sup\left\{\sum_{i=1}^n \Phi\left(\frac{G(E_i)}{\mu(E_i)}\right)\mu(E_i) : \{E_i\}_1^n \text{ disjoint}, E_i \in \Sigma_1 \cap E, n \ge 1\right\}, \tag{5}$$

if Ψ is not purely discontinuous, and Σ_1 is replaced by Σ in the opposite case. We set $I_\Phi(G) = I_\Phi(G, \Omega)$. If $I_\Phi(G) < \infty$, then G is said to have the Φ-*bounded variation* relative to μ on Σ. The space $A_\Phi(\mu)$ is called an *Orlicz space of additive set functions* on (Ω, Σ, μ), and is discussed later, in Chapter X.

If G is a real valued point function on $\Omega = I\!R$, setting $G(a, b) = (\Delta G)(a, b) = G(b) - G(a)$, and μ as Lebesgue measure, $I_\Phi(G)$ is termed a "generalized Hellinger" integral by Morse and Transue (1950). The above definition is an abstract analog of the latter concept.

It is clear that $A_\Phi(\mu)$ is a vector space of additive set functions and $N_\Phi(\cdot)$ of (4) is a norm so that $(A_\Phi(\mu), N_\Phi(\cdot))$ is a normed linear space. [Note that $I_\Phi(\cdot)$ is a convex functional.] If $\Phi(x) = |x|^p$, $p \geq 1$, then $A_\Phi(\mu)$ is the space of additive set functions of p-bounded variation of classical analysis. We shall show that $A_\Phi(\mu)$ is an adjoint space of a (certain) normed linear space from which one concludes that it is a Banach space.

Since the members of $A_\Phi(\mu)$ can be finitely additive, and since we intend to give an integral representation of the continuous linear functionals on $\mathcal{M}^\Phi$ and $L^\Phi(\mu)$ relative to such set functions, we need to recall integration of scalar functions with respect to them. Thus if $G : \Sigma \to I\!R$ (or $\mathbb{C}$) is additive and $f : \Omega \to I\!R$ (or $\mathbb{C}$) is a simple function, $f = \sum_{i=1}^{n} a_i \chi_{A_i}$, $A_i \in \Sigma$, then as usual

$$\int_\Omega f dG = \sum_{i=1}^{n} a_i G(A_i).$$

Next if f is a general measurable function, it is said to be integrable relative to G, provided there is a sequence of simple functions f_n converging to f in G-measure [i.e., for each $\varepsilon > 0$

$$\lim_{n \to \infty} |G|(|f - f_n| \geq \varepsilon) = 0$$

where $|G|(\cdot)$ is the variation of G which is an additive set function $(|G|(E) = I_\Phi(G, E)$ with $\Phi(x) = |x|$ in (5))], and satisfying the additional condition

$$\lim_{m,n \to \infty} \int_E |f_n - f_m| d|G| = 0, \qquad E \in \Sigma. \tag{6}$$

When this holds, we simply denote it as $\int_E f dG = \lim\limits_{n \to \infty} \int_E f_n dG$, $E \in \Sigma$. The usual measure theory arguments show that the integral is well defined and is independent of the $\{f_n\}_1^{\infty}$-sequence used, and is additive. It has many of the properties of the σ-additive G. However, the Lebesgue limit theorems are not valid without supplementary hypotheses. We shall state these conditions when we employ them. [A good account is in the standard work of Dunford-Schwartz (1958).]

We establish the following result and use it later:

Proposition 3. *Let Φ be a continuous Young function with $\Phi(x) = 0$ iff $x = 0$. If $f \in \tilde{L}^{\Phi}(\mu)$ and $G : E \mapsto \int_E f d\mu$, $E \in \Sigma_1 = \{A \in \Sigma : \mu(A) < \infty\}$, then $I_{\Phi}(G) < \infty$ and moreover*
   (a) $I_{\Phi}(G, E) = \int_E \Phi(f)d\mu$, $E \in \Sigma$;
   (b) $\lim\limits_{\mu(A) \to 0} I_{\Phi}(G, A) = 0$, and $I_{\Phi}(G) = 0$ iff $f = 0$ a.e.; and
   (c) *for each $\varepsilon > 0$, there is an $A \in \Sigma_1$ such that $I_{\Phi}(G, A^c) < \varepsilon$.*

Proof: Let us establish (a). The rest will follow from this. The necessary argument is a modification of that given in Morse and Transue ((1950), p. 596). Since μ is a measure and f is measurable, $f \in \tilde{L}^{\Phi}(\mu)$ implies for each $\{E_1, \dots, E_n\} \subset \Sigma_1$, E_i of positive μ-measure and disjoint [by Jensen's inequality (cf. Proposition III.3.5)], that

$$\sum_{i=1}^{n} \Phi\left(\int_{E_i} f \frac{d\mu}{\mu(E_i)}\right) \mu(E_i) \leq \sum_{i=1}^{n} \int_{E_i} \Phi(f)d\mu \leq \int_{\Omega} \Phi(f)d\mu.$$

Hence taking the supremum on the left over all such $\{E_i\}_1^n$, one gets

$$I_{\Phi}(G) \leq \int_{\Omega} \Phi(f)d\mu. \tag{7}$$

For the opposite inequality, since $I_{\Phi}(G) < \infty$ by (7), given $\varepsilon > 0$ there is a finite collection $\{E_1, \dots, E_{n_{\varepsilon}}\} \subset \Sigma_1$ such that $f_{\varepsilon} = \sum\limits_{i=1}^{n_{\varepsilon}} \frac{G(E_i)}{\mu(E_i)} \chi_{E_i}$ satisfies

$$\int_{\Omega} \Phi(f_{\varepsilon})d\mu = \sum_{i=1}^{n_{\varepsilon}} \Phi\left(\frac{G(E_i)}{\mu(E_i)}\right) \mu(E_i) \leq I_{\Phi}(G) \tag{8}$$

and $f_{\varepsilon} \to f$ in measure as $\varepsilon \searrow 0$. The latter from the fact that if $A = \bigcup\limits_{i=1}^{n} E_i$ ($\in \Sigma_1$), then $L^{\Phi}(A, \Sigma(A), \mu) \subset L^1(\mu)$ and $\int_{\Omega} \Phi(f_{\varepsilon} - f)d\mu \to 0$ as

$\varepsilon \searrow 0$, and so $f_\varepsilon \to f$ in measure. Hence by Fatou's inequality applied to (8) gives $\int_\Omega \Phi(f)d\mu \le I_\Phi(G)$, which together with (7) implies (a) if $E = \Omega$. But the same procedure with $f\chi_E$ replacing f proves (a) as stated.

(b) Since $\Phi(f) \in L^1(\mu)$ and $\Phi(x) = 0$ iff $x = 0$, (a) $\Rightarrow$ (b) at once.

(c) This is also a consequence of (a) since $\Phi(f) \in L^1(\mu)$ and

$$I_\Phi(G) = I_\Phi(G, A \cup A^c) = \int_{A \cup A^c} \Phi(f)d\mu = \int_A \Phi(f)d\mu + \int_{A^c} \Phi(f)d\mu$$
$$= I_\Phi(G, A) + I_\Phi(G, A^c),$$

so that for each $\varepsilon > 0$, there is an $A_\varepsilon \in \Sigma_1$ such that $I_\Phi(G, A_\varepsilon^c) < \epsilon$ by the standard properties of integrable functions. ∎

Remark. In the proof of (a) above, we only used the fact that Φ is a Young function, and no other restrictions used for (b) and (c) are needed. This fact is useful later.

The following property of any G in $A_\Phi(\mu)$ is useful.

Lemma 4. *If $G \in A_\Phi(\mu)$, and the complementary function Ψ of Φ is continuous (hence satisfies $\Psi(x) < \infty$ for all $x < \infty$), then* $\lim_{\mu(A) \to 0} G(A) = 0.$

Proof: Given $\varepsilon > 0$, choose $K \ge 1$ such that $2I_\Phi(G) < K\varepsilon$ which is possible since $I_\Phi(G) < \infty$. Let $A \in \Sigma$ be such that $2\Psi(K)\mu(A) < \varepsilon$ which is also possible since $\Psi(K) < \infty$. If $A_1, \ldots, A_n \in \Sigma$ is a partition of A such that $\mu(A_i) > 0$, let $x_i = G(A_i)/\mu(A_i)$, and $f_\varepsilon = \sum_{i=1}^n x_i \chi_{A_i}$ ($\in \mathcal{M}^\Phi$). Then

$$|G(A)| \le \sum_{i=1}^n |G(A_i)| = \sum_{i=1}^n |x_i|\mu(A_i) = \int_A K\left|\frac{|f_\varepsilon|}{K}\right|d\mu$$

$$\le \int_A \Phi\left(\frac{|f_\varepsilon|}{K}\right)d\mu + \int_A \Psi(K)d\mu, \text{ by Young's inequality,}$$

$$\le \frac{1}{K}\sum_{i=1}^n \Phi(x_i)\mu(A_i) + \Psi(K)\mu(A), \text{ since } \Phi \text{ is convex,}$$

$$\le \frac{1}{K}I_\Phi(G) + \Psi(K)\mu(A) < \frac{\varepsilon}{2} + \frac{\varepsilon}{2} = \varepsilon,$$

which implies the assertion. ∎

We have another simple property of the functional I_Φ.

Lemma 5. *The functional $I_\Phi : A_\Phi(\mu) \to \mathbb{R}^+$ is lower semi-continuous, i.e., if $G_n \in A_\Phi(\mu)$, $G_n \to G$ pointwise, then $\liminf_{n\to\infty} I_\Phi(G_n) \geq I_\Phi(G)$.*

Proof: Let $\pi = \{A_1, \ldots, A_n\} \subset \Sigma$, $0 < \mu(A_i) < \infty$, be a disjoint collection. Then $G_n(A_i) \to G(A_i)$ as $n \to \infty$, so that

$$S_\Phi^\pi(G_n) = \sum_{i=1}^{n} \Phi\left(\frac{G_n(A_i)}{\mu(A_i)}\right)\mu(A_i) \to S_\Phi^\pi(G), \text{ as } n \to \infty. \tag{9}$$

But since by hypothesis $G_n \to G$ pointwise on the metric space (Σ, ρ) where $\rho(A,B) = \arctan \mu(A \triangle B)$, and since $S_\Phi^\pi(\cdot)$ itself is lower semi-continuous for each π, we get the desired result (i.e., the Fatou inequality) on taking suprema of π over Σ_1, on both sides of (9). ∎

We are now ready to characterize $(\mathcal{M}^\Phi)^*$, the main result of this section.

Theorem 6. *Let (Φ, Ψ) be a pair of Young functions and $(M^\Phi, \mathcal{M}^\Phi)$ and $A_\Psi(\mu)$ be the spaces introduced in Definition III.4.2 and Definition 2 above on a measure space (Ω, Σ, μ). Then for each $x^* \in (\mathcal{M}^\Phi)^*$ there exists a unique $G \in A_\Psi(\mu)$ such that*

$$x^*(f) = \int_\Omega f dG, \qquad f \in \mathcal{M}^\Phi, \tag{10}$$

and if Φ is continuous with $\Phi(x) = 0$ only for $x = 0$, then $\mathcal{M}^\Phi = M^\Phi$, and G in (10) is σ-additive. Moreover, if either μ is σ-finite (or generally localizable, which is recalled below) or both (Φ, Ψ) are N-functions (and μ is unrestricted) then $(\mathcal{M}^\Phi)^* = L^\Psi(\mu)$ so that

$$x^*(f) = \int_\Omega fg d\mu, \qquad f \in M^\Phi, \tag{11}$$

for a unique $g \in L^\Psi(\mu)$. In both (10) and (11) we have

$$\|x^*\| = \sup\{|x^*(f)| : N_\Phi(f) \leq 1, f \in \mathcal{M}^\Phi\} = \|G\|_\Psi$$
$$= \sup\{|\int_\Omega f dG| : N_\Phi(f) \leq 1, f \in \mathcal{M}^\Phi\}, \tag{12}$$

or

$$\|x^*\| = \|g\|_\Psi. \tag{13}$$

Thus $(\mathcal{M}^{\Phi})^*$ or $(M^{\Phi})^*$ is isometrically isomorphic to $A_{\Psi}(\mu)$ or $L^{\Psi}(\mu)$.

Remark. Since Φ can be discontinuous, so that $\mathcal{M}^{\Phi} = L^{\infty}(\mu)$ is possible, in (10) G can only be finitely additive in which case the integral is taken in the sense defined after Definition 2. Even if $\Phi(x) = |x|$, so that $M^{\Phi} = \mathcal{M}^{\Phi}$ and G is σ-additive in (10), one cannot replace it by (11) unless μ is restricted, the best condition being *localizability*. Thus, if $\Sigma_0 = \{A \in \Sigma : \mu(A) < \infty\}$, then $\Sigma = \sigma(\Sigma_0)$ and for each $\mathcal{C} \subset \Sigma$ there is a $B \in \Sigma$ such that (i) $A \in \mathcal{C} \Rightarrow \mu(A - B) = 0$ and (ii) $B_0 \in \Sigma$, $\mu(A - B_0) = 0$ for all $A \in \mathcal{C}$, then $\mu(B - B_0) = 0$. Thus B is the supremum of $\mathcal{C}$ in Σ. It is easy to verify that every σ-finite measure is localizable, but not conversely. A Haar measure on a locally compact group is localizable but not σ-finite. (For a discussion and properties of localizability, which was originally introduced by I. E. Segal in the early 1950's, see, e.g. Rao (1987), p. 70.) An interesting aspect of this result is that it extends all the $(L^p(\mu))^*$-spaces, $1 \leq p \leq +\infty$, on a general measure space (Ω, Σ, μ).

Proof: We establish (10) in two parts: (i) The integral in (10) defines an element x^* in $(\mathcal{M}^{\Phi})^*$ for each $G \in A_{\Psi}(\mu)$, and (ii) every x^* in $(\mathcal{M}^{\Phi})^*$ is of the form (10) for a unique $G_{x^*} \in A_{\Psi}(\mu)$. Then (iii) we get (11), and (iv) (12) and (13) are deduced. The argument will be given in steps for clarity.

(i) Let $0 \leq f \in \mathcal{M}^{\Phi}$, and $\Sigma_1 = \{A \in \Sigma : \mu(A) < \infty\}$. For each $E \in \Sigma_1$, set $f_E = f\chi_E \in \mathcal{M}^{\Phi} \subset L^{\Phi}(\mu)$. Then there exist $0 \leq f_n \uparrow f_E$, f_n simple and $f_n \in \mathcal{M}^{\Phi}$ (each $f_n = 0$ on E^c). If $G \in A_{\Psi}(\mu)$, then G vanishes on μ-null sets. Let $f_n = \sum_{i=1}^{m} a_{ni}\chi_{E_{ni}}$ be a typical representation of the simple $f_n \geq 0$. Define $h = \sum_{i=1}^{m} b_{ni}\chi_{E_{ni}}$ where $b_{ni} = |G(E_{ni})|/\mu(E_{ni})$ if $\mu(E_{ni}) > 0$, and $= 0$ otherwise. Then

$$\left|\int_E f_n dG\right| = \left|\sum_{i=1}^{m} a_{ni}G(E_{ni})\right| \leq \sum_{i=1}^{m} a_{ni}|G(E_{ni})|, \quad E \in \Sigma_1,$$

$$= KN_{\Phi}(f_E)\int_E \frac{|f_n|}{N_{\Phi}(f_E)}\frac{h}{K}d\mu, \quad K > 0,$$

$$\leq KN_{\Phi}(f_E)\left[\int_E \Phi\left(\frac{f_n}{N_{\Phi}(f_E)}\right)d\mu + \int_E \Psi\left(\frac{h}{K}\right)d\mu\right],$$

by Young's inequality,

$$\leq KN_\Phi(f_E)\left[\int_E \Phi\left(\frac{f_n}{N_\Phi(f_E)}\right)d\mu + I_\Psi\left(\frac{G}{K},E\right)\right], \text{ by (5)},$$

$$\leq 2N_\Phi(f_E)N_\Psi(G_E), \text{ by (4) with } K = N_\Psi(G_E),$$

$$\leq 2N_\Phi(f)N_\Psi(G), \tag{14}$$

where $G_E = G \mid \Sigma(E)$, the restriction. Thus for each E in Σ_1 the left side of (14) is uniformly bounded by the right constant of (14).

(a) Suppose now that Φ is continuous. Then by Lemma 4, G is μ-continuous and since $E \in \Sigma_1$, G_E has finite total variation on E (because the Φ-variation on a set of finite measure is never smaller than its total variation). So G_E is σ-additive on $\Sigma(E)$ and hence by the monotone convergence theorem applied to the variation $v(G_E)$, it follows from standard measure theory that $\int_E f_n dG \to \int_E f_E dG = \int_E f dG$. Hence (14) and this last result imply

$$\left|\int_E f dG\right| \leq 2N_\Phi(f)N_\Psi(G), \qquad E \in \Sigma_1. \tag{15}$$

(b) Suppose Φ is discontinuous, so that $\Phi(x) < \infty$ for $0 \leq x < x_0$, and $\Phi(x) = +\infty$ for $x \geq x_0$. Then as seen in Proposition III.4.3, $L^\Phi(\mu) \subset L^\infty(\mu)$ and restricting these spaces to $E \in \Sigma_1$, $\mathcal{M}^\Phi = L^\Phi(\mu) = L^\infty(\mu)$. But then the Ψ-bounded variation of G_E reduces to the ordinary variation, $N_\Phi(f_E)$ and $\|f\|_\infty$ are equivalent. The same is true of $N_\Psi(G_E)$ and $(v(G_E))(E)$. In this case therefore we have (the integral of a finitely additive set function being as in (6))

$$\left|\int_E f_E dG\right| \leq \int_E |f_E| dv(G) \leq 2CN_\Phi(f_E)N_\Psi(G_E)$$

$$\leq 2CN_\Phi(f)N_\Psi(G), \tag{16}$$

where $0 < C < \infty$ is a constant depending only on $0 < x_0 < \infty$. Thus in both cases (15) and (16) hold for all $E \in \Sigma_1$ since the right side bounds are independent of E. It should be shown here that E can be replaced by Ω.

For the last conclusion we need to use another (unfortunately non-trivial) result from Measure Theory. Let $\lambda^f : E \mapsto \int_E f_E dG = \int_E f dG$,

$E \in \Sigma$. Then $\lambda^f(\cdot)$ is an additive set function for each $f \in \mathcal{M}^\Phi$, and $\lambda^{(\cdot)}(E)$ is linear on $\mathcal{M}^\Phi$ for each $E \in \Sigma_1$. Fix an f as before, and let $\alpha = \sup\{|\lambda^f(E)| : E \in \Sigma_1\}$. Then by (15) and (16), $0 \leq \alpha \leq CN_\Phi(f)N_\Psi(G) < \infty$. Here (Σ_1, λ^f) can be represented as follows: Consider (Ω, Σ). By the Stone representation theorem (cf., e.g., Rao (1987), p. 504) there exists a measurable space $(\tilde{\Omega}, \tilde{\Sigma})$, where $\tilde{\Omega}$ is a totally disconnected compact Hausdorff space with $\tilde{\Sigma}$ as its σ-algebra generated by the clopen (= closed-open) sets of Ω such that $\tilde{\Sigma}$ is isomorphic to Σ. This result is also true if Σ is replaced by the ring Σ_1 and then $\tilde{\Sigma}_1$ will be a σ-ring generated by the compact open sets of $\tilde{\Omega}$, and again there is an isomorphism. If h is this isomorphism, then the mapping $T : \lambda \mapsto \lambda \circ h^{-1}$ defined by $(T\lambda)(\tilde{E}) = \lambda(T^{-1}(\tilde{E}))$, $\tilde{A} \in \tilde{\Sigma}_1$ and λ is a bounded additive set function, makes $(T\lambda)(\cdot)$ σ-additive on $\tilde{\Sigma}_1$ and bounded, and further $(T\lambda)$ has a unique σ-additive extension to $\sigma(\Sigma_1)$. Moreover $v(T\lambda) = v(\lambda)$, so that the mapping T is an isometric isomorphism. (For a proof of this, see, e.g., Rao (1987), p. 509, or Dinculeanu (1967), p. 372 for a more general result on vector applications.) Then $\alpha = \sup\{T(v(\lambda^f))(\tilde{A}) : \tilde{A} \in \Sigma_1\}$, and since $T(v(\lambda^f))$ is σ-additive, there exists a sequence $\tilde{A}_n \in \Sigma_1$, $\tilde{A}_n \subset \tilde{A}_{n+1}$ such that if $\tilde{B} = \bigcup_{n=1}^{\infty} \tilde{A}_n$ ($\in \Sigma_1$) we get $(T(v(\lambda^f)))(\tilde{B}) = \alpha$. It follows from the by-now-familiar argument that $(T(v(\lambda^f)))(C_0) = 0$ for each $C_0 \in \tilde{\Sigma}$ with $C_0 \cap B = \emptyset$. Hence $(T(v(\lambda^f)))(\tilde{\Omega}) = \alpha$. The isomorphism theorem (recalled above) implies that $\lambda^f(\Omega)$ is defined, the integral $\lambda^f(\Omega)$ is additive in f and one has

$$|\lambda^f(\Omega)| = \left| \int_\Omega f dG \right| \leq CN_\Phi(f)N_\Psi(G).$$

It follows that, writing $f = f_1 - f_2 + i(f_3 - f_4)$, $f_j \geq 0, j = 1, 2, 3, 4$, one concludes that (10) defines a bounded linear functional on $\mathcal{M}^\Phi$ (even on $L^\Phi(\mu)$) for each $G \in A_\Psi(\mu)$. This proves (i).

(ii) We again start with $\Sigma_1 = \{A \in \Sigma : \mu(A) < \infty\}$. Let $x^* \in (\mathcal{M}^\Phi)^*$. Then using the fact that step functions are dense in $\mathcal{M}^\Phi$, we see that $x^*(\chi_E) = G(E)$ defines an additive set function on Σ which is zero iff $x^* = 0$. We first show that $G \in A_\Psi(\mu)$. Since $\chi_A = 0$ a.e. for μ-null sets A, $G(A) = 0$ for such A. If $E \in \Sigma$, and $E_i \subset E$ with $E_i \in \Sigma$, disjoint, $0 < \mu(E_i) < \infty$, $i = 1, \ldots, n$, define, for the nontrivial case $x^* \neq 0$ ($\psi =$ left derivative of Ψ),

$$f_n = \sum_{i=1}^{n} b_i \chi_{E_i}, \qquad b_i = \psi\left(\frac{G(E_i)}{K\mu(E_i)}\right), \qquad K > 0. \qquad (17)$$

Then by Proposition III.4.3, $f_n \in \tilde{L}^{\Phi}(\mu)$, and we have on taking K in (17) as $|x^*(f)| \leq K\|f\|_{\Phi}$, a positive constant,

$$1 + \int_{\Omega} \Phi(f_n)d\mu \geq \|f_n\|_{\Phi} \geq \frac{x^*(f_n)}{K}$$

$$= \sum_{i=1}^{n} b_i \left(\frac{G(E_i)}{K\mu(E_i)}\right)\mu(E_i)$$

$$= \sum_{i=1}^{n} \left[\Phi(b_i) + \Psi\left(\frac{G(E_i)}{K\mu(E_i)}\right)\right]\mu(E_i), \text{ by the}$$

equality in Young's inequality,

$$= \int_{\Omega} \Phi(f_n)d\mu + \sum_{i=1}^{n} \Psi\left(\frac{G(E_i)}{K\mu(E_i)}\right)\mu(E_i).$$

This implies on taking the supremum over Σ_1 and noting that $E \in \Sigma$ is arbitrary, that $I_\Psi(\frac{G}{K}) \leq 1$, so that $G \in A_\Psi(\mu)$.

But now by part (i), for each $f \in L^{\Phi}(\mu)$, we can define

$$y^*(f) = \int_{\Omega} f dG, \qquad (18)$$

so that y^* is a continuous linear functional. Also $y^*(f_n) = x^*(f_n)$ for each simple function. If $\mu(\Omega) < \infty$, then on using the fact that simple functions are norm dense in $\mathcal{M}^{\Phi}$, and since a continuous linear mapping on a normed vector space is uniformly continuous, we conclude that $x^* = y^*$ on $\mathcal{M}^{\Phi}$ itself. Here we invoked the classical result that two uniformly continuous functions agreeing on a dense subset of a metric space coincide on the whole space. Thus the representation (10) holds on Σ_1. But in the general case, we can apply the procedure of the last part of (i) word-for-word and conclude that (10) holds as stated. It remains to establish (11) and (12) (or (13)) under the additional conditions given in the statement.

(iii) If Φ is continuous, with $M^{\Phi} = \mathcal{M}^{\Phi}$, G in $A_\Psi(\mu)$ is σ-additive and μ-continuous. Since G is thus a signed measure, we can apply the

Radon-Nikodým theorem iff μ is localizable in which case $G(E) = \int_E g\,d\mu$ for a μ-unique measurable function g. (This is Segal's extension of the classical Radon-Nikodým theorem, cf., e.g., Rao (1987), p. 275.) Now by Proposition 3 above, $g \in L^{\Psi}(\mu)$. Thus $A_{\Psi}(\mu)$ can be identified with $L^{\Psi}(\mu)$ and (11) follows. The uniqueness assertion is an immediate consequence of Proposition III.4.10 for (11) and its obvious extension to (10) since $N_{\Psi}(G)$ is a norm on $A_{\Psi}(\mu)$, by (4). In case (Φ, Ψ) are N-functions but μ is arbitrary, then again each element G of $A_{\Psi}(\mu)$ is σ-additive and μ-continuous as before. But by definition $I_{\Psi}(G) < \infty$, and by (5) since $\Psi(x) = 0$ iff $x = 0$, we claim that G vanishes outside a σ-finite set of μ-measure. In fact, $I_{\Psi}(G, E) \leq I_{\Psi}(G, F)$ for $E \subset F$ and $I_{\Psi}(G, \cdot)$ is σ-subadditive. So by (5) there is a sequence $E_n \subset E_{n+1}$, $E_i \in \Sigma_1$ such that $I_{\Psi}(G) = \lim_{n \to \infty} I_{\Psi}(G, E_n)$. Let $\tilde{E} = \bigcup_{n=1}^{\infty} E_n$. Then $\tilde{\Sigma}$ is σ-finite for μ. If there is a $B \in \Sigma_1$, $B \cap \tilde{E} = \emptyset$ and $\mu(B) > 0$, then, since Φ is also an N-function, Ψ is strictly convex so that $I_{\Psi}(G, \tilde{E}) < I_{\Psi}(G, E_0)$ where $E_0 = \tilde{E} \cup B = \bigcup_{n=1}^{\infty}(E_n \cup B)$, $E_n \cup B \in \Sigma_1$, $n \geq 1$. Since both terms equal $I_{\Psi}(G) = I_{\Psi}(G, \Omega)$, this gives a contradiction. Thus $\mu(B) = 0$ and $G = G_{\tilde{E}}$. So $G_{\tilde{E}}(A) = G(\tilde{E} \cap A) = \int_{\tilde{E} \cap A} g\,d\mu$ for a μ-unique g which vanishes outside the σ-finite set $\tilde{E}$. Hence $g \in L^{\Psi}(\mu)$ as before and (11) holds in this case also. It should be noted that $\tilde{E}$ varies with G so that μ on Σ itself is not restricted.

(iv) Finally (12) and (13) give different (but equivalent) norms of x^* by using the different (but equivalent) $N_{\Phi}(\cdot)$ or $\|\cdot\|_{\Phi}$ respectively, and these isometries are immediate from Proposition III.4.10 again. This establishes the theorem in all its parts. ∎

The above proof can be simplified, and the use of finitely additive set functions and their integration can be avoided, if we assume that Φ is continuous, $\Phi(x) = 0$ iff $x = 0$, and μ is σ-finite. Further simplification is obtained if (Φ, Ψ) are N-functions and μ is a finite measure. The latter was essentially given by Krasnosel'skii and Rutickii (1961), and we state the relevant result for comparison and reference.

Theorem 7. *Let (Φ, Ψ) be a complementary Young pair and (Ω, Σ, μ) be a σ-finite measure space, Φ continuous, $\Phi(x) = 0$ iff $x = 0$. Then $(M^{\Phi})^* = L^{\Psi}(\mu)$ and for each $x^* \in (M^{\Phi})^*$, there is a unique $g_{x^*} \in L^{\Psi}(\mu)$ such that*

$$x^*(f) = \int_\Omega f g_{x^*} d\mu, \qquad f \in M^\Phi, \tag{19}$$

$$\|x^*\| = \sup\{|x^*(f)| : N_\Phi(f) \le 1, f \in M^\Phi\} = \|g_{x^*}\|_\Psi \tag{20}$$

$$\|x^*\|' = \sup\{|x^*(f)| : \|f\|_\Phi \le 1, f \in M^\Phi\} = N_\Psi(g_{x^*}). \tag{21}$$

Thus $\|\cdot\|$ and $\|\cdot\|'$ are equivalent norms for $(M^\Phi)^*$. In particular, if (Φ, Ψ) are complimentary N-functions and (Ω, Σ, μ) is σ-finite, then $(M^\Phi)^* = L^\Psi(\mu)$ and $(M^\Psi)^* = L^\Phi(\mu)$, the equalities being isometric isomorphisms, when one of the norms is used for M^Φ and the other for $L^\Psi(\mu)$.

Since the adjoint space of any normed linear space is complete, and since $N_\Phi(G)$ of (4) is a gauge norm of G and similarly

$$\|G\|_\Phi = \sup\{|\int_\Omega h \, dG| : N_\Psi(h) \le 1\} \tag{22}$$

is the analog of the Orlicz norm of G [the fact that $\|\cdot\|_\Psi$ is a norm follows from (10)], both are equivalent. Hence we have

Corollary 8. *For any Young function Φ, and measure space (Ω, Σ, μ), the spaces $(A_\Phi(\mu), N_\Phi(\cdot))$ and $(A_\Phi(\mu), \|\cdot\|_\Phi)$ are Banach spaces.*

This is significant since $L^\Phi(\mu)$ need not be complete if μ is only finitely additive, and even then $A_\Phi(\mu)$ turns out to be complete. This will be discussed further in Chapter X. Also note that if Φ is Δ_2-regular, then $M^\Phi = L^\Phi(\mu)$, hence we have the consequence:

Corollary 9. *If Φ is a Young function which is Δ_2-regular, and if (Ω, Σ, μ) is localizable (or σ-finite) or Ψ is continuous with $\Psi(x) = 0$ iff $x = 0$, then $(L^\Phi(\mu))^* = L^\Psi(\mu)$. The same result holds if Φ is such that $L^\Phi(\mu)$ has an absolutely continuous norm and the other conditions on Ψ and μ hold. The equalities are in terms of isometric isomorphisms.*

It should be noted that in the above isometrics one uses the gauge or Orlicz norms in $L^\Phi(\mu)$ and $L^\Psi(\mu)$ alternately, *but not the same norms in general*. However, if the same $N_\Phi(\cdot)$ or $\|\cdot\|_\Phi$ are used throughout, then we have topological equivalence in lieu of isometries. This may be improved in some cases under normalization, but impossible in the general theory.

Recall that if $\mathcal{X}$ is a Banach space and $\mathcal{X}^*$ its adjoint which is always a Banach space under the adjoint norm, let $(\mathcal{X}^*)^*$ be the second adjoint, simply denoted $\mathcal{X}^{**}$. If the image $\hat{\mathcal{X}}$ of $\mathcal{X}$ is all of $\mathcal{X}^{**}$, one says that $\mathcal{X}$ is a *reflexive* Banach space, and writes $\mathcal{X} = \mathcal{X}^{**}$ where equality is understood as isometric isomorphism (or topological equivalence if equivalent [but different] norms are used). We can characterize the reflexive Orlicz spaces using the preceding result.

Theorem 10. *Let (Ω, Σ, μ) be a measure space and (Φ, Ψ) be a pair of complementary Young functions. Then $L^\Phi(\mu)$ $[L^\Psi(\mu)]$ is reflexive iff $L^\Phi(\mu) = M^\Phi$ and $L^\Psi(\mu) = M^\Psi$, or equivalently both $L^\Phi(\mu)$ and $L^\Psi(\mu)$ have absolutely continuous norms. In particular this holds if both Φ and Ψ are Δ_2-regular.*

Proof: Let $L^\Phi(\mu) = M^\Phi$ and $L^\Psi(\mu) = M^\Psi$. Then by Theorem 6, both (Φ, Ψ) must be continuous and $\Phi(x) = 0$ iff $x = 0$. Also

$$(L^\Phi(\mu))^* = (M^\Phi)^* = L^\Psi(\mu) = M^\Psi.$$

Hence

$$(L^\Phi(\mu))^{**} = (M^\Psi)^* = L^\Phi(\mu).$$

This implies that $L^\Phi(\mu)$ is reflexive, and similarly $L^\Psi(\mu)$ is reflexive. [Actually the general Banach space theory tells us that $\mathcal{X}$ is reflexive iff its adjoint $\mathcal{X}^*$ is. So the statement about $L^\Psi(\mu)$ is a direct consequence.]

Conversely, let $L^\Phi(\mu)$ be reflexive. Then by Proposition 3 with each $g \in L^\Psi(\mu)$, $G(E) = \int_E g d\mu$ defines $G \in A_\Psi(\mu)$ and $N_\Psi(G) = N_\Psi(g)$ so that $L^\Psi(\mu) \subset A_\Psi(\mu)$ isometrically. Also $M^\Psi \subset L^\Psi(\mu)$ implies that

$$L^\Phi(\mu) \subset A_\Phi(\mu) = (M^\Psi)^* \subset (L^\Psi(\mu))^*$$

and hence also

$$(L^\Phi(\mu))^* \subset (L^\Psi(\mu))^{**}.$$

These two inclusions imply, on interchanging Φ and Ψ in the above,

$$L^\Phi(\mu) \subset (L^\Psi(\mu))^* \subset (L^\Phi(\mu))^{**} = L^\Phi(\mu),$$

the last equality by hypothesis. It follows that $L^\Phi(\mu)$ is the adjoint space of $L^\Psi(\mu)$ and since by the above quoted general theory, this is true iff $L^\Psi(\mu)$ is reflexive. Hence $L^\Psi(\mu)$ is the adjoint space of $L^\Phi(\mu)$. By Theorem 6, this implies both Φ, Ψ are continuous as noted above, and then $(M^\Phi)^* = L^\Psi(\mu), (M^\Psi)^* = L^\Phi(\mu)$. Since $M^\Phi \subset L^\Phi(\mu)$ and in general $(M^\Phi)^* = A_\Psi(\mu) \supset L^\Psi(\mu)$, it follows that

$$(M^\Phi)^* \subset (L^\Phi(\mu))^* = L^\Psi(\mu) \subset (M^\Phi)^*.$$

So $M^\Phi = L^\Phi(\mu)$ and similarly $M^\Psi = L^\Psi(\mu)$.

By Theorem III.5.5, the condition $M^\Phi = L^\Phi(\mu)$ is equivalent to the absolute continuity of the norm of $L^\Phi(\mu)$. Similarly $L^\Psi(\mu)$ has the analogous property for its norm. The last statement is immediate. ∎

As a consequence of this result and Theorem III.5.1, we have

Corollary 11. *Let (Φ, Ψ) be a Young complementary pair, and $A_\Psi(\mu)$ be the space of set functions on (Ω, Σ, μ) introduced in Definition 2. Then the separability of $A_\Psi(\mu)$ implies the reflexivity of $L^\Phi(\mu)$ [hence also of $L^\Psi(\mu)$] whenever $L^\Phi(\mu) = M^\Phi$.*

Proof: Since by Theorem 6, $A_\Psi(\mu) = (M^\Phi)^*$, the separability of the first one implies, by a general result in the abstract theory, that M^Φ is separable. But $L^\Psi(\mu) \subset A_\Psi(\mu)$, as noted already, and so $L^\Psi(\mu)$ is separable. By Theorem III.5.1, this implies $L^\Psi(\mu) = M^\Psi$. Since by hypothesis $L^\Phi(\mu) = M^\Phi$, the result follows from the preceding theorem. ∎

We remark that, in general, a separable space need not be reflexive and a reflexive space need not be separable.

Using the local conditions Δ_2 and ∇_2, we can deduce immediately from Theorem II.3.3 and Theorem 10 above the following result. A direct proof is also simple.

Corollary 12. *Let (Ω, Σ, μ) be a finite measure space and Φ be an N-function. Then $L^\Phi(\mu)$ is reflexive iff $\Phi \in \Delta_2 \cap \nabla_2$.*

With this analysis of M^Φ and $\mathcal{M}^\Phi$ spaces, we are ready to consider the adjoint space of $L^\Phi(\mu)$ itself without restrictions. We now turn to this study.

4.2. Continuous linear functionals on L^Φ: general theory

The work in the preceding section does not give much indication about the linear functionals that vanish on $\mathcal{M}^\Phi$ or M^Φ but not on all of $L^\Phi(\mu)$. This is the case when the quotient space $L^\Phi(\mu)/M^\Phi$ [or

$L^{\Phi}(\mu)/\mathcal{M}^{\Phi}$] is not $\{0\}$. When this happens, Theorem III.5.5 shows that such a quotient will be infinite dimensional! In fact, we know N-functions such as $\Phi : x \mapsto e^{x^2} - 1$ for which $(L^{\Phi}(\mu)/M^{\Phi})$ certainly exhibits this phenomenon. We thus develop the required theory to include all these, and many other cases and complete the solution of the problem of characterizing $(L^{\Phi}(\mu))^*$ on general measure spaces.

An important step in the characterization of $(L^{\Phi}(\mu))^*$ is to obtain a detailed structure of the adjoint of the quotient space $L^{\Phi}(\mu)/\mathcal{M}^{\Phi}$ and compare it with $(\mathcal{M}^{\Phi})^*$. Because of the particular space $\mathcal{M}^{\Phi}$ under consideration, it will be seen that $(L^{\Phi}(\mu))^*$ admits a direct sum decomposition into the two spaces mentioned above. From this one obtains the desired representation of $(L^{\Phi}(\mu))^*$. In the following we denote the annihilator of $\mathcal{M}^{\Phi}$ as $(\mathcal{M}^{\Phi})^{\perp}$ which, by definition, is the set of elements of $(L^{\Phi}(\mu))^*$ that vanish on $\mathcal{M}^{\Phi}$. The following statement about $(\mathcal{M}^{\Phi})^{\perp}$ is a consequence of the classical facts on annihilators in general Banach spaces.

Lemma 1. *The quotient space $\mathcal{N}^{\Phi} = L^{\Phi}(\mu)/\mathcal{M}^{\Phi}$ with its norm $d(\cdot)$ given by $d(\hat{f}) = \inf\{N_{\Phi}(f - h) : h \in \mathcal{M}^{\Phi}, \hat{f} = f + \mathcal{M}^{\Phi}\}$ is a Banach space, and the canonical mapping $\lambda : L^{\Phi}(\mu) \to \mathcal{N}^{\Phi}$ takes the open unit ball of $L^{\Phi}(\mu)$ onto the open unit ball of $\mathcal{N}^{\Phi}$. The correspondence $j : x^* \to z^*$ where $x^* \in (\mathcal{M}^{\Phi})^{\perp}$ and $z^* \in (\mathcal{N}^{\Phi})^*$ defined by $z^*(f + \mathcal{M}^{\Phi}) = x^*(f), f \in L^{\Phi}(\mu)$, is an isometric isomorphism of $(\mathcal{M}^{\Phi})^{\perp}$ onto $(\mathcal{N}^{\Phi})^*$, so that $\|j(x^*)\| = \|x^*\|$. [Here $\lambda(f) = \hat{f}, f \in L^{\Phi}(\mu)$.]*

The facts that $d(\cdot)$ is a norm, and if Φ is continuous and $\Phi(x) = 0$ only for $x = 0$, then $d(\cdot)$ remains unchanged for both $N_{\Phi}(\cdot)$ and $\|\cdot\|_{\Phi}$, were proved in Proposition III.5.3. That $\mathcal{N}^{\Phi}$ is a Banach space, and that $j((\mathcal{M}^{\Phi})^{\perp}) = (\mathcal{N}^{\Phi})^*$ are general facts about quotient Banach spaces. [Cf., e.g., Dunford-Schwartz (1958), II.4.18.] There is no specialization here.

Taking into account that $L^{\Phi}(\mu)$, in the real case, is a Banach lattice, we shall discuss the induced order structure of $\mathcal{N}^{\Phi}$ by the canonical mapping λ, and then show how a corresponding order is definable in $(\mathcal{N}^{\Phi})^*$. This will lead us to establish the crucial fact that the norm of $(\mathcal{N}^{\Phi})^*$ is additive on positive elements so that the space becomes an "abstract (L)-space" and a representation of its elements will give us a key element of the representation of $(L^{\Phi}(\mu))^*$. These ideas will now be implemented.

We introduce an order relation in $\mathcal{N}^{\Phi}$ induced by that of $L^{\Phi}(\mu)$. If $\hat{f}, \hat{g}$ are in $\mathcal{N}^{\Phi}$, $\hat{f} = f + \mathcal{M}^{\Phi}$, $\hat{g} = g + \mathcal{M}^{\Phi}$, then let $\hat{f} \leq \hat{g}$ to mean $f_1 \leq g_1$ a.e. for some $f_1 \in \hat{f}$ and $g_1 \in \hat{g}$. Similarly if $z^* \in (\mathcal{N}^{\Phi})^*$ we say that $z^* \geq 0$ iff $z^*(\hat{f}) \geq 0$ for $\hat{f} \geq 0$. With these notations we can establish the following lemma.

Lemma 2. *The space $\{\mathcal{N}^{\Phi}, d(\cdot), \leq\}$ introduced above is a Banach lattice. Moreover $\lambda(\tilde{L}^{\Phi}(\mu))$ $(\subset \mathcal{N}^{\Phi})$ is closed under lattice operations. Further the adjoint space $(\mathcal{N}^{\Phi})^*$ under the induced ordering "$\geq$" is also a Banach lattice, and each $z^* \in (\mathcal{N}^{\Phi})^*$ admits a Jordan type decomposition*

$$z^* = (z^*)^+ - (x^*)^-, (z^*)^+ \wedge (z^*)^- = \inf((z^*)^+, (z^*)^-) = 0, \quad (1)$$

where $(z^*)^+$ and $(z^*)^-$ are positive functionals so that $(z^*)^{\pm}(\hat{f}) \geq 0$ if $\hat{f} \geq 0$.

Proof: To see that "$\leq$" is a partial order in $\mathcal{N}^{\Phi}$, let $\hat{f}, \hat{g}$ be in $\mathcal{N}^{\Phi}$ such that $\hat{f} \leq \hat{g}$ as in the statement. Since "$\leq$" is evidently reflexive, let $\hat{f}, \hat{g}, \hat{h}$ be in $\mathcal{N}^{\Phi}$ such that $\hat{f} \leq \hat{g}$ and $\hat{g} \leq \hat{h}$. To see that $\hat{f} \leq \hat{h}$ must hold, let $f_0 \in \hat{f}$, $g_0, g_1 \in \hat{g}$, and $h \in \hat{h}$ be any representative members. Then by definition $f_0 \leq g_0$ a.e. and $g_1 \leq h$ a.e. If $g_2 = g_1 - g_0$, then again by definition of the coset $\hat{g}$, $g_2 \in \mathcal{M}^{\Phi}$ so that $f_2 = f_0 + g_2 \in \hat{f}$ and $f_2 \leq g_0 + g_2 = g_1 \leq h_1$, a.e. Thus $\hat{f} \leq \hat{h}$. Finally to verify that "$\leq$" is antisymmetric, let $\hat{u}, \hat{v}$ be a pair of elements of $\mathcal{N}^{\Phi}$ such that $\hat{u} \leq \hat{v}$ and $\hat{v} \leq \hat{u}$. To see that $\hat{u} = \hat{v}$, let $u_0, u_1 \in \hat{u}$ and $v_0, v_1 \in \hat{v}$ such that $u_0 \leq v_0$ and $v_1 \leq u_1$ a.e. If $w = v_1 - v_0$ ($\in \mathcal{M}^{\Phi}$), then $u_2 = u_0 + w \leq v_0 + w = v_1 \leq u_1$ a.e. Hence $0 \leq u_1 - u_2 = (u_1 - u_0) - w \in \mathcal{M}^{\Phi}$ and $0 \leq u_1 - v_1 \leq u_1 - u_2 \in \mathcal{M}^{\Phi}$, using the second set of inequalities. But $\mathcal{M}^{\Phi}$ is solid so that $u_1 - v_1 \in \mathcal{M}^{\Phi}$. It follows that u_1 and v_1 are in $\mathcal{M}^{\Phi}$ so that $\hat{u} = \hat{v}$. This shows that "$\leq$" is a partial order in $\mathcal{N}^{\Phi}$. If $\hat{f}, \hat{g} \in \mathcal{N}^{\Phi}$ and $f \in \hat{f}, g \in \hat{g}$, let $h = \sup(f, g)$ and $\hat{h} = h + \mathcal{M}^{\Phi}$. Then $\hat{f} \leq \hat{h}, \hat{g} \leq \hat{h}$ and one verifies that $h \in \mathcal{N}^{\Phi}$, so that $\mathcal{N}^{\Phi}$ is a vector lattice and under the norm $d(\cdot)$ which is monotone under this ordering (and Proposition III.5.3). Similarly $\inf(\hat{f}, \hat{g}) \in \mathcal{N}^{\Phi}$ and that the operations 'inf' and 'sup' (denoted '$\wedge$' and '$\vee$') are continuous under $d(\cdot)$. It follows from this and Lemma 1 that $\mathcal{N}^{\Phi}$ is a Banach lattice.

The general theory of Banach lattices shows that the ordering of $\mathcal{N}^{\Phi}$ induces a corresponding one in $(\mathcal{N}^{\Phi})^*$ as given in the statement. Moreover, under that ordering the adjoint space is also a Banach lattice.

But the decomposition (1) is again a general property of such spaces so that all the statements hold, since $\mathcal{N}^\Phi$ is shown to be a Banach lattice above. ∎

Another important aspect of $(\mathcal{N}^\Phi)^*$ is given by the following result, if we set $\hat{f}_E = f_E + \mathcal{M}^\Phi$ where $f_E = f\chi_E$, $E \in \Sigma$, and let z_E^* be defined by $z_E^*(\hat{f}) = z^*(\hat{f}_E)$, $\hat{f} \in \mathcal{N}^\Phi$ and $z^* \in (\mathcal{N}^\Phi)^*$.

Lemma 3. *Let $z^* \in (\mathcal{N}^\Phi)^*$ and $|z^*| = (z^*)^+ + (z^*)^-$, where $(z^*)^\pm$ are as in (1). Then the norms of z^* and $|z^*|$ are equal and*

$$\|z^*\| = \|(z^*)^+\| + \|(z^*)^-\| \quad (= \sup\{|z^*|(\hat{f}) : 0 \le \hat{f} \in \lambda(\tilde{L}^\Phi)\}). \quad (2)$$

More generally if $z_i^* \in (\mathcal{N}^\Phi)^*$, $i = 1, 2$ are any positive elements, then

$$\|z_1^* + z_2^*\| = \|z_1^*\| + \|z_2^*\|, \quad (3)$$

where the norm is given by (2). Further $\|z_{(\cdot)}^*\| : \Sigma \to \mathbb{R}^+$ is a purely finitely additive and bounded function in that if $0 \le \xi(\cdot) \le \|z_{(\cdot)}^*\|$ and $\xi(\cdot)$ is σ-additive, then $\xi \equiv 0$.

Proof: The equality between the various terms of (2) holds in any Banach lattice if $\lambda(\tilde{L}^\Phi(\mu))$ is replaced by its closed or open unit ball. Since $\lambda(\tilde{L}^\Phi(\mu))$ contains the open unit ball of $\mathcal{N}^\Phi$ and is contained in its closed unit ball, by Proposition III.5.4, it thus follows that the norm of $(\mathcal{N}^\Phi)^*$ is determined by the elements of $\lambda(\tilde{L}^\Phi(\mu))$. [This should be compared with the last part of Proposition III.5.3 also.] Since the middle equality in (2) is a special case of (3), let us establish the latter. For this, by the subadditivity of the norm, it suffices to show the superadditivity, when $z_i^* \ge 0$ and $\ne 0$.

Since by definition $\|z_i^*\| = \sup\{z_i^*(\hat{f}) : \hat{0} \le \hat{f} \in \lambda(\tilde{L}^\Phi(\mu))\} < \infty$, we have, for each $\varepsilon > 0$, elements $\hat{0} \le \hat{f}_i \in \lambda(\tilde{L}^\Phi(\mu))$, such that $\|z_i^*\| < z_i^*(\hat{f}_i) + \frac{\varepsilon}{2}$, $i = 1, 2$. By the preceding lemma $\lambda(\tilde{L}^\Phi(\mu))$ is a lattice so that $\hat{h} = \hat{f}_1 \vee \hat{f}_2 \in \lambda(\tilde{L}^\Phi(\mu))$ and hence

$$\|z_1^*\| + \|z_2^*\| - \varepsilon \le z_1^*(\hat{f}_1) + z_2^*(\hat{f}_2) \le (z_1^* + z_2^*)(\hat{h}) \le \|z_1^* + z_2^*\|.$$

This implies (3), since $\varepsilon > 0$ is arbitrary.

For the last part, because of (2), we may assume $z^* \ge 0$. Let $A_i \in \Sigma$ be disjoint and $\Omega = A_1 \cup A_2$. Since $z^* = z_{A_1 \cup A_2}^*$, by definition, we have

$$\|z^*\| = \|z_{A_1}^* + z_{A_2}^*\| = \|z_{A_1}^*\| + \|z_{A_2}^*\|,$$

by (3). Hence if $\nu(A) = \|z_A^*\|$, then $\nu : \Sigma \to I\!\!R^+$ is a bounded additive set function, and if $\mu(A) = 0$, then $\|z_A^*\| = 0$ so that $\nu(A) = 0$ also. We claim that $\nu(\cdot)$ is in fact purely finitely additive. Indeed, given $\varepsilon > 0$, there is an $\hat{0} \leq \hat{f}_0 \in \mathcal{N}^\Phi$ depending on ε, such that $|d(\hat{f}_0) - 1| < \varepsilon$ and $z^*(\hat{f}_0) = \|z^*\|$, as a consequence of the Hahn-Banach theorem. If $\tilde{\nu}(E) = z^*(\hat{f}_{0E}) = z^*(\hat{f}_0)$, then $\tilde{\nu}$ is a bounded additive set function and $\nu(\Omega) = \tilde{\nu}(\Omega)$, as above. Also $\tilde{\nu}(E) \leq \nu(E)(1 + \varepsilon)$ for all $E \in \Sigma$, so that $|\nu - \tilde{\nu}|(\Omega) < \varepsilon$. Now if $f_0 \in \lambda^{-1}(\hat{f}_0)$, λ being the canonical map, and $E_n = \{\omega : f_0(\omega) \geq n\} \in \Sigma$, then $E_n \downarrow \emptyset$ and $f_0 - f_{0E_n} \in \mathcal{M}^\Phi$ so that $z^*(\hat{f}_0) = z^*(\hat{f}_{0E_n})$ all n. Hence $\tilde{\nu}(E_n) = \tilde{\nu}(\Omega)$, $n \geq 1$, and so $\tilde{\nu}$ is purely finitely additive. However, as is well known, the set of purely finitely additive set functions forms a closed linear subspace of all bounded additive set functions on (Ω, Σ, μ). Since $|\nu - \tilde{\nu}|(\Omega) < \varepsilon$, it then follows that ν also must be purely finitely additive. ∎

A number of consequences of the preceding result will be recorded and they help motivate the ensuing discussion and development.

Remarks. 1. For $\tilde{\nu}$ and $\hat{f}_0 = f_0 + \mathcal{M}^\Phi$ in the preceding proof, let $E = \{\omega : |f_0(\omega)| \leq x\}$. Then $\tilde{\nu}(E \cap F) = z^*(\hat{f}_{0E \cap F}) = 0$ for every $0 \leq x < \infty$ and $F \in \Sigma$ since $f_{0E \cap F} \in \mathcal{M}^\Phi$.

2. If Φ jumps to $+\infty$ at some $x = x_0$, then $\mathcal{M}^\Phi = L^\Phi(\mu)$ so that $\mathcal{N}^\Phi = \{0\}$. Thus the latter is nontrivial only if $\Phi(x) < \infty$ for all $x < \infty$ and Φ does not satisfy any growth conditions such as the Δ_2-regularity. Since then for $f \in L^\Phi(\mu) - \mathcal{M}^\Phi$, $\Phi(\frac{f}{k})$ is integrable for some $k > 0$, we deduce that $E_a = \{\omega : |f(\omega)| > a\}$ has finite μ-measure for large enough a, and $\mu(E_a) \to 0$ as $a \to \infty$. In particular this is true for the f_0 of 1. above, since $d(\hat{f}_0) \geq 1$, $\frac{1}{k}f_0 \in \tilde{L}^\Phi(\mu)$ for some $k \geq 1$. Hence by the structure theorem there exist a sequence of step functions $f_n \to f_0$ pointwise and satisfying $|f_0 - f_n| \leq |f_0|$. So $f_n \in \mathcal{M}^\Phi$ and $\int_\Omega \Phi(\frac{f_0 - f_n}{k})d\mu \to 0$ on $n \to \infty$, by the dominated convergence. Thus given $\varepsilon > 0$, there exists a $g \in \mathcal{M}^\Phi$ ($g = f_{n_0}$ for some n_0) such that if $h = \frac{1}{k}(f_0 - g)$, then $\int_\Omega \Phi(h)d\mu < \varepsilon$. Therefore the set function ν ($= \tilde{\nu}$) corresponding to $z^* \in (\mathcal{N}^\Phi)^*$, in the nontrivial case, (i.e., $z^* \neq 0$) satisfies $\nu(E) = z^*(\hat{f}_{0E}) = z^*(\hat{h})$ where $E = \{\omega : |h(\omega)| > 0\}$. *It follows that the support of ν lies in the support of $h \in \tilde{L}^\Phi(\mu) - \mathcal{M}^\Phi$, which is of arbitrarily small μ-measure.* This fact is important in the characterization of $(\mathcal{N}^\Phi)^*$.

In representing the functionals of $(\mathcal{N}^\Phi)^*$, we first show that $\mathcal{N}^\Phi$ can be mapped isometrically onto a subspace of $L^\infty(\mu)$ and then obtain integral representations of the elements of $(\mathcal{N}^\Phi)^*$ in terms of the finitely additive set functions ν of the preceding lemma. Here is the outline of the desired construction, essentially following Andô (1960) in an important case, and especially Rao (1968), for the general conditions that we are using in this book.

Let $\hat{0} \leq \hat{f} \, (= f + \mathcal{M}^\Phi \,) \in \mathcal{N}^\Phi$ and $\beta = d(\hat{f}) > 0$. Then there is a $g \in \mathcal{M}^\Phi$ such that $f - g \geq 0$ and $f - g \in \hat{f}$. One may assume $f > 0$ a.e. If $\pi = (E_1, \dots, E_n)$ is a partition of Ω, let $f_E = f\chi_E$ as before and set $a_k = d(\hat{f}_{E_k}) \leq \beta$. For each $0 \leq \nu \in ba(\Omega, \Sigma, \mu)$, the space of finitely additive scalar set functions vanishing on μ-null sets with total variation as norm, consider $\sum_{k=1}^n a_k \nu(E_k)$. As the partitions are refined, these sums form a monotone bounded sequence of positive numbers converging to a limit, denoted by $\int_\Omega f d\nu$. Set $\tilde{f}_\pi = \sum_{i=1}^n a_i \chi_{E_i}$ so that $\tilde{f}_\pi \in L^\infty(\mu)$ and $\tilde{f}_\pi = 0$ iff $f \in \mathcal{M}^\Phi$. Further $\|\tilde{f}_\pi\|_\infty = \max_{1 \leq i \leq n}(a_i) = d(\hat{f})$, by Remark 2 above. Thus the mapping $\hat{f} \mapsto \tilde{f}_\pi$ is an isometry of $\mathcal{N}^\Phi$ on a subspace of $L^\infty(\mu)$ determined by such functions $\tilde{f}_\pi$. Now define a functional x^*:

$$x^*(\hat{f}) = \int_\Omega f d\nu, \qquad \hat{0} \leq \hat{f} \in \mathcal{N}^\Phi, 0 \leq f \in \hat{f}. \qquad (4)$$

Using standard arguments in abstract analysis and integration, one can verify that x^* is well defined and the integral is linear in ν. We actually can assert more as follows.

Lemma 4. *The functional x^* of (4) is in $(\mathcal{N}^\Phi)^*$, and for each $0 \leq \hat{f} \in \mathcal{N}^\Phi$, one has $0 \leq x^*(\hat{f}) \leq d(\hat{f})\nu(\Omega)$.*

Proof: From the definition of the integral we get $0 \leq x^*(\hat{f}) \leq \|\tilde{f}_\pi\|_\infty \nu(\Omega) = d(\hat{f})\nu(\Omega) < \infty$, for each $\hat{f} \geq 0$. Hence x^* will be in $(\mathcal{N}^\Phi)^*$ if it is shown that it is linear. Since $\widehat{af} = a\hat{f}$, it is clear that $x^*(a\hat{f}) = ax^*(\hat{f})$ for all $a \geq 0$. From the fact that $d(\cdot)$ is a lattice norm on $\mathcal{N}^\Phi$, and from the definition of (4), we get for $\hat{f}_i \geq \hat{0}$ in $\mathcal{N}^\Phi$, $x^*(\hat{f}_1 + \hat{f}_2) \leq x^*(\hat{f}_1) + x^*(\hat{f}_2)$. It is asserted that the opposite inequality also holds so that the additivity of x^* follows.

Let $0 \leq f_i \in \hat{f}_i$ such that f_1, f_2 have disjoint supports. Then from the definition of (4) it follows easily that $x^*(\hat{f}_1 + \hat{f}_2) = x^*(\hat{f}_1) + x^*(\hat{f}_2)$.

If $\hat{f}_i \geq \hat{0}$ are such that $f_i \in \hat{f}_i$ and $0 < f_1 \leq f_2$ a.e., so that $\hat{f}_1 \leq \hat{f}_2$, $0 \leq x^*(\hat{f}_1) \leq x^*(\hat{f}_2)$, and $\frac{f_1}{f_2} \in L^\infty(\mu)$ we may approximate this bounded function. For each $\varepsilon > 0$ there exists an $f_\varepsilon = \sum_{i=1}^{n} \alpha_i \chi_{E_i}$, $\alpha_i \geq 0$, E_i disjoint in Σ, such that $0 \leq \frac{f_1}{f_2} - f_\varepsilon < \frac{\varepsilon}{k}$ where $k = x^*(\hat{f}_2) > 0$. Then this can be written, with $E_0 = \Omega - \bigcup_{i=1}^{n} E_i$ and $\alpha_0 = 0$, as

$$f_\varepsilon \cdot f_2 \leq \left(\sum_{i=0}^{n} \alpha_i \chi_{E_i} + \frac{\varepsilon}{k} \right) f_2 \tag{5}$$

and hence using the result on the additivity of (4) for functions with disjoint supports

$$x^*(\hat{f}_2) + x^*(\hat{f}_1) \leq \sum_{i=0}^{n} x^*(\hat{f}_2 \chi_{E_i}) + \sum_{i=0}^{n} (\alpha_i + \frac{\varepsilon}{k}) x^*(\hat{f}_2 \chi_{E_i})$$

$$= \sum_{i=0}^{n} (1 + \alpha_i) x^*(\hat{f}_2 \chi_{E_i}) + \frac{\varepsilon}{k} x^*(\hat{f}_2)$$

$$= x^* \left(\sum_{i=0}^{n} (1 + \alpha_i) \hat{f}_2 \chi_{E_i} \right) + \frac{\varepsilon}{k} x^*(\hat{f}_2)$$

$$\leq x^*(\hat{f}_2 + \hat{f}_1) + \varepsilon, \quad \text{by (5)}.$$

Since $\varepsilon > 0$ is arbitrary, we get with the earlier case, $x^*(\hat{f}_1) + x^*(\hat{f}_2) = x^*(\hat{f}_1 + \hat{f}_2)$. If $0 \leq f_i \in \hat{f}_i$ are arbitrary we consider the result on $\Omega_1 = \{\omega : f_1(\omega) \leq f_2(\omega)\}$ and on $\Omega_2 = \Omega - \Omega_1$, use the above result and add. This gives the lemma. ∎

Since $\mathcal{N}^\Phi$ and its adjoint are vector lattices, one may decompose each of its elements into a linear combination of positive elements and apply the above result. This then extends the definition of (4) to all elements of $\mathcal{N}^\Phi$ and of $(\mathcal{N}^\Phi)^*$ for which (4) defines a bounded linear functional. We use the same symbol in the general case also.

For the characterization of $(\mathcal{N}^\Phi)^*$, one needs to select a suitable subclass of elements of $ba(\Omega, \Sigma, \mu)$. This is isolated in the following:

Definition 5. If (Φ, Ψ) is a complementary Young pair, then let $B_\Psi(\mu) = \{\nu \in ba(\Omega, \Sigma, \mu): \text{the support of } \nu \text{ is contained in some } f \in \tilde{L}^\Phi(\mu) - \mathcal{M}^\Phi\}$. For each $\nu \in B_\Psi(\mu)$, let $|\nu|(\Omega) = $ total variation of ν on Ω.

It can be verified that $(B_\Psi(\mu), |\cdot|)$ is a normed linear space. (Linearity is clear.) However, the following result implies that it is identifiable isometrically with $(\mathcal{N}^\Phi)^*$ so that it is a Banach space.

Theorem 6. *The space $(\mathcal{N}^\Phi)^*$ and $B_\Psi(\mu)$, with their respective norms, are isometrically isomorphic. The correspondence is given by*

$$x^*(\hat{f}) = \int_\Omega f\, d\nu, \qquad f \in \hat{f} \in \mathcal{N}^\Phi, x^* \in (\mathcal{N}^\Phi)^* \tag{6}$$

and

$$\|x^*\| = |\nu|(\Omega). \tag{7}$$

Here the integral in (6) is that defined for (4).

Proof: Since every element ν of $B_\Psi(\mu)$ and x^* of $(\mathcal{N}^\Phi)^*$ can be expressed as a linear combination of nonnegative elements [because $(\mathcal{N}^\Phi)^*$ is a Banach lattice and so is $ba(\Omega, \Sigma, \mu)$ when real elements are considered], we will assume that $\nu \geq 0$ in (6) for convenience. Then by Lemma 4, (6) defines a nonnegative x^* of $(\mathcal{N}^\Phi)^*$, and $\|x^*\| \leq \nu(\Omega)$. To see that (7) holds, we need to show that there is equality here. Let $\pi = \{E_1, \ldots, E_n\}$ be a partition of Ω ($E_i \in \Sigma$, and disjoint) such that $\nu(E_i) > 0$, $1 \leq i \leq n$. Then there exist $0 \leq f_i \in \tilde{L}^\Phi(\mu) - \mathcal{M}^\Phi$ with supports E_i, by the definition of elements of $B_\Psi(\mu)$. If $f = f_1 + \cdots + f_n$ $(= \max_{1 \leq i \leq n}(f_i))$, then $f \in \tilde{L}^\Phi(\mu) - \mathcal{M}^\Phi$, $d(\hat{f}_{E_i}) = 1$, and $d(\hat{f}) = \max_{1 \leq i \leq n}(d(\hat{f}_i)) = 1$. [This is a property of any "abstract (M)-space," and $\mathcal{N}^\Phi$ can be identified as a closed subspace of the adjoint of $(\mathcal{N}^\Phi)^*$ in which the norms of nonnegative terms are additive by Lemma 3. Thus $(\mathcal{N}^\Phi)^*$ is an abstract (L)-space and its adjoint is an abstract (M)-space. These are consequences of a general theory of such spaces established by S. Kakutani (1941). Except for the above fact, we do not need the general results.] Thus $\nu(\Omega) = \sum_{i=1}^n d(\hat{f}_{E_i})$ and refining the partitions on the left side one gets, by definition of the integral in (4), $x^*(\hat{f}) = \nu(\Omega)$. This and the inequality of Lemma 4 imply that $\|x^*\| = \nu(\Omega)$ and hence (7) is true.

In the reverse direction, let $x^* \in (\mathcal{N}^\Phi)^*$ and it suffices to consider $x^* \geq 0$, as noted at the beginning. Then define $\nu : E \mapsto x^*(\hat{f}_E)$, $\hat{f} \in \mathcal{N}^\Phi$, $d(\hat{f}) = 1$, such that $x^*(\hat{f}) = \|x^*\|$, i.e., $\nu(E) = \|x_E^*\|$, as was defined in the proof of Lemma 3 above. Then $\nu \in B_\Psi(\mu)$ as seen in that proof and discussed in the two remarks that followed it. Let $\tilde{x}^*$ be a functional defined by (4) for this ν. Then $\tilde{x}^* \geq 0$. It is to be shown that $x^* = \tilde{x}^*$. By definition of the integral (4), we have for partitions π of Ω,

$$\tilde{x}^*(\hat{f}) = \lim_\pi \sum_{i=1}^n d(\hat{f}_{E_i})\nu(E_i) = \lim_\pi \sum_{i=1}^n d(\hat{f}_{E_i})\|x^*_{E_i}\|$$
$$\geq x^*(\hat{f}), \qquad 0 \leq \hat{f} \in \mathcal{N}^\Phi, d(\hat{f}) \leq 1 \qquad (8)$$

Hence $x^* \leq \tilde{x}^*$. This implies

$$x^*(\hat{f}) \leq \tilde{x}^*(\hat{f}) \leq d(\hat{f})\nu(\Omega) \leq d(\hat{f})\|x^*\|, \qquad 0 \leq \hat{f} \in \mathcal{N}^\Phi. \qquad (9)$$

Taking the suprema on the left as $\hat{f}$ varies over $\lambda(\tilde{L}^\Phi(\mu))$, we get equality throughout so that $\|x^*\| = \|\tilde{x}^*\|$. Since $\tilde{x}^* - x^* \geq 0$, we have

$$\|\tilde{x}^*\| = \|(\tilde{x}^* - x^*) + x^*\| = \|\tilde{x}^* - x^*\| + \|x^*\|, \qquad \text{by (3),}$$

so that $\|\tilde{x}^* - x^*\| = \|\tilde{x}^*\| - \|x^*\| = 0$. Consequently $x^* = \tilde{x}^*$ for positive elements and hence in general. Thus representation (6) holds, and by the preceding paragraph (7) must then also be true.

Regarding uniqueness, if ν_1, ν_2 in $B_\Psi(\mu)$ represent x^*, then by the argument in the first paragraph and the linearity of $B_\Psi(\mu)$, we get $\|0\| = |\nu_1 - \nu_2|(\Omega)$ so that $\nu_1 = \nu_2$. Hence $(\mathcal{N}^\Phi)^*$ is identifiable isometrically with $B_\Psi(\mu)$ as desired. ∎

As a consequence of the representation (6), we have

Corollary 7. *For each $x^* \in (\mathcal{N}^\Phi)^*$ there correspond a $\nu \in B_\Psi(\mu)$ and an $\hat{f} \in \mathcal{N}^\Phi$ such that $d(\hat{f}) = 1$ and for each $E \in \Sigma$,*

$$\text{(a)} \quad x^*(\hat{f}_E) = x^*_E(\hat{f}) = \nu(E), \quad and \quad \text{(b)} \quad \|x^*_E\| = |\nu|(E). \qquad (10)$$

Proof: This follows from the fact that $\|x^*\| = x^*(\hat{f}), d(\hat{f}) = 1$ by the Hahn-Banach theorem, and then (a) and (b) can be deduced from (6) and (7). In fact $x^*(\hat{f}) = \int_\Omega f d\nu$. By (7), for $x^* > 0$, $x^*(\hat{f}) = \nu(\Omega) = \|x^*\|$. Here $\hat{f}$ need not be unique but $d(\hat{f}) = 1, \hat{f} \geq \hat{0}$. If $E \in \Sigma$, then $x^*_E(\hat{f}) = x^*(\hat{f}_E)$ defines $x^*_E \in (\mathcal{N}^\Phi)^*$. If $\nu_E(\cdot) = \nu(E \cap \cdot)$, the restriction, and if $\tilde{x}^*$ is defined by (6) for this ν_E, then $\|\tilde{x}\| = \nu_E(\Omega) = \nu(E)$. The same argument then shows that $\tilde{x}^* = x_E$ so that $\|x_E^*\| = \nu(E) = x^*(\hat{f}_E)$, as asserted. ∎

To present a representation of the elements of $(L^\Phi(\mu))^*$, we need finally a decomposition of its elements as given by the following.

Proposition 8. *Every element* $x^* \in (L^\Phi(\mu))^*$, *can be expressed uniquely as* $x^* = y^* + z^*$ *where* $y^* \in (\mathcal{M}^\Phi)^*$ *and* $z^* \in (\mathcal{M}^\Phi)^\perp$ *so that* $|y^*| \wedge |z^*| = \inf(|y^*|, |z^*|) = 0$. *Here* $|y^*| = (y^*)^+ + (y^*)^-$ *and similarly* $|z^*|$.

Proof: This result is a consequence of general results in normed vector lattices. However, we can present a simple proof using the work we have already completed. Indeed let Σ_1 be the δ-ring of sets of finite μ-measure in Σ. Define $G : \Sigma_1 \to \mathbb{R}$ by the relation $G(E) = x^*(\chi_E)$. Then by Theorem 1.6, $G \in A_\Psi(\mu)$. But as shown in part (i) of the proof of that theorem, $y^*(f) = \int_\Omega f dG$ is defined for all $f \in L^\Phi(\mu)$ and $y^* \in (L^\Phi(\mu))^*$. Restricting y^* to $\mathcal{M}^\Phi$, it is clear that $y^* \in (\mathcal{M}^\Phi)^*$, and if $z^* = x^* - y^*$, then $z^*(\mathcal{M}^\Phi) = 0$, i.e., $z^* \in (\mathcal{M}^\Phi)^\perp$, and is a continuous linear functional. Hence $x^* = y^* + z^*$ is a decomposition of the required type.

Regarding uniqueness, let $x^* = y_1^* + z_1^*$ be another such decomposition so that $y_1^*(f) = \int_\Omega f dG_1$, $f \in \mathcal{M}^\Phi$ and $G_1 \in A_\Psi(\mu)$. Then $y^* - y_1^* = z_1^* - z^*$ and hence

$$y^*(f) - y_1^*(f) = \int_\Omega f d(G - G_1) = z_1^*(f) - z^*(f) = 0.$$

Since $G - G_1 \in A_\Psi(\mu)$, Theorem 1.6 implies $\|0\| = \|G - G_1\|_\Psi$ so that $G = G_1$. Hence $y^* = y_1^*$ and then $z^* = z_1^*$. So the decomposition is unique. ∎

Note that if Φ is discontinuous, then $\mathcal{M}^\Phi = L^\Phi(\mu)$ is possible so that $z^* = 0$. In this event the representing measure G itself can be purely finitely additive and so in some cases the elements of $(\mathcal{M}^\Phi)^*$ may be further decomposed into "absolutely continuous" and "singular" types. We discuss this matter in the next section in more detail. However, if Φ is continuous but μ is unrestricted, then the above result takes a simpler form.

Corollary 9. *Let* (Φ, Ψ) *be a complementary N-pair. Then each element* x^* *of* $(L^\Phi(\mu))^*$ *can be expressed uniquely as*

$$x^*(f) = \int_\Omega fg d\mu + z^*(f) = y^*(f) + z^*(f), \quad f \in L^\Phi(\mu), \quad (11)$$

for a unique $g \in L^\Psi(\mu)$ and $z^* \in (M^\Phi)^\perp$, so $y^* \in (M^\Phi)^*$.

The representation for y^* given in (11) is now a consequence of the last part of Theorem 1.6, since now both Φ and Ψ are continuous.

Let us now introduce the following space of set functions.

Definition 10. Let $\mathcal{A}_\Phi(\mu) = A_\Phi(\mu) \oplus B_\Phi(\mu)$, a direct sum, with norm

$$N'_\Phi(G) = N_\Phi(G_1) + |\nu_1|(\Omega) \quad (12)$$

where $G = G_1 + \nu_1$, $G_1 \in A_\Phi(\mu), \nu_1 \in B_\Phi(\mu)$; these are given in Definitions 1.2 and 5 of this section. An equivalent norm is given by

$$\|G\|'_\Phi = \|G_1\|_\Phi + |\nu_1|(\Omega), \quad G = G_1 + \nu_1. \quad (13)$$

With all the preceding work we can present the main result of this section in the following theorem.

Theorem 11. *Let (Φ, Ψ) be a complementary pair of Young functions, $L^\Phi(\mu)$ and $\mathcal{A}_\Psi(\mu)$ be the corresponding spaces defined already on a general measure space (Ω, Σ, μ). Then $(L^\Phi(\mu))^*$ and $\mathcal{A}_\Psi(\mu)$ are isometrically isomorphic. More explicitly, for each $x^* \in (L^\Phi(\mu))^*$ there corresponds a unique $G \in \mathcal{A}_\Psi(\mu)$ $(G = G_1 + \nu_1 \in A_\Phi(\mu) \oplus B_\Phi(\mu))$ such that when $L^\Phi(\mu)$ is given the gauge norm,*

$$x^*(f) = \int_\Omega f dG \quad (= \int_\Omega f dG_1 + \int_\Omega f d\nu), \quad f \in L^\Phi(\mu), \quad (14)$$

and

$$\|x^*\| = \|G\|'_\Psi \quad (= \|G_1\|_\Psi + |\nu_1|(\Omega)) \quad (15)$$

where $\|G_1\|_\Psi = \sup\{|\int_\Omega f dG_1| : N_\Phi(f) \le 1\}$, $|\nu_1|(\Omega)$ is the total variation norm of ν_1. Here the integrals are defined for additive set functions as stated just preceding Proposition 1.3, and in (4) above. [If $L^\Phi(\mu)$ is given the Orlicz norm, then in (15) we shall have $N'_\Psi(G)$.]

The result is an immediate consequence of Proposition 8, Theorem 1.6 and Theorem 6 above.

Using Proposition III.4.3 and Corollary 9 above, we can record the following consequence of the above theorem for reference.

Corollary 12. *If* (Ω, Σ, μ) *is a finite measure space, then* (14) *and* (15) *are expressible for each* $x^* \in (L^\Phi(\mu))^*$, *uniquely as:*

$$x^*(f) = \int_\Omega f g d\mu + \int_\Omega f d\nu_0 + \int_\Omega f d\nu_1 \qquad f \in L^\Phi(\mu), \qquad (16)$$

and, with the gauge norm for $L^\Phi(\mu)$ *and Orlicz norm for* $L^\Psi(\mu)$,

$$\|x^*\| = \|g\|_\Psi + |\nu_0|(\Omega) + |\nu_1|(\Omega_1) \qquad (17)$$

where $g \in L^\Psi(\mu), \nu_1 \in B_\Psi(\mu)$ *and* ν_0 *is a purely finitely additive set function. Further,* $\nu_0 = 0$ *if* Φ *is continuous with* $\Phi(x) = 0$ *for* $x > 0$, *and* $\nu_1 = 0$ *if* Φ *is discontinuous.*

Proof: Since $\mu(\Omega) < \infty$, it follows from the support line property of a convex function that $A_\Psi(\mu) \subset ba(\Omega, \Sigma, \mu)$. Then by a classical theorem of Yosida and Hewitt (cf., e.g., Rao (1987), p. 182), each G_1 of $A_\Psi(\mu)$ can be uniquely decomposed into a σ-additive G_1' and a purely finitely additive ν_0, $G_1 = G_1' + \nu_0$. Since G_1 is μ-continuous, by the Radon-Nikodým theorem, $dG_1' = gd\mu$ and the fact that G_1' is in $A_\Psi(\mu)$ implies $g \in L_\Psi(\mu)$ as a simple consequence of Proposition 1.3. Hence (16) and (17) follow from (14) and (15). Note also that if Φ is discontinuous, then $\mathcal{M}^\Phi = L^\Phi(\mu)$ so that $\mathcal{N}^\Phi = \{0\}$ and $\nu_1 = 0$; and if Φ is continuous and $\Phi(x) > 0$ for $x > 0$, then $\mathcal{M}^\Phi = M^\Phi$. So by Lemma 1.4, G is μ-continuous and additive, hence it is σ-additive. Thus $\nu_0 = 0$ in this case as asserted. ∎

In some cases we can give conditions in order for an x^* to be in $(\mathcal{M}^\Phi)^\perp$. Observe that the norm of $x^* \in (L^\Phi(\mu))^*$ has two equivalent expressions accordingly as the equivalent gauge or Orlicz norms of $L^\Phi(\mu)$ are used. As is expected from Proposition III.5.3, both these norms of x^* in $(\mathcal{N}^\Phi)^*$ $(= (\mathcal{M}^\Phi)^\perp)$ will be the same. More precisely, we can state the following proposition.

Proposition 13. *Let* $x^* \in (L^\Phi(\mu))^*$, $\mathcal{N}^\Phi \neq \{0\}$ *and either* μ *be localizable (or* σ-*finite) or* $L^\Psi(\mu)$ *have an absolutely continuous norm (i.e.,* $M^\Psi = L^\Psi(\mu)$ *). If the complementary Young functions* (Φ, Ψ) *are continuous, then* $x^* \in (\mathcal{N}^\Phi)^*$ $(= (\mathcal{M}^\Phi)^\perp)$ *whenever* $\|x^*\|' = \|x^*\|$, *i.e.,*

$$\sup\{|x^*(f)| : \|f\|_\Phi \leq 1\} = \sup\{|x^*(f)| : N_\Phi(f) \leq 1\}. \qquad (18)$$

In the other direction, if $x^* \in (\mathcal{M}^\Phi)^\perp$, then (18) always holds.

Proof: Since by Proposition 8, when $x^* \in (\mathcal{M}^\Phi)^\perp$, so that $x^*(\mathcal{M}^\Phi) = 0$, we can apply Theorem 6 from which it follows that $\|x^*\| = |\nu|(\Omega)$ and this is the value obtained by using either side of (18) so that (18) holds. Note that if $\Phi(x) > 0$ for $x > 0$ and Φ is continuous, then Proposition III.5.3 implies (18) without any restrictions on μ. Since $\mathcal{N}^\Phi \neq \{0\}$, it follows that Φ is positive and continuous on an interval, so that in (14) and (15) which are valid in all cases only the last integral is nonvanishing and (18) holds always because no special use of either of the norms is used in the isometry (7).

For the converse, let (18) hold and assume the conditions there. If $x^* = y^* + z^*$ is the decomposition given by Proposition 8 in order for $z^* \in (\mathcal{M}^\Phi)^\perp$, then $\|x^*\| = \|x^*\|'$ implies $\|y^*\| = \|y^*\|'$. Since Φ is continuous, by Lemma 1.4, the representing measure G_1 of y^* is μ-continuous. But by the additional hypotheses on μ and on (Φ, Ψ) imply that G_1 is σ-additive and $\frac{dG_1}{d\mu} = g_1 \in L^\Psi(\mu)$. Then the assumed equality $\|y^*\| = \|y^*\|'$ now implies that $\|g_1\|_\Psi = N_\Psi(g_1)$. Since Ψ is continuous, this can hold only when $g_1 = 0$ a.e. by Proposition III.3.17. Hence $y^* = 0$ and $x^* = z^*$, as asserted. ∎

The restriction on μ was needed in deducing that $y^*(f) = \int_\Omega f dG_1$ $= \int_\Omega f g_1 d\mu$. However if (Φ, Ψ) are N-functions, this last representation holds by the second part of Theorem 1.6. Hence the preceding result takes the following symmetric form.

Proposition 14. *Let (Φ, Ψ) be a complementary pair of N-functions and $L^\Phi(\mu)$ be the corresponding Orlicz space on a measure space (Ω, Σ, μ). Then an x^* of $(L^\Phi(\mu))^*$ belongs to $(\mathcal{M}^\Phi)^\perp$ iff (18) holds.*

The proof is the same as the above one with slight simplifications and the details will not be repeated. The following companion of the above result has an interest for our study.

Proposition 15. *Let (Φ, Ψ), $L^\Phi(\mu)$, and M^Φ be as in the above proposition. Then every $x^* \in (M^\Phi)^*$ has a unique Hahn-Banach (i.e., norm preserving) extension to all of $L^\Phi(\mu)$.*

Proof: Let $x^* \in (M^\Phi)^*$. Then by the second part of Theorem 1.6,

$$x^*(f) = \int_\Omega f g d\mu, \qquad g \in L^\Psi(\mu), f \in M^\Phi, \qquad (19)$$

for a unique g. But the integral in (19) is well defined for all $f \in M^\Phi$ (even $L^\Phi(\mu)$) and by the Hölder inequality, it defines an $x_1^* \in (L^\Phi(\mu))^*$ such that $\|x^*\| = \|x_1^*\| = \|g\|_\Psi$. Let x_2^* be any other Hahn-Banach extension of x^* to $L^\Phi(\mu)$. Then $\|x^*\| = \|x_2^*\|$, and $x_2^*(f) = x^*(f)$ ($= x_1^*(f)$), $f \in M^\Phi$. So if $z_1^* = x_2^* - x_1^*$, then $z_1^* \in (M^\Phi)^\perp$ and $x_2^* = x_1^* + z_1^*$. Thus x_2^* admits a decomposition (must be unique) guaranteed by Proposition 8. Hence by Theorem 11 and the preceding condition one has

$$\|x^*\| = \|x_2^*\| = \|x_1^*\| + |\nu|(\Omega) = \|x^*\| + |\nu|(\Omega), \qquad (20)$$

where ν represents z_1^*. But (20) shows that $\nu = 0$. Hence $x_1^* = x_2^*$. ∎

Both the representation theory as well as its consequences of the preceding sections depended on the subspace $\mathcal{M}^\Phi$ and the quotient space $\mathcal{N}^\Phi$. We introduced M^Φ and N^Φ also. If $\Phi(x) = 0$ iff $x = 0$ and Φ is continuous, these satisfy $\mathcal{M}^\Phi = M^\Phi$ and hence $\mathcal{N}^\Phi = N^\Phi$. In the general theory $M^\Phi = \{0\}$ is possible so that one has to consider several cases to characterize $(L^\Phi(\mu))^*$. To have a unified treatment we used $\mathcal{M}^\Phi$. The price paid for this is that $\mathcal{M}^\Phi = L^\infty(\mu)$ is possible and hence integration relative to finitely additive set functions has to be considered at the early stages of the theory. There does not seem to be a shortcut. A second point to note is that the two norms in $L^\Phi(\mu)$ are equivalent but seldom equal. If $L^\Phi(\mu)$ is $L^p(\mu)$ then both norms agree. However, *when* (Φ, Ψ) *are not normalized, our definitions of* $\|\cdot\|_\Phi$ *and* $N_\Phi(\cdot)$ *do not give the exact classical Lebesgue norms, but each is a constant multiple of the latter, the two constants being* **distinct**. This accounts for Proposition III.3.17. We shall explain more on the normalization and its consequences in Chapter VII. Let us now turn to $(N^\Phi)^*$ for further analysis.

4.3. *Singular linear functionals*

We return once more to the study of functionals $x^* \in (L^\Phi(\mu))^*$, as a supplement to the theory of the last two sections. Since when Φ is discontinuous, $\mathcal{M}^\Phi = L^\infty(\mu)$ is possible and when Φ is continuous and strictly increasing then $M^\Phi = \mathcal{M}^\Phi$ always ($= L^1(\mu)$ is possible), the latter spaces having absolutely continuous norms, we should consider the elements of $(\mathcal{M}^\Phi)^\perp$ and analyze them. Thus by analogy with Definition III.4.13, we introduce 'absolutely continuous' and 'singular' linear functionals without restricting the basic measures, and these concepts agree with Zaanen's (1967) general definitions in Banach function lattices on σ-finite spaces.

Definition 1. Let Φ be a Young function and $L^\Phi(\mu)$ be the corresponding Orlicz space. Then $x^* \in (L^\Phi(\mu))^*$ is called *absolutely continuous* if for each $f \in L^\Phi(\mu)$ and each sequence $A_n \searrow \emptyset$, $A_n \in \Sigma$, one has $x^*(f\chi_{A_n}) \to 0$ as $n \to \infty$. Similarly an $x^* \in (L^\Phi(\mu))^*$ is called *singular* if it is expressed as a linear combination of (at most four) nonnegative elements $x_j^* [\in (L^\Phi(\mu))^*]$, then for any absolutely continuous functional $0 \le y_j^* \le x_j^*$ [i.e., $0 \le y_j^*(f) \le x_j^*(f)$, for all $0 \le f \in L^\Phi(\mu)$], one has $y_j^* \equiv 0$.

This definition is motivated by the concepts of absolutely continuous and singular set functions in the classical Yosida-Hewitt decomposition theory of finitely additive set functions. We shall see that the results 2.8, 2.9, 2.11–2.15 give at once the corresponding decompositions when Φ is strictly increasing and continuous, but there are also some $x^* \in (\mathcal{M}^\Phi)^*$, such as those in $(L^\infty(\mu))^*$, which admit analogous decompositions. We shall discuss all these in what follows.

It was already noted in Section 1, and used there, that $L^\Phi(\mu)$ is a Banach lattice (of equivalence classes of real functions) and the pointwise order of functions in this space induces a corresponding partial order in $(L^\Phi(\mu))^*$ which then becomes a Banach lattice. Consequently every x^* in the latter space admits a *Jordan decomposition*, $x^* = x_1^* - x_2^*$ where $x_i^* \ge 0$, meaning $x_i^*(f) \ge 0$, for all $0 \le f \in L^\Phi(\mu)$. Moreover, if we demand that $x_1^*(f) = \sup\{x^*(g) : 0 \le g \le f \in L^\Phi(\mu)\}$, and set $x_2^* = x_1^* - x^*$, then the obtained decomposition is the most efficient one in the sense that $x^* = y_1^* - y_2^*$ with $y_i^* \ge 0$ implies $x_i^* = y_i^*$, $i = 1, 2$. Thus by definition x^* is absolutely continuous iff each x_i^* in its decomposition is absolutely continuous, $i = 1, 2$. Similarly a continuous functional z^* ($= z_1^* - z_2^*$) is singular if its positive and negative parts have the same property. It follows also that $0 \le \ell_1 \le \ell_2$ and ℓ_2 is absolutely continuous implies that ℓ_1 has the same property. From this we easily deduce that the set of absolutely continuous functionals of $(L^\Phi(\mu))^*$ forms a solid subspace. We state these facts, which are true for Banach lattices, for reference, as follows:

Lemma 2. *Let $L^\Phi(\mu)$ be the (real) Orlicz space on (Ω, Σ, μ). Then the set $\mathcal{A} \subset (L^\Phi(\mu))^*$ of absolutely continuous elements forms a closed* *solid subspace.*

We can present a slight extension of Proposition 2.8 as:

Proposition 3. *The space $(L^{\Phi}(\mu))^*$ admits a direct sum decomposition as $(L^{\Phi}(\mu))^* = \mathcal{A} \oplus \mathcal{S}$ where $\mathcal{A}$ is the set of absolutely continuous elements and where $\mathcal{S}$ denotes the set of singular elements. Hence $x^* \in (L^{\Phi}(\mu))^*$ is uniquely representable as $x^* = y^* + z^*$ where $y^* \in \mathcal{A}$ and $z^* \in \mathcal{S}$.*

Proof: By the preceding discussion it suffices to consider $x^* \geq 0$, and show that $x^* = y^* + z^*$ for positive y^*, z^* of the desired type. The argument is an adaptation of that for the classical Yosida-Hewitt decomposition for set functions (cf., e.g., Rao (1987), p. 182 and Zaanen (1967), p. 467). Thus for $0 \leq x^* \in (L^{\Phi}(\mu))^*$ define for $0 \leq f \in L^{\Phi}(\mu)$

$$y^*(f) = \sup\{\ell(f) : 0 \leq \ell \leq x^*\}. \tag{1}$$

It is clear that $0 \leq y^*(f_1 + f_2) \leq y^*(f_1) + y^*(f_2)$ and $y^*(af_1) = ay^*(f_1)$ for $0 \leq f_i \in L^{\Phi}(\mu)$ and $a \geq 0$.

To prove the opposite inequality, given $\varepsilon > 0$, and $0 \leq f_i \in L^{\Phi}(\mu)$, choose ℓ_i such that $y^*(f_i) < \ell_i(f_i) + \frac{\varepsilon}{2}$, $i = 1, 2$. Then from the lattice properties of the spaces involved, we have $\ell = \ell_1 \vee \ell_2 = \max(\ell_1, \ell_2) \leq x^*$ and

$$0 \leq y^*(f_1) + y^*(f_2) \leq \ell(f_1) + \ell(f_2) + \varepsilon = \ell(f_1 + f_2) + \varepsilon \leq y^*(f_1 + f_2) + \varepsilon.$$

Hence $y^*(f_1) + y^*(f_2) = y^*(f_1 + f_2)$, and y^* is linear. It is clear that $0 \leq y^*(f_n) \leq x^*(f_n) \to 0$ as $f_n \downarrow 0$ and hence $y^* \in \mathcal{A}$. Let $z^* = x^* - y^*$. Then $z^* \geq 0$ and $z^* \in (L^{\Phi}(\mu))^*$. If it is not singular, then there is a nonvanishing $0 \leq y_1^* \leq z^* = x^* + y^*$, $y_1^* \in \mathcal{A}$. Then $y^* + y_1^* \leq x^*$ and $0 \leq y^* < y^* + y_1^* \leq x^*$ so that y^* is not the largest element contradicting the definition (1). Thus $y_1^* = 0$, so $x^* = y^* + z^*$ is a desired type of decomposition, with $y^* \in \mathcal{A}$ and $z^* \in \mathcal{S}$.

The uniqueness of decomposition is simple. If $x^* = y_1^* - z_1^* = y_2^* - z_2^*$, then $y_1^* - y_2^* = z_1^* - z_2^*$ and each is both absolutely continuous and singular at the same time. So each is zero so that $y_1^* = y_2^*$ and $z_1^* = z_2^*$, as asserted. ∎

Remark: If the Young function Φ is strictly increasing and continuous, then it is clear that $M^{\Phi} = \mathcal{M}^{\Phi}$ so that $(M^{\Phi})^* = \mathcal{A}$. If Φ is either discontinuous or $\Phi(x) = 0$ for $0 < x \leq x_0$ and $\Phi(x) > 0$ for $x \geq x_0$ and (perhaps) continuous, then $\mathcal{A} \subset (\mathcal{M}^\Phi)^*$ properly so that $(\mathcal{M}^\Phi)^\perp \subset \mathcal{S}$ with possibly a proper inclusion. Thus $(\mathcal{N}^\Phi)^* \subset \mathcal{S}$, $\mathcal{S}$ being the largest such space. Using the same type of argument as that employed for $(\mathcal{N}^\Phi)^*$, it is possible to show that $\mathcal{S}$ is also an abstract (L)-space.

A key idea for this proof is to reestablish Proposition 2.13, using a suitable form of Proposition III.5.4, directly without basing it on the representation theory. More precisely, we have:

Proposition 4. *If $0 \leq x^* \in \mathcal{S}$, then $\|x^*\| = \sup\{x^*(f) : 0 \leq f \in \tilde{L}^\Phi(\mu)\}$, where Φ is any Young function and (Ω, Σ, μ) is a measure space.*

Proof: We outline the main points of proof since the ideas are familiar, although the computations need care. [Complete details can be found in deJonge (1973).] Since the unit ball of $L^\Phi(\mu)$ is contained in $\tilde{L}^\Phi(\mu)$, under either of the $N_\Phi(\cdot)$- or $\|\cdot\|_\Phi$-norms, and $\tilde{L}^\Phi(\mu)$ is a larger set, we always have $\|x^*\| \leq \sup\{x^*(f) : 0 \leq f \in \tilde{L}^\Phi(\mu)\}$. Thus we need to obtain the opposite inequality (hence equality) using the fact that x^* is singular.

If Φ is continuous and strictly increasing, then $M^\Phi = \mathcal{M}^\Phi$, so that $\mathcal{S} = (\mathcal{N}^\Phi)^*$ using our earlier notations. In this case the result holds (i.e., equality is true) by our previous work. If $\Phi(x) = 0$ for $0 < x \leq x_0 < \infty$, and $= +\infty$ for $x > x_0$, then $L^\Phi(\mu) = L^\infty(\mu)$ and the norms are equivalent. Thus $\mathcal{S}$ is the set corresponding to the family of purely finitely additive functions on Σ so that by the classical work (Yosida-Hewitt decomposition corresponds to $\mathcal{A} \oplus \mathcal{S}$), the result is again true. If $\Phi(x) = 0$, $0 < x \leq x_0$, and Φ is continuous (so $\Phi(x) < \infty$ for all $x < \infty$), then for each $0 < f \in \tilde{L}^\Phi(\mu)$, and $\varepsilon > 0$, we have a step function $0 < f_\varepsilon \leq f$ such that $\int_\Omega \Phi(f - f_\varepsilon) < \varepsilon$. This is trivial if f is bounded, and in the general case the result follows from the fact that $f \geq f - f_\varepsilon \downarrow 0$ and the conclusion is drawn with the dominated convergence theorem. Hence, using Young's inequality,

$$N_\Phi(f - f_\varepsilon) \leq \|f - f_\varepsilon\|_\Phi < 1 + \varepsilon.$$

Since $x^*(f_\varepsilon) = 0$ (the fact that x^* is singular implies $x^*(h) = 0$ for all step functions), we have

$$0 \leq x^*(f) = x^*(f - f_\varepsilon) < (1 + \varepsilon)x^*\left(\frac{f - f_\varepsilon}{N_\Phi(f - f_\varepsilon)}\right) \leq (1 + \varepsilon)\|x^*\|. \quad (2)$$

This gives the desired inequality since $\varepsilon > 0$ is arbitrary. Thus it remains to consider the cases of discontinuous Φ such that $\Phi(x_0) > 0$, for some $0 < x_0 < \infty$ and $\Phi(x) = +\infty$ for $x > x_0$ with or without $\Phi(x_0) < \infty$. The necessary discussion follows.

First let Φ be a continuous Young function on the closed interval $[0, x_0]$ with $\Phi(x_0) > 0$, and $\Phi(x) = +\infty$ for $x > x_0$. Thus $\varphi(x_0) < \infty$ also. In this case consider $\Phi_1(x) = 0$, $0 \leq x \leq x_0$, and $\Phi(x) = +\infty$ for $x > x_0$ and $\Phi_2(x) = \Phi(x)$ for $0 \leq x \leq x_0$, and be a straight line with slope $\varphi(x_0)$ passing through $(x_0, \Phi(x_0))$. Thus this equation has the form $y = \varphi(x_0)x + \Phi(x_0) - \varphi(x_0)x_0$, $x \geq x_0$ where φ is the (left) derivative of Φ, as usual. Then Φ_i, $i = 1, 2$, are Young functions and $\Phi = \max(\Phi_1, \Phi_2)$. Hence $L^{\Phi}(\mu) = L^{\Phi_1}(\mu) \cap L^{\Phi_2}(\mu)$ as sets. But it is well known and easily verified that $\tilde{N}_{\Phi}(f) = \max(N_{\Phi_1}(f), N_{\Phi_2}(f))$ is a norm under which $L^{\Phi}(\mu)$ becomes a Banach space. [This is true even if one uses the equivalent norms $[(N_{\Phi_1}(f))^p + (N_{\Phi_2}(f))^p]^{\frac{1}{p}}$, $1 < p < \infty$.] On the other hand, it is seen that the gauge norm $N_{\Phi}(\cdot)$ of $L^{\Phi}(\mu)$ satisfies $N_{\Phi_2}(f) \geq \tilde{N}_{\Phi}(f)$, and under which $L^{\Phi}(\mu)$ is again complete. Consequently $N_{\Phi}(\cdot)$ and $\tilde{N}_{\Phi}(\cdot)$ are equivalent norms. [The same holds if Orlicz norms are used instead.] Although these norms are only equivalent and in general are not equal, it can be verified in the present case that $N_{\Phi}(\cdot) = \tilde{N}(\cdot)$ does indeed hold. [The verification is not hard and will be left to the reader.] From the general principles of adjoint norms, it can be inferred that $\tilde{N}_{\Phi}^*(\cdot) = N_{\Phi_1}^*(\cdot) + N_{\Phi_2}^*(\cdot)$. Hence for $0 \leq x^* \in (L^{\Phi_i}(\mu))^*$, one can find $0 \leq x_i^* \in (L^{\Phi_i}(\mu))^*$ such that $\tilde{N}_{\Phi}^*(x^*) = N_{\Phi_1}^*(x_1^*) + N_{\Phi_2}^*(x_2^*)$, or in the notation of $(L^{\Phi}(\mu), N_{\Phi}(\cdot))^* = ((L^{\Phi}(\mu))^*, \|\cdot\|_{\Psi})$, we get

$$\|x^*\|_{\Phi} = \|x_1^*\|_{\Phi_1} + \|x_2^*\|_{\Phi_2}. \tag{3}$$

These are classical representations, and are often used in such contexts. [See e.g. Calderón (1964), p. 114; Aronszajn and Gagliardo (1965), p. 57. This is also detailed in the Orlicz space context by deJonge (1973).] By the preceding paragraph, however, $\|x_i^*\| = \sup\{x_i^*(f) : 0 \leq f \in \tilde{L}^{\Phi_i}(\mu)\}$, $i = 1, 2$. Hence

$$\|x^*\|_{\Phi} = \sup\{x_1^*(f) : 0 \leq f \in \tilde{L}^{\Phi_1}(\mu)\} + \sup\{x_2^*(f) : 0 \leq f \in \tilde{L}^{\Phi_2}(\mu)\}$$
$$\geq \sup\{(x_1^* + x_2^*)(f) : 0 \leq f \in L^{\Phi}(\mu)\}. \tag{4}$$

This with the initial reduction shows that there is equality here.

We next consider $\Phi(x)$ to be continuous on $[0, x_0)$ for some $0 < x_0 < \infty$ and $= +\infty$ for $x > x_0$, with $\Phi(x_0)$ finite or not. Let $0 < x_n < x_{n+1} \to x_0$, such that $\Phi(x_1) > 0$. We reduce this case to the preceding one as follows. Let $\Phi_n(x) = \Phi(x)$, $0 < x \le x_n$, and $= +\infty$, $x > x_n$, $n \ge 1$. Then $\Phi_n(x) \searrow \Phi(x)$, as $n \to \infty$, for each x, and each $L^{\Phi_n}(\mu)$ is of the type considered above. Clearly $L^{\Phi_1}(\mu) \subset L^{\Phi_2}(\mu) \subset \cdots \subset L^{\Phi}(\mu) \subset L^{\infty}(\mu)$, and $N_{\Phi_i}(f) \ge N_{\Phi_{i+1}}(f) \to N_{\Phi}(f)$. A simple computation using the definition of gauges shows that all these spaces contain the same elements, and that these norms are equivalent on using the fact that $L^{\Phi_1}(\mu) = L^{\Phi}(\mu) \subset L^{\infty}(\mu)$. Thus if x^* is a positive singular linear functional on $L^{\Phi}(\mu)$, then one has

$$\|x^*\|_{\Phi_n} = \sup\{x^*(f) : 0 \le f \in \tilde{L}^{\Phi_n}(\mu)\}, \quad n \ge 1, \qquad (5)$$

by the last paragraph. On the other hand, if $0 \le \alpha < \|x^*\|_{\Phi} = \sup\{x^*(f) : 0 < f \in \tilde{L}^{\Phi}(\mu)\}$ where x^* is a singular linear functional, then there is an $0 < f_\alpha \in \tilde{L}^{\Phi}(\mu)$ such that $\alpha < x^*(f_\alpha)$. Consequently,

$$\alpha < x^*(f_\alpha) = x_0 x_n^{-1} x^*(x_0^{-1} x_n f_\alpha) \le x_0 x_n^{-1} \sup\{x^*(f) : 0 \le f \in \tilde{L}^{\Phi_n}(\mu)\}$$
$$= x_0 x_n^{-1} \|x^*\|_{\Phi_n} \le x_0 x_n^{-1} \|x^*\|_{\Phi}, \quad n \ge 1. \qquad (6)$$

Letting $n \to \infty$, we get $\alpha \le \|x^*\|_{\Phi}$. This with the initial inequality gives equality, as asserted. ∎

With this result, we can now prove the following.

Theorem 5. *S is an abstract (L)-space so that the norm is additive on its positive elements.*

Proof: Let $0 \le x_i^* \in S$, $i = 1, 2$. Since $\|x_1^* + x_2^*\| \le \|x_1^*\| + \|x_2^*\|$ is always true, we need to establish only the opposite inequality. Thus given $\varepsilon > 0$, there are $f_i \in \tilde{L}^{\Phi}(\mu)$, by Proposition 4, such that $x_i^*(f_i) > \|x_i^*\| - \frac{\varepsilon}{2}$. Since $h = \max(f_1, f_2) \in \tilde{L}^{\Phi}(\mu)$, we have $x_i^*(h) \ge x_i^*(f_i)$, $i = 1, 2$. Consequently

$$\|x_1^*\| + \|x_2^*\| - \varepsilon < x_1^*(f_1) + x_2^*(f_2) < (x_1^* + x_2^*)(h) \le \|x_1^* + x_2^*\|. \qquad (7)$$

Since $\varepsilon > 0$ is arbitrary, (7) and the earlier inequality implies the equality. ∎

Since $S \supset (\mathcal{N}^{\Phi})^*$ with possibly a strict inclusion (especially if Φ is not a N function), and since some elements of $(\mathcal{M}^{\Phi})^*$ may not be absolutely continuous, they can be decomposed into singular and absolutely continuous parts. However, for an integral representation of x^* in $(L^\Phi(\mu))^*$ the preceding work is needed. If $\mu(\Omega) < \infty$, then $L^\Phi(\mu) \subset L^1(\mu)$ so that $\mathcal{A}_\Psi(\mu) \subset ba(\Omega, \Sigma, \mu)$, we can directly employ the Yosida-Hewitt decomposition. In this case Corollary 2.12 gives the representation which makes all these statements self evident.

It should be observed that $N_\Phi(\cdot) = \tilde{N}_\Phi(\cdot)$ noted in the proof of Proposition 4 is based on using the gauge norms, and otherwise one has only equivalence for the Orlicz norm. It is thus of interest to present an expression for the norm $\|x^*\|' = \sup\{|x^*(f)| : \|f\|_\Phi \leq 1\}$, when Φ does not satisfy any growth restrictions. The following result, essentially due to Andô (1960), includes such an expression, and will also serve as a motivation for defining a so-called "F-norm" in Chapter X.

Proposition 6. *Let Φ be an N-function and $L^\Phi(\mu)$ be an Orlicz space. If $x^* \in (L^\Phi(\mu))^*$ and $x^* = y^* + z^*$ where y^* is absolutely continuous and z^* singular, and if g in $L^\Psi(\mu)$ corresponds to y^* (cf. Theorem 1.6), then*

$$\|x^*\|' = \inf\{\varepsilon > 0 : \rho_\Psi(\tfrac{g}{\varepsilon}) + \tfrac{1}{\varepsilon}\|z^*\| \leq 1\} \tag{8}$$

where $\|z^*\|$ is the norm of the singular functional which is unchanged when we use $N_\Phi(\cdot)$ or $\|\cdot\|_\Phi$ in the $L^\Phi(\mu)$ space (cf. Proposition 2.13).

Proof: Let $\varepsilon > 0$ and $f \in L^\Phi(\mu)$ such that $\|f\|_\Phi = 1$ and the right side of (8) holds. Then by Theorem III.3.13 there exists a $k > 0$ such that

$$1 = \|f\|_\Phi = \tfrac{1}{k}(1 + \rho_\Phi(kf)). \tag{9}$$

On the other hand, by Proposition 4 (cf. also Proposition 2.14), we have

$$|z^*(kf)| \leq \|z^*\|. \tag{10}$$

Hence

$$\frac{k}{\varepsilon}|x^*(f)| \leq \int_\Omega |\frac{kfg}{\varepsilon}|d\mu + \frac{1}{\varepsilon}|z^*(kf)|, \text{ by Corollary 2.9,}$$

$$\leq \rho_\Phi(kf) + \rho_\Psi(\frac{g}{\varepsilon}) + \frac{1}{\varepsilon}\|z^*\|, \text{ by Young's inequality}$$
$$\text{and (10),}$$

$$\leq 1 + \rho_\Phi(kf) = k, \text{ by the choice of } \varepsilon \text{ to satisfy (8),}$$
$$\text{and by (9).}$$

It follows that $|x^*(f)| \leq \varepsilon$ so that $\|x^*\|' \leq$ right side of (8).

For the opposite inequality, we may take $\|x^*\|' = 1$, $x^* \geq 0$ and show that the right side of (8) is at most one. If this is false, then there is a $\delta > 0$ such that the right side of (8) is $> 1 + \delta$ (and $\varepsilon \geq 1$ must hold there). Thus

$$\rho_\Psi(g) + \|z^*\| > 1 + \delta. \tag{11}$$

Hence by Proposition 4, as before, we can find $f_1 \in L^\Phi(\mu)$ such that

$$\rho_\Psi(g) + z^*(f_1) > 1 + \delta, \tag{12}$$

since $x^* \geq 0$ implies $y^* \geq 0$ and $z^* \geq 0$ as well. Also

$$1 = \|x^*\|' \geq \sup\{x^*(f) : \|f\|_\Phi \leq 1, 0 \leq f \in M^\Phi\},$$
$$= \sup\{y^*(f) : \|f\|_\Phi \leq 1, 0 \leq f \in M^\Phi\}, \text{ since } z^* \in (M^\Phi)^\perp,$$
$$= \sup\{\int_\Omega fg d\mu : \|f\|_\Phi \leq 1, 0 \leq f \in M^\Phi\} = N_\Psi(g), g \geq 0,$$

by Proposition III.4.10.

Thus $\rho_\Psi(g) \leq 1$. But by using the equality conditions in Young's inequality,

$$\rho_\Psi(g) = \int_\Omega [\psi(g)g - \Phi(\psi(g))] d\mu, \tag{13}$$

where ψ is, as usual, the (left) derivative of Ψ. Let $f = \psi(g)$. Then $f \in \tilde{L}^\Phi(\mu)$. By using Theorem 2.6, we can find for each $0 < \eta < \frac{\delta}{2}$, an $0 \leq h \in \tilde{L}^\Phi(\mu) - M^\Phi$ such that $\int_\Omega \Phi(h) d\mu < \eta$ [h will have arbitrarily small support] and $\nu(\Omega) < z^*(h) + \eta$. If $h_1 = \max(f, h)$ then

$$\rho_\Phi(h_1) = \int_\Omega \Phi(h_1) d\mu = \int_{[f \geq h]} \Phi(f) d\mu + \int_{[f < h]} \Phi(h) d\mu \leq \rho_\Phi(f) + \rho_\Phi(h)$$
$$\leq \rho_\Phi(f) + \eta. \tag{14}$$

From this, one has on using the positivity of x^*, and hence y^* and z^*,

$$x^*(h_1) - \rho_\Phi(h_1) = y^*(h_1) + z^*(h_1) - \rho_\Phi(h_1)$$
$$\geq y^*(f) + z^*(h) - \rho_\Phi(f) - \eta, \text{ by (14)},$$
$$\geq \rho_\Psi(g) + z^*(h) - \eta, \text{ by (13)},$$
$$> \rho_\Psi(g) + \nu(\Omega) - 2\eta, \text{ by choice of } h,$$
$$> 1 + \delta - 2\eta, \text{ by (11)}. \tag{15}$$

Let $\varepsilon = x^*(h_1) - \delta + 2\eta > 0$. Then (15) implies $1 + \rho_\Phi(\varepsilon h_2) < \varepsilon$, if we let $h_2 = h_1/\varepsilon$. But then the Orlicz norm of (εh_2) satisfies $\|\varepsilon h_2\|_\Phi < \varepsilon$ so that $\|h_2\|_\Phi < 1$ and hence

$$\|h_1\|_\Phi < \varepsilon = x^*(h_1) - \delta + 2\eta \leq \|h_1\|_\Phi - \delta + 2\eta. \tag{16}$$

Since $\delta > 2\eta$, the inequality in (16) is impossible. Thus (8) holds as stated. ∎

The interest in the form of (8) is seen when it is expressed as:

$$\|x^*\|' = \inf\{\varepsilon > 0 : \int_\Omega \Psi(\tfrac{g}{\varepsilon})d\mu \leq (1 - \tfrac{1}{\varepsilon}|\nu|(\Omega))\} \tag{17}$$

and thus the right side is also a function of $\varepsilon > 0$. A form of this definition, used as a "gauge" or an "F"-norm, will be employed in studying nonlocally convex Orlicz spaces in the last chapter. Of course, the result has independent interest here and serves as a comparison to the isometry property given in Theorem 2.11, using the gauge norm.

4.4. *The topologies $\sigma(L^\Phi(\mu), \mathcal{M}^\Psi)$ and $\sigma(L^\Phi(\mu), \mathcal{A}_\Psi(\mu))$*

In all the preceding work we have only considered the norm (or strong) topology of $L^\Phi(\mu)$, and then showed that the latter is a Banach space. But there are also other (weaker) topologies of considerable interest in important applications. Some of these will now be discussed here.

For each $g \in \mathcal{M}^\Psi$, the functional $x_g^*(f) = \int_\Omega fg\,d\mu$ defines an element x_g^* of $(L^\Phi(\mu))^*$ so that we may identify $\mathcal{M}^\Psi$ as a closed subset of the adjoint of $L^\Phi(\mu)$ by the mapping $j : g \mapsto x_g^*$. Further, for any $f \in L^\Phi(\mu)$ we have $x_g^*(f) = 0$, $g \in \mathcal{M}^\Psi$ implies $f = 0$, a.e. by Proposition III.4.10. Then $\mathcal{M}^\Psi$ [or any subspace of linear functions on $L^\Phi(\mu)$] with the above property is called a *total* subspace. The family of subsets

$$\mathcal{U}(x_1^*, \ldots, x_n^*; \varepsilon) = \{f \in L^\Phi(\mu) : |x_i^*(f)| < \varepsilon,$$
$$x_i^* \in j(\mathcal{M}^\Psi), i = 1, \ldots, n\}, \tag{1}$$

as n and $\varepsilon > 0$ vary, define a neighborhood base of '0' and hence a Hausdorff vector topology (since $\mathcal{M}^\Psi$ is total) on $L^\Phi(\mu)$, denoted $\sigma(L^\Phi(\mu), \mathcal{M}^\Psi)$. Also this is locally convex. This is called a weak topology for $L^\Phi(\mu)$, since one can verify that each closed (or open) set for the norm topology is also closed for $\sigma(L^{\Phi}(\mu), \mathcal{M}^{\Psi})$ but not conversely. Also since $\mathcal{M}^{\Psi} \subset L^{\Psi}(\mu)$, and proper subspace in general, it follows that $\sigma(L^{\Phi}(\mu), \mathcal{M}^{\Psi})$ is weaker than $\sigma(L^{\Phi}(\mu), L^{\Psi}(\mu))$ which in turn is weaker than $\sigma(L^{\Phi}(\mu), \mathcal{A}_{\Psi}(\mu))$ since $L^{\Psi}(\mu)$ is again embeddable as a proper subspace of $(L^{\Phi}(\mu))^* \cong \mathcal{A}_{\Psi}(\mu)$. The last topology is generally called *the weak topology* of $L^{\Phi}(\mu)$, and the former two are (for distinction) called the $\mathcal{M}^{\Psi}$- or ($L^{\Psi}(\mu)$-) topology of $L^{\Phi}(\mu)$. On the other hand, if Φ is an N-function, then $L^{\Phi}(\mu)$ is identifiable as $(\mathcal{M}^{\Psi})^*$ ($= (\mathcal{M}^{\Psi})^*$), so that $\sigma(L^{\Phi}(\mu), \mathcal{M}^{\Psi})$ is then called the *weak-star* topology of $L^{\Phi}(\mu)$ which is weaker than its weak topology where $\mathcal{M}^{\Psi}$ is replaced by $\mathcal{A}_{\Psi}(\mu)$. [Recall that a stronger topology has more open sets in the neighborhood base than the weaker one.] We include some properties of the $L^{\Phi}(\mu)$ spaces in these topologies for later applications.

If $\Phi(x) = |x|^p$, $p > 1$, then $L^{\Phi}(\mu) = \mathcal{M}^{\Phi} = M^{\Phi}$, and $L^{\Psi}(\mu) = \mathcal{M}^{\Psi} = M^{\Psi}$, and if $p = 1$, then $\Psi(x) = 0$ for $0 \leq x \leq 1$, and $= +\infty$ for $x > 1$, so that $L^{\Psi}(\mu) = \mathcal{M}^{\Psi} \neq M^{\Psi}$, and the classical theory shows that in the $\sigma(L^{\Phi}(\mu), \mathcal{M}^{\Psi})$ topology $L^{\Phi}(\mu)$ is complete when $1 \leq p < \infty$. However, if $p = 1$ then $L^{\Psi}(\mu)$ ($= L^{\infty}(\mu)$) is not complete in the $\sigma(L^{\Psi}(\mu), L^{\Phi}(\mu))$-topology, but it is sequentially complete in the $\sigma(L^{\Psi}(\mu), \mathcal{M}^{\Phi})$-topology. This indicates a difference between the topologies $\sigma(L^{\Phi}(\mu), \mathcal{M}^{\Psi})$, $\sigma(L^{\Phi}(\mu), L^{\Psi}(\mu))$ and $\sigma(L^{\Phi}(\mu), \mathcal{A}_{\Psi})$ as well. We need some information in these topologies, as they will be useful in a study of the integral (or kernel) operators on these spaces, and are important in several applications.

Let S denote the space of $\mathcal{M}^{\Psi}$, $L^{\Psi}(\mu)$ or $\mathcal{A}_{\Psi}(\mu)$. Then $L^{\Phi}(\mu)$ is said to be *sequentially complete* in $\sigma(L^{\Phi}(\mu), S)$ topology if every Cauchy sequence in this topology converges to an element in $L^{\Phi}(\mu)$. If the word "sequence" is replaced by "net" and the statement still holds, then the space is complete. However, this turns out to be too restrictive and the $L^{\Phi}(\mu)$ spaces seldom have this property so that we shall not consider it. Let us establish the following result in which both Φ, Ψ are not simultaneously discontinuous.

Theorem 1. *Let (Φ, Ψ) be complementary Young functions and (Ω, Σ, μ) be a measure space. Let $L^{\Phi}(\mu), \mathcal{M}^{\Psi}$ be the spaces as before. If μ is localizable, then $L^{\Phi}(\mu)$ is sequentially complete in $\sigma(L^{\Phi}(\mu), \mathcal{M}^{\Psi})$-topology. The same is true if Φ is continuous and vanishes only at the origin but now μ can be unrestricted. In particular the result holds* *if Φ (or Ψ) is an N-function and μ is unrestricted. Under the same hypotheses on Φ, Ψ and μ, $L^\Phi(\mu)$ is also sequentially complete for the $\sigma(L^\Phi(\mu), L^\Psi(\mu))$ topology.*

Proof: The argument is a modification of the classical $L^p(\mu)$-case to the present context. Here are the details.

Let $\{f_n, n \geq 1\} \subset L^\Phi(\mu)$ be a Cauchy sequence in $\sigma(L^\Phi(\mu), \mathcal{M}^\Psi)$-topology, so that for each $g \in \mathcal{M}^\Phi$, $\int_\Omega (f_n - f_m)g d\mu \to 0$ as $m, n \to \infty$, by definition of the topology (1). Hence the set $\{\int_\Omega f_n g d\mu, n \geq 1\}$ of numbers is bounded for each g. If we let $x_n^*(g) = \int_\Omega f_n g d\mu$, then $x_n^* : \mathcal{M}^\Psi \to \mathbb{R}$ is additive and (by Hölder's inequality) bounded so that $x_n^* \in (\mathcal{M}^\Psi)^*$ for each $n \geq 1$. This and the preceding line imply that the set $\{x_n^*, n \geq 1\}$ of continuous linear mappings from the Banach space $\mathcal{M}^\Psi$ into $\mathbb{R}$ is pointwise bounded (i.e., $\sup_n |x_n^*(g)| = K_g < \infty$ for each g in $\mathcal{M}^\Psi$). Hence by the classical uniform boundedness principle (even if the set if not countable) $\sup_n \|x_n\| < \infty$, and this translates by Theorem 1.6 to $\sup_n N_\Phi(f_n) = K < \infty$ (cf.,e.g., Rao (1987), pp. 291–2). Also since with each $g \in \mathcal{M}^\Psi$, $g_E = g\chi_E \in \mathcal{M}^\Psi$ (the space being solid), we have $\nu_n(E) = \int_\Omega f_n g_E d\mu = \int_E f_n g d\mu$, $E \in \Sigma$, with ν_n as σ-additive on Σ. The fact that f_n is Cauchy in this topology implies $\nu_n(E) \to \nu(E)$ for each $E \in \Sigma$, as $n \to \infty$. Since $\nu_n(\cdot)$ is a signed measure, it follows by a classical result of Nikodým (cf.,e.g., Rao (1987), p. 181) that $\nu : \Sigma \to \mathbb{R}$ is also σ-additive. At this point the additional hypotheses on (Φ, Ψ) or μ is needed.

By definition of ν_n, it is clear that ν_n is μ-continuous. Since $\nu_n^g(E) \to \nu^g(E)$, as shown above, it follows by the classical Vitali-Hahn-Saks theorem again (cf.,e.g., Rao (1987), p. 176) that ν^g is μ-continuous. Since μ is assumed localizable, we can apply the (general) Radon-Nikodým theorem to deduce that $\nu^g(E) = \int_E f_0 g d\mu$ for a μ-unique f_0. But each ν_n and hence ν satisfies $\nu_n, \nu \in \mathcal{A}_\Phi(\mu)$, which implies by Proposition 1.3(a) that $f_0 \in L^\Phi(\mu)$. This is in fact a simple consequence of the inverse Hölder inequality (Proposition 1.1). Thus $\int_\Omega f_n g d\mu \to \int_\Omega f_0 g d\mu$, $g \in \mathcal{M}^\Psi$, and $L^\Phi(\mu)$ is $\sigma(L^\Phi(\mu), \mathcal{M}^\Psi)$ sequentially complete.

Suppose that $\Phi(x) > 0$ for $x > 0$ and be continuous. Then in the above proof we note that each f_n vanishes outside a set E_n of σ-finite measure. If $E = \bigcup_{n \geq 1} E_n$, then E is σ-finite and the set $\{f_n, n \geq 1\}$ vanishes outside of E. Consequently all the ν_n vanish on E^c. But then on $\Sigma(E) = \{A \cap E : A \in \Sigma\}$, μ is σ-finite so that the classical Radon-Nikodým theorem applies, and the argument of the last paragraph is applicable. In particular, if Φ (or Ψ) is an N-function, then $\Phi(x) = 0$ iff $x = 0$ and Φ is continuous. So the preceding argument automatically includes this case.

Finally, one notes that in the above proof the properties of $\mathcal{M}^{\Psi}$ are used only to the extent that it is a complete space and it is norm determining for $L^{\Phi}(\mu)$ as well as it being solid. These are obviously present for $L^{\Psi}(\mu)$ and hence $L^{\Phi}(\mu)$ is $\sigma(L^{\Phi}(\mu), L^{\Psi}(\mu))$ sequentially complete if the same conditions of (Φ, Ψ) and μ are assumed. ∎

Remark. It should be noted that $\mathcal{M}^{\Phi}$ cannot be replaced by $\mathcal{A}_{\Psi}(\mu)$ because some of its elements are only finitely additive so that the proof breaks down soon after the norm boundedness of $\{f_n, n \geq 1\}$ is established. In fact, not only the proof but the result itself is false. For instance, $L^{\infty}(\mu)$ is not $\sigma(L^{\infty}(\mu), (L^{\infty}(\mu))^*)$ sequentially complete unless μ is an essentially trivial measure.

Some special cases will be recorded for ready reference.

Corollary 2. *If Φ is an N-function, Ψ its complementary function and (Ω, Σ, μ) is a measure space, then $L^{\Phi}(\mu)$, is $\sigma(L^{\Phi}(\mu), \mathcal{M}^{\Psi})$ as well as $\sigma(L^{\Phi}(\mu), L^{\Psi}(\mu))$-sequentially complete.*

Regarding the $\sigma(L^{\Phi}(\mu), \mathcal{A}_{\Psi}(\mu))$-topology which we may denote also by $\sigma(L^{\Phi}(\mu), (L^{\Phi}(\mu))^*)$, and call it *the weak topology*. Then we have:

Theorem 3. *Let Φ be a strictly increasing continuous Young function and (Ω, Σ, μ) be a measure space. Then $L^{\Phi}(\mu)$ is weakly sequentially complete iff $L^{\Phi}(\mu) = M^{\Phi}$. The last condition is essentially equivalent to Φ be Δ_2-regular.* [The word "essentially" is explained at the end of proof.]

Proof: Let $\{f_n, n \geq 1\} \subset L^{\Phi}(\mu)$ be a weak Cauchy sequence. We need to show that there exists an $f_0 \in L^{\Phi}(\mu)$ such that $f_n \to f_0$ weakly iff the given condition holds. Since $\Phi(x) > 0$ for $x > 0$, it follows that the support of each f_n is σ-finite. Hence the countable collection $\{f_n, n \geq 1\}$ itself vanishes outside of a set of σ-finite measure and hence we may restrict ourselves to such a set. In other words for this proof μ may be assumed σ-finite on (Ω, Σ), and we do so now.

Under this reduction, if $L^{\Phi}(\mu) = M^{\Phi}$, then $(L^{\Phi}(\mu))^* = L^{\Psi}(\mu)$, by Theorem 1.7. Consequently the weak topology of $L^{\Phi}(\mu)$ is precisely $\sigma(L^{\Phi}(\mu), L^{\Psi}(\mu))$, and the result follows from the last part of Theorem 1.

For the converse, let $\{f_n, n \geq 1\}$ be weakly convergent to f_0 in $L^{\Phi}(\mu)$ and suppose, if possible, that $M^{\Phi} \underset{\neq}{\subset} L^{\Phi}(\mu)$. Then $N^{\Phi} = L^{\Phi}(\mu)/M^{\Phi} \neq \{0\}$. (It is an infinite dimensional vector lattice by Theorem 3.5.) Hence there exists an $f_0 \geq 0$ in $L^{\Phi}(\mu) - M^{\Phi}$ with a σ-finite support. [Thus we may again assume that μ is σ-finite for this proof.] Let f_n be a sequence of step functions such that $f_n \nearrow f_0$, the convergence being pointwise. This is possible by the structure theorem of measurable functions. Also $f_n \in M^{\Phi}$, $n \geq 1$. We assert that $\{f_n, n \geq 1\}$ is a weak Cauchy sequence in $L^{\Phi}(\mu)$. Indeed, let $x^* \in (L^{\Phi}(\mu))^*$ and let $x^* = y^* + z^*$ where y^* is absolutely continuous and z^* singular, belonging to $(L^{\Phi}(\mu))^*$. Then $z^*(f_n) = 0$ so that $x^*(f_n) = y^*(f_n) = \int_{\Omega} f_n g d\mu$ for a unique $g \in L^{\Psi}(\mu)$ ($= (M^{\Phi})^*$) by Theorem 1.7. It follows that for $n > m$ (and the monotone convergence theorem),

$$|x^*(f_n - f_m)| = \left| \int_{\Omega} (f_n - f_m) g d\mu \right| \leq \int_{\Omega} (f_0 - f_n)|g| d\mu \to 0 \quad (2)$$

as $n \to \infty$. This establishes our assertion.

If $L^{\Phi}(\mu)$ is weakly complete, then $f_n \to \tilde{f}_0$ weakly for an $\tilde{f}_0 \in L^{\Phi}(\mu)$. On the other hand (2) shows that for any h in $L^{\Psi}(\mu)$ we have

$$\left| \int_{\Omega} (f_n - f_0) h d\mu \right| \leq \int_{\Omega} (f_0 - f_n)|h| d\mu \to 0 \quad (3)$$

as $n \to \infty$ by the monotone convergence theorem again, so that $f_n \to f_0$ in $\sigma(L^{\Phi}(\mu), L^{\Psi}(\mu))$. These two statements imply $f_0 = \tilde{f}_0$ since the topologies involved are Hausdorff and so the limits are unique. But since $f_0 \notin M^{\Phi}$, and the latter is a closed subspace of $L^{\Phi}(\mu)$, by the classical Hahn-Banach theorem there is an $x_0^* \in (L^{\Phi}(\mu))^*$ such that $x_0^*(M^{\Phi}) = 0$ and $x_0^*(f_0) = 1$. [Thus x_0^* is singular.] From this we get

$$1 = x_0^*(f_0) = x_0^*(f_0 - f_n) + x_0^*(f_n) = x_0^*(f_0 - f_n) \to 0 \quad (4)$$

since by assumption $f_n \to f_0 = \tilde{f}_0$ weakly. This contradiction shows that $M^\Phi = L^\Phi(\mu)$ must hold.

For the last statement, if Φ is Δ_2-regular, then $M^\Phi = L^\Phi(\mu)$, and on the other hand the equality $M^\Phi = L^\Phi(\mu)$ implies, when μ is diffuse that Φ must be Δ_2-regular. This is why we say "essentially" in the statement since in general $M^\Phi = L^\Phi(\mu)$ may hold without the Δ_2-regularity on Φ (regarding such a difference in these cases, see Theorem III.4.12(a)). This is the desired conclusion. ∎

If Φ is an N-function, then $(M^\Phi)^* = L^\Psi(\mu)$ by Theorem 1.6. Although $\sigma(M^\Phi, L^\Psi(\mu))$ is the weak topology of M^Φ, by the preceding proof if $0 \leq f_0 \in L^\Phi(\mu) - M^\Phi$, then $0 \leq f_n \uparrow f_0$, pointwise, implies $f_n \to f_0$ in $\sigma(M^\Phi, L^\Psi(\mu))$ so that $f_0 \notin M^\Phi$. Thus we have the following consequence.

Corollary 4. *If Φ is an N-function, then M^Φ is weakly sequentially complete iff $M^\Phi = L^\Phi(\mu)$, or essentially equivalently Φ is Δ_2-regular. Also the $\sigma(L^\Phi(\mu), \mathcal{M}^\Psi)$ completion of M^Φ is all of $L^\Phi(\mu)$ even if Φ is a strictly increasing Young function (and μ is unrestricted).*

Proof: The first part was already noted above. Regarding the last half, note that by (2), for each $0 \leq f \in L^\Phi(\mu) - M^\Phi$, there are simple functions $f_n \in M^\Phi$ (since Φ is continuous and $\Phi(x) > 0$ for $x > 0$) such that $0 \leq f_n \uparrow f$ pointwise and for each $g \in M^\Psi$

$$|\int_\Omega (f_n - f)g d\mu| \leq \int_\Omega (f - f_n)|g|d\mu \to 0$$

as $n \to \infty$ implying that $f_n \to f$ in $\sigma(L^\Phi(\mu), M^\Psi)$ topology. Since $L^\Phi(\mu)$ is a lattice, this implies the assertion. ∎

The point of the last part here is that although M^Φ is not sequentially complete for the weak topology, as implied by the first part, its weak*- (or $\sigma(L^\Phi(\mu), \mathcal{M}^\Psi)$-) closure fills up $L^\Phi(\mu)$.

In the following section we shall relate the convergence in this topology with convergence in measure for certain Orlicz spaces, after discussing weak compactness concepts.

We now present an application of the preceding results.

Theorem 5. *Let Φ be a Young function and $L^\Phi(\mu)$ be the corresponding Orlicz space on a measure space (Ω, Σ, μ) such that μ is diffuse on a set of positive measure. Then the following statements are mutually equivalent:*

(i) Φ *is* Δ_2-*regular;*

(ii) $L^{\Phi}(\mu)$ *does not contain a closed subspace isomorphic to* c_0, *the space of convergent sequences with limit zero, under uniform norm;*

(iii) $L^{\Phi}(\mu)$ *does not contain a closed subspace isomorphic to* ℓ^{∞}, *the space of bounded sequences, with uniform norm.*

Proof: We shall show that (i) $\Rightarrow$ (ii) $\Rightarrow$ (iii) without restrictions on μ, and that (iii) $\Rightarrow$ (i) under the diffuseness condition, as was done in Chapter III for several results.

(i) $\Rightarrow$ (ii)   This is the place we use the weak convergence result of Theorem 3. Indeed if (i) holds, then Φ is strictly increasing and $L^{\Phi}(\mu) = M^{\Phi}$, and hence $L^{\Phi}(\mu)$ is weakly sequentially complete. But by a classical result of Banach spaces, weak and strong closures of a convex set are the same (cf.,e.g., Dunford-Schwartz (1958), V.3.13). Hence if c_0 is isomorphic to a closed subspace of $L^{\Phi}(\mu)$, the latter is also weakly closed and hence complete. This implies by the assumed isomorphism, c_0 should be weakly sequentially complete. As is well known, the latter property is not shared by c_0 so that (ii) must be true as stated.

(ii) $\Rightarrow$ (iii)   This follows again for the same reason.

(iii) $\Rightarrow$ (i)   We establish this by contradiction. As noted many times before, we may assume for this proof that μ is diffuse on a set A of finite positive measure (because of the standing assumption of the finite subset property of μ). Thus we may and do take (Ω, Σ, μ) as a diffuse probability space, and suppose that Φ does not satisfy the Δ_2-condition. A subspace of $L^{\Phi}(\mu)$ which is isomorphic to ℓ^{∞} will now be constructed to show that (iii) is not valid, and that will complete the demonstration.

Thus, let $\{\Omega_n, n \geq 1\}$ be a disjoint measurable sequence such that $0 < \mu(\Omega_n) \leq 2^{-n}$, $\Omega = \bigcup_{n=1}^{\infty} \Omega_n$. In the detailed computation we employ the gauge norm here, and by the equivalence of $N_{\Phi}(\cdot)$ and $\|\cdot\|_{\Phi}$, the result holds for both norms. Since Φ is continuous, and does not satisfy the Δ_2-conditions, we may select a double sequence of numbers $\{x_{nk}, k \geq 1, n \geq 1\} \subset \mathbb{R}^+$ such that for each $n \geq 1$ and $k \geq 1$,

$$\Phi(x_{n1}) > [\mu(\Omega_n)]^{-1}, \text{ and } \Phi((1 + \tfrac{1}{k})x_{nk}) > 2^k \Phi(x_{nk}). \qquad (5)$$

Next for each $n \geq 1$, partition Ω_n into $\{\Omega_{nk}, k \geq 1\}$ such that Ω_{nk}'s are disjoint, and $\mu(\Omega_{nk}) = 2^{-n-k}(\Phi(x_{nk}))^{-1}$, $k \geq 1$. If we let $f_n = \sum_{k=1}^{\infty} x_{nk}\chi_{\Omega_{nk}}$, then f_n is measurable and

$$\rho_\Phi(f_n) = \sum_{k=1}^{\infty} \Phi(x_{nk})\mu(\Omega_{nk}) = \sum_{k=1}^{\infty} 2^{-n-k} = 2^{-n} < 1. \quad (6)$$

This shows that $N_\Phi(f_n) \leq 1$. To see there is equality here, let $\alpha > 1$. Then there is a $k_0 > 1$ such that for $k \geq k_0$, $1 + \frac{1}{k} < \alpha$, and hence using (5) we have

$$\rho_\Phi(\alpha f_n) = \sum_{k=1}^{\infty} \Phi(\alpha x_{nk})\mu(\Omega_{nk})$$
$$\geq \sum_{k \geq k_0} \Phi((1 + \tfrac{1}{k})x_{nk})\mu(\Omega_{nk}) > \sum_{k \geq k_0} 2^k \Phi(x_{nk})\mu(\Omega_{nk})$$
$$= \sum_{k \geq k_0} 2^{-n} = +\infty, \qquad n \geq 1. \quad (7)$$

This and (6) yield, since $\alpha > 1$ is arbitrary, $N_\Phi(f_n) = 1$, $n \geq 1$. Now for each $a = (a_1, a_2, \ldots) \in \ell^\infty$, define $f_a = \sum_{n=1}^{\infty} a_n f_n$ for the f_n constructed above and $\|a\|_\infty = \sup_n |a_n| > 0$. Then f_a is supported on Ω_n, and

$$\rho_\Phi\left(\frac{1}{\|a\|_\infty}f_a\right) = \sum_{n=1}^{\infty} \rho_\Phi\left(\frac{1}{\|a\|_\infty}a_n f_n\right) \leq \sum_{n=1}^{\infty} \rho_\Phi(f_n) \leq 1, \text{ by (6)}.$$

Thus $N_\Phi(f_a) \leq \|a\|_\infty$. For the opposite inequality, let $0 < \varepsilon < \|a\|_\infty$. Then there is an $n_0 \geq 1$ such that $|a_{n_0}| > \|a\|_\infty - \frac{\varepsilon}{2}$, and if α_ε is defined by

$$1 < \alpha_\varepsilon = \frac{\|a\|_\infty - \frac{\varepsilon}{2}}{\|a\|_\infty - \varepsilon} < \frac{|a_{n_0}|}{\|a\|_\infty - \varepsilon},$$

then by (7) for this α_ε, we get

$$\rho_\Phi\left(\frac{f_a}{\|a\|_\infty - \varepsilon}\right) = \sum_{n=1}^{\infty} \rho_\Phi\left(\frac{|a_n|}{\|a\|_\infty - \varepsilon}f_n\right) \geq \rho_\Phi\left(\frac{a_{n_0}}{\|a\|_\infty - \varepsilon}f_{n_0}\right) = +\infty.$$

Hence $\|a\|_\infty - \varepsilon < N_\Phi(f_a)$, and so $\|a\|_\infty = N_\Phi(f_a)$. Then the subspace $S = \{f_a : a \in \ell^\infty\}$ of $L^\Phi(\mu)$ is isomorphic to ℓ^∞ so that (iii) is violated. ∎

    This result indicates the important role played by the Δ_2-condition in the geometric properties of the Orlicz spaces, especially when the underlying measure space is diffuse. In the latter case, Φ being Δ_2-regular, the above theorem and a result of Rosenthal (1978) imply that $L^\Phi(\mu)$ is either reflexive or contains a subspace isomorphic to ℓ^1. Since in these cases the three weak topologies defined here specialize to give additional information, we turn to an analysis of compactness of subsets of $L^\Phi(\mu)$ in these structures. That will enable us to consider further applications having significant interest for the subject.

4.5. Weak compactness in L^Φ-spaces

It was already noted in the last section that $\sigma(L^\Phi(\mu),(L^\Phi(\mu))^*)$ is the weak- and, when Φ is an N-function, $\sigma(L^\Phi(\mu),M^\Psi)$ the weak*-topology of $L^\Phi(\mu)$. Here we discuss some compactness criteria for these topologies and establish a relation with uniformly integrable sets in these spaces. Actually we consider only weak sequential compactness. This however is the same as the general case because of a deep classical result in Banach space theory known as the Eberlein-Šmulian theorem according to which a subset A of a Banach space is relatively weakly compact iff it is relatively weakly sequentially compact. In other words, for a weakly closed set B, both concepts coincide. [For a proof, and related discussion, we refer the reader to Dunford-Schwartz (1958), Thm. V.6.1.] Another result of a similar nature from the general theory that we have occasions to employ (but it also helps to understand this work) is the following, called the Alaoglu-Bourbaki-Kakutani (ABK) theorem: If $\mathcal{X}$ is a Banach space and $\mathcal{X}^*$ its adjoint, then the closed ball $B = \{x^* \in \mathcal{X}^* : \|x^*\| \leq \alpha\}$ is compact in the $\sigma(\mathcal{X}^*,\mathcal{X})$ (or weak*-) topology of $\mathcal{X}^*$ (for a proof, cf., e.g., the preceding reference, Thm. V.4.2).

We start with the following criterion for compactness:

Theorem 1. *Let (Φ,Ψ) be a Young complementary pair in which Φ is continuous and strictly increasing. Then a subset $S \subset L^\Phi(\mu)$ is conditionally sequentially compact for $\sigma(L^\Phi(\mu),M^\Psi)$ iff (i) S is bounded in norm topology, and (ii) for each $g \in M^\Psi$ and any sequence $E_n \in \Sigma, E_n \searrow \emptyset$, as $n \to \infty$ one has $\int_{E_n} |fg|d\mu \to 0$ uniformly for f in S.*

Proof: Suppose that S is relatively sequentially compact in the $\sigma(L^\Phi(\mu),M^\Psi)$-topology. Then S is certainly bounded. Indeed, in the contrary case, for each $g \subset M^\Psi$ and $n \geq 1$, there will exist $f_n \in S$ such that we would have $|\int_\Omega f_n g d\mu| \geq n$. But by hypothesis $\{f_n, n \geq 1\}$ ($\subset S$) has a convergent subsequence f_{n_k} in this topology so that in particular, the set of numbers $\{|\int_\Omega f_{n_k} g d\mu|, k \geq 1\}$ is bounded and this violates our selection of the sequence f_n. Thus there is no such sequence and if $x_g^* \in (L^\Phi(\mu))^*$ corresponds to $g \in \mathcal{M}^\Psi$, then $\{x_g^*(f), f \in S\}$ is bounded for each $g \in \mathcal{M}^\Psi$, and since $\mathcal{M}^\Psi$ is complete, it follows by the uniform boundedness principle that S is bounded in norm (of $L^\Phi(\mu)$). So (i) holds. If (ii) is false, then for some $g_0 \in \mathcal{M}^\Psi$ and a sequence $E_n \searrow \emptyset, E_n \in \Sigma$, we must have $|\int_{E_n} f_n g_0 d\mu| \geq \varepsilon > 0$ for some $\varepsilon > 0$ and some $\{f_n, n \geq 1\} \subset S$. But again by the compactness condition of S, there exists a subsequence f_{n_k} such that $\int_\Omega f_{n_k} g_0 \chi_A d\mu \to$ a limit for each $A \in \Sigma$, as $k \to \infty$. If we let $\nu_k(A) = \int_\Omega f_{n_k} g_0 \chi_A d\mu$, then for each k, $\nu_k : \Sigma \to I\!R$ is μ-continuous and σ-additive, and since $\nu_k(A) \to \nu(A)$ for each $A \in \Sigma$, by the Nikodým classical Vitali-Hahn-Saks theorem (cf., e.g., Rao (1987), p. 181) $\nu(\cdot)$ is σ-additive and for any sequence $A_n \searrow \emptyset$, $\lim_{n\to\infty} \nu_k(A_n) = 0$ uniformly in k. In particular then $\nu_k(E_k) \to 0$ as $k \to \infty$, which contradicts our choice of g_0 and E_n. This shows that (ii) must also hold if S is conditionally $\sigma(L^\Phi(\mu), \mathcal{M}^\Psi)$-sequentially compact. As yet the continuity hypothesis on Φ is not used. However, we need it for the proof of the reverse direction.

Thus for the converse let the conditions (i) and (ii) hold. Now by (i) S is norm bounded. If $\{f_n, n \geq 1\} \subset S$ is any sequence, we need to show that it has a convergent subsequence in the given topology. Now using the fact that Φ is continuous and $\Phi(x) > 0$ for $x > 0$, we note that the $\{f_n, n \geq 1\}$ vanishes outside a set Ω_1 of σ-finite measure and hence for this proof we may assume that μ is σ-finite. To use another simplification note that the smallest σ-algebra $\widetilde{\Sigma} \subset \Sigma$ relative to which $\{f_n, n \geq 1\}$ is measurable, is countably generated since in $f_n^{-1}(\mathcal{B})$, $\mathcal{B}$ the Borel σ-algebra of $I\!R$, has that property and $n = 1, 2, \ldots$. Thus $(\Omega_1, \widetilde{\Sigma}, \mu)$ is a separable σ-finite measure space in which there is a sequence $\{A_n, n \geq 1\}$ of sets generating $\widetilde{\Sigma}$. Moreover $\{\chi_{A_n}, n \geq 1\} \subset \mathcal{M}^\Psi$, and is dense in the latter. Consequently $\int_\Omega f_n \chi_A d\mu$, $A \in \{A_n, n \geq 1\}$, is a bounded sequence of reals, and we proceed to select, by the Cantor diagonal process, a subsequence $\{g_n, n \geq 1\}$ from $\{f_n, n \geq 1\}$, which converges for each A_n. Thus if $\nu(A) = \int_A g_n d\mu$, $A \in \widetilde{\Sigma}_0$, the algebra generated by the A_n's, then $\nu(A) = \lim_{n\to\infty} \nu_n(A)$. But by Theorem 4.1 $L^\Phi(\mu)$ is $\sigma(L^\Phi(\mu), \mathcal{M}^\Psi)$ sequentially complete so that the $g_n \to g$ for some $g \in L^\Phi(\mu)$ is this topology. But this implies, from the initial reduction, that S is conditionally sequentially compact in $\sigma(L^\Phi(\mu), \mathcal{M}^\Psi)$-topology, as asserted. ∎

Remark. In the above proof we used only the fact that $\mathcal{M}^\Psi$ is norm determining for $L^\Phi(\mu)$. Hence the result, with the same proof, holds for the $\sigma(L^\Phi(\mu), L^\Psi(\mu))$-topology also. Unless Φ is Δ_2-regular, this topology is strictly weaker than the weak topology of $L^\Phi(\mu)$, which is $\sigma(L^\Phi(\mu), \mathcal{A}_\Psi(\mu))$, and the above assertions need not hold for the latter topology.

In the converse part of the proof, we used the strict continuity of Φ only in reducing the result for the σ-finite measures so that an earlier proposition on sequential completeness that relied on the Radon-Nikodým theorem was invoked. Thus if we restrict the measure space, the result can be stated with the same proof, for general Φ, Ψ as follows:

Corollary 2. *Let (Φ, Ψ) be a complementary pair of Young functions, (Ω, Σ, μ) be σ-finite, and $L^\Phi(\mu)$ be the Orlicz space. Then a set $S \subset L^\Phi(\mu)$ is conditionally sequentially compact for the topology $\sigma(L^\Phi(\mu), \mathcal{M}^\Psi)$ [or $\sigma(L^\Phi(\mu), L^\Psi(\mu))$] iff (i) S is bounded in this topology (hence also in norm), and (ii) for each $g \in \mathcal{M}^\Psi[L^\Psi(\mu)]$ and each $E_n \in \Sigma$, $E_n \searrow \emptyset$, one has $\lim_{n\to\infty} \int_{E_n} |fg| d\mu = 0$ uniformly in $f \in S$.*

The preceding result and its proof are a simple modification of the classical $L^1(\mu)$ case (cf., e.g., Dunford-Schwartz (1958), IV.8.9), and the statement admits a similar extension to numerous Banach function spaces, as shown in Luxemburg and Zaanen (1963).

With a further specialization of the Young functions and/or the measures, we may present simpler conditions for weak sequential compactness. The following, due to Andô (1962), is such a result. An extension of this to σ-finite μ will be indicated later.

Theorem 3. *Let Φ be an N-function and (Ω, Σ, μ) a finite measure space. Then a set $S \subset L^\Phi(\mu)$ is in $\sigma(L^\Phi(\mu), L^\Psi(\mu))$-topology conditionally sequentially compact, Ψ being complementary to Φ, iff*

$$\lim_{k\to 0} \frac{1}{k} \int_\Omega \Phi(kf) d\mu = 0 \tag{1}$$

uniformly in $f \in S$.

Proof: Suppose S is $\sigma(L^\Phi(\mu), L^\Psi(\mu))$-conditionally sequentially compact. Then as seen in the proof of Theorem 1 and before, it follows (as a consequence of the uniform boundedness principle) that S is norm bounded in $L^\Phi(\mu)$. For convenience we can assume that S is contained in the unit ball of $(L^\Phi(\mu), N_\Phi(\cdot))$ and hence $\rho_\Phi(f) \le 1$ for all $f \in S$. To see that (1) holds, suppose the contrary. Then there exist an $\varepsilon > 0$, $k_n \searrow 0$, and a sequence $\{f_n, n \ge 1\} \subset S$ such that

$$\frac{1}{k_n} \int_\Omega \Phi(k_n f_n)d\mu = \frac{1}{k_n}\rho_\Phi(k_n f_n) \ge \varepsilon, \ n \ge 1. \tag{2}$$

Since Φ is an N-function so that $\frac{\Phi(x)}{x} \to 0$ as $x \to 0$, we may select a subsequence of k_n (denoted by the same symbol) such that

$$0 < k_n \le 2^{-n}, \sum_{n=1}^{\infty} k_n \le 1 \text{ and } \frac{\Phi(nk_n)\mu(\Omega)}{k_n} \le \frac{\varepsilon}{2}, \ n \ge 1. \tag{3}$$

If φ is the (left) derivative of Φ, let $g = \sup_{n \ge 1} \varphi(k_n|f_n|)$. Then $g \in L^\Psi(\mu)$. In fact, using Young's equality,

$$\Psi(\varphi(x)) = x\varphi(x) - \Phi(x) \le \Phi(2x) - \Phi(x) \le \Phi(2x), \ x \ge 0, \tag{4}$$

we have

$$\int_\Omega \Psi(g)d\mu \le \sum_{n=1}^{\infty} \int_\Omega \Psi(\varphi(k_n|f_n|))d\mu$$

$$\le \sum_{n=1}^{\infty} \int_\Omega \Phi(2k_n f_n)d\mu, \text{ by (4)},$$

$$\le \sum_{n=1}^{\infty} 2k_n \int_\Omega \Phi(f_n)d\mu = 2\sum_{n=1}^{\infty} k_n \le 2, \text{ by (3)}.$$

If $\Omega_n = \{\omega : |f_n(\omega)| > n\}$, then using the fact that $L^\Phi(\mu) \subset L^1(\mu)$, we get $\mu(\Omega_n) \searrow 0$ so that $\Omega_n \searrow \emptyset$ a.e. Hence by Theorem 1, there is an $n_0 \ge 1$ such that for all $n \ge n_0$,

$$\int_{\Omega_n} |f_n g|d\mu < \frac{\varepsilon}{2}. \tag{5}$$

From this we obtain

$$\int_\Omega \Phi(k_n f_n) d\mu = \int_{\Omega_n} \Phi(k_n f_n) d\mu + \int_{\Omega_n^c} \Phi(k_n f) d\mu$$

$$\leq \int_{\Omega_n} [\Phi(k_n f_n) + \Psi(\varphi(k_n|f_n|))] d\mu + \Phi(n k_n) \mu(\Omega_n^c)$$

$$\leq \int_{\Omega_n} |k_n f_n \varphi(k_n|f_n|)| d\mu + k_n \frac{\varepsilon}{2},$$

by the Young equality and (3),

$$\leq k_n \int_{\Omega_n} |f_n g| d\mu + k_n \frac{\varepsilon}{2} < k_n \varepsilon, \text{ by (5).}$$

But this contradicts (2), so that (1) must hold.

For the converse let (1) be satisfied. Given a $g \in L^\Psi(\mu)$, let $r > 0$ be chosen such that $\rho_\Psi(rg) < \infty$. Then for any $\varepsilon > 0$, there is a $k_0 > 0$ such that (1) becomes

$$\sup\left\{\frac{1}{k_0} \int_\Omega \Phi(k_0 f) d\mu : f \in S\right\} < \frac{\varepsilon r}{2}. \tag{6}$$

By the integrability of $\Psi(rg)$, we can find a $\delta > 0$ such that for $E \in \Sigma$, $\mu(E) < \delta \Rightarrow$

$$\int_E \Psi(rg) d\mu < \frac{\varepsilon r k_0}{2}. \tag{7}$$

Hence

$$\sup_{f \in S} \int_E |fg| d\mu = \frac{1}{rk_0} \sup_{f \in S} \int_E |k_0 frg| d\mu$$

$$\leq \frac{1}{rk_0}\left\{\sup_{f \in S} \int_E \Phi(k_0 f) d\mu + \int_E \Psi(rg) d\mu\right\}$$

$$< \frac{1}{r}\left(\frac{\varepsilon r}{2}\right) + \frac{\varepsilon}{2} = \varepsilon, \text{ by (6) and (7).} \tag{8}$$

Taking a sequence of $E_n \searrow \emptyset$ with $\mu(E_1) < \delta$, (8) implies that the condition (ii) of Theorem 1 is satisfied, and hence S is conditionally sequentially compact in the given (weak) topology. ∎

Note that, for the converse above, Φ can be any continuous Young function with $\Phi(x) > 0$ for $x > 0$. The more restrictive condition that Φ be an N-function is not needed. Another consequence of the preceding results will be given. For this we recall a classical theorem of Banach. If $\mathcal{X}^*$ is the adjoint space of a Banach space $\mathcal{X}$, then the closed unit ball of $\mathcal{X}^*$ which is compact in the weak*-topology (by the ABK theorem recalled before) is a metric topology iff $\mathcal{X}$ is separable (cf., e.g., Dunford-Schwartz (1958), V.5.1), and hence it will be a compact metric space. But then its compact and weak sequential compact concepts are equivalent (cf., e.g., the same reference I.6.15). Using this result, we have the following in the context of Orlicz spaces which was essentially noted by Ren (1986).

Proposition 4. *Let* (Φ,Ψ) *be a complementary pair of N-functions and* (Ω, Σ, μ) *be a finite separable measure space. Then each closed norm bounded set in* $L^\Phi(\mu)$ *is* $\sigma(L^\Phi(\mu), L^\Psi(\mu))$*-sequentially compact if* $\Phi \in \nabla_2$. *On the other hand, if* μ *is diffuse, then the* ∇_2*-condition is also necessary for the asserted compactness.*

Proof: If $\Phi \in \nabla_2$ then by Theorem II.3.3, the complementary N-function $\Psi \in \Delta_2$. Hence $M^\Psi = L^\Psi(\mu)$ and since (Ω, Σ, μ) is separable, by Theorem III.5.1 $L^\Psi(\mu)$ is separable. Also $(L^\Psi(\mu))^* = L^\Phi(\mu)$ by Corollary 4.9. Hence by the above noted result of Banach and the ABK theorem, we deduce that any closed bounded set of $L^\Phi(\mu)$ is weak*-sequentially compact. Actually this can also be proved directly by the Cantor diagonal method, as in the proof of Theorem 1.

For the reverse direction, suppose if possible that $\Phi \notin \nabla_2$, and that any closed bound S in $L^\Phi(\mu)$ is weak*-sequentially compact where μ is a finite diffuse measure. Since by the definition of membership in ∇_2, for each $\varepsilon > 0$, there is a $\delta > 0, x_0 > 0$ such that for $x \geq x_0$ we have $\frac{1}{\delta}\Phi(\delta x) < \varepsilon\Phi(x)$, if $\Phi \notin \nabla_2$ then there exist $x_n \nearrow \infty$, $\varepsilon_0 > 0$, such that

$$n\Phi(\tfrac{x_n}{n}) > \varepsilon_0\Phi(x_n), \qquad n \geq 1. \tag{9}$$

We may take $\Phi(x_1)\mu(\Omega) \geq 1$. By diffuseness of μ, one can select $\Omega_n \in \Sigma$ such that $\Phi(x_n)\mu(\Omega_n) = 1$, $n \geq 1$. If $f_n = x_n\chi_{\Omega_n}$, then we get $N_\Phi(f_n) = 1$, and thus the set $S = \{f_n, n \geq 1\}$ is in the unit ball of $L^\Phi(\mu)$. Hence from (1) and (9) with $k_n = \frac{1}{n}$ there, we have

$$\limsup_{n\to\infty} \frac{1}{k_n} \frac{1}{k_n}\int_\Omega \Phi(k_n f_n)d\mu = \limsup_{n\to\infty}\frac{n\Phi(\frac{1}{n}x_n)}{\Phi(x_n)} \geq \varepsilon_0.$$

By Theorem 3 therefore S cannot be $\sigma(L^\Phi(\mu), L^\Psi(\mu))$-conditionally sequentially compact, contradicting the hypothesis. ∎

Since by Theorem III.5.1, for any N-function Ψ, M^Ψ is separable iff the measure space (Ω, Σ, μ) is separable, the first part of the above theorem can be stated as follows. Here one uses the fact that $(M^\Psi)^* = L^\Phi(\mu)$. This is a consequence of Theorem 1.6.

Corollary 5. *Let (Φ, Ψ) be a complementary pair of N-functions and (Ω, Σ, μ) be a separable measure space. Then each closed norm bounded set $S \subset L^\Phi(\mu)$ is $\sigma(L^\Phi(\mu), M^\Psi)$-sequentially compact.*

We can present a connection between the convergence in measure and a suitable weak convergence for an Orlicz space.

Proposition 6. *Let (Ω, Σ, μ) be a finite separable measure space and Φ be an N-function with Ψ as its complementary function. If $\{f_n, n \geq 1\}$ is a bounded set in $L^\Phi(\mu)$ and if $f_n \to f$ in μ-measure then it is also convergent in the $\sigma(L^\Phi(\mu), M^\Psi)$-topology.*

Proof: Since (Ω, Σ, μ) is separable and Ψ, being an N-function, is strictly increasing and continuous, we deduce that M^Ψ is a separable Banach space by Theorem III.5.1. Further $(M^\Psi)^* = L^\Phi(\mu)$, as noted already. Hence each bounded set is conditionally sequentially compact by the above corollary. Thus $f_{n_k} \to \tilde{f}$ in this topology for a subsequence $\{f_{n_k}, k \geq 1\}$. [$\tilde{f} \in L^\Phi(\mu)$ because of Theorem 4.1.] Note also that $f_n \to f$ in measure implies, $f_{n_k} \to f$ in the same sense. We now show using the finiteness of μ at this point that $f = \tilde{f}$ a.e. This will immediately imply that each convergence subsequence of $\{f_n, n \geq 1\}$ will have the same limit f, so that the whole sequence converges in the desired (weak) sense.

Since $f_n \to f$ in measure implies $f_{n_i} \to f$ a.e. for a subsequence and hence $|f_{n_i}| \to |f|$ a.e., we get by Fatou's lemma that $f \in L^\Phi(\mu)$. By the fact that $\mu(\Omega) < \infty$ and Φ is an N-function, it follows that the bounded set $\{f_n, n \geq 1\}$ of $L^\Phi(\mu)$ is uniformly integrable by Theorem I.2.2. Hence $\{f_{n_k}, k \geq 1, f, \tilde{f}\}$ also has the same property. Also $\int_\Omega (f_{n_k} - \tilde{f})g d\mu \to 0$ as $k \to \infty$ for each $g \in M^\Psi$. In particular let $\tilde{g} = \chi_{A_m}\text{sgn}(f - \tilde{f})$, where $A_m = \{\omega : |f - \tilde{f}|(\omega) \leq m\}$. Thus for each

$\varepsilon > 0$, there is a k_0 such that for $k \geq k_0$,

$$\left| \int_\Omega (f_{n_k} - \tilde{f}) \tilde{g} d\mu \right| < \frac{\varepsilon}{5}. \tag{10}$$

By Theorem I.2.2, for this $\varepsilon > 0$ there is a $\delta > 0$ such that $\mu(A) < \delta \Rightarrow$

$$\int_A |\tilde{f}| d\mu < \frac{\varepsilon}{5}, \quad \int_A |f_{n_k}| d\mu < \frac{\varepsilon}{5}, \quad k \geq 1. \tag{11}$$

Using the convergence in measure hypothesis choose k_1 such that $k \geq k_1$ implies

$$\mu(B_k) < \delta, \tag{12}$$

where $B_k = \{\omega : |f_{n_k} - f|(\omega) > \frac{\varepsilon}{5\mu(\Omega)}\}$. If $\tilde{k} = \max(k_0, k_1)$ then for $k > \tilde{k}$ we have, if $\delta < (\varepsilon/5m)$ (cf., (10)–(12)),

$$\int_\Omega |f - \tilde{f}| \chi_{A_m} d\mu \leq \left| \int_\Omega (f_{n_k} - \tilde{f}) \tilde{g} d\mu \right| + \int_\Omega |f - f_{n_k}| \chi_{A_m} d\mu$$

$$< \frac{\varepsilon}{5} + \int_{B_k} |f - f_{n_k}| \chi_{A_m} d\mu + \int_{B_k^c} |f - f_{n_k}| d\mu$$

$$< \frac{\varepsilon}{5} + \int_{A_m \cap B_k} |f - \tilde{f}| d\mu + \int_{B_k} |\tilde{f} - f_{n_k}| d\mu + \frac{\varepsilon}{5\mu(\Omega)} \mu(B_k^c)$$

$$< \frac{2\varepsilon}{5} + m \cdot \mu(A_m \cap B_k) + \int_{B_k} |\tilde{f}| d\mu + \int_{B_k} |f_{n_k}| d\mu$$

$$< \frac{2\varepsilon}{5} + m \cdot \frac{\varepsilon}{5m} + \frac{\varepsilon}{5} + \frac{\varepsilon}{5} = \varepsilon. \quad \blacksquare$$

We now include some comments on an extension of the weak compactness statements from finite to σ-finite measures. Let (Ω, Σ, μ) be σ-finite. Then it is well known that there is a finite measure ν or Σ which is equivalent to μ. Indeed, let $\{A_n, n \geq 1\}$ be a disjoint sequence of sets A_n ($\in \Sigma$) of finite μ-measure such that $\Omega = \bigcup_{n=1}^\infty A_n$, which is a consequence of the σ-finiteness of μ. Let $\nu : \Sigma \to I\!R^+$ be defined by the relation

$$\nu(E) = \sum_{n=1}^\infty \frac{2^{-n} \mu(A_n \cap E)}{1 + \mu(A_n)}. \tag{13}$$

It is clear that $\nu(\Omega) \le 1$ and that μ and ν are equivalent measures. Moreover, if $f \in L^\Phi(\mu)$ then

$$\int_\Omega \Phi(f)d\nu = \sum_{n=1}^\infty \int_{A_n} \frac{\Phi(f)d\mu}{2^n(1+\mu(A_n))} \le \sum_{n=1}^\infty 2^{-n} \int_{A_n} \Phi(f)d\mu$$

$$\le \int_\Omega \Phi(f)d\mu. \tag{14}$$

Hence $L^\Phi(\nu) \supset L^\Phi(\mu)$ and that the topology of $L^\Phi(\mu)$ is not weaker than that of $L^\Phi(\nu)$.

Suppose now that $S \subset L^\Phi(\mu)$ be a set such that (1) is satisfied, i.e.,

$$\lim_{k\to 0} \tfrac{1}{k}\rho_{\Phi,\mu}(kf) = \lim_{k\to 0} \frac{1}{k} \int_\Omega \Phi(kf)d\mu = 0, \tag{15}$$

uniformly for f in S. This implies that (cf.(14)) the same holds with ν in place of μ. But since ν is a finite measure, and (1) is satisfied for $S \subset L^\Phi(\nu)$, it follows by Theorem 3 that S is $\sigma(L^\Phi(\nu), M^\Psi)$-conditionally sequentially compact. Hence for each $\{f_n, n \ge 1\} \subset S$, there is a subsequence $f_{n_k} \to f_0$ in the stated topology, and $f_0 \in L^\Phi(\nu)$. Thus $\int_\Omega (f_{n_k} - f_{n_\ell})g d\nu \to 0$ for all $g \in L^\Psi(\nu)$. Since $L^\Psi(\mu) \subset L^\Phi(\nu)$, if we let $\xi = \frac{d\nu}{d\mu}$, then $\xi > 0$ a.e.$[\mu]$, and replacing g by $g\xi\chi_{[\xi\le 1]} = \tilde{g}$, one gets $\int_\Omega (f_{n_k} - f_{n_\ell})\tilde{g}d\mu \to 0$ and similarly replacing g by $g\xi\chi_{[\xi>1]} = \tilde{\tilde{g}}$, one can conclude that $\int_\Omega (f_{n_k} - f_{n_\ell})\tilde{\tilde{g}}d\mu \to 0$. This implies that $S \subset L^\Phi(\mu)$ is $\sigma(L^\Phi(\mu), L^\Psi(\mu))$-conditionally sequentially compact. On the other hand, if $S \subset L^\Phi(\mu)$ has the latter property (and $L^\Phi(\mu) \subset L^\Phi(\nu)$), it is also conditionally sequentially compact in $\sigma(L^\Phi(\nu), L^\Psi(\nu))$-topology since this topology of $L^\Phi(\mu)$ is not stronger than the former. But then by Theorem 3, $\lim_{k\searrow 0} \tfrac{1}{k}\rho_{\Phi,\nu}(kf) = 0$ uniformly in $f \in S$. From this using a similar argument one can conclude that $\tfrac{1}{k}\rho_{\Phi,\mu}(kf) \to 0$ uniformly in $f \in S$, as $k \searrow 0$. We omit the detail. A somewhat different method, still following Andô's idea, is found in M. Nowak (1986). The result can thus be stated as:

Proposition 7. *If Φ is an N-function and $S \subset L^\Phi(\mu)$ on a σ-finite space (Ω, Σ, μ), then S is conditionally sequentially compact in $\sigma(L^\Phi(\mu), L^\Psi(\mu))$-topology iff $\lim_{k\to 0} \tfrac{1}{k}\rho_{\Phi,\mu}(kf) = 0$ uniformly in $f \in S$.*

The point of this discussion is that, in many cases, results can be extended, with only a little more effort, from the finite to the σ-finite measure space. This becomes somewhat more involved if μ is not restricted (i.e., only assumed to have the [essentially unrestricted] finite subset property). We shall omit this line of analysis and study another aspect of our function spaces, useful in some applications.

Bibliographical Notes: Extending the $L^p(\mu)$, $1 \le p \le \infty$, theory the corresponding adjoint space to be considered is $(\mathcal{M}^{\Phi})^*$ on a general measure space. If Φ is an N-function and the measure μ is finite, or σ-finite, this has been considered by Morse and Transue (1950), and the general case is not a simple extension. The N-function case with Φ satisfying the Δ_2 condition with μ finite, was originally given by W. Orlicz (1932), who also noted in (1936) that Corollary 1.9 does not have the general form. Our treatment given in Section 1, follows Rao (1964). The general $(L^{\Phi}(\mu))^*$ was stated as an open problem in Krasnosel'skii and Rutickii's book. A key special case of this characterization was given by Andô (1960c), and the characterization without unnecessary restrictions was stated in Rao (1964) and the full detail appears in Rao (1968), whose treatment is followed in Section 2. A more detailed analysis of singular linear functionals, presented in Section 3, combines the works of Rao (1968), Zaanen (1967) and de Jonge (1973). Proposition 3.6 is essentially in Andô (1960c). The σ-finiteness of the underlying measure is always assumed in the latter works. In this context our treatment is based on the paper by Luxemburg and Zaanen (1963)., specialized from the Banach function space theory. A slightly restricted version of Theorem 4.3 appears in Kantorovitch and Akilov (1982). Similarly the first half of Corollary 4.4 is in Wang (1985) and Theorem 4.5 is in Turett (1976).

The discussion on weak topologies is an adaptation of the general Banach space theory to the Orlicz space context. We have been influenced from the preceding papers and especially from the monograph by Krasnosel'skii and Rutickii (1961). Theorem 5.3 is due to Andô (1962) and Proposition 5.4 is a modified version of that in Ren (1986). It is of some interest to remark that in most cases we aimed at a unified presentation, sometimes involving generalizations and other times simplifications of the known results, often reworking the existing proofs.

V

Comparison of Orlicz Spaces

The material in this chapter is essentially necessary for the following work. After discussing the extensiveness and comparison of the Orlicz spaces with the classical Lebesgue classes, we consider compact subsets and embedding problems of these spaces. Further, a somewhat detailed version of the pointwise, tensor and integral (generalization of convolution) products of Orlicz spaces, initiated in Section III.3, is presented. The latter will be used in a study of integral operators in the next chapter.

5.1. *The extensiveness of Orlicz spaces*

At the outset of a study of Orlicz spaces, it is natural to ask about the extensiveness of these spaces in relation to the familiar Lebesgue spaces. Essentially as in the latter classical case, inequivalent N-functions, but not all Young functions, determine distinct Orlicz spaces. On the other hand, one can find between a pair of distinct Orlicz spaces a nonreflexive Orlicz space. This would imply, in particular, that within a pair of reflexive Lebesgue spaces there is a nonreflexive Orlicz space which gives an idea of the existence in abundance of Orlicz spaces. They are more numerous than the Lebesgue spaces. We now present the precise details.

The first assertion shows that there is considerable flexibility in employing the Orlicz spaces in applications. This fact may be compared with the Lebesgue case where there is not such a possibility.

Theorem 1. *Let Φ be a continuous strictly increasing Young function, and $L^\Phi(\mu)$ be the associated Orlicz space on a measure space (Ω, Σ, μ). Then there exists (nonuniquely) a strictly convex Young function Φ_1 such that $L^\Phi(\mu) = L^{\Phi_1}(\mu)$ and the norms are equivalent, so that both spaces are isomorphic. Further if Φ is Δ_2-regular, so is Φ_1.*

Proof: Let Φ be a continuous Young function. Being convex, it admits a representation as

$$\Phi(x) = \int_0^x \varphi(t)dt, \tag{1}$$

where φ is the (left) derivative of Φ (cf. Theorem I.3.1). Define a new function $\tilde{\varphi}$ as $\tilde{\varphi}(t) = \varphi(t)(2 - e^{-t})$, $t \geq 0$. Then Φ_1 defined by

$$\Phi_1(x) = \int_0^{|x|} \tilde{\varphi}(t)dt, \tag{2}$$

is a strictly convex Young function since $\tilde{\varphi}$ (≥ 0) is strictly increasing. It is also evident that $\Phi(x) \leq \Phi_1(x) \leq 2\Phi(x)$, $x \in \mathbb{R}$. Hence if $L^\Phi(\mu)$ and $L^{\Phi_1}(\mu)$ are the corresponding spaces, then $f \in L^\Phi(\mu)$ iff $f \in L^{\Phi_1}(\mu)$ and the inequalities imply that $N_\Phi(f) \leq N_{\Phi_1}(f) \leq 2N_\Phi(f)$ holds. The fact that both Φ and Φ_1 are Δ_2-regular or not is immediate from the construction.

To see that Φ_1 is not the only such Young function, we now give a family of equivalent Young functions to Φ. Indeed let $\alpha > 1$. Consider φ_α given by

$$\varphi_\alpha(t) = \sum_{i=1}^{n-1} \frac{\varphi(i)}{\alpha^i} + \frac{\alpha^n + t - (n-1)}{\alpha^n}\varphi(t), \qquad t \geq 0, n \geq 1 \tag{3}$$

where the empty sum is taken as zero. If now we let

$$\Phi_\alpha(x) = \int_0^{|x|} \varphi_\alpha(t)dt, \tag{4}$$

then Φ_1 is a Young function such that $\Phi \leq \Phi_\alpha \leq \alpha\Phi$ for each $\alpha > 1$. Moreover Φ and Φ_α are simultaneously Δ_2-regular or not, and $L^\Phi(\mu) = L^{\Phi_\alpha}(\mu)$ with $N_\Phi(f) \leq N_{\Phi_\alpha}(f) \leq \alpha N_\Phi(f)$, as before. Thus all these (uncountably many) Orlicz spaces are isomorphic to the given $L^{\Phi}(\mu)$, with equivalent topologies. ∎

One of the consequences of this result is that from among the isomorphic Orlicz spaces $L^{\Phi_{\alpha}}(\mu)$ of $L^{\Phi}(\mu)$, one can often choose a space that is more useful in particular applications than the original one. Such advantage is not available in the Lebesgue classes. We shall illustrate this point later in Chapter VII when the geometric properties of Orlicz spaces are analyzed in detail.

If we restrict the measure space to be finite, then Φ can be more general, as seen from

Proposition 2. *Let* (Ω, Σ, μ) *be a finite measure space and* Φ *be a Young function such that* $\Phi(x) < \infty$ *for all* $|x| < \infty$ *(so* $\Phi(\mathbb{R}) \subset \mathbb{R}^{+}$ *). Then there is a strictly convex Young function* Φ_1 *(as in the preceding theorem), such that* $L^{\Phi}(\mu) = L^{\Phi_1}(\mu)$, *i.e., these spaces are isomorphic. (Of course* Φ_1 *is again not unique.)*

Proof: By hypothesis Φ is finite and convex, hence continuous, but $\Phi(x) = 0$ for $0 \leq x < x_1$ is possible. In this case, which is the only new possibility that is not covered by Theorem 1, there exists a continuous Young function Φ_0, satisfying $\Phi_0(x) > 0$ for $|x| > 0$, for which Φ forms the principal part (cf., e.g., Lemma III.4.1). Since $\mu(\Omega) < \infty$, both $L^{\Phi}(\mu)$ and $L^{\Phi_0}(\mu)$ have the same elements and their norms are equivalent. But by Theorem 1, $L^{\Phi_0}(\mu)$ is isomorphic to $L^{\Phi_1}(\mu)$ where Φ_1 is strictly convex. Hence $L^{\Phi}(\mu)$ and $L^{\Phi_1}(\mu)$ have equivalent norms, and contain the same elements. ∎

The remaining case not covered by the preceding two results is that Φ is discontinuous so that $\Phi(x) < \infty$ for $|x| \leq x_1 < \infty$ and $\Phi(x) = +\infty$, $|x| > x_1$. In this event only a somewhat weaker assertion holds and it and other geometric properties will be studied in Section VII.1. It turns out that with $\mu(\Omega) < \infty$, every $L^{\Phi}(\mu)$ is isomorphic to a strictly convex Banach space which need not always be an Orlicz space. Here by definition a Banach space $\mathcal{X}$ is *strictly convex* (or *rotund*) if, for any pair of distinct elements x, y of $\mathcal{X}$, its norm has the property that $\|\alpha x + (1 - \alpha)y\| < \alpha\|x\| + (1 - \alpha)\|y\|$ for each $0 < \alpha < 1$.

From the very definition of an Orlicz space, it follows that $L^{\Phi_1}(\mu)$ and $L^{\Phi_2}(\mu)$ on a given measure space (Ω, Σ, μ) are inequivalent if Φ_1 and Φ_2 are inequivalent and the spaces are not finite dimensional. Thus for Φ_1 and Φ_2 such that $\Phi_1 \prec \Phi_2$ or $\Phi_2 \prec \Phi_1$ is not satisfied, then $L^{\Phi_1}(\mu)$ and $L^{\Phi_2}(\mu)$ are distinct Orlicz spaces. However, when one of these relations holds (globally, or locally if $\mu(\Omega) < \infty$), then the corresponding inclusion relation is valid. We establish the following result.

Theorem 3. *Let $\Phi_i, i = 1, 2$ be a pair of (not necessarily complementary) Young functions. If $L^{\Phi_i}(\mu)$, $i = 1, 2$, are Orlicz spaces, then $\Phi_1 \prec \Phi_2 [\Leftrightarrow \Phi_2 \succ \Phi_1]$, so that $\Phi_1(x) \leq \Phi_2(bx)$, $x \geq x_0 \geq 0$ (and $x_0 = 0$ if $\mu(\Omega) = \infty$) for some $b > 0$, implies $L^{\Phi_1}(\mu) \supset L^{\Phi_2}(\mu)$. The converse implication holds whenever μ is diffuse on a set of positive μ-measure and Φ_i, $i = 1, 2$, are also strictly increasing and continuous.*

Proof: If $\Phi_1 \prec \Phi_2$ holds globally (so $x_0 = 0$ in the ordering), then the inclusion holds at once from the definition for any measure μ. Suppose then it is only a local ordering and $\mu(\Omega) < \infty$, and let $f \in L^{\Phi_2}(\mu)$ and $N_{\Phi_2}(f) > 0, x_0 > 0$. If $K = 1 + \Phi_1(x_0)\mu(\Omega)$, and $A = \{\omega : |f(\omega)| < x_0 b N_{\Phi_2}(f)\}$, then $\Phi_1 \prec \Phi_2$ gives

$$\int_\Omega \Phi_1\left(\frac{f}{bK N_{\Phi_2}(f)}\right) d\mu \leq \frac{1}{K} \int_\Omega \Phi_1\left(\frac{f}{b N_{\Phi_2}(f)}\right) d\mu$$

$$\leq \frac{1}{K}\left\{\Phi_1(x_0)\mu(A) + \int_\Omega \Phi_2\left(\frac{f}{N_{\Phi_2}(f)}\right) d\mu\right\}$$

$$\leq \frac{1}{K}\left\{\Phi_1(x_0)\mu(\Omega) + 1\right\} = 1. \tag{5}$$

Hence $N_{\Phi_1}(f) \leq bK N_{\Phi_2}(f)$ so that $f \in L^{\Phi_1}(\mu)$.

For the converse we need additional restrictions. Thus suppose that $L^{\Phi_1}(\mu) \supset L^{\Phi_2}(\mu)$ but $\Phi_1 \prec \Phi_2$ is not true. We shall show then that the inclusion is also not true. Since μ has the finite subset property, there is a finite positive $\Omega_0 \in \Sigma$ on which μ is diffuse and it suffices to show that there is a function supported by Ω_0 which is in $L^{\Phi_2}(\mu)$ but not in $L^{\Phi_1}(\mu)$. For this we may and do assume that (Ω, Σ, μ) is a finite diffuse space from now on. Then there is a sequence of numbers $a_n \nearrow \infty$ such that $\Phi_1(a_n) > \Phi_2(n^2 a_n 2^n)$, $n \geq 1$. This is possible since $\Phi_1 \prec \Phi_2$ is false by assumption. From the above inequality combined with the fact that Φ_2 is convex, we get

$$\Phi_1(a_n) > \Phi_2(n^2 a_n 2^n) \geq 2^n \Phi_2(n^2 a_n), \quad n \geq 1. \tag{6}$$

Now using the diffuseness of μ on Ω, we may find a disjoint sequence of measurable sets Ω_n such that $\mu(\Omega_n) > 0$ and

$$\mu(\Omega_n) = \frac{\Phi_2(a_1)\mu(\Omega)}{2^n\Phi_2(n^2a_n)}, \qquad n \geq 1. \tag{7}$$

This is possible, since $\Phi_2(a_n) > 0$, $\sum_{n=1}^{\infty} \mu(\Omega_n) < \mu(\Omega) < \infty$. Let $f_0 = \sum_{n=1}^{\infty} b_n\chi_{\Omega_n}$ where $b_n = na_n$. Then $f_0 \in L^{\Phi_2}(\mu)$ and in fact even in M^{Φ_2} since for any $\alpha > 0$,

$$\int_{\Omega} \Phi_2(\alpha f_0)d\mu = \sum_{n=1}^{\infty} \Phi_2(\alpha b_n)\mu(\Omega_n)$$

$$\leq \sum_{n=1}^{n_0} \Phi_2(\alpha b_n)\mu(\Omega_n) + \sum_{n>n_0} \Phi_2(nb_n)\mu(\Omega_n),$$

$$\text{where } n_0 \geq \alpha > 0,$$

$$\leq n_0\Phi_2(\alpha b_{n_0})\mu(\Omega) + \Phi_2(b_1)\mu(\Omega) < \infty, \text{ by } (7). \tag{8}$$

However, for any $\varepsilon > 0$, if $n_1 > \frac{1}{\varepsilon}$, then

$$\int_{\Omega} \Phi_1(\varepsilon f_0)d\mu > \sum_{n\geq n_1} \Phi_1(\varepsilon b_n)\mu(\Omega_n)$$

$$\geq \sum_{n\geq n_1} \Phi_1\left(\frac{b_n}{n}\right)\mu(\Omega_n) = \infty, \quad \text{by } (6) \text{ and } (7).$$

Since $\varepsilon > 0$ is arbitrary, $f_0 \notin L^{\Phi_1}(\mu)$, giving the desired conclusion. ∎

As an immediate consequence of the above result and of the computation given in (8), we deduce the following:

Corollary 4. *Let Φ_i, $i = 1, 2$, be a pair of continuous strictly increasing Young functions with $L^{\Phi_i}(\mu)$ and M^{Φ_i} ($\subset L^{\Phi_i}(\mu)$) being the corresponding spaces. If (Ω, Σ, μ) is a finite diffuse measure space, then $\Phi_1 \sim \Phi_2$ (i.e., $\Phi_1 \prec \Phi_2$ and $\Phi_2 \prec \Phi_1$) iff $L^{\Phi_1}(\mu) = L^{\Phi_2}(\mu)$ or equivalently $M^{\Phi_1} = M^{\Phi_2}$. The equalities in these spaces signify equivalence of norms, or topological isomorphisms but not necessarily isometries.*

In fact, by the theorem, we have $\Phi_1 \prec \Phi_2 \Leftrightarrow L^{\Phi_1}(\mu) \supset L^{\Phi_2}(\mu) \Rightarrow M^{\Phi_2} \subset L^{\Phi_1}(\mu)$. If $f \in M^{\Phi_2}$, then $\int_{\Omega} \Phi_2(\alpha f)d\mu < \infty$ for all $\alpha > 0$, and if $\Omega_1 = \{\omega : \alpha|f(\omega)| < x_0\}$, then $\Phi_1 \prec \Phi_2$ signifies

$$\int_\Omega \Phi_1(\alpha f)d\mu \leq \Phi_1(x_0)\mu(\Omega_1) + \int_{\Omega_1^c} \Phi_2(\alpha b f)d\mu < \infty$$

so that $f \in M^{\Phi_1}$. Thus $M^{\Phi_1} \supset M^{\Phi_2}$. Conversely, this last inclusion implies $M^{\Phi_2} \subset L^{\Phi_1}(\mu)$ which by the proof of the theorem gives $\Phi_1 \prec \Phi_2$. Similarly $\Phi_2 \prec \Phi_1$ yields the opposite inclusion at the same time, and this establishes the corollary.

These inclusions will be used in considering and comparing the compact subsets of both the spaces in their norm topologies. Such results complement the work of the last two sections of Chapter IV on weak sequential compactness, for applications with embedding theorems. We thus present this aspect in the next section.

5.2. *Compact subsets of Orlicz spaces*

Recall that an element f of $L^\Phi(\mu)$ has absolutely continuous norm if $N_\Phi(f\chi_{E_n}) \to 0$ for each $E_n \searrow \emptyset, E_n \in \Sigma$. A set $S \subset L^\Phi(\mu)$ is said to have *equi-absolutely continuous norms* (or *absolutely continuous norms uniformly in S*), if $\lim_{n\to\infty} \sup_{f\in S} N_\Phi(f\chi_{E_n}) = 0$ for each $E_n \searrow \emptyset$, $E_n \in \Sigma$. (Compare this with Theorem III.4.14.) Since M^Φ has absolutely continuous norm for   a continuous Φ, and $L^\Phi(\mu)$ has that same property iff $L^\Phi(\mu) = M^\Phi$, we always have $S \subset M^\Phi$ if S is with equi-absolutely continuous norms. But M^Φ is a complete metric space. So we may specialize a classical result stating that a set $S \subset M^\Phi$ is compact iff S is closed and sequentially compact or iff it is totally bounded [i.e., for each $\varepsilon > 0$ there exist a finite number of open balls of radius ε with centers in S, whose union covers S] and the closure $\bar{S}$ is complete. [For a proof, cf. Dunford-Schwartz (1958), p. 22.] Our specialization involves only μ and Φ. It is an extension of Vitali's convergence theorem.

We begin the analysis with a preliminary assertion which is related to the statement of Theorem III.4.12(a).

Lemma 1. *Let Φ be a continuous Young function with $\Phi(x) = 0$ iff $x = 0$. Let $M^\Phi \subset L^\Phi(\mu)$ be the subspace determined by the simple functions, where (Ω, Σ, μ) is a measure space. Then for $f_n \in M^\Phi$, $N_\Phi(f_n) \to 0$ iff $\rho_\Phi(kf_n) \to 0$ for each $k \geq 1$, as $n \to \infty$. In particular,* *if Φ is Δ_2-regular, then Φ-mean convergence and norm convergence are equivalent in $L^\Phi(\mu)$.*

Proof: If $\{f_n, n \geq 1\} \subset M^\Phi$ such that $N_\Phi(f_n) \to 0$, then for given $k > 0$, choose n_0 such that $kN_\Phi(f_n) \leq 1$, for all $n \geq n_0$. Hence by convexity

$$\int_\Omega \Phi(kf_n)d\mu \leq kN_\Phi(f_n) \int_\Omega \Phi\left(\frac{f_n}{N_\Phi(f_n)}\right)d\mu \leq kN_\Phi(f_n) \to 0, \qquad (1)$$

as $n \to \infty$. Thus $\rho_\Phi(kf_n) \to 0$ for any $k > 0$.

Conversely, given $\varepsilon > 0$, choose $k_0 > 1$ such that $\frac{1}{k_0} < \varepsilon$. Then there exists an n_0 such that $\rho_\Phi(k_0 f_n) \leq 1$ for $n \geq n_0$. Hence $N_\Phi(k_0 f_n) \leq 1$ so that $N_\Phi(f_n) \leq \frac{1}{k_0} < \varepsilon$. The last statement is simple. In fact it follows from Theorem III.4.12(a) itself. ∎

The following is the desired extension of a classical result (cf., Dunford and Schwartz (1958), III.6.15).

Theorem 2. *Let Φ be a strictly increasing continuous Young function and $L^\Phi(\mu)$ be the Orlicz space on a measure space (Ω, Σ, μ) with M^Φ as the corresponding subspace of functions of absolutely continuous norm. Then a set $S \subset M^\Phi$ is conditionally compact iff it is bounded and*

(i) (a) *each infinite set $\{f_n, n \geq 1\}$ from S contains a subsequence $\{f_{n_k}, k \geq 1\}$ converging in μ-measure to an element f of M^Φ;*

(b) *for each $\varepsilon > 0$, there is a set $A_\varepsilon \in \Sigma$ such that $\mu(A_\varepsilon) < \infty$ and*

$$\int_{A_\varepsilon^c} \Phi(f_{n_k})d\mu < \varepsilon, \qquad uniformly\ in\ k \geq 1;$$

(ii) *S has equi-absolutely continuous norms.*

Proof: First let $S \subset M^\Phi$ be conditionally compact. Hence it is conditionally sequentially compact also so that for each sequence $\{f_n, n \geq 1\} \subset S$, there is a norm convergent subsequence $\{f_{n_k}, k \geq 1\}$ with a limit $f \in M^\Phi$ since the latter is complete. But by the above lemma and Theorem III.4.12(a), this implies $f_{n_k} \to f$ in Φ-mean, i.e., $\rho_\Phi(f_{n_k}-f) \to 0$ as $k \to \infty$. Then by Markov's inequality $f_{n_k} \to f$ in measure. In fact, for any $\varepsilon > 0$

$$\mu\{\omega : |f_{n_k} - f|(\omega) \geq \varepsilon\} \leq \frac{1}{\Phi(\varepsilon)}\rho_\Phi(f_{n_k} - f) \to 0.$$

Thus (i)(a) holds. We next deduce that (ii) and (i)(b) are true simultaneously.

Since the conditional compactness in a metric space is equivalent to total boundedness, given $\varepsilon > 0$, there exist a finite set $g_1, \ldots, g_{n_0}$ in S and balls $B(g_i, \varepsilon)$, centers g_i and of radius ε, whose union covers S. But each g_i has an absolutely continuous norm. Since these are only finite, we get, for each sequence $A_n \in \Sigma$, $A_n \searrow \emptyset$, that $N_\Phi(g_i \chi_{A_n}) \to 0$, as $n \to \infty$, for all $1 \le i \le n_0$. Also by total boundedness given $f \in S$, there is an $i_0 \in \{1, 2, \ldots, n_0\}$ such that $N_\Phi(f - g_{i_0}) < \varepsilon$. Hence

$$N_\Phi(f \chi_{A_n}) \le N_\Phi\big((f - g_{i_0}) \chi_{A_n}\big) + N_\Phi(g_{i_0} \chi_{A_n}) < \varepsilon + N_\Phi(g_{i_0} \chi_{A_n}). \quad (2)$$

From this one has $\lim\limits_{\substack{n \to \infty \\ f \in S}} \sup N_\Phi(f \chi_{A_n}) \le \varepsilon$. Since $\varepsilon > 0$ is arbitrary, (ii) follows.

Regarding (i)(b), since $\Phi(g_i)$ is integrable we get, by the standard reasoning, for each $\varepsilon > 0$ there is an $A \in \Sigma$, $\mu(A) < \infty$ such that

$$\int_{A^c} \Phi(g_i) d\mu = \int_\Omega \Phi(g_i \chi_{A^c}) d\mu < \varepsilon, \qquad 1 \le i \le n_0. \quad (3)$$

Because of the absolute continuity of norm of M^Φ, one may assume that the A is so chosen that $N_\Phi(g_i \chi_{A^c}) < \frac{\varepsilon}{2}$. Hence for our sequence of (i), $N_\Phi(f_n - g_i) < \frac{\varepsilon}{2}$ for some i and

$$N_\Phi(f_n \chi_{A^c}) \le N_\Phi\big((f_n - g_i) \chi_{A^c}\big) + N_\Phi(g_i \chi_{A^c})$$
$$< \frac{\varepsilon}{2} + \frac{\varepsilon}{2} = \varepsilon, \qquad n \ge 1.$$

Since this is uniform in n, (i)(b) follows and S is bounded, so that the necessity is proved.

For the converse, let the given conditions (i), (ii) hold and S be bounded. Thus let $\{f_n, n \ge 1\} \subset S$ have a subsequence $f_{n_k} \to f$ in μ-measure. Then under (i) and (ii) we assert that $f_{n_k} \to f$ in norm. If this is shown, then every subsequence of S has a norm convergent subsequence so that S is conditionally sequentially compact. This gives what is desired since both the compactness conditions are equivalent in a metric space. We thus show the truth of our assertion.

Let $g_k = f_{n_k}$. Then by (ii) for each sequence $A_n \searrow \emptyset$, $A_n \in \Sigma$, $N_\Phi(g_k \chi_{A_n}) \to 0$ as $n \to \infty$, uniformly in $k \ge 1$. In particular, let

$B_{\ell k} = B^{\eta}_{\ell k} = \{\omega : |g_{\ell} - g_k|(\omega) \geq \Phi^{-1}(\eta)\}$ where $\eta > 0$ is arbitrarily fixed. Then $\mu(B_{\ell k}) \to 0$ as $k, \ell \to \infty$ by (i)(a). Also by (i)(b) for each $\varepsilon > 0$, and $\alpha > 0$, there exists an $A_{\varepsilon} \in \Sigma$, $\mu(A_{\varepsilon}) < \infty$, and $\rho_{\Phi}(\alpha g_k \chi_{A^c_{\varepsilon}}) < \varepsilon$ uniformly in k. Since $g_k \in M^{\Phi}$, this is equivalent to $N_{\Phi}(g_k \chi_{A^c_{\varepsilon}}) < c\varepsilon$ for some $0 < c < \infty$, independent of $\varepsilon > 0$ (cf. Lemma 1). Hence

$$N_{\Phi}(g_k - g_{\ell}) \leq N_{\Phi}\big((g_k - g_{\ell})\chi_{A_{\varepsilon}}\big) + N_{\Phi}(g_k \chi_{A^c_{\varepsilon}}) + N_{\Phi}(g_{\ell}\chi_{A^c_{\varepsilon}})$$
$$\leq N_{\Phi}\big((g_k - g_{\ell})\chi_{A_{\varepsilon} \cap B_{\ell k}}\big) + N_{\Phi}\big((g_k - g_{\ell})\chi_{A_{\varepsilon} - B_{\ell k}}\big) + 2c\varepsilon$$
$$\leq N_{\Phi}(g_k \chi_{B_{\ell k}}) + N_{\Phi}(g_{\ell}\chi_{B_{\ell k}}) + \eta N_{\Phi}(\chi_{A_{\varepsilon}}) + 2c\varepsilon. \qquad (4)$$

Since $\mu(B_{\ell k}) \to 0$ as $k, \ell \to \infty$, and by (ii) $N_{\Phi}(g_{\ell}\chi_{E_n}) \to 0$ for any $E_n \searrow \emptyset$, as $n \to \infty$ uniformly in ℓ, we get by letting $\ell, k \to \infty$,
$$\limsup_{k, \ell \to \infty} N_{\Phi}(g_{\ell} - g_k) \leq c\varepsilon + c'\eta, \quad \text{where } c' = N_{\Phi}(\chi_{A_{\varepsilon}}). \text{ Given } \varepsilon_0 > 0$$
choose $c\varepsilon + c'\eta < \varepsilon_0$ which is possible since $\varepsilon > 0$ and $\eta > 0$ are independently prescribable and c is independent of ε. Hence $\{g_{\ell}, \ell \geq 1\}$ is Cauchy so that S is conditionally compact. ∎

Note that if Φ is Δ_2-regular, then $M^{\Phi} = L^{\Phi}(\mu)$ and the result includes the classical case $L^p(\mu)$, $1 \leq p < \infty$. If $\mu(\Omega) < \infty$, then we may choose $A_{\varepsilon} = \Omega$ in (i)(b) and hence that condition may be omitted. Thus we have the following consequence.

Corollary 3. *If Φ is a continuous Young function with $\Phi(x) > 0$ for $x > 0$, and (Ω, Σ, μ) is a finite measure space, then a set $S \subset M^{\Phi}$ is conditionally compact iff it is bounded and (i)(a) and (ii) of the theorem are satisfied. If Φ verifies the Δ_2-condition, then M^{Φ} can be replaced by $L^{\Phi}(\mu)$ here, and (ii) is then equivalent to: (ii)′ given $\varepsilon > 0$, there is a $\delta_{\varepsilon} > 0$ such that $\mu(E) < \delta \Rightarrow \sup_{f \in S} \int_E \Phi(f)d\mu < \varepsilon$.*

The next result on the characterization was originally formulated by Kolmogorov (1931) for reflexive $L^p(\mu)$ spaces on a bounded set in the $\mathbb{R}^n$-space with μ as its Lebesgue measure. It was extended by several authors, and finally by Takahashi (1935) for the Orlicz spaces $L^{\Phi}(\mu)$ with Φ satisfying the Δ_2-regular condition and μ as the Lebesgue measure on $\mathbb{R}$. The following general form is due to Phillips (1940) in the $L^p(\mu)$ case, but it is closely related to the martingale convergence theory although that was not noticed until recently.

We present a general result, and later discuss its specialization to recover the original Kolmogorov formulation and its identification with a martingale. If (Ω, Σ, μ) is a measure space, $\pi = (A_1, \ldots, A_n)$ is a finite disjoint sequence of sets $A_i \in \Sigma$ of finite positive measure, and $f \in L^{\Phi}(\mu)$, define the π-averaged f, f_π, as:

$$f_\pi = \sum_{i=1}^{n} \frac{1}{\mu(A_i)} \left(\int_{A_i} f d\mu \right) \chi_{A_i}, \tag{5}$$

and let $U_\pi : f \mapsto f_\pi$ be the mapping, for each f in $L^{\Phi}(\mu)$. With this notation we have the following:

Theorem 4. *Let Φ be an increasing continuous Young function and (Ω, Σ, μ) be a measure space. If $S \subset M^{\Phi}$ is a bounded set, then S is conditionally compact iff $U_\pi f \to f$ uniformly for f in S, i.e., $N_\Phi(U_\pi f - f) \to 0$ as π is refined, the limit holding uniformly for f in S.*

Proof: Since U_π is clearly linear on $L^{\Phi}(\mu)$ into itself, we assert that it is a contraction. Indeed if $\pi = (A_1, \ldots, A_n)$, as given, then

$$\Phi(f_\pi) = \sum_{i=1}^{n} \Phi\left(\frac{1}{\mu(A_i)} \int_{A_i} f d\mu \right) \chi_{A_i}, \text{ since } A_i \text{ are disjoint}$$

$$\text{and } \Phi \text{ is increasing.}$$

Integrating and noting that Φ is convex,

$$\int_\Omega \Phi(U_\pi f) d\mu = \int_\Omega \Phi(f_\pi) d\mu$$

$$= \sum_{i=1}^{n} \Phi\left(\frac{1}{\mu(A_i)} \int_{A_i} f d\mu \right) \mu(A_i)$$

$$\leq \sum_{i=1}^{n} \frac{1}{\mu(A_i)} \left(\int_{A_i} \Phi(f) d\mu \right) \mu(A_i), \text{ by Jensen's inequal-}$$

$$\text{ity (cf. Proposition III.3.5),}$$

$$= \int_{\cup_i A_i} \Phi(f) d\mu \leq \int_\Omega \Phi(f) d\mu. \tag{6}$$

Replacing f by $\frac{f}{N_\Phi(f)}$ in (6) we get $N_\Phi(U_\pi f) \leq N_\Phi(f)$, so that U_π is a contraction. It is clear that U_π has a finite dimensional range. Hence U_π

maps bounded sets into (bounded hence) conditionally compact sets of a finite dimensional space. But if f is a simple function, then $\Phi(x) > 0$ for $x > 0$ implies that the support of f has finite measure and hence can be included in a π_i. Thus by the refinement order among these π's (i.e., $\pi' \succ \pi$ iff each element of π is a.e. a union of some elements of π' so that '$\succ$' is a partial ordering), it follows that if $\pi_i \succ \pi$ then $U_{\pi_i} f = f$ for such f. Since simple functions are dense in M^Φ, and $\{U_\pi\}_{\pi \in \Pi}$ are uniformly bounded (contractions), it follows that $U_\pi f \to f$ in norm for each $f \in M^\Phi$. The conclusion of the theorem is now an immediate consequence of the next lemma, valid for any Banach space.

Lemma 5 (Phillips). *Let $\mathcal{X}$ be a Banach space and $\{U_\alpha, \alpha \in I\}$ be a net of uniformly bounded linear operators on $\mathcal{X}$ into itself, I being a partially ordered set. If for each $x \in \mathcal{X}$, $\lim_\alpha U_\alpha x = x$, then this limit exists uniformly for x in any compact set $S \subset \mathcal{X}$. Conversely, if the limit exists uniformly for x in a bounded set $S \subset \mathcal{X}$ and moreover $\{U_\alpha x : x \in S\}$ is conditionally compact for each $\alpha \in I$, then S is conditionally compact.*

Proof: If S is compact, it is totally bounded so that given $\varepsilon > 0$, there exists a finite collection $\{x_1, \ldots, x_n\}$ in S such that the balls $B(x_i, \varepsilon)$ cover S. Hence by hypothesis there is an α_ε such that $\alpha \geq \alpha_\varepsilon \Rightarrow \|U_\alpha x_i - x_i\| < \varepsilon$ and then for each $x \in S$, there is an i_0, $1 \leq i_0 \leq n$ such that $\|x - x_{i_0}\| < \varepsilon$ so that

$$\|U_\alpha x - x\| \leq \|U_\alpha (x - x_{i_0})\| + \|U_\alpha x_{i_0} - x_{i_0}\| + \|x_{i_0} - x\|$$
$$\leq (\|U_\alpha\| + 1)\varepsilon + \varepsilon. \tag{7}$$

Since $\|U_\alpha\| \leq K$, $\alpha \in I$, by hypothesis, (7) is at most $(K + 2)\varepsilon$. Hence the uniform convergence of $U_\alpha x \to x$ follows for $x \in S$, since ε is arbitrary.

For the converse, let $\varepsilon > 0$ be given and choose α_ε such that $\alpha \geq \alpha_\varepsilon$ implies $\|U_\alpha x - x\| < \varepsilon$ for all $x \in S$. Since $U_\alpha S$ is conditionally compact by hypothesis, it is totally bounded so that there are $x_i \in S$, $i = 1, 2, \ldots, m$, such that the balls $B(x_i, \varepsilon)$ cover $U_\alpha S$. Hence $\|U_\alpha x - x_i\| < \varepsilon$ for some x_i, and each $x \in S$. Thus $\|x - x_i\| \leq \|x - U_\alpha x\| + \|U_\alpha x - x_i\| < 2\varepsilon$, and S is totally bounded so that it is conditionally compact. This establishes the lemma, and hence also the theorem. ■

We shall now present the earlier stated specializations. In the preceding theorem the fact that $\Phi(\cdot)$ is strictly increasing and continuous was needed in the necessity conclusion that every step function in $L^\Phi(\mu)$ has a support of finite μ-measure. Then a step function is a simple function and $U_\pi f = f$ for such f. This will no longer be true if Φ is discontinuous or if $\Phi(x) = 0$, for $0 \le x < a$, and simple functions are not necessarily dense in the $L^\Phi(\mu)$ or in $L^\infty(\mu)$. However by Proposition III.4.3, if $\mu(\Omega) < \infty$, then $M^\Phi = L^\Phi(\mu)$ in such a case, and in particular $L^\infty(\mu)$ is included. Thus Theorem 4 holds in this general context also if μ is a finite measure with the same proof. We state this for reference as follows.

Corollary 6. *If (Ω, Σ, μ) is a finite measure space and Φ is any Young function, let $M^\Phi \subset L^\Phi(\mu)$ be the corresponding subspace determined by step functions. Then a set $S \subset M^\Phi$ is conditionally compact iff it is bounded and $U_\pi f \to f$ uniformly in $f \in S$ as π is refined, where U_π is defined as in (5).*

Let us recall the concept of a martingale. If (Ω, Σ, μ) is a σ-finite measure space and $\{\mathcal{B}_\alpha, \alpha \in I\}$ is an increasing net of σ-subalgebras of Σ such that $\mu|\mathcal{B}_\alpha$ is σ-finite [in particular this condition is automatic if $\mu(\Omega) < \infty$], let f_α be a $\mathcal{B}_\alpha$-measurable μ-integrable function. Then $\{f_\alpha, \mathcal{B}_\alpha, \alpha \in I\}$ is called a *martingale* indexed by I, if for each $\alpha < \beta$ in I we have

$$\int_A f_\alpha d\mu = \int_A f_\beta d\mu, \qquad A \in \mathcal{B}_\alpha. \tag{8}$$

A family $\{f_\alpha, \alpha \in I\}$ is called *terminally uniformly integrable* if for each $\varepsilon > 0$, there is an $\alpha(\varepsilon) \in I$ such that $\{f_\alpha, \alpha > \alpha(\varepsilon)\}$ is bounded in $L^1(\mu)$ and

$$\lim_{\mu(A)\to 0} \int_A |f_\alpha| d\mu = 0, \quad \text{uniformly in } \alpha > \alpha(\varepsilon), \tag{9}$$

and there is a set $A_0 \in \Sigma$, $\mu(A_0) < \infty$, satisfying

$$\int_{A_0^c} |f_\alpha| d\mu < \varepsilon, \quad \text{uniformly in } \alpha > \alpha(\varepsilon). \tag{10}$$

The last condition may be omitted if $\mu(\Omega) < \infty$, taking $A_0 = \Omega$. A standard result in modern martingale theory states that a terminally

uniformly integrable martingale $\{f_\alpha, \mathcal{B}_\alpha, \alpha \in I\}$ converges in the norm
of $L^1(\mu)$. [This result appears, e.g., in Rao (1981), p. 209, for finite
measures and an abstract version of it is in Section 5.2 of the same
reference.] If $I = \{1, 2\}$, with $\mathcal{B}_2 = \Sigma$ and $\mathcal{B}_1 = \mathcal{B}$, then (8) says that
for any $f \in L^1(\mu)$, and $\tilde{f}$ satisfying

$$\int_A \tilde{f} d\mu = \int_A f d\mu, \qquad A \in \mathcal{B}, \tag{11}$$

gives $\{\tilde{f}, f\}$ as a trivial "martingale." That the desired $\tilde{f}$ always exists for each f is a consequence of the Radon-Nikodým theorem since
$\nu(A) = \int_A f d\mu$ ($\mu|\mathcal{B}$ is σ-finite) implies ν is μ-continuous and $\tilde{f} = \frac{d\nu}{d\mu}$.
Such an $\tilde{f}$, denoted $E^\mathcal{B}(f)$, is called a *conditional expectation* of f rel-
ative to $\mathcal{B}$. Thus if $\mathcal{B} = \sigma(\pi)$ in (5), then U_π constructed there is the
conditional expectation, without using the Radon-Nikodým theorem,
relative to the simple-minded $\mathcal{B}$. Thus $\{f_\pi, \pi \in \Pi\}$ is a martingale relative to the ordered set Π. Further it is easily seen that the martingale
is uniformly integrable. This is a consequence of (6). Thus the special
computations given for Theorem 4 are simply a proof of the norm convergence of such a directed martingale in M^Φ or $\mathcal{M}^\Phi$, and then Phillips'
lemma completed the proof. We now indicate Kolmogorov's version of
the original special, but pioneering, result. It must be observed that
the concept of a martingale was not available in 1931. This was introduced by J. Ville and P. Lévy in the late 1930's and the convergence
theory was developed by J. L. Doob, and E. S. Andersen and B. Jessen
and others in the 1940's and later.

Let $\Omega = \mathbb{R}^n$, $\Sigma =$ Borel σ-algebra, and $\mu =$ Lebesgue measure.
For each $x \in \mathbb{R}^n$ and $r > 0$, let $B_r(x)$ be the ball of radius r and center
x in $\mathbb{R}^n$. For each $f \in L^1(\mu)$, let

$$f_r(x) = \frac{1}{\mu(B_r(x))} \int_{B_r(x)} f(t) d\mu(t). \tag{12}$$

Then f_r is called a *Steklov function*. If $\mathcal{B}(\frac{1}{r})$ is the σ-algebra generated
by the balls $\{B_r(x), x \in \Omega\}$, then $\mathcal{B}(\frac{1}{r}) \subset \mathcal{B}(\frac{1}{r'}) \subset \Sigma$ for $r < r'$ and
$\{f_r, \mathcal{B}(\frac{1}{r}), r > 0\}$ forms a martingale. The mapping $f \mapsto f_r$ corresponds
to $E^{\mathcal{B}(\frac{1}{r})}(\cdot)$ when $\Omega \subset \mathbb{R}^n$ is a bounded set. Then the Kolmogorov-
Takahashi result is that the set $S \subset L^\Phi(\mu)$ is conditionally compact

if it is bounded and $\lim_{r \to 0} E^{B(\frac{1}{r})}(f) = f$ uniformly for $f \in S$. Thus this result is a special case of Theorem 4, although for the particular space at hand one can prove it by examining the properties of Steklov functions, as these authors did.

Finally we present another characterization of compactness, following F. Riesz, if $\Omega = I\!R^n$ using the group structure of the space. From the standard measure theory, it is known that the translation operator τ_s, defined by $(\tau_s f)(x) = f(x + s)$, is norm-preserving and continuous on $L^p(\Omega, \Sigma, \mu)$, $0 < p < \infty$ where μ is the Lebesgue measure. (cf., e.g., Rao (1987), p. 194.) Thus $\|\tau_s f\|_p = \|f\|_p$ and $\|\tau_s f - f\|_p \to 0$ as $s \to 0$. The same argument applies to the Orlicz space case and shows that $N_\Phi(\tau_s f) = N_\Phi(f)$ for all $f \in L^\Phi(\mu)$. Next using the fact that continuous functions with compact supports are dense in M^Φ, the $L^p(\mu)$-argument extends to M^Φ to establish that $N_\Phi(\tau_s f - f) \to 0$ as $s \to 0$ for each $f \in M^\Phi$ because of the translation invariance of the Lebesgue measure μ, and the continuity of Φ with $\Phi(x) = 0$ iff $x = 0$. These two facts of the norm together imply that M^Φ is a *homogeneous Banach function space* (these two are the defining conditions of the latter concept).

We can now state the analog of F. Riesz's compactness condition for Orlicz spaces as follows:

Theorem 7. *Let Φ be a strictly increasing continuous Young function and $(\Omega, \mathcal{B}, \mu)$ be the Lebesgue measure space with $\Omega = I\!R^n$. Then a subset $S \subset M^\Phi$ is conditionally compact iff it is bounded and*

(i) $\lim_{s \to 0} N_\Phi(\tau_s f - f) = 0$, *and*

(ii) *for each $\varepsilon > 0$, there is a compact cube $C_a = \underset{i=1}{\overset{n}{\times}} [-a, a]^i$ such that*

$$\lim_{a \to \infty} N_\Phi(f \chi_{C_a^c}) = 0,$$

uniformly in $f \in S$. In particular if Ω is the n-torus [or a bounded set isomorphic to some torus], then condition (ii) above is superfluous.

A proof of this result in the classical Lebesgue case is given in Dunford and Schwartz ((1958), IV.8.21) which with all the details is quite tedious. The same ideas and work admit an extension to the present case. The finite measure aspect is somewhat simpler although is still long, as given in Krasnosel'skii and Rutickii's monograph using the special properties of Steklov functions. We shall not include these details here. However, the above theorem is a simple consequence of the following general (and abstract) result of G. E. Šilov (1951) at least when Ω is bounded. In fact Šilov's theorem is valid for compact abelian groups with μ as the Haar measure, since M^Φ is a homogeneous function space. Let us present the desired result.

Theorem 8 (Šilov). *Let H be a homogeneous Banach function space on a compact abelian group $(\Omega, \mathcal{B}, \mu)$. Then a set $S \subset H$ is conditionally compact iff it is bounded and its elements are equi-continuous for translation, in the sense that for each $\varepsilon > 0$, there is a neighborhood $\mathcal{U}_\varepsilon$ of the identity of Ω, such that $\|\tau_h f - f\| < \varepsilon$, for $h \in \mathcal{U}_\varepsilon$, uniformly in $f \in S$. $[(\tau_h f)(g) = f(g+h)$, as usual, '+' being the group operation.]*

It is seen that in M^Φ, the translation τ_h satisfies the hypothesis of the above theorem, and since condition (ii) of Theorem 7 is automatic in this case, the latter result follows from Šilov's theorem. The locally compact case considered in Theorem 7, which may be proved by the methods of Dunford and Schwartz (1958), can perhaps be deducible from an extension of Šilov's result for the locally compact abelian group case. Such a result does not seem to be in the literature. The σ-compact case may be modeled after the work of the $I\!\!R^n$-method, but the details of this, as well as its possible extension for general locally compact groups, are still subjects for future research.

If Φ does not satisfy a Δ_2-regularity condition, then $L^\Phi(\mu)$ is not a homogeneous Banach space. In fact this will be true, as seen from the analog of our work in Chapter III, iff $L^\Phi(\mu)$ has an absolutely continuous norm, and this holds iff $L^\Phi(\mu) = M^\Phi$. Hence one may ask for the related characterization of compact sets in $L^\Phi(\mu)$ in general. This means when $L^\Phi(\mu)/\mathcal{M}^\Phi \neq \{0\}$ so that the latter space is infinite dimensional; it is an abstract (M)-space. But by the classical characterization of such a space as an $L^\infty(\tilde{\mu})$ on a compact topological triple $(\tilde{\Omega}, \tilde{\Sigma}, \tilde{\mu})$, we may apply Corollary 3 above and get conditions for compact subsets of this quotient space, and then translate them to $L^\Phi(\mu)$. However, the necessary effort does not seem worthy of presentation here, especially since they are not very useful for applications. With somewhat similar ideas and the quotient norms, conditions are given by Y. P. Wu (1982). We here omit further considerations along these lines.

Before proceeding further it must be emphasized that the results for Orlicz spaces cannot be obtained directly from the Lebesgue spaces. This is because for any $0 \neq f \in L^{\Phi}(\mu)$, the naive definition $\|f\| = \Phi^{-1}[\int_{\Omega} \Phi(f)d\mu]$ never gives a norm if $\Phi(x) \neq c|x|^p$ for some $p \geq 1$. This point was noted by A. Zygmund in 1935, and recently a characterization of it was given by Zaanen (1981). We include this analysis as an illustration of the above remark.

Proposition 9. *Let (Ω, Σ, μ) be a diffuse measure space and $\mathcal{X}$ be a Banach function space on it, i.e., $(\mathcal{X}, \|\cdot\|)$ is a Banach space in $L^0(\mu)$, the space of finite μ-measurable scalar functions. Suppose that $\chi_E \in \mathcal{X}$ for each $E \in \Sigma$, $\mu(E) < \infty$. If $\Phi : \mathbb{R} \to \mathbb{R}^+$ is a strictly increasing, continuous function with $\Phi(1) \neq 0$, $\Phi(-x) = \Phi(x)$, then*

$$\|\cdot\| : f \mapsto \Phi^{-1}[\int_{\Omega} \Phi(f)d\mu] \tag{13}$$

is a norm iff $\Phi(x) = \Phi(1)|x|^p$ for some $p \geq 1$.

Proof: Since the "if" part is trivial, we only need to consider the converse. Thus let (13) be a norm and since μ has the finite subset property (by our standing assumption), choose an $s_0 > 0$ and a set $E_0 \in \Sigma$ of positive finite measure such that $\Phi(s_0)\mu(E_0) = \Phi(1)$. This is possible by the diffuseness of μ and the hypothesis on Φ. Then for each $s > s_0$, we can find $E_s \in \Sigma$ such that $0 < \mu(E_s) < \mu(E_0)$ and $\Phi(s)\mu(E_s) = \Phi(1)$. Hence for $s \geq s_0$ and $t > 0$, one has:

$$s\|\chi_{E_s}\| = \|s\chi_{E_s}\| = \Phi^{-1}[\Phi(s)\mu(E_s)] = 1, \quad s \geq s_0, \tag{14}$$

and similarly

$$ts\|\chi_{E_s}\| = \Phi^{-1}[\Phi(ts)\mu(E_s)] = \Phi^{-1}\left[\Phi(ts)\frac{\Phi(1)}{\Phi(s)}\right]. \tag{15}$$

From (15) we get, since $s\|\chi_{E_s}\| = 1$ by (14),

$$\Phi(ts)\Phi(1) = \Phi(t)\Phi(s), \quad s \geq s_0, t > 0. \tag{16}$$

This leads to the classical functional equation related to Cauchy's identity. In fact, let $f(v) = \log \Phi(e^v) - \log \Phi(1)$, so that (16) gives

$$f(v + s) = f(v) + f(s), \quad v \in \mathbb{R}, s \geq \alpha = \max(1, \log s_0). \tag{17}$$

This is the desired equation, so that $f(n\alpha) = nf(\alpha)$, and then the well known unique continuous solution is $f(w) = pw$ where $p = f(\alpha)/\alpha$. Thus if we set $x = e^w$, then $\Phi(x) = \Phi(1)x^p$, $x \geq 0$. Since (13) can be a norm only if $p \geq 1$, and $\Phi(-x) = \Phi(x)$, this implies the necessity. ∎

The following consequences are of interest.

Corollary 10. *If (Ω, Σ, μ) is as in the theorem and Φ is an N-function, then $N_\Phi(f) = \Phi^{-1}[\int_\Omega \Phi(f)d\mu]$ iff $\Phi(u) = |u|^p$, $p > 1$. Similarly (13) is equal to $\|f\|_\Phi$ iff $\Phi(u) = p^{-p}(p-1)^{p-1}|u|^p$ for some $p > 1$.*

We next turn to the general theory, and consider in detail the ideas started in Section 1 of embedding of different Orlicz spaces.

5.3. *Embedding theorems*

Using various order relations of the Young functions discussed at length in Chapter II, we consider the corresponding relations in the Orlicz spaces under different conditions on the basic measure functions. Since all the order relations of Chapter II on the Young (or N-) functions are local and the function spaces are on finite measure spaces, we will concentrate on the latter. But the results extend for the global relations with infinite measures with a small extra effort, as we illustrate. Further, many of the converses of the inclusions imply the order relations among the Young functions whenever the measures are diffuse on a set of positive measure, thereby reducing these converses to finite diffuse measure spaces. This strategy used many times before will be employed frequently here. Let us turn to details.

Recall that for a pair of Young functions $\Phi_i, i = 1, 2$, $\Phi_1 \bowtie \Phi_2$ if for each $\varepsilon > 0$ there is a $K > 0$ and an $x_0 \geq 0$, such that $\Phi_2(\frac{x}{\varepsilon}) \leq K\Phi_1(x)$, $x \geq x_0$. This is a local ordering, and $x_0 = 0$ for the global ordering. We now state:

Theorem 1. *Let Φ_1, Φ_2 be continuous Young functions and $L^{\Phi_i}(\mu)$, $i = 1, 2$ be Orlicz spaces on (Ω, Σ, μ). Consider the conditions:*
   (i) $\Phi_1 \bowtie \Phi_2$;
   (ii) $L^{\Phi_1}(\mu) \subset M^{\Phi_2}$;
   (iii) $\lim\limits_{\rho_{\Phi_1}(f) \to 0} N_{\Phi_2}(f) = 0$;
   (iv) *Each norm bounded set in $L^{\Phi_1}(\mu)$ is modular bounded in $L^{\Phi_2}(\mu)$.*

Then (i) *implies* (ii), (iii) *as well as* (iv). *On the other hand if* μ *is diffuse on a set of positive* μ*-measure and* $\Phi_i(x) > 0$ *for* $x > 0$*, then all four conditions are equivalent.*

Proof: Let (i) hold. We first verify each of the other three conditions. Thus let $0 \neq f \in L^{\Phi_1}(\mu)$, and $k > 0$ be arbitrary. If $\varepsilon = (kN_{\Phi_1}(f))^{-1} > 0$, then by definition of the ordering, there exist $K > 0$ and $x_0 \geq 0$ [$= 0$ if we are in the global case], such that

$$\Phi_2(kN_{\Phi_1}(f)x) = \Phi_2\left(\frac{x}{\varepsilon}\right) \leq K\Phi_1(x), \qquad x \geq x_0. \tag{1}$$

Set $\Omega_1 = \{\omega : |f(\omega)| < x_0 N_{\Phi_1}(f)\}, \Omega_2 = \Omega_1^c$ (with $\Omega_1 = \emptyset$ if $x_0 = 0$). Then by (1),

$$\int_\Omega \Phi_2(kf)d\mu \leq \Phi_2(kx_0 N_{\Phi_1}(f))\mu(\Omega_1) + K\int_{\Omega_2} \Phi_1\left(\frac{f}{N_{\Phi_1}(f)}\right)d\mu$$

$$\leq \Phi_2\left(\frac{x_0}{\varepsilon}\right)\mu(\Omega) + K < \infty.$$

Since $k > 0$ is arbitrary, this implies that $f \in M^{\Phi_2}$ so that (ii) holds.

To see that (i) $\Rightarrow$ (iii), let $\varepsilon > 0$, $K > 0$ and $x_0 \geq 0$ be as in the ordering (with $x_0 = 0$ in the global case). Let $0 < \lambda < 1$ be chosen such that $\Phi_2(\lambda)\mu(\Omega) < \frac{1}{2}$ in the case of local ordering and $\lambda = 0$ in the global case. Then putting $K_1 = \max\{\Phi_2\left(\frac{x}{\varepsilon}\right)/\Phi_1(x) : x \in [\lambda\varepsilon, x_0]\}$, and $K_2 = \max(K, K_1)$, we have $\Phi_2\left(\frac{x}{\varepsilon}\right) \leq K_2\Phi_1(x), x \geq \lambda\varepsilon$. If $\delta = \frac{1}{2K_2} > 0$, then for any $f \in L^{\Phi_1}(\mu)$ with $\rho_{\Phi_1}(f) < \delta$, and $\Omega_1 = \{\omega : |f(\omega)| < \lambda\varepsilon\}$ [or $= \emptyset$ when $\lambda\varepsilon = 0$], we get

$$\int_\Omega \Phi_2\left(\frac{f}{\varepsilon}\right)d\mu \leq \Phi_2(\lambda)(\Omega_1) + K_2\int_{\Omega_1^c} \Phi_1(f)d\mu < \frac{1}{2} + K_2\delta = 1. \tag{2}$$

Hence $N_{\Phi_2}(f) \leq \varepsilon$, so that $\lim_{\rho_{\Phi_1}(f)\to 0} N_{\Phi_2}(f) = 0$, and (iii) holds.

To prove that (i) $\Rightarrow$ (iv), let $S \subset L^{\Phi_1}(\mu)$ be norm bounded and (i) hold. Then there is an $\alpha > 0$ such that S is contained in a ball of radius α. Taking $\varepsilon = \frac{1}{\alpha}$ in the ordering, and $K > 0, x_0 \geq 0$ as the other two constants [with $x_0 = 0$ in the global case], we get $\Phi_2\left(\frac{x}{\varepsilon}\right) \leq K\Phi_1(x)$, $x \geq x_0$. This inequality immediately implies that each $N_{\Phi_1}(\cdot)$-bounded set is necessarily $\rho_{\Phi_2}(\cdot)$-bounded, i.e., (iv) must hold. In other words, (iv) is also equivalent to $\lim_{\rho_{\Phi_2}(f)\to\infty} N_{\Phi_1}(f) = +\infty$.

We now consider the reverse implications assuming the diffuseness of μ. By the finite subset property of μ, there is a set $\Omega_0 \in \Sigma$, $0 < \mu(\Omega_0) < \infty$ and μ is diffuse on Ω_0. If (i) is false, then we shall derive a contraction. For this procedure, we may as well assume that (Ω, Σ, μ) is a finite diffuse measure space to simplify writing. With this, if (ii) holds and (i) does not, then there are $\varepsilon_0 > 0$ and $a_n \nearrow \infty$ such that

$$\Phi_2\left(\frac{a_n}{\varepsilon_0}\right) > 2^n \Phi_1(a_n), \qquad n \geq 1, \tag{3}$$

since the Φ_i are strictly increasing and $\rightthreetimes$ is violated. Let Ω_n be a disjoint sequence of measurable sets such that

$$\mu(\Omega_n) = \Phi_1(a_1)\mu(\Omega)/2^n \Phi_1(a_n), \qquad n \geq 1. \tag{4}$$

Then $f_0 = \sum\limits_{n=1}^{\infty} \frac{a_n}{\varepsilon_0} \chi_{\Omega_n} \in L^{\Phi_1}(\mu)$, and

$$\rho_{\Phi_1}(\varepsilon_0 f_0) = \sum_{n=1}^{\infty} \Phi_1(a_n)\mu(\Omega_n) = \Phi_1(a_1)\mu(\Omega) < \infty.$$

But

$$\rho_{\Phi_2}(f_0) = \sum_{n=1}^{\infty} \Phi_2\left(\frac{a_n}{\varepsilon_0}\right)\mu(\Omega_n) > \sum_{n=1}^{\infty} 2^n \Phi_1(a_n)\mu(\Omega_n) = +\infty.$$

Hence we have a contradiction to (ii). So (i) $\Leftrightarrow$ (ii).

Similarly if (iii) holds but not (i), then with the choice of $\varepsilon_0 > 0$, a_n as in (3) and taking, for convenience, $\Phi_2(\frac{a_1}{\varepsilon_0})\mu(\Omega) \geq 1$, we can choose $\Omega_n \in \Sigma$, satisfying $\mu(\Omega_n) = [\Phi_2(\frac{a_n}{\varepsilon_0})]^{-1}$, $n \geq 1$. This implies for the $f_n = a_n \chi_{\Omega_n}$, $\rho_{\Phi_1}(f_n) = \Phi_1(a_n)\mu(\Omega_n) < 2^{-n} \to 0$ as $n \to \infty$, and $N_{\Phi_2}(f_n) = a_n N_{\Phi_2}(\chi_{\Omega_n}) = \varepsilon_0 > 0$, by definition of $\mu(\Omega_n)$ above and Corollary III.4.7 for $N_{\Phi_2}(\chi_E)$. But then (iii) is contradicted.

Finally if (i) does not hold so that we again have (3), then for (iv) we can choose $\Omega_n \in \Sigma$, such that $\mu(\Omega_n) = [\Phi_1(a_n)]^{-1}$, and $\Phi_1(a_1)\mu(\Omega) \geq 1$. If $f_n = \frac{a_n}{\varepsilon_0}\chi_{\Omega_n}$, then

$$\rho_{\Phi_2}(f_n) = \Phi_2\left(\frac{a_n}{\varepsilon_0}\right)\mu(\Omega_n) > 2^n \to \infty \quad \text{as } n \to \infty.$$

But $N_{\Phi_1}(f_n) = \frac{a_n}{\varepsilon_0}N_{\Phi_1}(\chi_{\Omega_n}) = \frac{1}{\varepsilon_0} < \infty$, so that (iv) cannot hold. Thus each of (ii)–(iv) is equivalent to (i) under the present hypothesis. ∎

Since $\Phi \rightarrowtail \Phi$ iff Φ is Δ_2-regular, the following consequence of the theorem will complement and illuminate Corollary III.4.15.

Corollary 2. *If Φ is a continuous strictly increasing Young function and (Ω, Σ, μ) is a measure space with μ being diffuse on a set of positive measure, then $L^\Phi(\mu) = M^\Phi$ or equivalently norm bounded sets and modular bounded sets of $L^\Phi(\mu)$ are the same; or the norm and modular convergences are equivalent in $L^\Phi(\mu)$ iff $\Phi \rightarrowtail \Phi$ or iff Φ is Δ_2-regular.*

We recall that a bounded linear operator A between a pair of Banach spaces $\mathcal{X}, \mathcal{Y}$ is *conditionally (sequentially) compact* in a topology $\mathcal{T}$, if A maps each bounded subset of $\mathcal{X}$ into a *conditionally (sequentially) compact* subset in the topology $\mathcal{T}$ of $\mathcal{Y}$. Thus if the inclusion map $i : \mathcal{X} \to \mathcal{Y}$ has the stated compactness property, then the embedding is called *compact* in the same sense. The preceding theorem merely gives conditions for the continuous embedding of $L^{\Phi_1}(\mu)$ into M^{Φ_2}. We now consider compact embeddings.

Recall that for a pair of continuous Young functions $\Phi_1 \succ \Phi_2$ if for each $\varepsilon > 0$, there is an $\alpha > 0$ and $x_0 \geq 0$ ($= 0$ for the global ordering), we have $\frac{1}{\alpha}\Phi_2(\alpha x) \leq \varepsilon \Phi_1(x)$, $x \geq x_0$.

Theorem 3. *Let Φ_i be a pair of N-functions and $L^{\Phi_i}(\mu)$, $i = 1, 2$, be Orlicz spaces on a measure space (Ω, Σ, μ). Consider the following:*

(i) $\Phi_1 \succ \Phi_2$;

(ii) $\displaystyle \lim_{N_{\Phi_1}(f) \to 0} \frac{1}{N_{\Phi_1}(f)} \rho_{\Phi_2}(f) = 0$;

(iii) $\displaystyle \lim_{N_{\Phi_2}(f) \to \infty} \frac{1}{N_{\Phi_2}(f)} \rho_{\Phi_1}(f) = +\infty$;

(iv) *if τ is an embedding map of $L^{\Phi_1}(\mu)$ into $L^{\Phi_2}(\mu)$, then τ is $\sigma(L^{\Phi_2}(\mu), L^{\Psi_2}(\mu))$-compact.*

Then (i) implies (globally if $\mu(\Omega) = +\infty$, and locally if $\mu(\Omega) < \infty$) (ii), (iii) and [(ii) $\Rightarrow$] (iv). On the other hand, if μ is diffuse on a set of positive measure, then all conditions (i)–(iv) are equivalent.

Proof: (i) $\Rightarrow$ (ii) Since Φ_2 satisfies $\lim_{x \to 0} \frac{\Phi_2(x)}{x} = 0$, given $\varepsilon > 0$, let $x_0 \geq 0$ be the "lower" constant in the ordering between Φ_1 and Φ_2. Define $\Omega_1 = \{\omega : |f(\omega)| < x_0 N_{\Phi_1}(f)\}$, if $x_0 > 0$, $f \not\equiv 0$, and $= \emptyset$ otherwise. Note that if the ordering is global, then $x_0 = 0$ and $\Omega_1 = \emptyset$ and in fact the following holds for any continuous Young functions with

$\Phi_i(x) > 0$ for $x > 0$. Thus for $f \in L^{\Phi_1}(\mu)$ with $N_{\Phi_1}(f) < \alpha$,

$$\int_\Omega \Phi_2(f)d\mu = \int_{\Omega_1} \Phi_2(f)d\mu + \int_{\Omega_1^c} \Phi_2\left(\frac{N_{\Phi_1}(f)\alpha|f|}{\alpha N_{\Phi_1}(f)}\right)d\mu$$

$$\leq \Phi_2(x_0 N_{\Phi_1}(f))\mu(\Omega_1) + \varepsilon N_{\Phi_1}(f)\int_{\Omega_1^c} \Phi_1\left(\frac{f}{N_{\Phi_1}(f)}\right)d\mu,$$

by (i),

$$\leq \Phi_2(x_0 N_{\Phi_1}(f))\mu(\Omega_1) + \varepsilon N_{\Phi_1}(f).$$

Thus in both local and global ordering cases,

$$\lim_{N_{\Phi_1}(f)\to 0} \frac{\rho_{\Phi_2}(f)}{N_{\Phi_1}(f)} \leq \varepsilon, \tag{5}$$

since Φ_2 is an N-function (used in the local case). This gives (ii) because $\varepsilon > 0$ is arbitrary.

(i) $\Rightarrow$ (iii). Take $\varepsilon = \frac{1}{2\lambda}$, $0 < \lambda < \infty$, $\alpha > 0$ in the ordering of (i), and set $r = \frac{2}{\alpha}(1 + \Phi_2(\alpha x_0)\mu(\Omega))$ if $\mu(\Omega) < \infty$, and $= \frac{2}{\alpha}$ if $\mu(\Omega) = +\infty$ since then $x_0 = 0$. If $f \in L^{\Phi_2}(\mu)$ such that $N_{\Phi_2}(f) > r$, let $\Omega_1 = \{\omega : |f(\omega)| < x_0\}$. Then

$$\rho_{\Phi_1}(f) \geq \int_{\Omega_1^c} \Phi_1(f)d\mu \geq \frac{2\lambda}{\alpha}\int_{\Omega_1^c} \Phi_2(\alpha f)d\mu$$

$$\geq \frac{2\lambda}{\alpha}(\rho_{\Phi_2}(\alpha f) - \Phi_2(\alpha x_0)\mu(\Omega_1))$$

$$\geq \frac{2\lambda}{\alpha}(\alpha N_{\Phi_2}(f) - \frac{\alpha r}{2}) > \lambda N_{\Phi_2}(f).$$

Since $\lambda > 0$ is arbitrary this implies (iii) in both cases that $\mu(\Omega) \leq \infty$.

(ii) $\Rightarrow$ (iv). Suppose τ is an embedding map so that for each bounded set S of $L^{\Phi_1}(\mu)$, $\tau(S)$ is bounded in $L^{\Phi_2}(\mu)$. Thus let $a = \sup\{N_{\Phi_1}(f) : f \in S\} < \infty$. If $\alpha_n \searrow 0$, so that $N_{\Phi_1}(\alpha_n f) \leq \alpha_n a \to 0$, we see that (ii) implies

$$\frac{1}{\alpha_n}\int_\Omega \Phi_2(\alpha_n f)d\mu \leq a\frac{1}{\alpha_n N_{\Phi_1}(f)}\int_\Omega \Phi_2(\alpha_n f)d\mu \to 0, \tag{6}$$

as $n \to \infty$, uniformly in $f \in S$, $f \neq 0$, by (ii). But this implies by Proposition IV.5.7 that $\tau(S)$ is conditionally sequentially $\sigma((L^{\Phi_2}(\mu), L^{\Psi_2}(\mu))$-compact.

We next turn to the converse for which we can assume that (Ω, Σ, μ) is a finite diffuse measure, and show that if (i) is not true then none of the others can hold for some f_n and this will imply the equivalences in the statement.

Since it is assumed that (i) does not hold, there exist an $\varepsilon_0 > 0$ and a sequence $a_n \nearrow \infty$ such that

$$n \, \Phi_2\left(\frac{a_n}{n}\right) > \varepsilon_0 \Phi_1(a_n), \quad n \geq 1. \tag{7}$$

Since these are N-functions, we also have $\frac{\Phi_i(x)}{x} \to 0$ as $x \to 0$ and $\frac{\Phi_i(x)}{x} \to \infty$ as $x \to \infty$. We use these properties below. It may be assumed that $\Phi_1(a_1)\mu(\Omega) \geq 1$ and then one may choose $\Omega_n \in \Sigma$ such that $\Phi_1(a_n)\mu(\Omega_n) = 1$, $n \geq 1$ by diffuseness of μ. If we set

$$g_n = \Phi_2^{-1}\left(\frac{\varepsilon_0}{n}\Phi_1(a_n)\right)\chi_{\Omega_n},$$

then by (7)

$$N_{\Phi_1}(g_n) = \frac{1}{a_n}\,\Phi_2^{-1}\left(\frac{\varepsilon_0}{n}\Phi_1(a_n)\right) < \frac{1}{n} \to 0, \tag{8}$$

as $n \to \infty$. The set $S = \{f \in L^{\Phi_1}(\mu) : N_{\Phi_1}(f) = 1\}$ is bounded and $f_n = \frac{g_n}{N_{\Phi_1}(g_n)} \in S$. But (8) also implies, if $\xi_n = N_{\Phi_1}(g_n)$, that

$$\frac{1}{\xi_n}\int_\Omega \Phi_2(\xi_n f_n)d\mu > n\int_{\Omega_n} \Phi_2(g_n)d\mu = \varepsilon_0 > 0, \quad n \geq 1,$$

by definition of Ω_n and a_n. This and Theorem IV.5.3 imply that S is not conditionally $\sigma(L^{\Phi_2}(\mu), L^{\Psi_2}(\mu))$-sequentially compact. Thus (i) $\Rightarrow$ (ii) $\Rightarrow$ (iv) $\Rightarrow$ (i).

Regarding (iii) $\Rightarrow$ (i), if (i) is false then there is the possibility of (7) above. That implies

$$\lim_{n\to\infty}\frac{n}{a_n}\Phi_2\left(\frac{a_n}{n}\right) \geq \varepsilon_0 \lim_{n\to\infty}\frac{1}{a_n}\Phi_1(a_n) = +\infty$$

so that $\frac{a_n}{n} \to \infty$ and $\Phi_2\left(\frac{a_n}{n}\right) \to \infty$, as these are N-functions. This time, choose $\Omega_n \in \Sigma$, and $n_0 \geq 1$, such that (by the diffuseness of μ again) $\Phi_2\left(\frac{a_n}{n}\right)\mu(\Omega) \geq 1$, and $\mu(\Omega_n)\Phi_2\left(\frac{a_n}{n}\right) = 1$, $n \geq n_0$. If we set $f_n = a_n\chi_{\Omega_n}$, then $N_{\Phi_2}(f_n) = n \to \infty$, and by (7),

$$\frac{1}{N_{\Phi_2}(f_n)}\int_\Omega \Phi_1(f_n)d\mu = \frac{1}{n}\Phi_1(a_n)\mu(\Omega_n) < \frac{1}{\varepsilon_0}, \quad n \geq n_0.$$

This shows (iii) also is not valid, so that (i) $\Leftrightarrow$ (iii) holds. ∎

By definitions of the order relation '$\succ$' and the growth condition ∇_2, we note that $\Phi \in \nabla_2$ iff $\Phi \succ \Phi$. Hence the above result implies the following consequences at once.

Corollary 4. *Let Φ be an N-function and (Ω, Σ, μ) be a measure space with μ being diffuse on a set of positive measure. Then the following are equivalent conditions:*

(i) $\Phi \in \nabla_2$;

(ii) $\lim\limits_{N_\Phi(f) \to 0} \frac{1}{N_\Phi(f)} \int\limits_\Omega \Phi(f) d\mu = 0$;

(iii) $\lim\limits_{N_\Phi(f) \to \infty} \frac{1}{N_\Phi(f)} \int\limits_\Omega \Phi(f) d\mu = +\infty$;

(iv) *each bounded set in $L^\Phi(\mu)$ is conditionally $\sigma(L^\Phi(\mu), L^\Psi(\mu))$-sequentially compact.*

We now strengthen the topology from $\sigma(L^\Phi(\mu), L^\Psi(\mu))$ to the usual weak compactness, i.e., to $\sigma(L^\Phi(\mu), (L^\Psi(\mu))^*)$, and characterize the resulting (weak) embedding assertion.

Theorem 5. *Let Φ_i be a pair of N-functions, and (Ω, Σ, μ) be a measure space. Then the embedding mapping τ of $L^{\Phi_1}(\mu)$ into $L^{\Phi_2}(\mu)$ is weakly compact if the following two conditions are satisfied: (a) $\Phi_1 \succ \Phi_2$ and (b) $\Phi_1 \rightarrowtail \Phi_2$. On the other hand, if μ is diffuse on a set of positive measure then (a) and (b) are also necessary for the weak compactness of τ.*

Proof: We first show that τ is a weakly compact embedding, so that for each bounded set $S \subset L^{\Phi_1}(\mu)$, $\tau(S)$ is a conditionally weakly compact set in $L^{\Phi_2}(\mu)$. By the Eberlein-Šmulian theorem, recalled is Section 4.5, this is equivalent to considering weak sequential compactness and hence only this will be established in both directions. Thus let S be bounded and conditions (a) and (b) hold. By (a) and Theorem 3, $\tau(S)$ is conditionally sequentially $\sigma(L^{\Phi_2}(\mu), L^{\Psi_2}(\mu))$-compact. So each sequence $\{g_n, n \geq 1\} \subset \tau(S)$ has a convergent subsequence (denoted by itself) in this topology with limit $g \in L^{\Phi_2}(\mu)$, since it is complete in the latter sense (cf. Theorem IV.4.1). Hence, as $n \to \infty$,

$$\int\limits_\Omega g_n h d\mu \to \int\limits_\Omega g h d\mu, \qquad h \in L^{\Psi_2}(\mu). \tag{9}$$

But (b) implies, by Theorem 1, that $L^{\Phi_1}(\mu) \subset M^{\Phi_2}$. [We can omit the display of τ since both spaces are on (Ω, Σ, μ) and are thus contained in $L^0(\Omega, \Sigma, \mu)$.] If $x^* \in (L^{\Phi_2}(\mu))^*$, then $x^* = y^* + z^*$ uniquely where $y^* \in (M^{\Phi_2})^*$ and $z^* \in (M^{\Phi_2})^\perp$, by Proposition IV.3.3. Since $\{g_n, n \geq 1\}$ is in M^{Φ_2}, $z^*(g_n) = 0$, for all n. Also, when sequences in M^{Φ_2} are considered (Φ_2 being strictly increasing and continuous), we may assume that μ is σ-finite as already noted in the proof of Theorem IV.4.1. But then with each y^* in $(M^{\Phi_2})^*$, corresponds an element h in $L^{\Psi_2}(\mu)$ so that (9) gives the $\sigma(L^{\Phi_2}(\mu), (L^{\Phi_2}(\mu))^*)$-sequential convergence. This clearly implies our assertion.

For the converse, assume the additional conditions also. Suppose if possible that $\tau(S)$ is conditionally weakly sequentially compact but both (a) and (b) are not satisfied. Under the present hypothesis Theorem 3(iv) is satisfied so that (i), which is our (a) here, must hold. Hence possibly (b) is not true.

We assert that, since (a) implies the ordering $\Phi_1 \succ \Phi_2$, neither Φ_1 nor Φ_2 is Δ_2-regular. In fact, let Φ_1 be Δ_2-regular. Then by the relation $\Phi_1 \succ \Phi_2$, there exist $0 < \alpha < \infty$ and $x_0 \geq 0$, such that $\Phi_2(x) \leq \Phi_1(\alpha x)$, $x \geq x_0$ ($x_0 = 0$ in the global case). By the Δ_2-regularity of Φ_1, for each $\varepsilon > 0$, there is a $0 < \beta < \infty$ and $x_1 \geq 0$ such that $\Phi_1(\frac{\alpha}{\varepsilon} x) \leq \beta \Phi_1(x)$, $x \geq x_1$. Then $\Phi_2(\frac{u}{\varepsilon}) \leq \Phi_1(\frac{\alpha u}{\varepsilon}) \leq \beta \Phi_1(u)$, $u \geq \max(\varepsilon x_0, x_1)$. Thus $\Phi_1 \bowtie \Phi_2$. Replacing Φ_2 by Φ_1 in Δ_2, we get the same relation again. This proves our assertion. So if (b) is false neither Φ_1 nor Φ_2 can be Δ_2-regular. Thus there exist $\varepsilon_0 > 0$, $a_n \nearrow \infty$ such that

$$\Phi_2\left(\frac{a_n}{\varepsilon_0}\right) > 2^n \Phi_1(a_n), \qquad n \geq 1. \tag{10}$$

Again as before, to produce a set $S \subset L^{\Phi_1}(\mu)$ which is not conditionally weakly sequentially compact in $L^{\Phi_2}(\mu)$, we may assume that (Ω, Σ, μ) is a finite diffuse measure space. Then we can choose disjoint sets $\Omega_n \in \Sigma$ such that

$$\mu(\Omega) \Phi_1(a_1) \geq 1, 2^n \Phi_1(a_n)\mu(\Omega_n) = 1, \qquad n \geq 1. \tag{11}$$

Letting $f_n = \sum_{i \geq n} a_i \chi_{\Omega_i}$, and $S = \{f_n, n \geq 1\}$, we have

$$\int_\Omega \Phi_1(f_n)d\mu = \sum_{i \geq n} \Phi_1(a_i)\mu(\Omega_i) = \frac{1}{2^{n-1}}, \qquad n \geq 1, \tag{12}$$

so that $N_{\Phi_1}(f_n) \leq 1$ and S is a bounded subset of $L^{\Phi_1}(\mu)$. Since $\Phi_1 \succ \Phi_2$, S is also bounded in $L^{\Phi_2}(\mu)$. We assert that S is not compact in the desired topology. For this we construct an $x^* \in (L^{\Phi_2}(\mu))^*$ such that $x^*(f_n) \not\to$ a limit. Thus consider

$$\int_\Omega \Phi_2\left(\frac{f_n}{\varepsilon_0}\right)d\mu = \sum_{i\geq n}\Phi_2\left(\frac{a_i}{\varepsilon_0}\right)\mu(\Omega_i) \geq \sum_{i\geq n}1 = +\infty, \qquad n \geq 1.$$

Therefore $S \subset L^{\Phi_2}(\mu) - M^{\Phi_2}$ and $N_{\Phi_2}(f_n) \geq \varepsilon_0$. Let $\mathcal{X}_0$ be the linear span of $\{f_n, n \geq 1\}$, so that each $f \in \mathcal{X}_0$ is of the form $f = \sum_{k=1}^m b_k f_k$ for some $b_k \in \mathbb{R}$. Then f may be expressed as

$$f = \sum_{i=m}^\infty \left(\sum_{j=1}^m b_j\right)a_i \chi_{\Omega_i}. \tag{13}$$

If $\sum_{j=1}^m b_j \neq 0$, then, using the fact that $|f| \geq \sum_{i\geq m} |\sum_{j=1}^m b_j|a_i\chi_{\Omega_i}$, we get

$$\int_\Omega \Phi_2\left(\frac{f}{\varepsilon_0|\sum_{j=1}^m b_j|}\right)d\mu \geq \sum_{i\geq m}\Phi_2\left(\frac{a_i}{\varepsilon_0}\right)\mu(\Omega_i) = +\infty.$$

Hence

$$N_{\Phi_2}(f) \geq \varepsilon_0|\sum_{j=1}^m b_j| > 0 \tag{14}$$

in this case. We define a functional x^* on $\mathcal{X}_0$ as:

$$x^*(f) = \sum_{j=1}^m b_j, \tag{15}$$

if f in $\mathcal{X}_0$ has the representation (13). To see that x^* is well-defined let $f = \sum_{j=1}^{m'} c_j f_j$ be another representation. Then comparing this with (13) and using the fact that $0 < \mu(\Omega_i) < \infty$, Ω_i disjoint, we deduce that $m' = m$, and then $\sum_{j=1}^m b_j = \sum_{j=1}^m c_j$ so that (15) is independent of the way f is represented, and is well-defined. It is clear that x^* is linear on $\mathcal{X}_0$ and $|x^*(f)| \leq |\sum_{j=1}^m b_j| \leq \frac{1}{\varepsilon_0}N_{\Phi_2}(f)$, by (14). Hence $x^* \in \mathcal{X}_0^*$, and since $\mathcal{X}_0 \subset L^{\Phi_2}(\mu)$, it has a norm preserving extension to $L^{\Phi_2}(\mu)$, denoted by the same symbol, with the Hahn-Banach theorem. This x^* is our candidate. Set $x^*(f) = 0$ if $\sum_{j=1}^m b_j = 0$ in (15) for definiteness.

Since by (12) and Markov's inequality $f_n \to 0$ in measure, if $\{f_n, n \geq 1\} \subset S$ is conditionally weakly sequentially compact, then there is a convergent subsequence, say $\{f_{n_k}, k \geq 1\}$. Since only countable sets are involved here, the σ-algebra generated by $\{f_n, n \geq 1\}$ is countably generated. Hence for this argument we may assume (Ω, Σ, μ) itself is separable. But then by Proposition IV.5.6, $f_{n_k} \to 0$ in $\sigma(L^{\Phi_2}(\mu), M^{\Psi_2})$ topology, and hence $f_{n_k} \to 0$ weakly. However $x^*(f_n) = 1$ since $b_k = 1$ in (13). So $x^*(f_{n_k}) = 1 \nrightarrow 0$, contradicting the preceding fact. Consequently S has no weakly convergent subsequence. This means condition (b) must hold for the weakly compact embedding, establishing the converse. ∎

Taking $\Phi_1 = \Phi_2 = \Phi$ in the above theorem so that $\Phi \in \nabla_2 \cap \Delta_2$, we get the following consequence.

Corollary 6. *If $L^{\Phi}(\mu)$ is an Orlicz space on (Ω, Σ, μ) with μ being diffuse on a set of positive measure and Φ an N-function, then a norm bounded subset in $L^{\Phi}(\mu)$ is conditionally weakly compact iff $\Phi \in \nabla_2 \cap \Delta_2$.*

This result strengthens Corollary IV.5.5, and, in view of Corollary IV.1.12 implying that $L^{\Phi}(\mu)$ is reflexive, it expresses the general fact that in a reflexive Banach space, bounded sets are conditionally weakly (sequentially) compact. In fact if $\Phi \in \nabla_2 \cap \Delta_2$, we may omit the diffuseness condition on μ.

As a final item of this section, let us include a characterization of uniform absolute continuity of norms of sets in an Orlicz space when it is embeddable in another. Recall now, from Definition II.2.1, that a Young functions Φ_1 is essentially stronger than another Φ_2 ($\Phi_1 \succ\!\!\succ \Phi_2$) if there is an $x_0 \geq 0$ ($x_0 = 0$ if this ordering is global) such that $\Phi_2(x) \leq \Phi_1(ax)$, $x \geq x_0$ ($= x_0(a)$) for every $a > 0$.

Proposition 7. *Let Φ_i, $i = 1, 2$ be N-functions and (Ω, Σ, μ) be a finite diffuse measure space. If $L^{\Phi_i}(\mu)$, $i = 1, 2$ are the corresponding Orlicz spaces, then each of the following conditions is equivalent to the others:*

(i) $\Phi_1 \succ\!\!\succ \Phi_2$;

(ii) *for each $\varepsilon > 0$, there is a $K_\varepsilon > 0$ such that*

$$\|f\|_{\Phi_2} \leq \varepsilon \|f\|_{\Phi_1} + K_\varepsilon \|f\|_1, \qquad f \in L^{\Phi_1}(\mu); \qquad (16)$$

(iii) *each norm bounded set S in $L^{\Phi_1}(\mu)$ has a uniformly $L^{\Phi_2}(\mu)$ absolutely continuous norm, i.e.,* $\displaystyle\lim_{\mu(A)\to 0} \|f\chi_A\|_{\Phi_2} = 0$ *uniformly in $f \in S$.*

Proof: (i) $\Rightarrow$ (ii). By Theorem II.2.2 (b), (i) implies $\Psi_2 \succ\!\!\succ \Psi_1$ where Ψ_i is the complementary function to Φ_i, $i = 1,2$. Hence given $\varepsilon > 0$, there is a $K_\varepsilon > 0$ such that

$$\Psi_1\left(\frac{y}{\varepsilon}\right) \le \Psi_2(y), \qquad y \ge K_\varepsilon. \tag{17}$$

If $g \in L^{\Psi_2}(\mu)$ such that $\rho_{\Psi_2}(g) \le 1$, and $\Omega_1 = \{\omega : |g(\omega)| \ge K_\varepsilon\}$, then

$$\int_\Omega \Psi_1\left(\frac{g}{\varepsilon}\chi_{\Omega_1}\right)d\mu \le \int_{\Omega_1} \Psi_2(g)d\mu \le \rho_{\Psi_2}(g) \le 1, \qquad \text{by (17)}.$$

This implies for each $f \in L^{\Phi_1}(\mu)$,

$$\left|\int_\Omega fg\,d\mu\right| \le \varepsilon \int_\Omega \frac{|fg|}{\varepsilon}\chi_{\Omega_1}\,d\mu + \int_{\Omega_1^c} |fg|d\mu \le \varepsilon\|f\|_{\Phi_1} + K_\varepsilon \int_\Omega |f|d\mu.$$

Taking the supremum over all such g, we get (16).

(ii) $\Rightarrow$ (iii). Let $S \subset L^{\Phi_1}(\mu)$ be a bounded set. We may assume that the bound is one, i.e., S is in the unit ball, so that $\rho_{\Phi_1}(f) \le 1$ for $f \in S$. Since $\mu(\Omega) < \infty$, and Φ_1 is an N-function, it follows by Theorem I.2.2, that $S \subset L^{\Phi_1}(\mu) \subset L^1(\mu)$, is uniformly integrable so that for each $\varepsilon > 0$, there is a $\delta > 0$ such that $\mu(E) < \delta \Rightarrow$

$$\sup_{f\in S} \int_E |f|d\mu < \frac{\varepsilon}{K_\varepsilon}. \tag{18}$$

But by (ii), (16) holds so that for all $f \in S$,

$$\|f\chi_E\|_{\Phi_2} \le \varepsilon\|f\chi_E\|_{\Phi_1} + K_\varepsilon\|f\,\chi_E\|_1 \le \varepsilon \cdot 1 + K_\varepsilon \cdot \frac{\varepsilon}{K_\varepsilon} = 2\varepsilon.$$

This shows (iii) is true.

(iii) $\Rightarrow$ (i). We now use the fact that μ is diffuse also. If (i) is not true, i.e., $\Phi_1 \succ\!\!\succ \Phi_2$ is not satisfied, then there exist an $\varepsilon_0 > 0$ and a sequence $a_n \nearrow \infty$ such that

$$\Phi_1^{-1}(a_n) > \varepsilon_0 \Phi_2^{-1}(a_n), \qquad n \geq 1. \tag{19}$$

We can choose sets $\Omega_n \in \Sigma$, such that $a_n\mu(\Omega_n) = 1$, $n \geq 1$ and that $a_1\mu(\Omega) \geq 1$. Let $f_n = \Phi_1^{-1}(a_n)\chi_{\Omega_n}$, $n \geq 1$. Then $N_{\Phi_1}(f_n) = 1$, so that $S = \{f_n, n \geq 1\}$ is a bounded set in $L^{\Phi_1}(\mu)$. However, since $N_{\Phi_2}(f_n) = \Phi_1^{-1}(a_n)N_{\Phi_2}(\chi_{\Omega_n}) > \varepsilon_0$ by (19), implying $\|f_n\chi_{\Omega_n}\|_{\Phi_2} \geq N_{\Phi_2}(f_n\chi_{\Omega_n}) = N_{\Phi_2}(f_n) > \varepsilon_0$, (iii) is contradicted. ∎

5.4 Products of Orlicz spaces

Motivated by the Hölder inequality and its extensions, we can consider products of functions when the individual members belong to different Orlicz spaces. Under what conditions does the product belong to such a space again? We have seen in Theorems III.3.7 and III.3.9 two types of products, namely pointwise and convolution. There are others. Here a more detailed investigation will be undertaken. The basic work of Krasnosel'skii and Rutickii, which was extended primarily by T. Andô (1960_b) and R. O'Neil (1968), is of primary interest here.

Following the notation of Proposition III.1.7(b), we denote by $L^{\Phi_1}(\mu) \cdot L^{\Phi_2}(\mu)$, the set $\{f \cdot g : f \in L^{\Phi_1}(\mu), g \in L^{\Phi_2}(\mu)\}$. Further, if $(\Omega_i, \Sigma_i, \mu_i) = 1, 2$ are measure spaces and $(\Omega, \Sigma, \mu) = (\Omega_1 \times \Omega_2, \Sigma_1 \otimes \Sigma_2, \mu_1 \otimes \mu_2)$ also denoted $(\Omega_1, \Sigma_1, \mu_1) \otimes (\Omega_2, \Sigma_2, \mu_2)$, is their product measure space (cf., e.g., Rao (1987), Sec. 6.1), then letting $h(\omega) = f(\omega_1)g(\omega_2)$, $\omega = (\omega_1, \omega_2) \in \Omega$, we may again write $L^{\Phi_1}(\mu_1) \otimes L^{\Phi_2}(\mu_2) = \overline{\mathrm{sp}}\{h = f \cdot g : f \in L^{\Phi_1}(\mu_1), g \in L^{\Phi_2}(\mu_2)\}$. For these two products where the first is pointwise and the second is tensor, we can establish the following pair of companion results, the first one being a slight refinement of Theorem III.3.7.

Theorem 1. *Let Φ_1, Φ_2 be Young functions and $L^{\Phi_i}(\mu)$ be the Orlicz spaces on a measure space (Ω, Σ, μ). If Φ_3 is another Young function, consider the following conditions:*

(i) $\Phi_3(\alpha xy) \leq \Phi_1(x) + \Phi_2(y)$ *for some $\alpha > 0$, and $x, y \geq x_0 \geq 0$;*

(ii) *there is a constant $C > 0$, such that*

$$N_{\Phi_3}(fg) \leq C\, N_{\Phi_1}(f)\, N_{\Phi_2}(g), \quad f \in L^{\Phi_1}(\mu), g \in L^{\Phi_2}(\mu);$$

(iii) $L^{\Phi_1}(\mu) \cdot L^{\Phi_2}(\mu) \subset L^{\Phi_3}(\mu)$.

Then (i) $\Rightarrow$ (ii) $\Rightarrow$ (iii) and if μ is diffuse on a set of positive measure and Φ_i, $i = 1, 2, 3$ are N-functions, then (iii) $\Rightarrow$ (i) so that all are equivalent conditions. [As usual if $\mu(\Omega) = +\infty$, $x_0 = 0$ in (i).]

We now present the corresponding tensor product statement, and prove both results thereafter. The similarity and differences will be illuminated.

Theorem 2. *Let Φ_i, $i = 1,2,3$ be Young functions and $L^{\Phi_i}(\mu_i)$, $i = 1,2$, be Orlicz spaces on measure spaces $(\Omega_i, \Sigma_i, \mu_i)$, $i = 1,2$, and $L^{\Phi_3}(\mu)$ be the space on the product space $(\Omega, \Sigma, \mu) = (\Omega_1, \Sigma_1, \mu_1) \otimes (\Omega_2, \Sigma_2, \mu_2)$. Assume that the product measure μ also has the finite subset property. Consider the conditions:*

(i) $\Phi_3(\alpha x y) \leq \Phi_1(x) \cdot \Phi_2(y)$ *for some $\alpha > 0$, and $x, y \geq x_0 \geq 0$;*

(ii) *there is a constant $C > 0$ such that*

$$N_{\Phi_3}(f \cdot g) \leq C N_{\Phi_1}(f) N_{\Phi_2}(g), \quad f \in L^{\Phi_1}(\mu_1), g \in L^{\Phi_2}(\mu_2);$$

(iii) $L^{\Phi_1}(\mu_1) \otimes L^{\Phi_2}(\mu_2) \subset L^{\Phi_3}(\mu)$.

Then (i) $\Rightarrow$ (ii) $\Rightarrow$ (iii), *and if μ is diffuse on a set of positive finite μ-measure, Φ_i, $i = 1,2,3$ are N-functions, then* (iii) $\Rightarrow$ (i) *so that all are equivalent conditions. [As usual $x_0 = 0$ if $\mu_i(\Omega_i) = +\infty$, for $i = 1$ or 2 in* (i).*]*

We prove Theorem 1, and then show that the conditions of (i) are locally equivalent, and, in general, they both are derived from one inequality.

Proof of Theorem 1: (i) $\Rightarrow$ (ii). Let $0 \neq f \in L^{\Phi_1}(\mu)$, $0 \neq g \in L^{\Phi_2}(\mu)$, for nontriviality. Define $\widetilde{\Omega}_1 = \{\omega : |f(\omega)| < x_0 N_{\Phi_1}(f)\}$, $\widetilde{\Omega}_2 = \{\omega : |g(\omega)| < x_0 N_{\Phi_2}(g)\}$ and $\widetilde{\Omega}_i = \emptyset$ if $x_0 = 0$. Then we get

$$\int_\Omega \Phi_3\left(\frac{\alpha f g}{N_{\Phi_1}(f) N_{\Phi_2}(g)}\right) d\mu = \left(\int_{\widetilde{\Omega}_1 \cap \widetilde{\Omega}_2} + \int_{\widetilde{\Omega}_1 \cap \widetilde{\Omega}_2^c} + \int_{\widetilde{\Omega}_2 \cap \widetilde{\Omega}_1^c} + \int_{(\widetilde{\Omega}_1 \cup \widetilde{\Omega}_2)^c}\right) \Phi_3(\ ) d\mu$$

$$\leq \Phi_3(\alpha x_0^2) \mu(\Omega) + \int_{\widetilde{\Omega}_1 \cap \widetilde{\Omega}_2^c} \Phi_3\left(\frac{\alpha x_0 g}{N_{\Phi_2}(g)}\right) d\mu + \int_{\widetilde{\Omega}_2 \cap \widetilde{\Omega}_1^c} \Phi_3\left(\frac{\alpha x_0 f}{N_{\Phi_1}(f)}\right) d\mu$$

$$+ \int_\Omega \Phi_1\left(\frac{f}{N_{\Phi_1}(f)}\right) d\mu + \int_\Omega \Phi_2\left(\frac{g}{N_{\Phi_2}(g)}\right) d\mu, \quad \text{by (i)},$$

$$\leq \mu(\Omega)[\Phi_3(\alpha x_0^2) + \Phi_1(x_0) + \Phi_2(x_0)] + 4, \text{ using (i) again,}$$

$$= K \quad \text{(say)}. \tag{1}$$

From (1) we deduce at once that $N_{\Phi_3}(fg) \leq C N_{\Phi_1}(f) N_{\Phi_2}(g)$ for some constant C depending only on α, x_0 and $\mu(\Omega)$ when $x_0 > 0$ and only on $\alpha > 0$ when $x_0 = 0$. This establishes (ii).

(ii) $\Rightarrow$ (iii). This follows from definition of the norm.

(iii) $\Rightarrow$ (i). Under the additional hypothesis, we intend to show that if (i) is not true then (iii) cannot hold for some pair of functions f, g. So as before, it is enough to show this under the further assumption that the (Ω, Σ, μ) is a finite diffuse measure space and Φ_i's are N-functions.

If (i) is false, then there exist $0 < a_n \nearrow \infty, 0 < b_n \nearrow \infty$ such that

$$\Phi_3\left(\frac{a_n b_n}{n 2^n}\right) > \Phi_1(a_n) + \Phi_2(b_n), \quad n \geq 1.$$

Hence using the convexity of Φ_3 we have

$$\Phi_1(a_n) + \Phi_2(b_n) < \frac{1}{2^n}\Phi_3\left(\frac{a_n b_n}{n}\right), \quad n \geq 1. \tag{2}$$

We may assume $(\Phi_1(a_1) + \Phi_2(b_1))\, \mu(\Omega) \geq 1$, and choose disjoint $\Omega_n \in \Sigma$ such that (because of diffuseness of μ on Ω)

$$2^n(\Phi_1(a_n) + \Phi_2(b_n))\mu(\Omega_n) = 1, \quad n \geq 1. \tag{3}$$

Let f_0 and g_0 be defined as:

$$f_0 = \sum_{n=1}^{\infty} a_n \chi_{\Omega_n}, \quad g_0 = \sum_{n=1}^{\infty} b_n \chi_{\Omega_n}.$$

Then

$$\int_\Omega \Phi_1(f_0)d\mu + \int_\Omega \Phi_2(g_0)d\mu = \sum_{n=1}^{\infty}(\Phi_1(a_n) + \Phi_2(b_n))\mu(\Omega_n) = 1$$

so that $f_0 \in L^{\Phi_1}(\mu)$, $g_0 \in L^{\Phi_2}(\mu)$. However, given $\varepsilon > 0$, choose $n_0 > \frac{1}{\varepsilon}$. Then we have

$$\int_\Omega \Phi_3(\varepsilon f_0 g_0)d\mu = \sum_{n=1}^{\infty} \Phi_3(\varepsilon a_n b_n)\mu(\Omega_n) \geq \sum_{n \geq n_0}\Phi_3\left(\frac{a_n b_n}{n}\right)\mu(\Omega_n) = +\infty,$$

by (2) and (3). Since $\varepsilon > 0$ is arbitrary, $f_0 g_0 \notin L^{\Phi_3}(\mu)$, so (iii) fails. ∎

We present a few specializations of this result before turning to the proof of the second theorem. The next result is due to Krasnosel'skii and Rutickii.

Corollary 3. *Let Φ_i be N-functions, $i = 1, 2, 3$. Suppose that one of the following conditions is satisfied:*

(i) *there is a complementary N-pair (Φ, Ψ) such that*

$$\Phi_1 \succ \Phi_3 \circ \Phi \quad and \quad \Phi_2 \succ \Phi_3 \circ \Psi, \tag{4}$$

(ii) *for the above (Φ, Ψ), if $\Phi_3 \in \Delta'$, suppose that*

$$\Phi_1 \succ \Phi \circ \Phi_3 \quad and \quad \Phi_2 \succ \Psi \circ \Phi_3. \tag{5}$$

Then $L^{\Phi_1}(\mu) \cdot L^{\Phi_2}(\mu) \subset L^{\Phi_3}(\mu)$, where (Ω, Σ, μ) is a measure space.

Proof: It suffices to note that both (4) and (5) imply (i) of Theorem 1. In fact by (4), and from the definition of '$\succ$,' we have

$$\Phi_3(\Phi(2\beta x)) \leq \Phi_1(x), \quad \Phi_3(\Psi(2\beta y)) \leq \Phi_2(y), \quad x, y \geq x_0 \geq 0,$$

for some $\beta > 0$ and $x_0 \geq 0$. Hence if we set $\alpha = \beta^2$,

$$\begin{aligned}
\Phi_3(\alpha x y) &\leq \Phi_3(\Phi(\beta x) + \Psi(\beta y)), \text{ by Young's integrality,} \\
&\leq \frac{1}{2}\Phi_3(\Phi(2\beta x)) + \frac{1}{2}\Phi_3(\Psi(2\beta y)), \text{ by the convexity of} \\
&\qquad \Phi_3, \Phi, \text{ and } \Psi, \\
&\leq \Phi_1(x) + \Phi_2(y), \quad x, y \geq x_0 \geq 0,
\end{aligned}$$

which is (i) of Theorem 1.

Next let (ii) be true. Then by Lemma II.3.8 and (5) we get

$$\begin{aligned}
\Phi_3(cxy) &\leq \Phi_3(x)\Phi_3(y) \\
&\leq \Phi[\Phi_3(x)] + \Psi[\Phi_3(y)], \text{ by Young's inequality,} \\
&\leq \Phi_1(bx) + \Phi_2(by), \quad x, y \geq x_0 \geq 0, \tag{6}
\end{aligned}$$

for some $c > 0$, $b > 0$, and $x_0 \geq 0$. Let $\alpha = \frac{c}{b^2}$, $u = bx$, $v = by$ and $u_0 = bx_0$. Then (ii) reduces to (i) of Theorem 1. So the result holds. ∎

Since (i) of Theorem 1, plays an important role in these results, we now present several equivalent (local) forms of it from which one can also see the close relations between the two theorems.

Lemma 4. *Let Φ_i, $i = 1, 2, 3$ be N-functions with Ψ_i as the corresponding complementary functions. Then condition (i) of Theorem 1 is equivalent to each of the following:*

(a) $\displaystyle \limsup_{x \to \infty} \frac{\Phi_1^{-1}(x)\Phi_2^{-1}(x)}{\Phi_3^{-1}(x)} < \infty;$

(b) $\alpha_1 xyz \leq \Phi_1(x) + \Phi_2(y) + \Psi_3(z)$, $x, y, z \geq x_1 \geq 0$;

(c) $\Psi_1(\alpha_2 yz) \leq \Phi_2(y) + \Psi_3(z)$, $y, z \geq x_2 \geq 0$;

(d) $\Psi_2(\alpha_3 xz) \leq \Phi_1(x) + \Psi_3(z)$, $x, z \geq x_3 \geq 0$, *where $\alpha_i > 0$ are some constants.*

Proof: If (i) of Theorem 1 holds, then for some $\alpha > 0$ and $x_0 \geq 0$ we have

$$\alpha xy \leq \Phi_3^{-1}(\Phi_1(x) + \Phi_2(y)), \qquad x, y \geq x_0.$$

Letting $x = \Phi_1^{-1}(u)$, $y = \Phi_2^{-1}(u)$ we get $\alpha \Phi_1^{-1}(u) \Phi_2^{-1}(u) \leq \Phi_3^{-1}(2u) \leq 2\Phi_3^{-1}(u)$, $u \geq u_0 = \max(\Phi_1(x_0), \Phi_2(x_0)) \geq 0$. This implies (a). Conversely, by (a), we can find $\alpha > 0, u_0 \geq 0$ such that

$$\Phi_1^{-1}(u)\Phi_2^{-1}(u) \leq \frac{1}{\alpha}\Phi_3^{-1}(u), \qquad u \geq u_0.$$

If we let $x = \Phi_1^{-1}(u), y = \Phi_2^{-1}(u)$, this becomes

$$\alpha xy \leq \Phi_3^{-1}\left(\frac{1}{2}(\Phi_1(x) + \Phi_2(y))\right) \leq \Phi_3^{-1}(\Phi_1(x) + \Phi_2(y)).$$

Hence $\Phi_3(\alpha xy) \leq \Phi_1(x) + \Phi_2(y)$, $x, y \geq \max\left(\Phi_1^{-1}(u_0), \Phi_2^{-1}(u_0)\right)$. This shows that (i)$\Leftrightarrow$(a).

(i) $\Rightarrow$ (b), since by Young's inequality

$$\alpha xyz \leq \Phi_3(\alpha xy) + \Psi_3(z) \leq \Phi_1(x) + \Phi_2(y) + \Psi_3(z), \qquad x, y, z \geq x_1 \geq 0.$$

Conversely, if (b) is assumed then to use the equality in Young's inequality, take $z = \varphi_3(\alpha_1 xy)$, where φ_3 is the (left) derivative of Φ_3, we have

$$\Phi_3(\alpha_1 xy) = \alpha_1 xyz - \Psi_3(z) \leq \Phi_1(x) + \Phi_2(y).$$

For (b) $\Rightarrow$ (c), let $x = \psi_1(\alpha_1 yz)$ so that with equality in Young's inequality,

$$\Psi_1(\alpha_1 yz) = \alpha_1 xyz - \Phi_1(x) \leq \Phi_2(y) + \Psi_3(z),$$

by (b), for $y, z \geq x_2 > 0$. Thus (c) holds. Conversely, if (c) is assumed by Young's inequality,

$$\alpha_2 xyz - \Phi_1(x) \leq \Psi_1(\alpha_2 yz) \leq \Phi_2(y) + \Psi_3(z), \qquad y, z \geq x_2 \geq 0.$$

Thus (b)$\Leftrightarrow$(c); (b)$\Leftrightarrow$(d) is similar. ∎

In view of the lemma, we have

Corollary 5. *If* Φ_i, $i = 1, 2, 3$ *are* N-*functions and* (Ω, Σ, μ) *is a finite diffuse measure space, then* $L^{\Phi_1}(\mu) \cdot L^{\Phi_2}(\mu) \subset L^{\Phi_3}(\mu)$ *iff any one of the equivalent conditions* (b)–(d) *of the lemma holds.*

We now turn to the proof of Theorem 2.

Proof of Theorem 2: The argument is a modifications of that given for Theorem 1. Thus to show (i) $\Rightarrow$ (ii), let $0 \neq f \in L^{\Phi_1}(\mu_1)$, $0 \neq g \in L^{\Phi_2}(\mu_2)$, the other case being trivial. Define $\widetilde{\Omega}_1$ and $\widetilde{\Omega}_2$ exactly as in that proof, these being taken as empty sets if $x_0 = 0$. Now, on expressing $\Omega = \Omega_1 \times \Omega_2$ in terms of the measurable rectangles $\widetilde{\Omega}_i \times \widetilde{\Omega}_j$ or $\widetilde{\Omega}_i \times \widetilde{\Omega}_j^c$, $i, j = 1, 2$, and replacing f, g by $\frac{f}{N_{\Phi_1}(f)}, \frac{g}{N_{\Phi_2}(g)}$, we take $N_{\Phi_1}(f) = 1 = N_{\Phi_2}(g)$ and consider

$$\int_{\Omega} \Phi_3(\alpha f g) d\mu \leq \Phi_3(\alpha x_0^2) \mu(\widetilde{\Omega}_1 \times \widetilde{\Omega}_2) + \int\int_{\widetilde{\Omega}_1 \widetilde{\Omega}_2^c} \Phi_3(\alpha x_0 g) d\mu_1 d\mu_2$$

$$+ \int\int_{\widetilde{\Omega}_i^c \widetilde{\Omega}_2} \Phi_3(\alpha x_0 f) d\mu_1 d\mu_2 + \int\int_{\Omega_1 \Omega_2} \Phi_1(f) \Phi_2(g) d\mu_1 d\mu_2$$

$$\leq \{\Phi_3(\alpha x_0^2)\mu(\Omega) + \Phi_1(x_0)\mu_1(\Omega_1) + \Phi_2(x_0)\mu_2(\Omega_2)\} + 1. \quad (7)$$

It follows that $N_{\Phi_3}(fg) \leq C = \frac{K}{\alpha}$ where K is the right side of (7). This gives (ii). Note that if $\mu(\Omega) = +\infty$, then $x_0 = 0$ so that the result follows at once with $C = \frac{1}{\alpha}$. However, we need the additional assumption, in this case, that μ has the finite subset property. Otherwise the double and the product integrals need not agree. [Cf., e.g., Rao (1987), Exercise 6.2.4 and 6.2.8. In particular if μ_1, μ_2 are σ-finite this is automatic.]

(ii) $\Rightarrow$ (iii) is a consequence of the definition.

(iii) $\Rightarrow$ (i). Under the additional diffuseness condition, we shall show that, contrapositively, if (i) is false then the inclusion of (iii) also fails. We may assume now, for this purpose, all the measure spaces are finite and both μ_i, $i = 1, 2$, are diffuse. [Note that the diffuseness of μ implies that of both μ_1, μ_2.]

Since (i) does not hold, there exist $a_n \nearrow \infty$, $b_n \nearrow \infty$ such that

$$\Phi_3\left(\frac{a_n b_n}{n}\right) > 4^n \Phi_3\left(\frac{a_n b_n}{n4^n}\right) > 4^n \Phi_1(a_n)\Phi_2(b_n), \quad n \geq 1. \quad (8)$$

We may assume $\Phi_1(a_1)\mu_1(\Omega_1) \geq 1, \Phi_2(b_1)\mu_2(\Omega_2) \geq 1$ and choose disjoint $\widetilde{\Omega}_{1n} \in \Sigma_1, \widetilde{\Omega}_{2n} \in \Sigma_2$ such that ($a_1 > 0, b_1 > 0$, can be taken)

$$\mu_1(\widetilde{\Omega}_{1n})2^n\Phi_1(a_n) = 1 \quad \text{and} \quad \mu_2(\widetilde{\Omega}_{2n})2^n\Phi_2(b_n) = 1, \quad n \geq 1. \quad (9)$$

If we define $f_0 = \sum_{n=1}^{\infty} a_n \chi_{\widetilde{\Omega}_{1n}}$, $g_0 = \sum_{n=1}^{\infty} b_n \chi_{\widetilde{\Omega}_{2n}}$, then (9) implies that $N_{\Phi_1}(f_0) \leq 1, N_{\Phi_2}(g_0) \leq 1$. However,

$$f_0 g_0 \geq \sum_{n \geq 1} a_n b_n \chi_{\widetilde{\Omega}_{1n}} \chi_{\widetilde{\Omega}_{2n}},$$

and for any $\lambda > 0$ if $n_0 > \frac{1}{\lambda}$, then

$$\int_\Omega \Phi_3(\lambda f_0 g_0)d\mu \geq \sum_{n \geq n_0} \Phi_3\left(\frac{a_n b_n}{n}\right)\mu_1(\widetilde{\Omega}_{1n})\mu_2(\widetilde{\Omega}_{2n})$$

$$\geq \sum_{n \geq n_0} 4^n \Phi_1(a_n)\Phi_2(b_n)\mu_1(\widetilde{\Omega}_{1n})\mu_2(\widetilde{\Omega}_{2n}) = +\infty,$$

by (8) and (9). Since $\lambda > 0$ is arbitrary, $f_0 g_0 \notin L^{\Phi_3}(\mu)$ so that (iii) cannot hold. ∎

We now present some consequences of this result in the following:

Corollary 6. *Let* $(\Omega_i, \Sigma_i, \mu_i)$, $i = 1, 2$, *be measure spaces such that their product* (Ω, Σ, μ) *also has a set* $A \in \Sigma$, $0 < \mu(A) < \infty$ *on which* μ *is diffuse. Let* Φ_i, $i = 1, 2, 3$ *be N-functions and* $L^{\Phi_i}(\mu_i)$, $i = 1, 2$ *and* $L^{\Phi_3}(\mu)$ *be the corresponding Orlicz spaces. Then* $L^{\Phi_3}(\mu_1) \otimes L^{\Phi_3}(\mu_2) \subset L^{\Phi_3}(\mu)$ *iff* $\Phi_3 \in \Delta'$. *Further* $\Phi_i \succ \Phi_3$, $i = 1, 2$ *and* $\Phi_3 \in \Delta'$ *are sufficient for the inclusion* $L^{\Phi_1}(\mu_1) \otimes L^{\Phi_2}(\mu_2) \subset L^{\Phi_3}(\mu)$.

Proof: By Lemma II.3.8, $\Phi_3 \in \Delta'$ iff $\Phi_3(\alpha x y) \leq \Phi_3(x)\,\Phi_3(y)$, $x, y \geq x_0$, for some $\alpha > 0$ and $x_0 \geq 0$. Hence taking $\Phi_1 = \Phi_2 = \Phi_3$, the first result follows from Theorem 2. For the second assertion, $\Phi_i \succ \Phi_3$, $i = 1, 2$, imply by Theorem 1.3 $L^{\Phi_i}(\mu_i) \subset L^{\Phi_3}(\mu_i)$, $i = 1, 2$. Hence, by the first part since $\Phi_3 \in \Delta'$,

$$L^{\Phi_1}(\mu_1) \otimes L^{\Phi_2}(\mu_2) \subset L^{\Phi_3}(\mu_1) \otimes L^{\Phi_3}(\mu_2) \subset L^{\Phi_3}(\mu),$$

the last inclusion uses the Δ' condition (locally or globally), and the first one follows from the definition of the tensor product. ∎

Regarding the above products one can raise a converse question; namely, given a pair of Orlicz spaces, when are they embeddable in another Orlicz space. If so, will there be a smallest (largest) such space? A comprehensive solution of this problem is provided by the following result, based on the work of Andô (1960$_b$).

Theorem 7. *Let Φ_1, Φ_2 be a pair of Young functions and (Ω, Σ, μ) be a finite measure space with μ diffuse on a set of positive measure. Then*

(i) *$L^{\Phi_1}(\mu) \cdot L^{\Phi_2}(\mu)$ is embeddable in some Orlicz space iff $\Phi_1 \succ\!\!\succ \Psi_2$ where Ψ_2 is complementary to Φ_2. Moreover when this condition holds (i.e., when an embedding is available) it is possible to find a minimal (or smallest) Orlicz space that contains the product.*

(ii) *Also there exist an Orlicz space $L^\Phi(\mu)$ such that $L^{\Phi_1}(\mu) \cdot L^\Phi(\mu)$ is embeddable in the given $L^{\Phi_2}(\mu)$ iff $\Phi_1 \succ\!\!\succ \Phi_2$. When this holds, it is possible to find a maximal (or largest) such $L^\Phi(\mu)$ space.*

Proof: (i) This will be deduced from Theorem 1. Thus suppose that there is a Young function Φ such that $L^{\Phi_1}(\mu) \cdot L^{\Phi_2}(\mu) \subset L^\Phi(\mu)$ so that the former is embedded in the latter Orlicz space. Then by Theorem 1(i) and its equivalent conditions given by Lemma 4(d), there exist an $\alpha > 0$ and a constant $x_0 \geq 0$ such that (Ψ's being complementary to the Φ's)

$$\Psi_2(\alpha x z) \leq \Phi_1(x) + \Psi(z), \qquad x, z \geq x_0. \tag{10}$$

Given $\varepsilon > 0$, choose $z_0 \geq x_0$ such that $\alpha z_0 \geq \frac{1}{\varepsilon}$. Then (10) implies

$$\Psi_2\left(\frac{x}{\varepsilon}\right) \leq \Psi_2(\alpha x z_0) \leq \Phi_1(x) + \Psi(z_0), \qquad x \geq x_0. \tag{11}$$

If we take $x_1 \geq x_0$ such that for $x \geq x_1 \Rightarrow \Phi_1(x) \geq \Psi(z_0)$, then from (11) we get $\Psi_2(\frac{x}{\varepsilon}) \leq 2\Phi_1(x)$, $x \geq x_1$ which means $\Phi_1 \succ\!\!\succ \Psi_2$.

Conversely, let the latter ordering be true. Then by Theorem II.2.2(b), $\Psi_1 \prec\!\!\prec \Phi_2$. These conditions can be expressed alternatively as follows. There exist N-functions R_1 and R_2 such that $\Phi_1 \succ \Psi_2 \circ R_1$, $\Phi_2 \succ \Psi_1 \circ R_2$. Let Q_i be the complementary function to R_i, and set $Q = \max(Q_1, Q_2)$. Then Q is an N-functions, and we produce the desired Φ and $L^\Phi(\mu)$ from this Q for the embeddability. Indeed let R be the complementary N-function to Q. Then, since Q dominates Q_i, we get $R \prec R_i$, $i = 1, 2$, so that the earlier relations between Φ_i and R_i become $\Phi_1 \succ \Psi_2 \circ R$, $\Phi_2 \succ \Psi_1 \circ R$. This means that there exist $\alpha > 0, x_0 \geq 0$ such that

$$\Psi_1(R(\alpha x)) \leq \Phi_2(x), \qquad \Psi_2(R(\alpha x)) \leq \Phi_1(x), \qquad x \geq x_0.$$

Define

$$\Phi(x) = \int\limits_0^{|x|} \frac{R(\alpha\sqrt{t})}{\sqrt{t}} dt, \qquad x \in \mathbb{R}.$$

Then Φ is a candidate. In fact, for $x \geq y \geq x_0$, we have

$$\Phi_1(x) + \Phi_2(y) \geq \Psi_2(R(\alpha x)) + \Phi_2(y) \geq y R(\alpha x),$$

by Young's inequality,

$$= xy \frac{R(\alpha x)}{x} \geq xy \frac{R(\alpha \sqrt{xy})}{\sqrt{xy}} \geq \Phi(xy).$$

It follows from Theorem 1 (whose measure theoretical assumptions are satisfied here), we see that $L^{\Phi_1}(\mu) \cdot L^{\Phi_2}(\mu)$ is embeddable in an Orlicz space, establishing (i) in both directions.

Under the embeddability hypothesis, we show that there is a minimal space in that it is contained in every Orlicz space into which the given product is embeddable. Thus by (i), $\Phi_1 \twoheadrightarrow \Psi_2$ or equivalently $\Phi_2 \twoheadrightarrow \Psi_1$. Define $Q : I\!\!R \to I\!\!R^+$ by the equation

$$Q(x) = \sup\{\Psi_1(xy) - \Phi_2(y) : y \geq 0\}, \qquad x \in I\!\!R. \qquad (12)$$

Then $Q(x) = Q(-x)$, $Q(0) = 0$, $Q(I\!\!R) \subset I\!\!R^+$, and since $\Psi_1(xy)$ is convex in x for each y, $Q(\cdot)$ is also convex. Moreover

$$\lim_{x \to \infty} \frac{Q(x)}{x} \geq \lim_{x \to \infty} \left(\frac{\Psi_1(x)}{x} - \frac{\Phi_2(1)}{x} \right) = +\infty.$$

Thus Q is an N-function which satisfies $\Psi_1(xy) \leq \Phi_2(y) + Q(x)$, by definition. Hence by Lemma 4(c) and Theorem 1, we see that $L^{\Phi_1}(\mu) \cdot L^{\Phi_2}(\mu)$ is embeddable into $L^R(\mu)$, where R is complementary to Q.

We now claim that $L^R(\mu)$ is the minimal space. Indeed if $L^{\Phi}(\mu)$ is any other space into which the product space is embeddable then by Lemma 4(c) and the converse of Theorem 1(iii), for some $b > 0, x_0 \geq 0$,

$$\Psi_1(xy) \leq \Psi(bx) + \Phi_2(y), \qquad x \geq x_0. \qquad (13)$$

But by definition of Q in (12), (13) implies that $Q \prec \Psi$ so that $R \succ \Phi$, and $L^R(\mu) \subset L^{\Phi}(\mu)$. Thus $L^R(\mu)$ is minimal.

(ii) If there is a Φ such that $L^{\Phi_1}(\mu) \cdot L^{\Phi}(\mu)$ is embeddable in the given Orlicz space $L^{\Phi_2}(\mu)$, then by Theorem 1(i) there exist $\alpha > 0$, and $x_0 \geq 0$ such that

$$\Phi_2(\alpha xy) \leq \Phi_1(x) + \Phi(y), \qquad x, y \geq x_0. \qquad (14)$$

For each $\varepsilon > 0$, choose $y_0 \geq x_0$ such that $\alpha y_0 \geq \frac{1}{\varepsilon}$. Then (14) becomes

$$\Phi_2\left(\frac{x}{\varepsilon}\right) \leq \Phi_2(\alpha x y_0) \leq \Phi_1(x) + \Phi(y_0), \qquad x \geq x_0.$$

This is equivalent to the relation $\Phi_1 \succ\!\!\succ \Phi_2$.

Conversely, if the last relation holds, then by (i) there is an N-function R such that $L^{\Phi_1}(\mu) \cdot L^{\Psi_2}(\mu) \subset L^R(\mu)$. This means that by Theorem 1,

$$R(\alpha x y) \leq \Phi_1(x) + \Psi_2(y), \qquad x, y \geq x_0, \tag{15}$$

for some $\alpha > 0$ and $x_0 \geq 0$. If Q is complementary to R, then by Lemma 4, (15) is equivalent to the inequality (for some $\alpha' > 0$)

$$\Phi_2(\alpha' x y) \leq \Phi_1(x) + Q(y), \qquad x, y \geq x_1,$$

so that $L^{\Phi_1}(\mu) \cdot L^Q(\mu) \subset L^{\Phi_2}(\mu)$, and the condition is sufficient.

Regarding a maximal space, since $\Phi_2 \prec\!\!\prec \Phi_1$, by (i) there is a minimal Orlicz space $L^R(\mu)$ such that $L^{\Phi_1}(\mu) \cdot L^{\Psi_2}(\mu) \subset L^R(\mu)$. If Φ is an N-function such that $L^{\Phi_1}(\mu) \cdot L^{\Phi}(\mu) \subset L^{\Phi_2}(\mu)$, then by Theorem 1(i) and Lemma 4(d), we also have $L^{\Phi_1}(\mu) \cdot L^{\Psi_2}(\mu) \subset L^{\Psi}(\mu)$ and $L^R(\mu) \subset L^{\Psi}(\mu)$ by the minimality of $L^R(\mu)$. This implies $L^{\Phi}(\mu) \subset L^Q(\mu)$, Q being the complementary N-function to R and is in fact defined by (12); i.e., the last part of (i) as: $Q(x) = \sup\{\Phi_2(xy) - \Phi_1(y) : y \geq 0\}$. This proves (ii) also. ∎

In these embeddings one may ask for conditions in order that the large space required to be an Orlicz space of absolutely continuous norm, i.e., $L^{\Phi}(\mu)$ of the type $\{f : \int_\Omega \Phi(\alpha f) d\mu < \infty$ for all $\alpha > 0\}$. This is obtained by demanding that the corresponding inequalities of Theorem 1(i) be valid for all $\alpha > 0$ instead of for some $\alpha > 0$. We thus state a precise version whose proof is the same as that given there with simple modification. We omit the repetition. This was also noted with a detailed (modified) argument by S. W. Wang (1963).

Proposition 8. *Under the hypothesis of Theorem 1, for the N-function Φ_i $i = 1, 2, 3$, $L^{\Phi_1}(\mu) \cdot L^{\Phi_2}(\mu) \subset M^{\Phi_3}$ iff for each $\varepsilon > 0$, there is a $K(= K_\varepsilon) > 0$ and $x_0(= x_0(\varepsilon)) \geq 0$ such that*

$$\Phi_3\left(\frac{xy}{\varepsilon}\right) \leq K(\Phi_1(x) + \Phi_2(y)), \qquad x, y \geq x_0.$$

It is possible to give the corresponding formulations for the tensor products. We present this for comparison and later use. The next result states that an embedding of a tensor product of Orlicz spaces is always possible. This will illuminate Theorem 2(iii), and is in sharp contrast with Theorem 7(i) of which it is a companion.

Theorem 9. *Let $(\Omega_i, \Sigma_i, \mu_i)$ be finite measure spaces such that their product (Ω, Σ, μ) has a set of positive diffuse measure. Then*

(i) *for any given pair of N-functions Φ_1, Φ_2, there is an N-function Φ such that $L^{\Phi_1}(\mu_1) \otimes L^{\Phi_2}(\mu_2) \subset L^{\Phi}(\mu)$;*

(ii) *for any pair Φ_1, Φ_2 as above, there exist an N-function Φ_0 such that $L^{\Phi_1}(\mu_1) \otimes L^{\Phi_0}(\mu_2) \subset L^{\Phi_2}(\mu)$ iff $\Phi_1 \rtimes \Phi_2$;*

(iii) *for all Φ_0 satisfying (ii) there is a maximal Φ, in the sense that $L^{\Phi_0}(\mu_2) \subset L^{\Phi}(\mu_2)$; and*

(iv) *for any pair Φ_1, Φ_2 verifying (i), there is a minimal Q in the sense that $L^{\Phi_1}(\mu_1) \otimes L^{\Phi_2}(\mu_2) \subset L^Q(\mu)$ and any other Q_1 satisfying this inclusion also verifies $L^Q(\mu) \subset L^{Q_1}(\mu)$.*

Proof: (i) This assertion is established by a construction. If Ψ_1, Ψ_2 are the complementary functions of Φ_1, Φ_2, let Ψ be defined by

$$\Psi(u) = \frac{\Psi_1(u)\Psi_2(u)}{|u|} = \frac{1}{|u|} \int_0^{|u|} \psi_1(t)dt \int_0^{|u|} \psi_2(\tau)d\tau, \quad \text{by Cor. I.3.2,}$$

$$= \psi_1(t_u^*) \int_0^{|u|} \psi_2(\tau)d\tau, \quad 0 < t_u^* < |u| \text{ a.e.,}$$

by the mean-value theorem for Lebesgue integrals. Since $t_u^* \le t_{u'}^*$ for $|u| \le |u'|$, it follows that $\psi_1(t_u^*)\psi_2(\tau) \ge 0$ and non-decreasing, so that $\Psi(\cdot)$ is convex. Since Ψ_i are N-functions, it is clear that $\frac{\Psi(u)}{u} \to 0$ (or $+\infty$) as $u \to 0$ (or $u \to +\infty$) so that Ψ is an N-function. Let Φ be complementary to Ψ. Since for $x, y \ge 1$, we have

$$\Psi(xy) = \frac{\Psi_1(xy)}{y} \cdot \frac{\Psi_2(xy)}{x} \ge \Psi_1(x)\Psi_2(y), \tag{16}$$

because of convexity of the Ψ_i. Hence there is an $x_0 \ge 0$ such that, by Proposition II.3.9,

$$\Phi\left(\frac{xy}{2}\right) \le \Phi_1(x)\Phi_2(y), \qquad x, y \ge x_0.$$

This inequality by Theorem 2 implies the embedding assertion.

(ii) Suppose that there is a Φ_0 such that $L^{\Phi_1}(\mu_1) \otimes L^{\Phi_0}(\mu_2) \subset L^{\Phi_2}(\mu)$. Then by Theorem 2, $\Phi_2(\alpha xy) \leq \Phi_1(x)\Phi_0(y)$, $x, y \geq x_0$ for some $\alpha > 0$ and $x_0 \geq 0$. Hence for any $\varepsilon > 0$, let $y_0 \geq \max(x_0, \frac{1}{\alpha\varepsilon})$, and we have from the preceding inequality

$$\Phi_2\Big(\frac{x}{\varepsilon}\Big) \leq \Phi_2(\alpha x y_0) \leq \Phi_1(x) \cdot \Phi_0(y_0) = K\Phi_1(x), \qquad x \geq x_0$$

with $K = \Phi_0(y_0)$. Thus $\Phi_1 \dashv \Phi_2$ holds.

Conversely let the latter ordering relation hold. If we define Φ_0 as:

$$\Phi_0(y) = \sup\Big\{ \frac{\Phi_2(xy)}{\Phi_1(x)} : x \geq 0 \Big\}, \tag{17}$$

then Φ_0 is an N-function, and $\Phi_2(xy) \leq \Phi_1(x)\Phi_0(y)$ for $x, y \geq 1$. So by Theorem 2 again, we get the desired inclusion, giving (ii).

(iii) Regarding the choice of a maximal Φ_0 when there is one $\bar{\Phi}_0$ satisfying (ii), by Theorem 2, there is a $b > 0$ and an $x_1 \geq 0$ such that $\Phi_2(xy) \leq \Phi_1(x)\bar{\Phi}_0(by)$, $x, y \geq x_1$. By definition of Φ_0 of (17), $\bar{\Phi}_0 \succ \Phi_0$. By Theorem 1.3, $L^{\bar{\Phi}_0}(\mu_2) \subset L^{\Phi_0}(\mu_2)$ so that Φ_0 defines the essential maximum space.

(iv) We only sketch the argument which is analogous to the corresponding one of Theorem 7(i). Let Φ be an N-function [which exists by (i)] such that

$$L^{\Phi_1}(\mu_1) \otimes L^{\Phi_2}(\mu_2) \subset L^{\Phi}(\mu). \tag{18}$$

Then by Theorem 2(i), using an equivalent relation from Proposition II.3.9, we get for some $\alpha > 0$ and $x_0 \geq 0$,

$$\Psi(\alpha xy) \geq \Psi_1(x)\,\Psi_2(y), \qquad x, y \geq x_0, \tag{19}$$

where the Ψ's are the corresponding complementary N-functions. Taking $x_0 = 1$ here for convenience, (19) can be expressed as

$$\Psi(\alpha x) \geq \Psi_1(y)\,\Psi_2\Big(\frac{x}{y}\Big), \qquad 1 \leq y \leq x. \tag{20}$$

Define $\overline{Q}$ by the equation

$$\overline{Q}(x) = \sup\Big\{ \Psi_1(y)\Psi_2\Big(\frac{x}{y}\Big) : 1 \leq y \leq x \Big\}. \tag{21}$$

A direct computation that $\overline{Q}$ is convex, is messy. But using the so-called first mean-value theorem for Lebesgue integrals, employed in the proof of (i) above, we see first that $\Psi_1(y)\Psi_2(\frac{x}{y})$ is convex, and hence [the supremum being a convex functional] $\overline{Q}$ is also convex. It is clear that $\frac{\overline{Q}(x)}{x} \nearrow \infty$ as $x \nearrow \infty$. But it may not be an N-function. Let Q be an N-function with $\overline{Q}$ as its principal part (cf. Lemma III.4.1). Then the complementary function R of Q satisfies the inequality of Lemma 4(a) in place of Φ_3 there. However for any $L^{\Phi}(\mu)$ containing $L^{\Phi_1}(\mu_1) \otimes L^{\Phi_2}(\mu_2)$, we must have $\Psi(\alpha x) \geq Q(x)$ for some $\alpha > 0$ and $x \geq 1$ by definition of $\overline{Q}$, and hence Q. This means $\Phi \prec R$, and so $L^R(\mu)$ is the minimal space into which the tensor product space $L^{\Phi_1}(\mu_1) \otimes L^{\Phi_2}(\mu_2)$ can be embedded. ∎

We now present the following result of Krasnosel'skii and Rutickii as a consequence of the above theorem.

Corollary 10. *Let $(\Omega_i, \Sigma_i, \mu_i)$, $i = 1, 2$, be finite measure spaces such that their product (Ω, Σ, μ) has a set in Σ of positive diffuse μ-measure. If (Φ_i, Ψ_i), $i = 1, 2$, are complementary N-pairs, let $\Psi_3 = \Psi_2 \circ \Psi_1$ and $\Psi_4 = \Psi_1 \circ \Psi_2$. If Φ_3 and Φ_4 are the complementary N-functions to Ψ_3 and Ψ_4, then*

$$L^{\Phi_1}(\mu_1) \otimes L^{\Phi_2}(\mu_2) \subset L^{\Phi_3}(\mu) \cap L^{\Phi_4}(\mu). \tag{22}$$

Proof: For the given Ψ_1, Ψ_2 define Ψ by the equation

$$\Psi(x) = \frac{\Psi_1(x)\Psi_2(x)}{|x|},$$

and let $x_1 > 0$ be chosen such that $\frac{\Psi_1(x_1)}{x_1} > 1$. Then

$$\Psi(x) = \frac{\Psi_1(x)}{|x|} \cdot \Psi_2(x) < \Psi_2(\Psi_1(x)) = \Psi_3(x),$$

by the convexity of Ψ_2 and $x \geq x_1$. Thus $\Psi_3 \succ \Psi$ so that $\Phi_3 \prec \Phi$. By Theorem V.1.3, $L^{\Phi}(\mu) \subset L^{\Phi_3}(\mu)$. By the preceding theorem we get

$$L^{\Phi_1}(\mu_1) \otimes L^{\Phi_2}(\mu_2) \subset L^{\Phi}(\mu) \subset L^{\Phi_3}(\mu).$$

The second part is similar. It shows that the tensor product is contained in $L^{\Phi_4}(\mu)$, from which the conclusion (22) follows. ∎

We next consider a different type of product using the integration of functions on product measure spaces. This will extend the convolution operator considered before from the special topological group spaces to general measure spaces. The final section of this chapter is devoted to this aspect.

5.5. Remarks on integral products

The third type of products we consider here is through integration and this will be used in the study of certain bounded linear operators on Orlicz spaces. The following result is due to R. O'Neil (1968). It and its consequences, as well as specializations due to the same author, will occupy most of this section.

Theorem 1. *Let* $(\Omega_i, \Sigma_i, \mu_i)$, $i = 1, 2$, *be measure spaces and* Φ_i, $i = 1, 2, 3$, *be Young functions with* $\Phi_i(x) = 0$ *iff* $x = 0$. *Suppose that there is an* $\alpha > 0$ *such that*

$$\Phi_1^{-1}(xy)\, \Phi_2^{-1}(y) \le \alpha y \Phi_3^{-1}(x), \qquad x, y \ge 0. \tag{1}$$

If (Ω, Σ, μ) *is the product measure space with the finite subset property, and if* $f \in L^{\Phi_1}(\mu)$, $g \in L^{\Phi_2}(\mu_2)$ *define* h *by the equation*

$$h(\omega_1) = \int_{\Omega_2} f(\omega_1, \omega_2)\, g(\omega_2)\, d\,\mu_2(\omega_2). \tag{2}$$

Then $h \in L^{\Phi_3}(\mu_1)$ *and in fact*

$$N_{\Phi_3}(h) \le \alpha\, N_{\Phi_1}(f)\, N_{\Phi_2}(g). \tag{3}$$

Moreover, if μ *is diffuse on a set of positive (sigma) finite measure then the failure of* (1) *implies the failure of* (3) *and even* h *may fail to be in* $L^{\Phi_3}(\mu_1)$, *so that* (1)–(3) *are the best possible.*

The proof of this result depends on an auxiliary lemma which we present in a slightly more general form because of its use in applications. Its proof, being somewhat long, will be given in steps. The lemma contains the essentials of the above theorem.

Lemma 2. *Let* Φ_i, $i = 1, 2, 3$, *be Young functions such that the left continuous* Φ_3 *satisfies* $\Phi_3(x) = 0$ *iff* $x = 0$, *and for some* $\alpha > 0$ *and* $\beta > 0$,

$$\Phi_1^{-1}(xy)\, \Phi_2^{-1}(y) \le \alpha\, y\, \Phi_3^{-1}(\beta x), \qquad x \ge 0, y \ge 0. \tag{4}$$

If $(\Omega_i, \Sigma_i, \mu_i)$ is as in the theorem, with (Ω, Σ, μ) as their product having the finite subset property, let $f : \Omega \to \mathbb{R}$ and $g : \Omega_2 \to \mathbb{R}$ be measurable, $N_{\Phi_2}(g) \le 1$, and $h : \omega_1 \mapsto \int_{\Omega_2} f(\omega_1, \omega_2) \, g(\omega_2) \, d\mu_2(\omega_2)$, $\omega_1 \in \Omega_1$. Then

$$\int_{\Omega_1} \Phi_3\left(\frac{h}{\alpha}\right) d\mu_1 \le \beta \int_{\Omega} \Phi_1(f) d\mu. \tag{5}$$

Proof: We start with a simple inequality and extend it successively to the general case, giving (5).

Step 1. If $x \ge 0$, $y > 0$ satisfying $\Phi_2(y) > 0$, then it is true that

$$xy \le \alpha \Phi_2(y) \, \Phi_3^{-1}\left(\frac{\beta \Phi_1(x)}{\Phi_2(y)}\right). \tag{6}$$

This follows from:

$$xy \le \Phi_1^{-1}(\Phi_1(x)) \, \Phi_2^{-1}(\Phi_2(y))$$

$$= \Phi_1^{-1}\left(\frac{\Phi_2(y) \cdot \Phi_1(x)}{\Phi_2(y)}\right) \Phi_2^{-1}(\Phi_2(y))$$

$$\le \alpha \Phi_2(y) \, \Phi_3^{-1}\left(\beta \frac{\Phi_1(x)}{\Phi_2(y)}\right), \qquad \text{by (4).}$$

Step 2. Suppose $\Phi_2(y) > 0$ for all $y > 0$, and let $B = \{\omega_2 : g(\omega_2) \ne 0\}$. For each $\omega_2 \in B$, define $x = |f(\omega_1, \omega_2)|, y = |g(\omega_2)|$. Then by (6)

$$|f(\omega_1, \omega_2) g(\omega_2)| \le \alpha \Phi_2(g(\omega_2)) \Phi_3^{-1}\left(\beta \frac{\Phi_1(f(\omega_1, \omega_2))}{\Phi_2(g(\omega_2))}\right).$$

Now since $N_{\Phi_2}(g) \le 1$, by hypothesis, we may integrate the above and use Jensen's inequality (cf. Prop. III.3.5) with the finite measure $\Phi_2(g) d\mu_2$ and suitable normalization. Thus since $0 < \int_B \Phi_2(g) d\mu_2$ $= \gamma \le 1$,

$$\Phi_3\left(\frac{h(\omega_1)}{\alpha}\right) \le \Phi_3\left(\gamma \int_B \Phi_3^{-1}\left(\beta \frac{\Phi_1(f(\omega_1, \omega_2))}{\Phi_2(g(\omega_2))}\right) \Phi_2(g(\omega_2)) \frac{d\mu_2(\omega_2)}{\gamma}\right)$$

$$\le \gamma \int_B \Phi_3\left[\Phi_3^{-1}\left(\beta \frac{\Phi_1(f(\omega_1, \omega_2))}{\Phi_2(g(\omega_2))}\right)\right] \Phi_2(g(\omega_2)) \frac{d\mu_2(\omega_2)}{\gamma}$$

$$\le \beta \int_B \Phi_1(f(\omega_1, \omega_2)) \, d\mu_2(\omega_2), \text{ since } \Phi_3(\Phi_3^{-1}(x)) \le x.$$

Integrating this we get

$$\int_{\Omega_1} \Phi_3\left(\frac{h}{\alpha}\right)d\mu_1 \le \beta \int_{\Omega_1}\int_{\Omega_2} \Phi_1(f(\omega_1,\omega_2))d\mu_2 d\mu_1 = \beta \int_{\Omega} \Phi_1(f)d\mu,$$

since μ has the finite subset property by hypothesis and a from of the Fubini-Tonelli theorem applies (cf., e.g., Rao (1987), Exercise 6.2.8). Thus (5) holds in this case.

Step 3. Let $\Phi_2^{-1}(0) = x_0 > 0$ and suppose that there is an $x_1 > x_0$ such that $\Phi_2(x) < \infty$ for $x_0 \le x \le x_1$, and that B of Step 2 is σ-finite. Then again (5) holds.

For, given $\varepsilon > 0$ for the set B of Step 2, since $B_1 = \{\omega_2 \in B : \Phi_2(g(\omega_2)) = 0\}$ is σ-finite for μ_2, we can find a measurable function $0 < u(\cdot) < \Phi_2(x_1)$, supported by B_1 such that $\int_{B_1} ud\mu_2 < \varepsilon$. Then the function $g_1 = |g|\chi_{B-B_1} + \Phi_2^{-1}(u)\chi_{B_1}$ satisfies $|g| \le g_1$, both with the same support B, and

$$\int_{\Omega_2} \Phi_3(g_1)d\mu_2 = \int_{B-B_1} \Phi_3(g)d\mu_2 + \int_{B_1} \Phi_2(\Phi_2^{-1}(u))d\mu_2$$
$$\le \int_{\Omega_2} \Phi_3(g)d\mu_2 + \int_{\Omega_2} ud\mu_2 \le 1 + \varepsilon.$$

If $h_1 = \int_{\Omega_2} |f|g_1 d\mu_2$, then $|h| \le h_1$ a.e. and f, g_1, and h_1 satisfy the hypothesis of Step 2 with $\alpha(1 + \varepsilon)$ in place of α. Hence we get

$$\int_{\Omega_1} \Phi_3\left(\frac{h}{\alpha(1+\varepsilon)}\right)d\mu_1 \le \int_{\Omega_1} \Phi_3\left(\frac{h_1}{\alpha(1+\varepsilon)}\right)d\mu_1$$
$$\le \beta \int_{\Omega_1}\int_{\Omega_2} \Phi_1(f)d\mu_2 d\mu_1 = \beta \int_{\Omega} \Phi_1(f)d\mu.$$

Letting $\varepsilon \searrow 0$, and using the monotone convergence theorem, (5) follows.

Step 4. If $\Phi_2^{-1}(0) = x_0 > 0$ and g has σ-finite support, then again (5) holds. (We reduce this to Step 3.)

For, since Φ_2 and Φ_2^{-1} are left continuous by definition, we may find $y_n \searrow 0$ such that $\Phi_2^{-1}(y_n) = x_n \searrow x_0$. We now employ a familiar argument, already used in the proof of Proposition IV.3.4. Thus let Φ_{2n} be a Young function defined as

$$\Phi_{2n}^{-1}(y) = [x_0 - \frac{1}{n} + \frac{y}{y_n}(x_n - x_0 + \frac{1}{n})]\chi_{[0,y_n]}(y) + \Phi_2^{-1}(y)\chi_{(y_n,\infty)}(y).$$

Then $\Phi_{2n}^{-1}(y)$, which is linear on $[0, x_n]$ of slope at most $\frac{x_n}{x_0}$, satisfies $\Phi_{2n}^{-1}(y) \le \frac{x_n}{x_0}\Phi_2^{-1}(y)$, and also $x = \Phi_2^{-1}(y) \le \frac{x_n}{x_0-\frac{1}{n}}\Phi_{2n}^{-1}(y)$. Hence for all $x \in [0, \infty)$, the latter inequality gives

$$\Phi_{2n}\left(\frac{(x_0 - \frac{1}{n})x}{x_n}\right) \le \Phi_2(x). \tag{7}$$

If we let $g_n = \frac{x_0-\frac{1}{n}}{x_n}g$, then $N_{\Phi_{2n}}(g_n) \le 1$. Define

$$h_n(\omega_1) = \int_{\Omega_2} f(\omega_1, \omega_2)g_n(\omega_2)d\mu_2 = \frac{x_0 - \frac{1}{n}}{x_n}h(\omega_1). \tag{8}$$

To see that Φ_1, Φ_{2n} and Φ_3 satisfy the hypothesis of Step 3, consider

$$\Phi_1^{-1}(xy)\,\Phi_{2n}^{-1}(y) \le \frac{x_n}{x_0}\Phi_1^{-1}(xy)\,\Phi_2^{-1}(y) \le \frac{\alpha x_n}{x_0}y\,\Phi_3^{-1}(\beta x),$$

using (4). Hence with α replaced by $\frac{\alpha x_n}{x_0}$ is Step 3, we get if $\alpha_n = \frac{\alpha x_n}{x_0}$, then

$$\int_{\Omega_1} \Phi_3\left(\frac{x_0 - \frac{1}{n}}{x_n} \cdot \frac{x_0}{\alpha x_n}h\right)d\mu_1 = \int_{\Omega_1} \Phi_3\left(\frac{h_n}{\alpha_n}\right)d\mu_1,$$

$$\le \beta \int_{\Omega_1}\int_{\Omega_2} \Phi_1(f)d\mu_2 d\mu_1 \le \beta \int_{\Omega} \Phi_1(f)d\mu.$$

Letting $n \to \infty$, and using Fatou's lemma this proves (5) in this case also.

   Step 5. Let $\Phi_2^{-1}(0) = x_0 > 0$, and the support of g is not necessarily σ-finite. We will reduce this to Step 4 so that (5) again holds. For, since $\Phi_2^{-1}(y)$ is nondecreasing, (4) and the present hypothesis imply

$$\alpha y\Phi_3^{-1}(\beta x) \ge \Phi_1^{-1}(xy)\,\Phi_2^{-1}(y) \ge x_0\,\Phi_1^{-1}(xy), \qquad x, y \ge 0. \tag{9}$$

Taking $x = 1$, and then $\alpha_0 = \alpha\, \Phi_3^{-1}(\beta)/x_0 > 0$, we get from (9) that $\Phi_1^{-1}(y) \le \alpha_0 y$ so that

$$0 \le x \le \Phi_1^{-1}(\Phi_1(x)) \le \alpha_0 \Phi_1(x).$$

It follows that

$$\int\!\!\int_{\Omega_1\Omega_2} |f| d\mu_2 d\mu_1 \le \alpha_0 \int\!\!\int_{\Omega_1\Omega_2} \Phi_1(f(\omega_1,\omega_2)) d\mu_2 d\mu_1 = \alpha_0 \int_{\Omega} \Phi_1(f) d\mu. \quad (10)$$

Since (5) is trivial if the right side is infinite, we may assume that it is finite. Then (10) implies (by the Fubini-Stone theorem) that f is also μ-integrable on using the fact that μ has the finite subset property. Thus f has σ-finite support, say, B. Hence for the rest of the proof we may assume Ω itself is σ-finite, by replacing it with $\Omega \cap B$ if necessary. Then $(\Omega_i, \Sigma_i, \mu_i)$ will also be σ-finite. If $F(\omega_1) = \{\omega_2 \in \Omega_2 : (\omega_1, \omega_2) \in B\}$, then $F(\omega_1) \in \Sigma_2$ and by the σ-finiteness of μ_2 (so it is "localizable") the collection $\{F(\omega_1), \omega_1 \in \Omega_1\} \subset \Sigma_2$ has a measurable supremum, i.e., there exits an $F \in \Sigma_2$ such that $\mu_2(F(\omega_1) - F) = 0$, $\omega_1 \in \Omega_1$, and if $\widetilde{F}$ is another set with the same property, then $\mu_2(F - \widetilde{F}) = 0$, i.e., F is the smallest such set (cf., e.g., Rao (1987), p. 79). Since Ω_2 is σ-finite, there exist $\Omega_{2n} \uparrow \Omega_2$, $\mu_2(\Omega_{2n}) < \infty$, so that $\mu_2(F \cap \Omega_{2n}) < \infty$ and $F = \bigcup_{n=1}^{\infty} (\Omega_{2n} \cap F)$. Thus F is σ-finite. Define h_1 by the equation

$$h_1(\omega_1) = \int_F f(\omega_1, \omega_2)\, g(\omega_2) d\mu_2. \quad (11)$$

Then $h_1 = h$ a.e., by definition of F. Since the support of $g\,\chi_F$ is σ-finite, and $h = h_1$ a.e., we have reduced the problem to that of Step 4, and hence (5) holds in all cases. ∎

We presented all details to show what additional technical conditions are needed to drop the σ-finiteness assumptions. Note that the product measure μ now is required to have the finite subset property which is *stronger* than demanding that the component measures μ_i have the same property. With this work we are ready to present the

Proof of Theorem 1: By hypothesis $N_{\Phi_1}(f) = k_1 < \infty$ and $N_{\Phi_2}(g) = k_2 < \infty$. We may assume for nontriviality that both $k_i > 0$, $i = 1, 2$.

Then using the fact that $N_{\Phi_2}(\frac{g}{k_2}) = 1$, and (1) [so $\beta = 1$ in (4)] we get by Lemma 2,

$$\int\limits_{\Omega_1} \Phi_3\left(\frac{h}{k_1 k_2 \alpha}\right) d\mu_1 \leq \int\limits_{\Omega_1}\int\limits_{\Omega_2} \Phi_1\left(\frac{f}{k_1}\right) d\mu_2 d\mu_1 = \int\limits_{\Omega} \Phi_1\left(\frac{f}{k_1}\right) d\mu \leq 1.$$

It follows that $N_{\Phi_3}(h) \leq k_1 k_2 \alpha$, giving (3).

For the converse part, let μ be diffuse on a set of positive (finite) measure. If (1) is false then there exist $\alpha_0 > 0$, $x_0 > 0$, $y_0 > 0$ such that

$$\Phi_1^{-1}(x_0 y_0)\, \Phi_2^{-1}(y_0) > \alpha_0\, y_0\, \Phi_3^{-1}(x_0) > 0. \tag{12}$$

By diffuseness of μ (and hence of μ_1, μ_2) there exist sets $A_1 \in \Sigma_1$, $B_1 \in \Sigma_2$ such that $\mu_1(A_1) = \frac{1}{x_0}$, $\mu_2(B_1) = \frac{1}{y_0}$ and if $C_1 = A_1 \times B_1$ ($\in \Sigma$), let $g = \Phi_2^{-1}(y_0) \chi_{B_1}$, $f = \Phi_1^{-1}(x_0 y_0) \chi_{C_1}$. Then $N_{\Phi_2}(g) = 1$, $N_{\Phi_1}(f) = 1$ and setting $h(\omega_1) = 0$ for $\omega_1 \in A_1^c$, we have for $\omega_1 \in A_1$,

$$h(\omega_1) = \int\limits_{\Omega_1} f(\omega_1, \omega_2)\, g(\omega_2) d\mu_2 = \Phi_1^{-1}(x_0 y_0)\, \Phi_2^{-1}(y_0)\, \chi_{A_1}(\omega_1) \frac{1}{y_0}$$

$$> \alpha_0\, \Phi_3^{-1}(x_0) \chi_{A_1}(\omega_1),$$

by (12). Hence $N_{\Phi_3}(h) > \alpha_0\, \Phi_3^{-1}(x_0)\, N_{\Phi_3}(\chi_{A_1}) = \alpha_0$, and (3) fails.

Although $h \in L^{\Phi_3}(\mu_1)$ here, we can use a more detailed construction of h as an elementary (instead of simple) function with f, g still in $\mathcal{M}^{\Phi_1}$ and $\mathcal{M}^{\Phi_2}$, but $h \notin L^{\Phi_3}(\mu_1)$. We omit this detail (but see the proof of Theorem 4), since the essential negative statement is demonstrated. ∎

Let us present a version of the above theorem for finite measures, since it is sufficient for some applications and is simplier to prove. The preceding key lemma takes the following form.

Lemma 3. *Let Φ_i, $i = 1, 2, 3$, be Young functions and $(\Omega_i, \Sigma_i, \mu_i)$, $i = 1, 2$, be finite measure spaces with (Ω, Σ, μ) as their product. If $\Phi_3(x) = 0$ implies $x = 0$, let there exist $\alpha > 0$, $\beta > 0$, $x_0 \geq 0$ such that*

$$\Phi_1^{-1}(xy)\, \Phi_2^{-1}(y) \leq \alpha y\, \Phi_3^{-1}(\beta x), \qquad x, y > x_0. \tag{13}$$

If $f : \Omega \to \mathbb{R}$, $g : \Omega_2 \to \mathbb{R}$ are measurable and $N_{\Phi_2}(g) \leq 1$, then for each $E \in \Sigma$, the function h defined by

$$h(\omega_1) = \int\limits_{\Omega_2} \chi_E(\omega_1, \omega_2)\, f(\omega_1, \omega_2)\, g(\omega_2) d\mu_2, \tag{14}$$

satisfies

$$\int_{\Omega_1} \Phi_3\left(\frac{h}{\alpha(1+x_0\mu_2(\Omega_2))}\right)d\mu_1$$

$$\leq \beta x_0\mu(E) + \frac{\beta}{1+x_0\mu_2(\Omega_2)}\int_E \Phi_1(f)d\mu. \qquad (15)$$

We use this lemma in proving the following specialization of Theorem 1, and then present its details.

Theorem 4. *Let* $(\Omega_i, \Sigma_i, \mu_i)$, $i = 1, 2$, *be finite measure spaces and* (Ω, Σ, μ) *their product. Let* Φ_i, $i = 1, 2, 3$, *be Young functions such that* $\Phi_i(x) > 0$ *for* $x > 0$ *and*

$$\Phi_1^{-1}(xy)\,\Phi_2^{-1}(y) \leq \alpha y\,\Phi_3^{-1}(x), \qquad x, y \geq x_0. \qquad (16)$$

for some $\alpha > 0$, $x_0 \geq 0$. *Then for each* $f \in L^{\Phi_1}(\mu)$, $g \in L^{\Phi_2}(\mu_2)$ *if* h *is defined by* (2), *we have* $h \in L^{\Phi_3}(\mu_1)$, *and in fact*

$$N_{\Phi_3}(h) \leq \alpha C\, N_{\Phi_1}(f)\, N_{\Phi_2}(g) \qquad (17)$$

where C *is a constant depending on* x_0, $\mu_i(\Omega_i)$, $i = 1, 2$, *only. If further* μ *is diffuse on a set of positive measure, the converse holds in the sense that the failure of* (16) *entails the failure of* (17) *for some* f, g.

Proof. For convenience we take $N_{\Phi_1}(f) \leq 1$, $N_{\Phi_2}(g) \leq 1$, since otherwise we can normalize. Let $\alpha_1 = 1 + x_0\mu_2(\Omega_2) \geq 1$. Then by the preceding lemma, in which we let $\beta = 1$ and $E = \Omega_1 \times \Omega_2$, with $\alpha_2 = 1 + x_0\mu_1(\Omega_1)\mu_2(\Omega_2) \geq 1$,

$$\int_{\Omega_1} \Phi_3\left(\frac{h}{\alpha\alpha_1\alpha_2}\right)d\mu_1 \leq \frac{1}{\alpha_2}\int_{\Omega_1} \Phi_3\left(\frac{h}{\alpha\alpha_1}\right)d\mu_1$$

$$\leq \frac{1}{\alpha_2}[x_0\mu_1(\Omega_1)\mu_2(\Omega_2)] + \frac{1}{\alpha_1\alpha_2}\int_\Omega \Phi_1(f)d\mu, \text{ by (15)},$$

$$\leq \frac{1}{\alpha_2}[x_0\mu_1(\Omega_1)\mu_2(\Omega_2)] + \frac{1}{\alpha_1\alpha_2} = 1 - \frac{1}{\alpha_2} + \frac{1}{\alpha_1\alpha_2} \leq 1.$$

Hence there is a constant $C > 0$ depending on the α's such that $N_{\Phi_3}(h) \leq \alpha C$. In fact, $C = \alpha_1\alpha_2$ will suffice.

For the converse, with the diffuseness hypothesis, we can assume that $(\Omega_i, \Sigma_i, \mu_i)$, $i = 1, 2$, are diffuse. If (16) does not hold, then for given $x_n \nearrow \infty$, $\alpha_n \nearrow \infty$, [e.g., $x_n(\mu_1(\Omega_1) \wedge \mu_2(\Omega_2)) = 2^n$, and $\alpha_n = 4^{2n}$, $n \geq 1$], we can choose $u_n \geq x_n$, $v_n \geq x_n$ such that

$$\Phi_1^{-1}(u_n v_n)\Phi_2^{-1}(v_n) > \alpha_n v_n \Phi_3^{-1}(u_n), \quad n \geq 1, \tag{18}$$

and we find disjoint sequences, $A_n \in \Sigma_1, B_n \in \Sigma_2$ such that

$$u_n \mu_1(A_n) = 1 \quad \text{and} \quad 2^n v_n \mu_2(B_n) = 1,$$

so that $\sum\limits_{n=1}^{\infty} \mu_1(A_n) \leq \mu_1(\Omega_1)$, $\sum\limits_{n=1}^{\infty} \mu_2(B_n) \leq \mu_2(\Omega_2)$. Now set

$$g = \sum_{n=1}^{\infty} 2^{-n} \Phi_2^{-1}(v_n) \chi_{B_n}, \quad f = \sum_{n=1}^{\infty} 2^{-n} \Phi_1^{-1}(u_n v_n) \chi_{A_n} \chi_{B_n}.$$

Then $N_{\Phi_2}(g) \leq 1$, $N_{\Phi_1}(f) \leq 1$, but

$$h = \int_{\Omega_2} fg d\mu_2 \geq \chi_{A_n} 4^{-n} \Phi_1^{-1}(u_n v_n) \Phi_2^{-1}(v_n) \mu_2(B_n)$$

and hence by (18),

$$N_{\Phi_3}(h) \geq N_{\Phi_3}(8^{-n} \alpha_n \Phi_3^{-1}(u_n) \chi_{A_n}) = 2^n.$$

Thus (17) cannot hold. In fact $h \notin L^{\Phi_3}(\mu_1)$. ∎
    We now complete the

Proof of Lemma 3: To follow the simple argument of Steps 1 and 2, of Lemma 2, which suffices, let $x_0 > 0$, and $x, y \geq 0$. Then the inequality of Step 1 of that lemma becomes

$$\begin{aligned} xy &\leq \Phi_1^{-1}(\Phi_1(x)) \, \Phi_2^{-1}(\Phi_2(y) + x_0) \\ &\leq \Phi_1^{-1}\left(\frac{\Phi_1(x) + x_0}{\Phi_2(y) + x_0} \cdot (\Phi_2(y) + x_0)\right) \Phi_2^{-1}(\Phi_2(y) + x_0) \\ &\leq \alpha(\Phi_2(y) + x_0) \, \Phi_3^{-1}\left(\beta \frac{\Phi_1(x) + x_0}{\Phi_2(y) + x_0}\right), \quad \text{by (13).} \end{aligned} \tag{19}$$

If we let $\alpha_1 = 1 + x_0 \mu_2(\Omega_2)$, then $\alpha_1 \geq \int_{\Omega_2} [\Phi_2(g) + x_0] d\mu_2$. Now take $x = |f(\omega_1, \omega_2)|$, $y = |g(\omega_2)|$ in (19). Then using the Jensen inequality as in the argument of Step 2 of the proof of Lemma 2, we get

$$\Phi_3\left(\frac{h(\omega_1)}{\alpha\alpha_1}\right) \leq \Phi_3\left[\int_{\Omega_2} \frac{\Phi_2(y)+x_0}{\alpha_1}\chi_E(\omega_1,\omega_2)\Phi_3^{-1}\left(\beta\frac{\Phi_1(x)+x_0}{\Phi_2(y)+x_0}\right)d\mu_2(\omega_2)\right]$$

$$\leq \frac{1}{\alpha_1}\int_{\Omega_2}(\Phi_2(y)+x_0)\Phi_3\left[\chi_E(\omega_1,\omega_2)\Phi_3^{-1}\left(\beta\frac{\Phi_1(x)+x_0}{\Phi_2(y)+x_0}\right)\right]d\mu_2,$$

by Proposition III.3.5,

$$\leq \frac{\beta}{\alpha_1}\int_{\Omega_2}\chi_E(\omega_1,\omega_2)(\Phi_1(x)+x_0)d\mu_2. \tag{20}$$

Integrating (20) on Ω_1 and applying the Fubini theorem, we get the desired inequality (15). ∎

Since in the above theorems h can be expressed as $h = Tg$ so that $T : L^{\Phi_2}(\mu_2) \to L^{\Phi_3}(\mu_1)$ is given by

$$(Tg)(\omega_1) = \int_{\Omega_2} f(\omega_1,\omega_2)\,g(\omega_2)d\mu_2, \tag{21}$$

our considerations give the best conditions for the continuity of the linear operator T. As a consequence of Theorem 4, we can obtain the following somewhat more general version of an important theorem from Krasnosel'skii and Rutickii ((1961), p. 145):

Theorem 5. *Let* $(\Omega_i, \Sigma_i, \mu_i)$, $i = 1, 2$, *be finite measure spaces and* Φ_i, $i = 1, 2, 3$ *be continuous Young functions such that the complementary functions* Ψ_1, Ψ_2 *are also continuous. If* (Ω, Σ, μ) *is the product measure space,* $f \in L^{\Phi_1}(\mu)$, $g \in L^{\Phi_2}(\mu_2)$ *and* $h = Tg$ *is given by* (21) *then T is continuous, in the sense that*

$$N_{\Phi_3}(Tg) = N_{\Phi_3}(h) \leq K\,N_{\Phi_1}(f)\,N_{\Phi_2}(g), \tag{22}$$

for a fixed constant $K > 0$, *when one of the following three conditions on* Φ_i's *holds:*

(i) $\Phi_3 \circ \Psi_2 \prec \Phi_1$

(ii) $\Psi_2 \circ \Phi_3 \prec \Phi_1$

(iii) Ψ_1 *satisfies the* Δ' *condition and* $\Psi_2 \prec \Phi_1$ *as well as* $\Phi_3 \prec \Phi_1$.

Remark. Since the measures are finite, all these order relations are local. Moreover, this result motivates a further analysis on the structure of the operators T of (21). That will be taken up in the next chapter.

Proof. If (i) holds, then by definitions of '$\prec$', we get for $x, y \geq x_0 \geq 1$,

$$\Phi_1(\alpha xy) \geq \Phi_3(\Psi_2(xy)) \geq \Phi_3(x\Psi_2(y)) \geq \Phi_3(x)\Psi_2(y), \qquad (23)$$

by the convexity of Φ_3 and Ψ_2 if $x_0 \geq 1$ is large such that $\Psi_2(x_0) \geq 1$. Similarly, if (ii) holds, we choose $x_0 \geq 1$ so that $\Phi_3(x_0) \geq 1$ and hence

$$\Phi_1(\alpha xy) \geq \Psi_2(\Phi_3(xy)) \geq \Psi_2(x\Phi_3(y)) \geq \Psi_2(x)\Phi_3(y). \qquad (24)$$

If (iii) holds, since $\Psi_1 \in \Delta'$ is equivalent to the statement

$$\Phi_1(u)\Phi_1(v) \leq \Phi_1(\alpha uv), \qquad \alpha > 0, \; u, v \geq u_0 \geq 0,$$

by Theorem II.3.11, we get

$$\Phi_3(x)\Psi_2(y) \leq \Phi_1(\alpha x)\Phi_1(\alpha y) \leq \Phi_1(\alpha^2 xy), \qquad x, y \geq x_0. \qquad (25)$$

Thus the result will follow if we establish it under the equivalent condition (23), (24) or (25). This will be deduced from Theorem 4.

Since Φ_i's are continuous, they are strictly increasing for large values, so that for $x \geq x_0$,

$$\Phi_1\big(\Phi_1^{-1}(x)\big) = x = \Phi_1^{-1}\big(\Phi_1(x)\big).$$

The same is true of Ψ_2 and Φ_3. Hence for large u, v ,

$$uv = \Phi_3\big(\Phi_3^{-1}(u)\big) \cdot \Psi_2\big(\Psi_2^{-1}(v)\big) \leq \Phi_1\big(\alpha\,\Phi_3^{-1}(u)\Psi_2^{-1}(v)\big), \text{ by (23)}.$$

This implies

$$\Phi_1^{-1}(uv) \leq \alpha\,\Phi_3^{-1}(u)\Psi_2^{-1}(v), \qquad u, v \geq u_0 > 0. \qquad (26)$$

Hence

$$\Phi_1^{-1}(uv)\Psi_2^{-1}(v) \leq \alpha\,\Phi_3^{-1}(u) \cdot \Phi_2^{-1}(v)\Psi_2^{-1}(v) \leq 2\alpha v\Phi_3^{-1}(u), \qquad (27)$$

by Proposition II.1.1. Thus (27) reduces to (16). Hence (22) follows from (17), as desired. ∎

It is, of course, possible to present some special results when the underlying measure space is purely atomic which will be in a direction opposite to that of Krasnosel'skii and Rutickii monograph where the treatment is restricted to a bounded set in $I\!R^n$ with Lebesgue measure. We do not treat such a specialization, but the following sample result on the embedding will give the flavor.

Proposition 6. *Let Φ_i be Young functions with $\Phi_i(x) > 0$ for $x > 0$, $i = 1, 2, 3$. Let ℓ^{Φ_i} be the corresponding sequence Orlicz space. Then for each $f \in \ell^{\Phi_1}$ and $g \in \ell^{\Phi_2}$, the function h defined by*

$$h(m) = \sum_{n=1}^{\infty} f(m, n)\, g(n), \qquad m \geq 1, \tag{28}$$

belongs to ℓ^{Φ_3} iff there is an $\alpha > 0$ and $u_0 > 0$ such that

$$\Phi_1^{-1}(uv)\Phi_2^{-1}(v) \leq \alpha v \Phi_3^{-1}(u), \qquad 0 \leq u, v \leq u_0. \tag{29}$$

When this condition is satisfied, then

$$N_{\Phi_3}(h) \leq \alpha K N_{\Phi_1}(f) N_{\Phi_2}(g) \tag{30}$$

holds for some positive constant K $(= K(u_0))$.

This is the analog of Theorem 1, and the proof running on similar (and slightly simpler) lines will not be included here. The details are given by O'Neil (1968). Note that this also will imply the continuity of the operator $T : g \mapsto Tg = h$ defined by (28) which is again an "integral operator". Various classes of such operators to be compact or factorizable or interpolatable, and their applications will be studied in the next chapter. So further discussion is omitted here.

Bibliographical notes. Unlike the Lebesgue class, the isomorphism results for the Orlicz spaces show the differences in the (strict) convexity properties vividly. Theorem 1.1 and Proposition 1.2 are taken from Rao (1968$_a$). The others in Section 1 are simple modifications of the classical work given in Krasnosel'skii and Rutickii's book. Similarly the results of Section 2 are generalizations of the classical work given in Dunford and Schwartz (1958, Section IV.8), and are motivated by the original theorems of Kolmogorov (1931) and Takahashi (1935).

The embedding results are also found in the book by Krasnosel'skii and Rutickii (1961), and several extensions are given by Andô (1960), Rutickii (1963) and completed by Ren (1986$_a$) (see also Rutickii (1962) and Ren (1963)). Our treatment is based on these works, but is freed from the assumptions of finiteness of measures and in particular the Lebesgue measures and to some extent also from the Φ_i being N-functions. Theorem 3.5 and Proposition 3.7 are based on Ren's paper (1986$_b$); see also Rabinovich (1968) and Ren (1987). Most of Section 4 is essentially based on Andô's (1960$_b$) paper. The generalizations given are motivated by O'Neil's (1968) treatment of the subject. On the other hand Section 5 follows O'Neil's (1968) paper closely. The pointwise and tensor products were originally treated in Andô (1960$_b$), but the integral products receive a detailed analysis in the work of O'Neil for general measure spaces and with minimal restrictions on the Young functions. Here we sought to present a unified (and somewhat simplified) treatment in the text. Various applications and sharper forms of some of these basic results will occupy the rest of this book.

VI

Analysis of Linear Operators

This chapter is devoted to the structure and analysis of classes of (mostly) linear operators between subspaces of Orlicz spaces, and some applications. We first consider the compactness, extension, and factorization properties as well as characterizations of linear integral operators on them. Then there follow various generalizations of the Riesz-Thorin interpolation theorem to linear and multilinear operators and also the Marcinkiewicz's interpolation theorem in the context of Orlicz spaces. A few applications are made to fractional integration and packing problems, and then we discuss briefly the existence of (Schauder) bases in M^Φ and L^Φ spaces.

6.1. Integral operators

If $\mathcal{X}, \mathcal{Y}$ are Banach spaces then, as usual, we denote by $B(\mathcal{X}, \mathcal{Y})$ the class of all bounded linear operators from $\mathcal{X}$ into $\mathcal{Y}$. Under the operator (or supremum) norm, $B(\mathcal{X}, \mathcal{Y})$ is a Banach space over the same scalar field as $\mathcal{X}$ and $\mathcal{Y}$. Also recall that $T \in B(\mathcal{X}, \mathcal{Y})$ is *compact* if T maps bounded sets of $\mathcal{X}$ into conditionally compact sets in the (norm) topology of $\mathcal{Y}$. It is *weakly compact* if T maps bounded sets of $\mathcal{X}$ into conditionally weakly compact, or equivalently (by the Eberlein-Šmulian theorem) into conditionally weakly sequentially compact sets of $\mathcal{Y}$. By a classical theorem of V. Gantmacher and M. Nakamura, every element of $B(\mathcal{X}, \mathcal{Y})$ is weakly compact if either $\mathcal{X}$ or $\mathcal{Y}$ is reflexive. Here we study the structure of $B(\mathcal{X}, \mathcal{Y})$ when $\mathcal{X}$ and $\mathcal{Y}$ are Orlicz spaces and

T is moreover an integral operator. To make this precise we introduce some notation and define these special classes from $B(\mathcal{X}, \mathcal{Y})$. If $\mathcal{X} = \mathcal{Y}$, we set $B(\mathcal{X}, \mathcal{X}) = B(\mathcal{X})$.

Thus let (Φ_i, Ψ_i), $i = 1, 2, 3$ be Young's complementary pairs and $(\Omega_i, \Sigma_i, \mu_i)$, $i = 1, 2, 3$ be measure spaces such that $(\Omega_3, \Sigma_3, \mu_3)$ is the product space $(\Omega_1 \times \Omega_2, \Sigma_1 \otimes \Sigma_2, \mu_1 \otimes \mu_2)$ with the finite subset property. We simply write $L^{\Phi_i}(\mu_i)$ for $L^{\Phi_i}(\Omega_i, \Sigma_i, \mu_i)$ and denote the tensor product, as usual, by $L^{\Phi_1}(\mu_1) \otimes L^{\Phi_2}(\mu_2)$. If $k : \Omega_1 \times \Omega_2 \to I\!R$ (or $\mathbb{C}$) which is measurable for Σ_3, then $k(\cdot, \cdot)$ is called a *kernel* and the associated operator T_k, an induced *kernel (or integral) operator* whenever $T_k f \in L^{\Phi_2}(\mu_2)$ for all $f \in L^{\Phi_1}(\mu_1)$ where

$$(T_k f) : \omega_2 \mapsto (T_k f)(\omega_2) = \int_{\Omega_1} k(\omega_1, \omega_2) f(\omega_1) d\mu_1(\omega_1), \quad \omega_2 \in \Omega_2. \quad (1)$$

The first questions considered are the continuity, compactness, and related properties of T_k under suitable conditions on Φ_i and μ_i. Also considered is the next question: which continuous linear $T : L^{\Phi_1}(\mu_1) \to L^{\Phi_2}(\mu_2)$ arise as kernel operators for some k of the type given above? We then treat, in the next section, the factorization (and extension) of such operators before turning to the interpolation problems on general (not necessarily kernel type) operators on these spaces.

Let us start with a modified version of Theorem V.5.4.

Proposition 1. *Let* $L^{\Phi_i}(\mu_i)$, $i = 1, 2, 3$ *be Orlicz spaces on finite measure spaces* $(\Omega_i, \Sigma_i, \mu_i)$ *and* N*-functions* Φ_i. *If* $k \in L^{\Psi_3}(\mu_3)$ *consider the following conditions for* T_k *given by* (1):

  (a) $\Phi_3(\alpha xy) \leq \Phi_1(x)\Psi_2(y)$, $x, y \geq x_0 \geq 0, \alpha > 0$;
  (b) $L^{\Phi_1}(\mu_1) \otimes L^{\Psi_2}(\mu_2) \subset L^{\Phi_3}(\mu_3)$,
  (c) $T_k \in B(L^{\Phi_1}(\mu_1), L^{\Phi_2}(\mu_2))$.

Then (a) $\Rightarrow$ (b) *and* (a) $\Rightarrow$ (c). *If, moreover,* μ_3 *is diffuse on a set of positive measure, then* (a) $\Leftrightarrow$ (b) $\Leftrightarrow$ (c) *so that all are equivalent.* [*We use* Ψ_i *as the complementary Young function to* Φ_i, *as usual.*]

Proof: (a) $\Leftrightarrow$ (b). Since the measures μ_i are finite, we have by Theorem V.4.2,

$$N_{\Phi_3}(fg) \leq C N_{\Phi_1}(f) N_{\Psi_2}(g), \qquad f \in L^{\Phi_1}(\mu_1), g \in L^{\Psi_2}(\mu_2). \quad (2)$$

Thus (b) follows.

(a) $\Rightarrow$ (c). Since (a) implies (2), we can employ Hölder's inequality with (2) to deduce (c), using Fubini's theorem, as follows.

$$\left| \int_{\Omega_2} (T_k f) g \, d\mu_2 \right| = \left| \int_{\Omega_2}\!\int_{\Omega_1} k(\omega_1, \omega_2) f(\omega_1) g(\omega_2) \, d\mu_1(\omega_1) \, d\mu_2(\omega_2) \right|$$

$$= \left| \int_{\Omega_3} k(fg) \, d\mu_3 \right|$$

$$\leq N_{\Psi_3}(k) \|(fg)\|_{\Phi_3}$$

$$\leq 2 C N_{\Psi_3}(k) N_{\Phi_1}(f) N_{\Psi_2}(g), \qquad \text{by (2)}.$$

Taking the supremum on the left relative to $\|g\|_{\Psi_2} \leq 1$ and then with $N_{\Phi_1}(f) \leq 1$, we get $\|T_k\| \leq 2 C N_{\Psi_3}(k) < \infty$, which is (c).

If now the measure μ_3, hence μ_i, is diffuse on a set of positive measure, we may assume for convenience that $(\Omega_i, \Sigma_i, \mu_i)$ themselves are diffuse. Since (Φ_i, Ψ_i) are complementary N-functions, by Proposition II.2.9, we have that (a) is equivalent to:

$$\Psi_1(x)\Phi_2(y) \leq \Psi_3(\beta x y), \qquad \text{for some } \beta > 0, \ x, y \geq y_0 \geq 0, \quad (3)$$

so that putting $\Psi_1(x) = v, \Phi_2(y) = u$,

$$\Psi_3^{-1}(uv)\Phi_1^{-1}(v) \leq \beta \Phi_1^{-1}(v)\Psi_1^{-1}(v)\Phi_2^{-1}(u) \leq 2\beta v \Phi_2^{-1}(u). \quad (4)$$

Hence by the last part of Theorem V.5.4, (a) and (c) are equivalent, since in the presence of diffuseness (4) and (a) are equivalent. We can also conclude by the now-familiar indirect argument that (a) and (b) must be equivalent also. This is the assertion of the theorem. ∎

Specializing this for the Lebesgue spaces we get immediately:

Corollary 2. *Let $1 < p_i, q_i < \infty$, $p_i^{-1} + q_i^{-1} = 1$, and $(\Omega_i, \Sigma_i, \mu_i)$ to be finite measure spaces which are diffuse. Then for each $k \in L^{q_3}(\mu_3)$ the integral operator T_k is in $B(L^{p_1}(\mu_1), L^{p_2}(\mu_2))$ iff $p_3 \leq \min(p_1, q_2)$.*

In what follows we consider characterizations of integral (or kernel) operators that are compact. We denote by $B_c(\mathcal{X}, \mathcal{Y})$ the subspace of $B(\mathcal{X}, \mathcal{Y})$ which has compact elements where $\mathcal{X}, \mathcal{Y}$ are Banach spaces.

The following auxiliary result, due to K. Vala (1964), will be of interest in this study, and it is a kind of dominated convergence statement for compact operators. This is a slight extension of Lemma V.2.5.

Lemma 3. *Let $\{T_\alpha, \alpha \in I\} \subset B_c(\mathcal{X}, \mathcal{Y})$ be a directed family (i.e., I is a directed set) such that (i) $\lim_\alpha T_\alpha(x) = T(x)$, $x \in \mathcal{X}$ (i.e., the net converges pointwise in $\mathcal{Y}$) and (ii) there is an $S \in B_c(\mathcal{X}, \mathcal{Y})$ such that $\|T_\alpha(x)\|_\mathcal{Y} \leq \|S(x)\|_\mathcal{Y}$, $x \in \mathcal{X}, \alpha \in I$. Then T is compact and $\lim_\alpha \|T_\alpha - T\| = 0$, the latter norm is that of $B(\mathcal{X}, \mathcal{Y})$, i.e., uniform norm.*

Proof: If U is the unit ball of $\mathcal{X}$, then $S(U)$ is conditionally compact in $\mathcal{Y}$ since S is a compact operator. Hence it has a finite covering which may be chosen as follows. Given $\varepsilon > 0$, select bounded sets $A_1, \ldots, A_n$ such that $\bigcup_{j=1}^{n} A_j \supset U$ and if $x_1, x_2 \in A_k$ then $\|S(x_1) - S(x_2)\|_\mathcal{Y} < \frac{\varepsilon}{3}$. This is possible since $S(U)$ is totally bounded. If $x_j \in A_j$, then (i) of the hypothesis implies that there is $\alpha_j \in I$ such that for $\alpha, \alpha' \geq \alpha_j$, one has $\|T_\alpha(x_j) - T_{\alpha'}(x_j)\|_\mathcal{Y} < \frac{\varepsilon}{3}$. By the directedness of the index I, there is an $\alpha_0 \geq \alpha_i$, $1 \leq i \leq n$, such that for $\alpha, \alpha' \geq \alpha_0$, and any $x \in U \subset \bigcup_{j=1}^{n} A_j$ so that $x \in A_j$ for some j, one has

$$\|T_\alpha(x) - T_{\alpha'}(x)\|_\mathcal{Y} \leq \|T_\alpha(x) - T_\alpha(x_j)\|_\mathcal{Y} + \|T_\alpha(x_j) - T_{\alpha'}(x_j)\|_\mathcal{Y}$$
$$+ \|T_{\alpha'}(x_j) - T_{\alpha'}(x)\|_\mathcal{Y}$$
$$\leq 2\|S(x - x_j)\|_\mathcal{Y} + \|T_{\alpha'}(x_j) - T_{\alpha'}(x_j)\|_\mathcal{Y}$$
$$< 2 \cdot \frac{\varepsilon}{3} + \frac{\varepsilon}{3} = \varepsilon, \tag{5}$$

since S dominates all T_α in norm, and the particular choice of the A_j is used. Taking supremum over all x in the unit ball U, (5) implies that $T_\alpha \to T$ in the operator norm. However, it is a classical result that $B_c(\mathcal{X}, \mathcal{Y})$ is a closed subspace of $B(\mathcal{X}, \mathcal{Y})$, and this implies $T \in B_c(\mathcal{X}, \mathcal{Y})$. ∎

We first give a characterization of compact operators in $B(\mathcal{X}, M^\Phi)$ and then specialize the result. For the former, we recall that an "averaging," or conditional expectation, was defined for Theorem V.2.4 as U_π for each finite partition $\pi = (A_1, \ldots, A_n)$ from Σ, $0 < \mu(A_i) < \infty$, A_j disjoint, on $L^\Phi(\mu)$, where

$$U_\pi f = \sum_{j=1}^{n} \frac{1}{\mu(A_j)} \left(\int_{A_j} f \, d\mu \right) \chi_{A_j}, \qquad f \in L^\Phi(\mu). \tag{6}$$

It was already shown in the proof of that theorem that $N_\Phi(U_\pi f) \leq N_\Phi(f)$, for each such π. The collection of all such partitions π, ordered by refinement and denoted Π is a directed set.

As a consequence of Theorem V.2.4, we have the following proposition.

Proposition 4. *For each Banach space $\mathcal{X}$, continuous Young function Φ and Orlicz space $L^{\Phi}(\mu)$ on a measure space (Ω, Σ, μ), one has T ($\in B(\mathcal{X}, M^{\Phi})$) is compact iff $\lim_{\pi \in \Pi} \|U_{\pi}T - T\| = 0$.*

Proof: Since for each $x \in \mathcal{X}$, $f_x = Tx \in M^{\Phi}$, the given condition implies that $N_{\Phi}(U_{\pi}f_x - f_x) \to 0$, as π varies over Π, uniformly in x in bounded sets B of $\mathcal{X}$. But U_{π} having a finite dimensional range is compact and $\|U_{\pi}\| \leq 1$. Then since $\lim_{\pi \in \Pi} U_{\pi} = T$, Theorem V.2.4 implies that $T(B)$ is conditionally compact in M^{Φ} so that $T \in B_c(\mathcal{X}, M^{\Phi})$. [The fact that simple functions are dense in M^{Φ} is essential here.]

For the converse, note that $N_{\Phi}(U_{\pi}Tx) \leq N_{\Phi}(Tx)$, and taking $S = T$ in the preceding lemma, T being compact, we conclude from it that $\|U_{\pi}T - T\| \to 0$, since $T(B)$ is conditionally compact by hypothesis so that $U_{\pi}f \to f$ uniformly for $f \in T(B)$ by the last part of Theorem V.2.4, as $\pi \nearrow$ in Π. ∎

Taking $\mathcal{X}$ as an Orlicz space, we can obtain the following result as a consequence of Propositions 1 and 4 above. The necessity given as the second part is due to Z. D. Ren (1983).

Theorem 5. *Let Φ_i be N-functions, $L^{\Phi_i}(\mu_i)$ be Orlicz spaces on finite measure spaces $(\Omega_i, \Sigma_i, \mu_i)$, $i = 1, 2, 3$ with $(\Omega_3, \Sigma_3, \mu_3)$ as the product of the other two spaces. Then for each $k \in M^{\Psi_3}(\mu_3)$ the kernel operator T_k is in $B_c(L^{\Phi_1}(\mu_1), M^{\Phi_2})$ whenever*

$$\Phi_3(\alpha xy) \leq \Phi_1(x)\Psi_2(y), \qquad x, y \geq x_0 \geq 0, \alpha > 0, \tag{7}$$

so that

$$L^{\Phi_1}(\mu_1) \otimes L^{\Psi_2}(\mu_2) \subset L^{\Phi_3}(\mu_3). \tag{8}$$

If on the other hand μ_3 is diffuse on a set of positive measure the equivalent conditions (7) and (8) are also necessary for T_k to be compact for each $k \in M^{\Psi_3}$.

Proof: That $(7) \Rightarrow (8)$ is clear and then by Proposition 1, T_k is continuous on $L^{\Phi_1}(\mu_1)$ into $L^{\Phi_2}(\mu_2)$. But since $k \in M^{\Psi_3}$, and Ψ_3 is an N-function so that simple functions are dense in this space, there exist simple $k_n \in M^{\Psi_3}$ satisfying $N_{\Psi_3}(k - k_n) \to 0$. Since $T_{k_n}f$ of

(1) is therefore in M^{Φ_2}, it follows from the completeness of M^{Φ_2}, the Hölder inequality of Orlicz spaces, and Fubini's theorem that $T_k \in B(L^{\Phi_1}(\mu_1), M^{\Phi_2})$. Moreover, $N_{\Psi_3}(k_n) \leq C_0 < \infty$ so that $\|T_{k_n}\| \leq C_1 < \infty$, all n, and each T_{k_n} is compact since it has a finite dimensional range. From this and Theorem V.2.4 (converse part), we infer that $\|U_\pi T_k - T_k\| \to 0$ so that by Proposition 4, $T_k \in B_c(L^{\Phi_1}(\mu_1), M^{\Phi_2})$.

For the converse, let μ_3 be diffuse. Then by Proposition 1, both (7) and (8) are equivalent. Suppose these conditions do not hold. Then we construct a $k_1 \in M^{\Psi_3}$ such that T_{k_1} is not compact, in fact, not even bounded, which would imply that (7) ($\Leftrightarrow$ (8)) is necessary.

Thus suppose (7) did not hold. We may assume μ_3 on Ω_3 and hence μ_i on Ω_i are diffuse. Then there exist numbers, $0 < c_n \uparrow \infty, 0 < d_n \uparrow \infty$, such that $\Phi_1(c_1)\mu_1(\Omega_1) > 1$, $\Psi_2(d_1)\mu_2(\Omega_2) > 1$, and

$$\Phi_3\left(\frac{c_n d_n}{n4^n}\right) > \Phi_1(c_n)\Psi_2(d_n), \qquad n \geq 1.$$

This is equivalent to

$$\Psi_3\left(\frac{n4^n \Phi_1(c_n)\Psi_2(d_n)}{c_n d_n}\right) < \Phi_1(c_n)\Psi_2(d_n), \qquad n \geq 1. \tag{9}$$

Let $\{A_n, n \geq 1\}, \{B_n, n \geq 1\}$ be mutually disjoint sets from Σ_1 and Σ_2 such that (by diffuseness of μ_i again)

$$2^n \Phi_1(c_n)\mu_1(A_n) = 1 \text{ and } 2^n \Psi_2(d_n)\mu_2(B_n) = 1. \tag{10}$$

Define k_1 as

$$k_1 = \sum_{n=1}^{\infty} \frac{4^n \Phi_1(c_n)\Psi_2(d_n)}{c_n d_n} \chi_{A_n \times B_n}. \tag{11}$$

We assert that $k_1 \in M^{\Psi_3}$ but T_{k_1} is not bounded.

Let $0 < \lambda < \infty$ be arbitrary and choose $n_0 \geq 1$ such that n_0 is the smallest integer such that $n_0 \geq \lambda$. Define

$$D(\lambda) = \sum_{n=1}^{n_0-1} \Psi_3\left(\frac{\lambda 4^n \Phi_1(c_n)\Psi_2(d_n)}{c_n d_n}\right)\mu_1(A_n)\mu_2(B_n).$$

From the choices (9)–(11) we get

$$\int_{\Omega_3} \Psi_3(\lambda k_1) d\mu_3 \leq D(\lambda) + \sum_{n=n_0}^{\infty} \Psi_3\left(\frac{n4^n \Phi_1(c_n)\Psi_2(d_n)}{c_n d_n}\right)\mu_1(A_n)\mu_2(B_n)$$

$$< D(\lambda) + \sum_{n \geq n_0} 4^{-n} < \infty.$$

Since $0 < \lambda < \infty$ is arbitrary, this gives $k_1 \in M^{\Psi_3}$. If we now let

$$f = \sum_{n=1}^{\infty} c_n \chi_{A_n}, \quad g = \sum_{n=1}^{\infty} d_n \chi_{B_n}, \tag{12}$$

then $\|f\|_{\Phi_1} \le 1 + \rho_{\Phi_1}(f) = 2$, and $\rho_{\Psi_2}(g) = 1$. Hence for T_{k_1},

$$\|T_{k_1}\| = \sup\{\|T_{k_1} h\|_{\Phi_2} : \|h\|_{\Phi_1} \le 1\} \ge \frac{1}{2}\|T_{k_1} f\|_{\Phi_2}$$

$$\ge \frac{1}{2} \left| \int_{\Omega_2} (T_{k_1} f)(w_2) g(w_2) d\mu_2(w_2) \right|$$

$$= \frac{1}{2} \int_{\Omega_1} \int_{\Omega_2} k_1(w_1, w_2) f(w_1) g(w_2) d\mu_2(w_2) d\mu_1(w_1)$$

$$\ge \frac{1}{2} \sum_{n=1}^{\infty} 1 = +\infty.$$

Thus $T_{k_1} \notin B(L^{\Phi_1}(\mu_1), M^{\Phi_2})$ as asserted. ∎

Taking $\Phi_1 = \Psi_2 = \Phi_3 = \Phi$, and $\Phi_2 = \Psi_3 = \Psi$ in the preceding result, we get:

Corollary 6. *Let (Φ, Ψ) be a pair of complementary N-functions, μ_3 diffuse, and $k \in M^{\Psi}(\mu_3)$. Then $T_k \in B_c(L^{\Phi}(\mu_1), M^{\Psi}(\mu_2))$ iff $\Phi \in \Delta'$.*

To complete the picture on integral operators, we construct, following D. V. Salehov (1963), an example of an integral operator $T_{k_0} \in B_c(L^{\Phi_1}(\mu_1))$ even though $k_0 \notin L^{\Psi}(\mu_1 \otimes \mu_1)$ so that the structure of integral operators on Orlicz spaces is more involved.

Example 7. *Let Φ_1 and Ψ be a pair of N-functions. Then there exists a measurable $k_0 : [0, 1] \times [0, 1] \to \mathbb{R}$ such that $k_0 \notin L^{\Psi}(\mu \otimes \mu)$ but $T_{k_0} \in B_c(L^{\Phi_1}(\mu))$, where μ is the Lebesgue measure on $[0, 1]$ and T_{k_0} is the kernel operator on $L^{\Phi_1}(\mu)$.*

Construction. Let Φ be complementary to the N-function Ψ and choose integers $1 < m_1 < m_2 < \cdots$ such that $\Phi^{-1}(m_n) > n \cdot 2^n$, $n \ge 1$. Consider squares on the diagonal of the unit square $[0, 1]^2$ defined by:

$$I_j^{(n)} = \{(w_1, w_2) : \frac{j-1}{m_n} < w_i < \frac{j}{m_n}, i = 1, 2\}, I^{(n)} = \bigcup_{j=1}^{m_n} I_j^{(n)}.$$

Then $\mu \otimes \mu(I^{(n)}) = \left(\frac{1}{m_n}\right)^2 \cdot m_n = \frac{1}{m_n}$. Next define k_n as

$$k_n(w_1, w_2) = \begin{cases} m_n & \text{if } (w_1, w_2) \in I^{(n)} \\ 0 & \text{if not,} \end{cases} \tag{13}$$

and consider the kernel operator $T_{k_n} : L^{\Phi_1}(\mu) \to L^{\Phi_1}(\mu)$, as in (1). This is in $B(L^{\Phi_1}(\mu))$ since

$$\|T_{k_n}\| = \sup\{\|T_{k_n}f\|_{\Phi_1} : \|f\|_{\Phi_1} \le 1\}$$

$$= \sup\{\sup\left[\int\int_{I^{(n)}} m_n|f(w_1)g(w_2)|d\mu(w_1)d\mu(w_2) : N_{\Psi_1}(g) \le 1\right] :$$

$$\|f\|_{\Phi_1} \le 1\}$$

$$\le \sup\{\sup\left[\sum_{j=1}^{m_n}\left(\int\int_{I_j^{(n)}} m_n\Phi_1(f(w_1))d\mu(w_1)d\mu(w_2)\right.\right.$$

$$\left.\left. + \int\int_{I_j^{(n)}} m_n\Psi_1(g(w_2))d\mu(w_2)d\mu(w_1)\right) : N_{\Psi_1}(g) \le 1\right] :$$

$$\|f\|_{\Phi_1} \le 1\}, \text{ by Young's inequality and (13)},$$

$$= \sup\{\sup\left[\left(\int_0^1 \Phi_1(f)d\mu(w_1) + \int_0^1 \Psi_1(g)d\mu(w_2)\right):N_{\Psi_1}(g) \le 1\right] :$$

$$\|f\|_{\Phi_1} \le 1\} \le 2.$$

Now let $k = \sum_{n=1}^{\infty} 2^{-n}k_n$, and $T_k = \sum_{n=1}^{\infty} 2^{-n}T_{k_n}$. Then $\|T_k\| \le 2$. Since each T_{k_n} is compact and $\|T_k - \sum_{j=1}^{n} 2^{-j}T_{k_j}\| \to 0$ as $n \to \infty$, it follows that $T_k \in B_c(L^{\Phi_1}(\mu))$. However, by the choice of m_n, and (13), $\|k\|_{\Psi} \ge 2^{-n}\|k_n\|_{\Psi} = 2^{-n}\Phi^{-1}(m_n) > n$ so that $k_0 = k \notin L^{\Psi}(\mu \otimes \mu)$. $\blacksquare$

The preceding example and propositions raise the question of characterizing integral operators whose kernels are in a given Orlicz space based on a product measure space. This problem is deeper than the direct parts considered thus far since its solution explicitly depends on a vector Radon-Nikodým theorem. The following motivational remarks will illustrate this situation.

Let $(\Omega_i, \Sigma_i, \mu_i)$ be finite measure spaces, $L^{\Phi_i}(\mu_i)$, $i = 1, 2$, be Orlicz spaces on them and $T : L^{\Phi_1}(\mu_1) \to L^{\Phi_2}(\mu_2)$ be a continuous linear operator. Then $\nu(E) = T(\chi_E)$, $E \in \Sigma$, is an element of the Banach space $L^{\Phi_2}(\mu_2)$, and in fact $\nu : \Sigma \mapsto L^{\Phi_2}(\mu_2)$ is an additive function which vanishes on μ_1-null sets. Moreover, if $\Phi_2 \in \Delta_2$, then for each $\ell \in (L^{\Phi_2}(\mu_2))^*$, we get

$$\ell(T(\chi_{E_n})) = \int_{\Omega_2} (T\chi_{E_n}) h_\ell d\mu_2, \qquad (14)$$

for a unique $h_\ell \in L^{\Psi_2}(\mu_2)$. From this it can be concluded that $\ell(T(\chi_{E_n})) = \ell(\nu(E_n)) \to 0$ as $n \to \infty$ if $E_n \searrow \emptyset$ for each such ℓ, so that $\nu(\cdot)$ is weakly and hence, by a classical theorem of Pettis in abstract analysis, strongly σ-additive, i.e., in the norm topology of $L^{\Phi_2}(\mu_2)$. Also $\nu \ll \mu_1$. It is then true in some (but *not* all) cases that ν can be expressed as

$$T(\chi_E) = \nu(E) = \int_E g d\mu_1, \qquad E \in \Sigma_1, \qquad (15)$$

so that

$$T(f) = \int_{\Omega_1} f g d\mu_1 \in L^{\Phi_2}(\mu_2). \qquad (16)$$

Comparing this with (1), we see that one can expect to obtain precise conditions on the Φ_i and T (hence on ν) in order that (16) is valid as well as T to be an integral or kernel operator with g as the appropriate kernel. The necessary work is involved and we shall present a simple but useful part of it here.

To state a vector Radon-Nikodým theorem, it is necessary to recall the concepts of variations of a vector measure. Thus if (Ω, Σ) is a measurable space and $\mathcal{X}$ a Banach space, then $\nu : \Sigma \to \mathcal{X}$ σ-additive in the norm topology (which is known to be equivalent to $x^*(\nu) : \Sigma \to$ scalars, being σ-additive for each $x^* \in \mathcal{X}^*$), then it is of *finite variation on* $E \in \Sigma$ if $(\Sigma(E) = \{A \cap E : A \in \Sigma\})$

$$|\nu|(E) = \sup \left\{ \sum_{i=1}^n \|\nu(A_i)\|_{\mathcal{X}} : A_i \in \Sigma(E), \text{ disjoint} \right\} < \infty. \qquad (17)$$

Also ν is said to have *finite semi-variation on* E if

$$\|\nu\|(E) = \sup \left\{ \left\| \sum_{i=1}^n a_i \nu(A_i) \right\|_{\mathcal{X}} : |a_i| \le 1, A_i \in \Sigma(E) \right.$$

$$\left. \text{disjoint}, a_i \in \mathbb{C} \right\} < \infty. \qquad (18)$$

It is clear that $\|\nu\|(E) \le |\nu|(E)$, but a strict inequality is possible if $\mathcal{X}$ is not finite dimensional. It can be shown that $\|\nu\|(\Omega) < \infty$ always, and it is true that $|\nu|(E) = +\infty$ for all E for which $\|\nu\|(E) > 0$ for certain $\mathcal{X}$. Even if $|\nu|(\Omega) < \infty$, and $\nu \ll \mu$ in the sense that $\mu(A) = 0$ implies $\nu(A) = 0$ in $\mathcal{X}$, then for infinite dimensional $\mathcal{X}$, ν need not be an indefinite integral of a measurable $f : \Omega \to \mathcal{X}$ relative to μ. The definition of this integral and conditions under which ν is an indefinite integral of such an f constitute a vector Radon-Nikodým theory and this is intimately related to the geometry of the Banach space $\mathcal{X}$. We shall indicate a few results of interest for our immediate application, to give a glimpse of the subject.

If (Ω, Σ, μ) is a localizable measure space, $f : \Omega \to \mathcal{X}$ is called *strongly* or *μ-measurable*, whenever (i) $f(\Omega - N) \subset \mathcal{X}$ is separable $\mu(N) = 0$, and (ii) $f^{-1}(G)$ is a μ-measurable set for each open set $G \subset \mathcal{X}$. The second condition may be replaced by $x^*(f) : \Omega \to \mathbb{C}$ is μ-measurable for each $x^* \in \mathcal{X}^*$. [This alternate condition is called *weak measurability* of f.] It may be verified that if $\mathcal{X}$ is a separable Banach space, then the weak and strong measurabilities of $f : \Omega \to \mathcal{X}$ are equivalent (but not otherwise). A μ-measurable $f : \Omega \to \mathcal{X}$ is *Bochner integrable* if there is a sequence $f_n : \Omega \to \mathcal{X}$ of simple functions converging to f in μ-measure and if $\int_\Omega \|f_n - f_m\|_{\mathcal{X}} d\mu \to 0$ as $m, n \to \infty$.

Then we set $\int_\Omega f d\mu = \lim_{n\to\infty} \int_\Omega f_n d\mu$, where $\int_A f_n d\mu = \sum_{i=1}^n a_i^n \mu(A \cap A_i^n)$ if $f_n = \sum_{i=1}^n a_i^n \chi_{A_i^n}$. It can be shown that the integral $\int_\Omega f d\mu$ is well defined, and many properties of the Lebesgue integration theory are valid for it if we replace the absolute value of the scalars with the norm of $\mathcal{X}$ in the general case. For instance, the Orlicz space $L^\Phi(\mu, \mathcal{X})$ of $\mathcal{X}$-valued strongly measurable functions on (Ω, Σ, μ) is defined as:

$$L^\Phi(\mu, \mathcal{X}) = \{f : \Omega \to \mathcal{X} \mid \int_\Omega \Phi\Big(\frac{\|f(w)\|_{\mathcal{X}}}{\alpha}\Big) d\mu < \infty \text{ for some } \alpha > 0\}$$

and the norms are then

$$N_\Phi(f) = \inf\{\alpha > 0 : \int_\Omega \Phi\Big(\frac{\|f(w)\|_{\mathcal{X}}}{\alpha}\Big) d\mu \le 1\}, \tag{19}$$

and

$$\|f\|_\Phi = \sup\{\Big|\int_\Omega \langle f(w), g(w)\rangle d\mu\Big| : N_\Psi(g) \le 1\}, \tag{20}$$

where $g : \Omega \to \mathcal{X}^*$ is strongly measurable and $\langle f(w), g(w) \rangle$ is the duality pairing, sometimes also written as $g(w)(f(w))$, following $x^*(x)$. It may be verified that (19) and (20) are equivalent norms under which $L^\Phi(\mu, \mathcal{X})$ is a Banach space. In case $\Phi \in \Delta_2$, it may be shown that simple functions are dense in it. To state the desired Radon-Nikodým theorem, we introduce:

Definition 8. Let (Ω, Σ, μ) be a finite or localizable measure space and $\mathcal{X}$ a Banach space. Then $\mathcal{X}$ is said to have the *Radon-Nikodým, or R-N, property* relative to μ, if for each measure $\nu : \Sigma \to \mathcal{X}$ having *finite variation* (i.e., $|\nu|(\Omega) < \infty$ as in (17)) and $\nu \ll \mu$, one has $\nu(A) = \int_A f d\mu$, $A \in \Sigma$, for a μ-unique Bochner integrable $f : \Omega \to \mathcal{X}$. The space $\mathcal{X}$ is then said to have the *R-N property*, if it has that property relative to each such measure space (Ω, Σ, μ) or such μ on Σ.

A classical result that suffices for our purposes here is:

Theorem 9 (Dunford-Pettis-Phillips). *If $\mathcal{X}$ is a separable dual space (meaning $\mathcal{X} = \mathcal{Y}^*$ for some Banach space $\mathcal{Y}$) or a reflexive space, then it has the R-N property.*

We shall not include a proof of this result. Our general reference for vector measure theory is N. Dinculeanu (1967), or Dunford-Schwartz (1958) or Hille-Phillips (1957) where all the preceding statements are also proved in detail. An "elementary" treatment of the Bochner integral is given by J. Mikusinski (1978). See also J. J. Uhl and J. Diestel (1977) for a contemporary treatment.

To use the *R-N* property, we need to ensure the finiteness of the variation of the vector measure derived from a bounded operator. Motivated by the total variation, (17), we introduce a more stringent norm on operators (as in Dinculeanu (1967)):

$$\|T\| = \sup\Big\{ \sum_{i=1}^n \|a_i T(\chi_{A_i})\|_\mathcal{X} : \|f\| \le 1, f = \sum_{i=1}^n a_i \chi_{A_i}, A_i \in \Sigma \Big\}. \quad (21)$$

It is easily seen that $\|\!\cdot\!\|$ is a norm functional. The definition of the usual operator norm $\|T\|$ thus corresponds to the semi-variation given in (18). Just as in the case of (17) and (18), one sees that $\|T\| \le \|T\|$, with strict inequality in many cases. The following two results are essentially due to Uhl (1971) and are an extension of some earlier work of Zaanen (cf. (1953), pp. 176–179). The next result will be recast on general measure spaces in Section VII.5, after the differentiation of norms is covered.

Theorem 10. *Let Φ be a continuous Young function, $\Phi(x) = 0$ iff $x = 0$, and X be a reflexive or a separable dual Banach space. Let (Ω, Σ, μ) be a localizable (or σ-finite) measure space with $T : M^\Phi \to X$ as a linear operator where M^Φ is the usual subspace of $L^\Phi(\mu)$ determined by its simple functions. Then $\|T\|_\Phi < \infty$ iff there is a strongly measurable $g : \Omega \to X$ such that $\|g\|(\cdot) \in L^\Psi(\mu)$ and*

$$Tf = \int_\Omega fgd\mu, \qquad f \in (M^\Phi, N_\Phi(\cdot)), \tag{22}$$

with

$$\|T\|_\Phi = \|g\|_\Psi. \tag{23}$$

If moreover, Φ is an N-function with the complementary Ψ in Δ_2, then T is also compact.

Proof: The argument is an extension of the corresponding part of Theorem IV.1.6. Suppose at first that $\mu(\Omega) < \infty$, and let $\nu : \Sigma \to X$ be defined by $\nu(E) = T(\chi_E)$. Since $\nu(\cdot)$ is evidently additive, consider $E_n \in \Sigma, E_n \searrow \emptyset$. Then

$$\|\nu(E_n)\| = \|T(\chi_{E_n})\| \le \|T\|N_\Phi(\chi_{E_n}) \to 0$$

as $n \to \infty$ since $\chi_{E_n} \in M^\Phi$ and Φ is continuous by hypothesis. This implies that $\nu(\cdot)$ is σ-additive. Since also $\lim_{\mu(A)\to 0} N_\Phi(\chi_A) = 0$ for the same reason, it follows that $\nu \ll \mu$. We now observe that ν has finite variation when $\|T\| < \infty$. This is obvious from (21). Since X has the R-N property, by Theorem 9, there is a strongly measurable $g : \Omega \to X$, such that $\int_\Omega \|g\|d\mu < \infty$, and

$$T(\chi_A) = \nu(A) = \int_A gd\mu = \int_\Omega \chi_A gd\mu, \tag{24}$$

so that for each simple function $f = \sum_{i=1}^{n} a_i \chi_{A_i}$, we get

$$Tf = \int_\Omega fgd\mu. \tag{25}$$

We need to show that $\|g\| \in L^\Psi(\mu)$ and then (22) and (23) follow.

Since by the classical theory $|\nu|(A) = \int_A \|g\|d\mu$, $A \in \Sigma$, it follows from (21) that, for any $f = \sum_{i=1}^{n} a_i \chi_{A_i}$, $N_\Phi(f) \leq 1$,

$$\sum_{i=1}^{n} \|a_i T(\chi_{A_i})\| = \sum_{i=1}^{n} |a_i| \left\| \int_{A_i} gd\mu \right\| \leq \sum_{i=1}^{n} |a_i| |\nu|(A_i) \leq \|T\|_\Phi < \infty. \quad (26)$$

Taking suprema on the left, we get by (21) that it is precisely $\|T\|_\Phi$ so that

$$\|T\|_\Phi = \sup \left\{ \int_\Omega |f| \|g\| d\mu : N_\Phi(f) \leq 1 \right\} = \|g\|_\Psi < \infty, \quad (27)$$

which is (23) in this case. Since simple functions are dense in M^Φ ($\Phi(x) = 0$ iff $x = 0$, since Φ is continuous), (25) implies that (22) holds, and the uniqueness of g is immediate.

If now (Ω, Σ, μ) is localizable, then for each $A \in \Sigma_0 = \{A \in \Sigma, \mu(A) < \infty\}$, then the preceding result implies that there is a unique g_A such that

$$\nu(A \cap \Sigma) = \int_E g_A d\mu,$$

and by the localizability of μ, the collection $\{g_A, A \in \Sigma\}$ determines a unique (strongly) measurable $g : \Omega \to \mathcal{X}$ such that $g_A = g\chi_A$, $A \in \Sigma$, by the standard theory (cf., e.g., Rao (1987), p. 275). But then by the preceding, $\|g_A\|_\Psi \leq \|T\|_\Phi$ so that g actually vanishes outside a σ-finite set and we can infer the truth of (22). From this we can use the same argument leading to (27) to conclude that (23) also holds. Thus the representation is established whenever $\|T\|_\Phi < \infty$.

For the converse, suppose T is given by (22). Then by the Hölder inequality, which is valid for the Bochner integral with essentially the same proof as in the Lebesgue case, we get, on taking f as a simple function and using (21), that $\|T\|_\Phi \leq \|g\|_\Psi < \infty$. But then we can apply the necessity proof to conclude that (23) holds and the main statement follows as asserted.

Finally suppose that $\Psi \in \Delta_2$ and Φ is an N-function, in addition to the rest of the hypothesis. Then there exists a sequence of simple $g_n : \Omega \to \mathcal{X}$ such that $g_n \to g$ a.e., $\|g_n\| \leq \|g\|$, and since now $M^\Psi = L^\Psi(\mu)$, it follows that $\int_\Omega \Psi(\|g_n - g\|\alpha^{-1})d\mu \to 0$ as $n \to \infty$ for each

$\alpha > 0$; hence $N_\Psi(g_n - g) \to 0$. Consequently $T_n(f) = \int_\Omega f g_n d\mu$ gives T_n as a bounded linear operator of finite range in $\mathcal{X}$ so that it is compact. But since $\|T - T_n\| \leq \|T - T_n\| = \|g - g_n\|_\Psi \leq 2N_\Psi(g_n - g) \to 0$, we conclude that T, the uniform limit of compact operators, is compact. ∎

Note: We observe that not every operator with $\|T\| < \infty$ is compact, and also not every compact operator satisfies $\|T\| < \infty$ on Banach spaces.

If $T : \mathcal{X} \to \mathcal{Y}$ is a bounded operator between the Banach spaces, $\mathcal{X}$ and $\mathcal{Y}$, then there exists a uniquely defined $S \in B(\mathcal{Y}^*, \mathcal{X}^*)$ such that $(Sy^*)(x) = y^*(Tx)$, $x \in \mathcal{X}, y^* \in \mathcal{Y}^*$. It is denoted by $S = T^*$, and termed the *adjoint* operator of T. It is not difficult to verify that $\|T\| = \|T^*\|$, which follows from definition. But a classical result of J. Schauder states that $T \in B_c(\mathcal{X}, \mathcal{Y})$ iff $T^* \in B_c(\mathcal{Y}^*, \mathcal{X}^*)$. A proof of this can be found in Dunford and Schwartz (1958; p. 485). Regarding the second norm, $\|\cdot\|$, such a statement cannot be made in general. Taking $\mathcal{X}$ to be an Orlicz space in the preceding theorem, one can obtain the following special assertion, from Uhl (1971). We state it without proof, since we do not use it below.

Proposition 11. *Let $L^\Phi(\mu)$ be a reflexive Orlicz space on (Ω, Σ, μ). If T_k is a kernel operator given by (1) and is in $B(L^\Phi(\mu))$, then for its adjoint $(T_k)^*$, $\|(T_k)^*\|_\Psi < \infty$ iff the function $w_1 \mapsto \|k(w_1, \cdot)\|_\Psi$ is in $L^\Phi(\mu)$. When this holds, $\|T_k\|_\Phi = \|(T_k)^*\|_\Psi$.*

Proceeding along the same lines of thought, Uhl (1970) has also given the following characterization of kernel operators which we state in a slightly restricted form without proof to indicate the possibilities.

Proposition 12. *Let $g : \Omega \to \mathcal{X}$ be a strongly measurable function on (Ω, Σ, μ) as before. If $\Phi(x) > 0$ for $x > 0$ and Φ is a continuous Young function, then T in $B(M^\Phi, \mathcal{X})$ is a kernel operator $(= T_g)$ with g as kernel when and only when $x^*(g) \in L^\Psi(\mu)$ for all $x^* \in \mathcal{X}^*$. In this case we can have $\|T\| = \sup\{\|x^*(g)\|_\Psi : \|x^*\| \leq 1\}$.*

We move on to another aspect of linear operators, namely their extension and factorization properties, on those spaces which will help understand their structure better.

6.2. *Extension and factorization of linear operators*

In the study of compactness of integral operators on Orlicz spaces, the $L^2(\mu)$-space plays an important role. The following motivational comment due to J. J. Uhl (included in Rao (1967), p. 135) is of interest.

Lemma 1. *Let* (Φ, Ψ) *be a pair of complementary Young functions and* $(L^{\Phi}(\mu), L^{\Psi}(\mu))$ *be the corresponding Orlicz spaces on* (Ω, Σ, μ). *Then any one of the following inclusions implies all the others.*

(i) $L^{\Phi}(\mu) \supset L^{\Psi}(\mu)$;

(ii) $L^2(\mu) \supset L^{\Psi}(\mu)$;

(iii) $L^{\Phi}(\mu) \supset L^2(\mu)$.

Proof: (i) $\Leftrightarrow$ (ii). The first one is immediate since $f \in L^{\Psi}(\mu) \subset L^{\Phi}(\mu)$ implies $\left| \int_{\Omega} f \cdot f d\mu \right| \leq \|f\|_{\Phi} N_{\Psi}(f) < \infty$ by the Hölder inequality, so that $f \in L^2(\mu)$. Conversely, if $f \in L^{\Psi}(\mu) \subset L^2(\mu)$, define $\ell_f(g) = \int_{\Omega} fg d\mu$ ($\in$ scalars), for all $g \in L^{\Psi}(\mu)$ ($\subset L^2(\mu)$). Hence by the inverse Hölder inequality (cf. Proposition IV.1.1) $f \in L^{\Phi}(\mu)$ so that $L^{\Psi}(\mu) \subset L^{\Phi}(\mu)$.

(ii) $\Leftrightarrow$ (iii). If $f \in L^2(\mu)$, then for all $g \in L^{\Psi}(\mu)$, $\left| \int_{\Omega} fg d\mu \right| < \infty$. Hence $f \in L^{\Phi}(\mu)$ by the same proposition used above. Conversely, if $f \in L^{\Psi}(\mu)$, then for all $g \in L^{\Phi}(\mu)$, we have by the Hölder inequality, that $\left| \int_{\Omega} fg d\mu \right| < \infty$. In particular if $g \in L^2(\mu)$ ($\subset L^{\Phi}(\mu)$), then $\left| \int_{\Omega} fg d\mu \right| < \infty$ for all $g \in L^2(\mu)$ so that $f \in L^2(\mu)$ by the inverse CBS (or Hölder) inequality. Hence $L^{\Psi}(\mu) \subset L^2(\mu)$. In these equivalences, μ is unrestricted. ∎

As an immediate consequence of this result and of Theorem V.1.3, we have the following:

Corollary 2. *Let* Φ *be an* N-*function and* (Ω, Σ, μ) *be a measure space such that* μ *is diffuse on a set of positive* μ-*measure. Then the equivalent inclusions of the theorem hold iff the following relations between* Φ, *its complementary function* Ψ, *and* Φ_0 *where* $\Phi_0(x) = x^2$ *hold (globally when* $\mu(\Omega) = +\infty$ *):*

$$\text{(i)} \quad \Phi \prec \Psi, \quad \text{(ii)} \quad \Phi_0 \prec \Psi, \quad \text{(iii)} \quad \Phi \prec \Phi_0. \tag{1}$$

In particular $\Phi \prec \Phi_0 \prec \Psi$ *iff* $L^{\Phi}(\mu) \supset L^2(\mu) \supset L^{\Psi}(\mu)$.

Note that, since $M^{\Phi_0} = L^2(\mu)$, and if $\Phi_0 \succ \Phi$ which implies $M^{\Phi} \supset M^{\Phi_0}$ in the above, one can replace $L^{\Phi}(\mu)$ by M^{Φ}.

We now establish a result on the continuous (and compact) extension of linear operators on certain Orlicz spaces "through the L^2-spaces" satisfying conditions as in the preceding two propositions. That will give immediately how the corresponding operators can be factorized.

Theorem 3. *Let Φ be an N-function and (Ω, Σ, μ) a measure space such that $M^\Phi \supset L^2(\mu) \supset L^\Psi(\mu)$, where Ψ is complementary to Φ. Let $T \in B(L^2(\mu))$ (respectively $T \in B_c(L^2(\mu))$), and let it satisfy the condition that $(Tf, g) = (f, Tg)$ and $(Tf, f) \geq 0$ for all $f, g \in L^2(\mu)$. Then there is a $\widetilde{T} \in B(M^\Phi, L^\Psi(\mu))$ $[\widetilde{T} \in B_c(M^\Phi, L^\Psi(\mu))]$ such that $\widetilde{T}f = Tf$ for all $f \in L^2(\mu)$ iff there is an $H \in B(L^2(\mu), L^\Psi(\mu))$ $[H \in B_c(L^2(\mu), L^\Psi(\mu))]$ such that $Hf = T^{\frac{1}{2}}f$ for all $f \in L^2(\mu)$, where $T^{\frac{1}{2}}$ is the unique square root of the operator T (which exists), i.e., $(Tf, f) = (T^{\frac{1}{2}}f, T^{\frac{1}{2}}f)$ for $f \in L^2(\mu)$.*

Remark: An operator S in a Hilbert space $\mathcal{H}$ satisfying $(Tf, g) = (f, Sg)$ for all f, g in $\mathcal{H}$ is denoted by $S = T^*$ and is called the *Hilbert adjoint* of T. In case $S = T$ then T is *self-adjoint*. Moreover, a self-adjoint operator T in a Hilbert space is termed *positive (semi-)definite* whenever $(Tf, f) \geq 0$ $[= 0$ can hold for some $f \neq 0$ also$]$. It should be noted that the Hilbert adjoint and the (Banach) adjoint defined prior to Proposition 1.11 are generally *not* the same. One often omits the qualification Hilbert or Banach when there is no possibility of misunderstanding. Also the operator $\widetilde{T}$ is called a continuous *extension* of T on the respective spaces shown. Similarly, H is a continuous *extension* of $T^{\frac{1}{2}}$, the square root of the positive semi-definite operator T on L^2.

Proof of Theorem: Let $T \in B(L^2(\mu))$ and be positive semi-definite. Suppose there is a $\widetilde{T} \in B(M^\Phi(\mu), L^\Psi(\mu))$ such that $\widetilde{T} \mid L^2(\mu) = T$ where $L^2(\mu) \subset M^\Phi(\mu)$. Let $T^{\frac{1}{2}}$ be the square root of T on $L^2(\mu)$ which is well known to exist always. If f is any simple function, then $f \in L^2(\mu)$ and since $Tf = \widetilde{T}f$ $(\in L^\Psi(\mu))$ we have

$$\|T^{\frac{1}{2}}f\|_2^2 = (Tf, f) = \int_\Omega (Tf)f d\mu = \int_\Omega (\widetilde{T}f)f d\mu,$$

$$\leq \|\widetilde{T}f\|_\Psi \|f\|_\Phi \leq \|\widetilde{T}\| \|f\|_\Phi^2 \leq C\|f\|_\Phi^2. \tag{2}$$

Hence $T^{\frac{1}{2}}$ maps all simple functions of $L^2(\mu)$ into $L^\Psi(\mu)$ and is bounded so that $T^{\frac{1}{2}}$ is also uniformly continuous by (2). Since $L^2(\mu)$ and $L^\Psi(\mu)$

are complete, and are metric spaces, $T^{\frac{1}{2}}$ has a unique uniformly continuous extension to the closure of this set which is $L^2(\mu)$, using the principle of extension by (uniform) continuity in complete metric spaces.
Hence there is an $H \in B(L^2(\mu), L^\Psi(\mu))$ such that $Hf = T^{\frac{1}{2}}f$ for all
$f \in L^2(\mu)$.

Suppose next that the given T is also compact. We assert that
H is also compact, since by the preceding paragraph H is continuous. As noted in the last section (prior to Proposition 1.11) by
the classical Schauder theorem H is compact iff H^* is, where $H^* \in$
$B((L^\Psi(\mu))^*, (L^2(\mu))^*) = B((L^\Psi(\mu))^*, L^2(\mu))$. But by the same argument $H^{**} : (L^2(\mu))^{**} = L^2(\mu) \to (L^\Psi(\mu))^{**}$ is continuous and
$(H^*)^* = H^{**} = H$ on $L^2(\mu)$. This means since $H(L^2(\mu)) \subset M^\Psi$,
we have $H^{**} = H : L^2(\mu) \to L^\Psi(\mu)$ also. Thus the compactness of
$H = (H^*)^*$ is equivalent to showing that of $H_1 = H^* \mid M^\Phi$, the restriction. This is because $H_1^* : (L^2(\mu))^* = L^2(\mu) \to (M^\Phi)^* = L^\Psi(\mu)$,
and $H_1^* = H^{**} = H$. We now connect T and H_1.

Since by hypothesis T has a compact extension $\widetilde{T}$ $(\in B_c(M^\Phi, L^\Psi(\mu)))$
each bounded set $\{f_n, n \geq 1\}$ of M^Φ is taken into $\{\widetilde{T}f_n, n \geq 1\}$ which
is sequentially compact. Hence there is a subsequence $\{\widetilde{T}f_{n_i}, i \geq 1\}$
which converges to some element in $L^\Psi(\mu)$. In particular, if our f_n is
in $L^2(\mu)$ ($\subset M^\Phi$), then $\widetilde{T}f_n = Tf_n$ so that

$$\|T(f_{n_i} - f_{n_j})\|_\Psi = \|\widetilde{T}(f_{n_i} - f_{n_j})\|_\Psi \to 0 \qquad \text{as } i, j \to \infty. \quad (3)$$

Consequently

$$\|H_1(f_{n_i} - f_{n_j})\|_2^2 = (T(f_{n_i} - f_{n_j}), (f_{n_i} - f_{n_j}))$$
$$\leq \|T(f_{n_i} - f_{n_j})\|_\Psi \|f_{n_i} - f_{n_j}\|_\Phi$$
$$\leq 2 \sup_n \|f_n\|_\Phi \|T(f_{n_i} - f_{n_j})\|_\Psi \to 0, \quad (4)$$

since the set is bounded. However simple functions and hence $L^2(\mu)$
itself form a dense set in M^Φ. So given a bounded set $\{g_n, n \geq 1\} \subset M^\Phi$,
we can find $\{\tilde{g}_n, n \geq 1\} \subset L^2(\mu)$ such that $\|g_n - \tilde{g}_n\|_\Phi < \frac{1}{n}$. Thus
$\|\tilde{g}_n\|_\Phi \leq \|g_n\|_\Phi + 1$ and $\{\tilde{g}_n, n \geq 1\}$ is also bounded in M^Φ. Replacing
f_n by $\tilde{g}_n$ in (3) and (4), we have

$$\|H_1(g_{n_i} - g_{n_j})\|_2 \leq \|H_1(g_{n_i} - \tilde{g}_{n_i})\|_2 + \|H_1(\tilde{g}_{n_i} - \tilde{g}_{n_j})\|_2 + \|H_1(\tilde{g}_{n_j} - g_{n_j})\|_2$$
$$\leq \|H_1\|\left[\frac{1}{n_i} + \frac{1}{n_j}\right] + \|H_1(\tilde{g}_{n_i} - \tilde{g}_{n_j})\|_2$$
$$\to 0, \qquad \text{as } n_i, n_j \to \infty.$$

Hence H_1, so that H itself, is compact as desired.

We next prove the converse parts. If $T^{\frac{1}{2}}$ has a continuous extension from $L^2(\mu)$ to $L^\Psi(\mu)$, then the restriction of its adjoint, say H_1 from M^Φ into $L^2(\mu)$ is also continuous. Hence for each f, simple, $T^{\frac{1}{2}}f = Hf$ and $f \in M^\Phi \cap L^2(\mu)$ so that

$$\|Tf\|_\Psi = \|T^{\frac{1}{2}}(T^{\frac{1}{2}}f)\|_\Psi = \|H(T^{\frac{1}{2}}f)\|_\Psi \le \|H\| \cdot \|T^{\frac{1}{2}}f\|_2$$
$$\le \|H\|\|H_1\|\|f\|_\Phi \le C_0\|f\|_\Phi. \qquad (5)$$

Since such functions are dense in M^Φ, it follows that, with the principle of extension by continuity, there is a $\widetilde{T} \in B(M^\Phi, L^\Psi(\mu))$, $\widetilde{T}f = Tf, f \in L^2(\mu)$.

Suppose finally that the extension of $T^{\frac{1}{2}}$ which is H is also compact. Then we can employ an identical argument of the preceding part (interchanging H and $\widetilde{T}$ there) and conclude that the extension $\widetilde{T}$ which exists by (5) is compact. Thus the parenthetical statements are also true. ∎

The preceding theorem in its direct part is due to Krasnosel'skii and Rutickii (1961) and the converse parts were considered by C. X. Wu (1962) and H. A. Ye (1982). Both these sets of authors considered the result for bounded Lebesgue measure spaces, and the generality given here follows immediately from the simple Lemma 1 (or Corollary 2).

The above theorem enables one to present a factorization of the extendable operator which is positive definite for its restriction to a subspace. More precisely, the following factorization assertion holds:

Theorem 4. *Let* (Ω, Σ, μ), *N-function* Φ, *and the inclusions* $M^\Phi \supset L^2(\mu) \supset L^\Psi(\mu)$ *be as in Theorem 3. If* T *is a positive (semi-) definite operator on* $L^2(\mu) \to L^2(\mu)$, *then its continuous extension (guaranteed by Theorem 3)* $\widetilde{T}: M^\Phi \to L^\Psi(\mu)$ *factors as* $\widetilde{T} = HH^*$ *where* $H \in B(L^2(\mu), L^\Psi(\mu))$ *is an extension of* $T^{\frac{1}{2}}$ *and* $H^* \in B(M^\Phi, L^2(\mu))$ *is the restriction of the adjoint of* H *to the subspace* M^Φ *of* $(L^\Psi(\mu))^*$ *into* $L^2(\mu)$. *If the operator* T *is compact to start with, the extension* $\widetilde{T}$ *and the components* H *and* H^* *are also compact. In other words, the following diagram explains the situation:*

$$L^\Psi(\mu) \hookrightarrow L^2(\mu) \xrightarrow{T} L^2(\mu) \hookrightarrow M^\Phi(\mu) \Rightarrow \begin{array}{ccc} M^\Phi & \xrightarrow{\widetilde{T}} & L^\Psi(\mu) \\ H^* \searrow & & \nearrow H \\ & L^2(\mu) & \end{array}$$

Here ↪ *denotes imbedding, and the second diagram commutes so that*
$\widetilde{T} = HH^*$.

We omit the proof since it is simply a restatement of Theorem 3 in a different terminology.

In case the measure space is diffuse, then the conditions on the inclusions of spaces can be replaced by $\Phi \prec \Phi_0 \prec \Psi$, as in Corollary 2. This is the form used in the monograph of Krasnosel'skii and Rutickii, in lieu of $M^\Phi \supset L^2(\mu) \supset L^\Psi(\mu)$. However, our statements do not need to assume the diffuseness of μ which is unnecessary here. Note also that the converse parts of Theorem 4 state that $\widetilde{T}$ is continuous or compact when H (or H^*) has respectively that property.

A number of interesting characterizations of integral operators through factorizations have been given by Grothendieck (1955) who relates it to the theory of tensor products of spaces. We shall not consider it here as it leads to a special study of integral operators.

6.3. *Interpolation of linear operators*

Instead of restricting to integral and related operators, as was done in the preceding two sections, we shall consider the possibility of continuously extending bounded linear (and later briefly sublinear) operators between classes of spaces "intermediate" to a given pair of Orlicz spaces. The corresponding result in the Lebesgue case, with the complex method, is the celebrated M. Riesz-G. O. Thorin interpolation theorem. In this section we present various Orlicz space extensions of this powerful classical result.

To motivate the generalization, let us recall the basic formulations of the Riesz interpolation on $L^p(\mu)$-spaces. If $1 \le p_0 < p_1 \le \infty$, let p_s be a number in the interval (p_0, p_1) defined as ($\frac{1}{\infty}$ is set $= 0$)

$$\frac{1}{p_s} = \frac{1-s}{p_0} + \frac{s}{p_1}, \qquad 0 \le s \le 1. \tag{1}$$

The space $L^{p_s}(\mu)$, for $\mu(\Omega) < \infty$, satisfies the inclusion relations $L^{p_0}(\mu) \supset L^{p_s}(\mu) \supset L^{p_1}(\mu)$ for all 's' above. Since $L^{p_0} = L^{p_s}$ for $s = 0$ and $L^{p_1} = L^{p_s}$ for $s = 1$ for any measure μ, one terms $L^{p_s}(\mu)$, an intermediate space for the couple $[L^{p_0}(\mu), L^{p_1}(\mu)]$ even when $\mu(\Omega) = +\infty$. The problem is to extend a given operator T which is bounded on both these spaces, to $L^{p_s}(\mu)$ and to find its bound. We now obtain the corresponding results on Orlicz spaces. For the most part we follow Rao

(1966, Sections 2 and 3) in this treatment and then present further
extensions due to other authors. The following definition which corresponds to (1) was independently given by Calderón (1964) and Rutickii
(1963).

Definition 1. Let Φ_1, Φ_2 be a pair of Young functions, and $0 \leq s \leq 1$ be fixed. Then Φ_s is the uniquely defined inverse of

$$\Phi_s^{-1} = (\Phi_1^{-1})^{1-s}(\Phi_2^{-1})^s, \tag{2}$$

where Φ_i^{-1} is the inverse of Φ_i; Φ_s is called an *intermediate function*.

We now establish some preliminary facts about Φ_s that follow from
the properties of Φ_i, $i = 1, 2$. They will be needed later.

Lemma 2. *Let (Φ_i, Ψ_i), $i = 1, 2$ be a pair of complementary
Young functions with Φ_i strictly increasing and continuous. Then for
each $0 < s < 1$, Φ_s and Ψ_s are Young functions [and N-functions if
Φ_1 and Φ_2 are] satisfying the Young inequality even though they need
not be complementary. Moreover, if $\Phi_1 \succ \Phi_2$, or $\Phi_1 \succcurlyeq \Phi_2$, then Φ_s
lies between Φ_1 and Φ_2 for the same ordering. If $\Phi_i \in \Delta_2$ [or Δ'] for
$i = 1, 2$, then $\Phi_s \in \Delta_2$ [or Δ'].*

Proof: The fact that Φ_s and Ψ_s are Young or N-functions follows
from definition when we observe that Φ_s [respectively Ψ_s] is convex.
The latter is a consequence of the concavity of Φ_i^{-1} and of the Hölder
inequality. Indeed, for $\sum_{i=1}^n \alpha_i = 1$, $\alpha_i > 0$,

$$\Phi_s^{-1}\left(\sum_{i=1}^n \alpha_i x_i\right) \geq \left(\sum_{i=1}^n \alpha_i \Phi_1^{-1}(x_i)\right)^{1-s}\left(\sum_{i=1}^n \alpha_i \Phi_2^{-1}(x_i)\right)^s$$

$$\geq \sum_{i=1}^n \alpha_i(\Phi_1^{-1}(x_i))^{1-s}(\Phi_2^{-1}(x_i))^s = \sum_{i=1}^n \alpha_i \Phi_s^{-1}(x_i),$$

so that $\Phi_s(\cdot)$ is convex.

To establish the Young inequality, let $t_1 = \Phi_s(x_1), t_2 = \Psi_s(x_2)$.
Then we need to verify $x_1 x_2 \leq t_1 + t_2$, and this follows from

$$\Phi_s^{-1}(t_1)\Psi_s^{-1}(t_2) = \{[\Phi_1^{-1}(t_1)\Psi_1^{-1}(t_2)]^{1-s}\}\{[\Phi_2^{-1}(t_1)\Psi_2^{-1}(t_2)]^s\}$$
$$\leq (t_1 + t_2)^{1-s}(t_1 + t_2)^s = t_1 + t_2 \tag{3}$$

since (Φ_i, Ψ_i) is a complementary pair for $i = 1, 2$. However, (3) does not imply that (Φ_s, Ψ_s) is a complementary pair.

If $\Phi_i \in \Delta_2$ then there exist $k > 0$ and $x_0 \geq 0$ such that

$$\Phi_i(2x) \leq k\Phi_i(x), \qquad i = 1, 2, x \geq x_0,$$

so that

$$2\Phi_i^{-1}(y) \leq \Phi_i^{-1}(ky), \quad y \geq y_0 = \max(\Phi_1(x_0), \Phi_2(x_0)).$$

Multiplying the inequalities for $i = 1, 2$, one gets

$$2\Phi_s^{-1}(y) \leq [\Phi_1^{-1}(ky)]^{1-s}[\Phi_2^{-1}(ky)]^s = \Phi_s^{-1}(ky), \quad y \geq y_0 \geq 0.$$

Hence $\Phi_s \in \Delta_2$ and this holds globally if it holds for each Φ_i.

A similar computation establishes the statement about Δ'.

Regarding the order relations, consider $\Phi_1 \succ \Phi_2$ so that $\Phi_2(x) \leq \Phi_1(bx)$, $x \geq x_0 \geq 0$ and some $b > 0$. Hence $\Phi_1^{-1}(y) \leq b\Phi_2^{-1}(y)$. Then

$$\Phi_s^{-1}(y) \geq (\Phi_1^{-1}(y))^{1-s}\left(\frac{1}{b}\Phi_1^{-1}(y)\right)^s = \frac{1}{b^s}\Phi_1^{-1}(y), \qquad y \geq y_0,$$

and so $\Phi_s(b^{-s}x) \leq \Phi_1(x)$, $x \geq x_0$, and $b^{-s} > 0$. Thus $\Phi_1 \succ \Phi_s$. Similarly one can verify that $\Phi_s \succ \Phi_2$. The other statements are obtained in exactly the same way. ∎

Since Ψ_s is not necessarily complementary to Φ_s, let Ψ_s^+ be complementary to Φ_s. Then it is clear that $\Psi_s^+(x) \leq \Psi_s(x)$, for all $x \geq 0$, by (3). This implies immediately that $N_{\Psi_s^+}(f) \leq N_{\Psi_s}(f)$. We shall show that, in fact, both these norms are equivalent.

Lemma 3. *Let* (Φ_i, Ψ_i), Ψ_s *and* Ψ_s^+ *be as defined above. Then the Orlicz spaces* $L^{\Psi_s}(\mu), L^{\Psi_s^+}(\mu)$, *on a measure space* (Ω, Σ, μ) *have the same elements (and hence are identical as sets) and*

$$N_{\Psi_s^+}(f) \leq N_{\Psi_s}(f) \leq 2N_{\Psi_s^+}(f), \quad f \in L^{\Psi_s}(\mu). \tag{4}$$

Proof: It suffices to establish the inequality

$$\Psi_s^+(x) \leq \Psi_s(x) \leq \Psi_s^+(2x), \qquad x \geq 0, \tag{5}$$

from which (4) and all the other statements of the lemma follow at once. For this we employ Proposition II.1.1(ii), on observing that the result there holds for all Young functions even though it was stated there only for N-functions. The proof clearly shows this. We thus have

$$a < \Phi_s^{-1}(a)(\Psi_s^+)^{-1}(a) \leq 2a, \text{ and } a < \Phi_i^{-1}(a)\Psi_i^{-1}(a) \leq 2a, \quad (6)$$

for $i = 1, 2$, and for all $a > 0$. Hence for $a > 0$

$$(\Psi_s^+)^{-1}(a) \leq \frac{2a}{\Phi_s^{-1}(a)} < 2(\Psi_1^{-1}(a))^{1-s}(\Psi_2^{-1}(a))^s = 2\Psi_s^{-1}(a). \quad (7)$$

Set $a = \Psi_s(x)$ in (7). It follows that $\Psi_s(x) \leq \Psi_s^+(2x)$. This yields (5) since the first inequality was already noted. ∎

Remark: Another argument without directly using the above noted proposition was originally given in the last reference, where (4) was proved on using the closed graph theorem. But the present proof is more elementary. It should be noted that for an arbitrary Young function Φ we use the relations for its inverse Φ^{-1}, which is nondecreasing and right continuous on $[0, \infty]$ with $\Phi^{-1}(+\infty) = +\infty$,

$$\Phi(\Phi^{-1}(x)) \leq x, \text{ and } \Phi^{-1}(\Phi(u)) \geq u, \quad \Phi(\Phi^{-1}(x) + \varepsilon) > x, \quad (8)$$

for $x, u \geq 0$ and $\varepsilon > 0$. As already noted by O'Neil (1968, p. 20), these hold even for generalized Young functions when $\Phi^{-1}(y)$ is taken as the Dedekind cut determined by $\{x : \Phi(x) \leq y\}$ and $\{x : y < \Phi(x)\}$.

We now present the Orlicz space extension of the Riesz-Thorin convexity theorem for complex spaces. It is ironic that the bound for the interpolated operators is larger for the real spaces than that for the complex spaces. For instance, the latter is half that of the real case for the spaces considered. So mainly the complex case will be considered. This is based on a consequence of the maximum modulus principle of complex analysis, known as the three-lines-theorem—since the bound on an intermediate line depends only on the bounds of the (analytic) function on the two boundary lines of a strip. For convenience, we include a short argument of this result.

Proposition 4 (Three-Lines-theorem). *Let $S = \{z = x + iy : 0 < x < 1\}$ be the (open) strip in $\mathbb{C}$. If f is an analytic function on S which is bounded and continuous on the closure $\overline{S}$ of S, let the bound of f on the vertical line at $x = s$ in S be M_s, i.e., $M_s = \sup\{|f(s + iy)| : y \in \mathbb{R}\}$. Then one has*

$$M_s \leq M_0^{1-s} M_1^s, \qquad 0 \leq s \leq 1. \quad (9)$$

Proof: For each $\widetilde{M}_0 > M_0$ and $\widetilde{M}_1 > M_1$, we need to show that $\tilde{f}(z) = f(z)/(\widetilde{M}_0)^{1-z}(\widetilde{M}_1)^z$ is bounded by 1 on S. By hypothesis on f, we note that $\tilde{f}$ is analytic and bounded by 1 on $s = 0$ and 1, and is bounded on $\overline{S}$, say by B_0. We assert that $B_0 = 1$ will also work for $\tilde{f}$. This may be deduced from the *maximum modulus principle* as follows.

Let $\tilde{f}_\varepsilon(z) = \tilde{f}(\varepsilon)/(1+\varepsilon z)$ for any $\varepsilon > 0$. Then $\tilde{f}_\varepsilon$ is analytic in S, and $|\tilde{f}_\varepsilon(z)| \leq 1$ on the boundary of S and $\leq B_0/\varepsilon|y|$ on S. Thus if we consider a rectangle I with vertices $\pm\frac{iB_0}{\varepsilon}$ and $\left(1 \pm \frac{iB_0}{\varepsilon}\right)$, then $|\tilde{f}_\varepsilon|$ is dominated by 1 on the boundary of I so that the same bound holds in the interior also. But outside I, $|\tilde{f}_\varepsilon|$ cannot be larger than this bound. Hence it follows that $|\tilde{f}(z)| < (1 + \varepsilon|z|)$ on S. Since $\varepsilon > 0$ is arbitrary, $|\tilde{f}(z)| \leq 1$ must hold on S. ∎

We are now ready to prove the desired result.

Theorem 5. *Let* (Ω, Σ, μ) *and* $(S, \mathcal{S}, \nu)$ *be measure spaces,* Φ_i, Q_i *be continuous Young functions* $\Phi_i(x) > 0$ *for* $x > 0$, *and* $L^{\Phi_i}(\mu) = M^{\Phi_i}$, $i = 1, 2$, *where* $L^{\Phi_i}(\mu), L^{Q_i}(\nu)$ *are complex Orlicz spaces. Let* Φ_s *and* Q_s *be the intermediate Young functions given by Definition 1,* $0 \leq s \leq 1$. *If* $T : L^{\Phi_i}(\mu) \to L^{Q_i}(\nu)$ *is defined on both spaces and is bounded by* M_1 *and* M_2, *i.e.,*

$$\|Tf\|_{Q_i} \leq M_i\|f\|_{\Phi_i}, \qquad f \in L^{\Phi_i}(\mu), \, i = 1, 2, \qquad (10)$$

then T *is defined on* $L^{\Phi_s}(\mu)$ *into* $L^{Q_s}(\nu)$ *for each* $0 < s < 1$ *and*

$$\max(\|Tf\|_{Q_s}, \|Tf\|_{Q_s^+}) \leq M_1^{1-s}M_2^s\|f\|_{\Phi_s}, \quad f \in L^{\Phi_s}(\mu). \qquad (11)$$

[If Φ_i' *and* Q_i' *have jumps, we may have to replace the* M_i *by a somewhat larger constant* M_i' *in (11).]*

In case the spaces $L^{\Phi_i}(\mu)$ and $L^{Q_i}(\nu)$ are real, and the rest of the hypothesis holds, then (10) implies (11) with the right side constant replaced by $2M_1^{1-s}M_2^s$, i.e., by doubling the bound.

Remark: The same result holds if we use the gauge norms in the spaces $L^{\Phi_i}(\mu)$ and $L^{Q_i}(\nu)$. The corresponding bounds M_1, M_2 will be replaced by K_1, K_2 that go with the other norm.

Proof: To simplify writing, let $\alpha_i(t) = \Phi_i^{-1}(t)$, and $\alpha_s = \alpha_1^{1-s}\alpha_2^s$, so that $\alpha_s(\cdot)$ is a concave nondecreasing function with $\alpha_s(0) = 0$, $0 \leq s \leq 1$. If R_i is complementary to Q_i, $\beta_i = R_i^{-1}$, let $\beta_s = \beta_1^{1-s}\beta_2^s$, $0 \leq s \leq 1$.

Let Φ_s and R_s be the inverses of α_s and β_s. Since by hypothesis simple functions are dense in $L^{\Phi_i}(\mu)$, the same fact holds for $L^{\Phi_s}(\mu)$ also by Lemma 2. It therefore suffices to establish (11) for all simple functions, and $0 < s < 1$; the general case then follows from the principle of extension by (uniform) continuity for such spaces.

In what follows we employ the Orlicz norm in $L^{\Phi_i}(\mu), L^{Q_i}(\nu)$ and the gauge norm in the respective complementary spaces. First assume that Φ'_i, Q'_i are continuous so that the same holds for Φ'_s and Q'_s. Let $f \in L^{\Phi_s}(\mu)$ be simple but complex valued, such that $\|f\|_{\Phi_s} = 1$. Let $g \in L^{R_s}(\nu)$ be simple and $N_{R_s}(g) \leq 1$. Since $f \in L^{\Phi_1}(\mu) \cap L^{\Phi_2}(\mu)$, Tf is defined and takes values in $L^{Q_i}(\nu)$, $i = 1, 2$, and by (10)

$$\|Tf\|_{Q_i} \leq M_i\|f\|_{\Phi_i}, \qquad i = 1, 2.$$

Since Φ'_s is continuous and f is simple, it follows from Theorem III.3.15 (see its proof) that there is a $k^*_s \geq 1$ satisfying

$$\int_\Omega \Psi^+_s(\Phi'_s(k^*_s f))d\mu = 1, \tag{12}$$

where Ψ^+_s is the complementary Young function of Φ_s. Using the equality conditions in Young's inequality, one has

$$1 = \|f\|_{\Phi_s} = \frac{1}{k^*_s}(1 + \int_\Omega \Phi(k^*_s f)d\mu) \tag{13}$$

(cf. Proposition III.3.16). This implies

$$\int_\Omega \Phi_s(k^*_s f)d\mu = k^*_s - 1 < \infty. \tag{14}$$

[Note: $2N_{\Phi_s}(f) \geq \|f\|_{\Phi_s} \geq N_{\Phi_s}(f)$, so that $\frac{1}{2} \leq N_{\Phi_s}(f) \leq 1$.] It also follows that k^*_s is continuous in $0 \leq s \leq 1$.

For the above f and g consider the expression

$$I = \int_S (Tf)g \, d\nu, \tag{15}$$

where the integral for complex valued functions is defined, as usual, for the real and imaginary parts and added. The integral, defining a complex function, can be extended to the strip $B = \{z = x + iy \in \mathbb{C} : 0 \leq x \leq 1\}$. We now use the classical procedure of G. O. Thorn's adapted to the Orlicz spaces as follows. Let $f(\cdot) = |f|(\cdot)e^{i\theta(\cdot)}$ and $g(\cdot) = |g|(\cdot)e^{i\theta'(\cdot)}$. Define F_z, G_z as

$$F_z = \frac{1}{k_s^*}\alpha_z(\Phi_s(k_s^*|f|))e^{i\theta}, \quad G_z = \beta_z(R_s(|g|))e^{i\theta'}, \qquad (16)$$

where α_z, β_z are given at the beginning of the proof when z is replaced by s, and $\alpha_s(0) = 0 = \beta_s(0)$. Since F_z and G_z of (16) are simple, the following integral is an extension of (15):

$$I(z) = \int_S (TF_z) \cdot G_z d\nu. \qquad (17)$$

Under the present hypothesis, $\alpha_s(\Phi_s)$ and $\beta_s(R_s)$ are identity functions so that $|I| = |I(s)|$. We assert that $I(z)$ is bounded and continuous on the strip B, and analytic in its interior. This is seen as follows. Writing the simple functions f, g in the canonical form as

$$f = \sum_{i=1}^{n} a_i \chi_{E_i}, \quad g = \sum_{j=1}^{m} b_j \chi_{F_j}, \quad a_j = |a_j|e^{i\theta_j}, \quad b_j = |b_j|e^{i\theta_j'},$$

we have for (17),

$$I(z) = \frac{1}{k_s^*}\sum_{j=1}^{n}\sum_{j'=1}^{m} e^{i(\theta_j + \theta_{j'}')}\alpha_z(\Phi_s(k_s^*|a_j|))\beta_z(R_s(|b_{j'}|))\int_S (T\chi_{E_j}) \cdot \chi_{F_{j'}} d\nu,$$

which is linear in exponentials of the form d^z, $d > 0$, with

$$d^z = \alpha_1(\Phi_s(k_s^*|a_j|)^{1-z})\alpha_2(\Phi_s(k_s^*|a_j|)^z)\beta_1(R_s(|b_{j'}|)^{1-z})\beta_2(R_s(|b_{j'}|)^z).$$

Thus $I(\cdot)$ is analytic in the interior of the strip B, and is continuous on B. We now obtain its bounds on the boundaries of B.

If $x = 0$, $TF_z \in L^{Q_1}(\nu)$, $G_z \in L^{R_1}(\nu)$ so that we get

$$|I(z)| \leq \|TF_z\|_{Q_1} N_{R_1}(G_z) \leq M_1\|F_z\|_{\Phi_1} N_{R_1}(G_z), \qquad (18)$$

by the Hölder inequality (Section III.3, expression (16)). The right side of (18) can be simplified for $x = 0$ as follows:

$$\int_\Omega \Phi_1(k_s^*|F_z|)d\mu = \int_\Omega \Phi_1(|\alpha_{iy}\Phi_s(k_s^*|f|)|)d\mu = \int_\Omega \Phi_s(k_s^*|f|)d\mu = k_s^* - 1.$$

Since $k_s^* \to k_0^*$ as $s \to 0$, it results from Fatou's lemma that

$$\int_\Omega \Phi_1(k_0^*|F_z|)d\mu \le k_0^* - 1 < \infty,$$

and

$$\|F_z\|_{\Phi_1} = \frac{1}{k_0^*}[1 + \int_\Omega \Phi_1(k_0^*|F_z|)d\mu] \le 1.$$

But we have $N_{R_1}(G_z) = N_{R_s}(g) \le 1$, from definition. Hence (18) becomes $|I(z)| \le M_1$ for $x = 0$. One can employ a similar computation for $z = 1 + iy$, to get $|I(z)| \le M_2$. It then follows from the Three-Line theorem (cf. Proposition 4) that for $z = s + iy$, $0 < s < 1$,

$$|I| = |I(s + iy)| \le M_1^{1-s}M_2^s\|f\|_{\Phi_s}, \qquad 0 < s < 1. \qquad (19)$$

By the density of simple functions in $L^{\Phi_s}(\mu)$, (18) also holds for all $f \in L^{\Phi_s}(\mu)$. Taking the supremum on all f with $\|f\|_{\Phi_s} = 1$, one gets (11) at once in this case.

In the general case, we employ Theorem V.1.1. By this result there exist $\widetilde{\Phi}_i$ and $\widetilde{Q}_i$ satisfying the hypothesis of the above case already considered and $L^{\Phi_i}(\mu) = L^{\widetilde{\Phi}_i}(\mu)$ with equivalent norms. A similar comment applies to the second space. With the equivalent norms, the given M_i will be replaced by $\widetilde{M}_i$ and hence we go back to the change of the latter constants. Thus if $M_i' = \max(M_i, \widetilde{M}_i)$, $i = 1, 2$, then (11) holds with these new constants and the result holds as stated.

Finally if the $L^{\Phi_i}(\mu)$ and $L^{Q_i}(\nu)$ are real spaces, and T is a linear operator as given, then we complexify the given spaces ($L_c^\Phi(\mu) = L^\Phi(\mu) + iL^\Phi(\mu)$ etc.) and define $\widetilde{T}(f_1 + if_2) = Tf_1 + iTf_2$ for $f_j \in L^\Phi(\mu)$ so that $\tilde{f} = f_1 + if_2 \in L_c^\Phi(\mu)$. Then $\widetilde{T}$ has the same properties as before, except that $\|\widetilde{T}\tilde{f}\|_{\Phi_i} \le 2M_i\|\tilde{f}\|_\Phi$, and hence we get (11) by the preceding with twice the constants. With this modification the result holds. ∎

Let us present some consequences and generalizations of the preceding result. Because of the last part, we only consider complex spaces without further comments. The bounding constants of (11) in our applications are unimportant, and so we use the same symbols indifferently (both for real and complex cases, as well as for the possibility of jumps in Φ_i' and Q_i').

A trite generalization is obtained if Φ_i does not satisfy any growth condition, but then one restricts to M^{Φ_i}—the subspaces of $L^{\Phi_i}(\mu)$ determined by simple functions. The proof remains unchanged, and hence one gets the following:

Corollary 6. *Let Φ_i, Q_i, $i = 1, 2$ be Young functions as in the theorem but without assuming any growth conditions. Then $T \in B(M^{\Phi_i}, L^{Q_i}(\nu))$, $i = 1, 2$ implies that T has an extension to $B(M^{\Phi_s}, L^{Q_s}(\nu))$, $0 < s < 1$, and the bounds as in (11) hold on the space M^{Φ_s}.*

To illustrate this interpolation method, we consider a convolution operator which complements and sharpens results of the type given in Theorem V.5.1. The result should also be compared with Theorem III.3.9, since both these may be termed *Young's integral inequalities.* The following is a simplified version of a result in (Rao (1967$_b$), Thm. 2) and is a consequence of the above theorem. The proof in the L^p-case is originally due to A. Weil (1938). Recall that if $f, g \in L^1(I\!R) \cap L^\infty(I\!R)$, then their convolution $f * g$ is defined as:

$$(f * g)(x) = \int_{I\!R} f(x - t)g(t)dt, \qquad x \in I\!R. \qquad (20)$$

For the following statement and proof, $I\!R$ may be replaced by a locally compact unimodular group with Haar measure, but this will not be discussed here.

Corollary 7. *Let Φ_i, $i = 1, 2, 3$ be Δ_2-regular N-functions, and Φ_0 be given by $\Phi_0(t) = |t|$. Let $L^{\Phi_i}(I\!R)$, $i = 1, 2, 3$ be the corresponding Orlicz spaces on the Lebesgue line. Suppose there exists an s, $0 < s < 1$, such that $\Phi_1^{-1} = (\Phi_0^{-1})^{1-s}(\Psi_2^{-1})^s$, $\Phi_3^{-1} = (\Phi_2^{-1})^{1-s}(\Psi_0^{-1})^s$, where the Ψ_i are the complementary Young functions to Φ_i, $0 \le i \le 3$. Then for any $f \in L^{\Phi_1}(I\!R), g \in L^{\Phi_2}(I\!R)$, the convolution $f * g$ is well defined and moreover,*

$$\|f * g\|_{\Phi_3} \le C\|f\|_{\Phi_1}\|g\|_{\Phi_2} \qquad (21)$$

where $0 < C < \infty$ is an absolute constant.

Proof: First we observe that for $f \in L^{\Phi_0}(I\!R)$ and $g \in L^{\Phi_2}(I\!R)$, then $f * g \in L^{\Phi_2}(I\!R)$. In fact, if $f \in L^{\Phi_0}(I\!R)$ and g is a continuous function with compact support, then $(x, y) \mapsto f(x - y)g(y)$ is jointly measurable and integrable in y so that $(f * g)(\cdot)$ exists. If now h is a simple function, then we have by Tonelli's theorem

$$\int_{I\!R} |f| * |g|(x)|h(x)|dx = \int_{I\!R} |f(y)| \Big[\int_{I\!R} |g|(x-y)|h(x)|dx \Big] dy$$

$$\le \|f\|_1 \|g\|_{\Phi_2} N_{\Psi_2}(h) < \infty, \tag{22}$$

by Hölder's inequality. Taking supremum over all such simple h satisfying $N_{\Psi_2}(h) \le 1$, (22) yields $\|(|f| * |g|)\|_{\Phi_2} \le \|f\|_1 \|g\|_{\Phi_2} < \infty$, as a consequence of Proposition III.4.10. Since $|f * g| \le |f| * |g|$ a.e., this implies $\|f * g\|_{\Phi_2} \le \|f\|_1 \|g\|_{\Phi_2}$ and hence $T : L^{\Phi_0}(I\!R) \to L^{\Phi_2}(I\!R)$ given by $Tf = f * g$ is a bounded linear operator with bound $\|g\|_{\Phi_2}$.

Next we note that for each $h \in L^{\Phi_2}(I\!R)$ and $k \in L^{\Psi_2}(I\!R)$, the function $h(x - \cdot)k(\cdot)$ is in $L^{\Phi_0}(I\!R)$ for almost all $x \in I\!R$ and by Hölder's inequality,

$$|(h * k)(x)| \le \|h\|_{\Phi_2} N_{\Psi_2}(k) < \infty, \tag{23}$$

so that if $h = g \in L^{\Phi_2}(I\!R)$ and $f = k \in L^{\Psi_2}(I\!R)$, then $f * g \in L^\infty(I\!R)$. Hence the mapping $Tf = f * g$ is a bounded linear operator from $L^{\Psi_2}(I\!R)$ into $L^{\Psi_0}(I\!R) = L^\infty(I\!R)$, with bound $\|g\|_{\Phi_2}$. If for some s, $0 < s < 1$, Φ_1 and Φ_3 satisfy the given hypothesis, so that they are "intermediate" to (Φ_0, Ψ_2) and (Φ_2, Ψ_0), we can apply the interpolation theorem to deduce (21). The constant $0 < C < \infty$ comes from using the gauge and Orlicz norms indifferently. ∎

Remarks: By Theorem V.1.1, the Φ_i may be replaced by equivalent Young (or N-) functions in the preceding work. Also if $\Phi_1(t) = |t|^p$, $\Phi_2(t) = |t|^q$, and $\Phi_3(t) = |t|^r$, then it is seen that $s = pq/(p-1)$ and $r^{-1} = p^{-1} + q^{-1} - 1$, $0 < s < 1$ obtain whenever $p^{-1} + q^{-1} \ge 1$. For Orlicz spaces there is more freedom. However in a given problem it is easier to check the condition on s in the L^p-case than in the L^Φ-case of the Corollary.

A natural question now is: Can we replace the known Φ_0 by a more general N-function and obtain analogous inequalities of the type (21)? The answer depends on a knowledge and characterization of all the "intermediate spaces" between the given $L^{\Phi_1}(\mu)$ and $L^{\Phi_2}(\mu)$. We discuss this problem since it illuminates the preceding one. First let us state precisely the concepts of intermediate and interpolation spaces. Thus a pair of Banach spaces $(\mathcal{X}_1, \mathcal{X}_2)$ is termed an *interpolation couple* if there is a topological vector space $\mathcal{Z}$ (not necessarily normed) such that $\mathcal{X}_1$ and $\mathcal{X}_2$ can be continuously embedded in $\mathcal{Z}$. For instance, if $\mathcal{X}_i = L^{\Phi_i}(\mu)$, then $\mathcal{Z} = L^0(\mu)$, the space of scalar measurable functions on (Ω, Σ, μ) with convergence in μ-measure as topology which makes $\mathcal{Z}$ a (non-locally convex) Fréchet space (for a proof of this classical fact, cf., e.g., Rao (1987), Prop. 3.2.6), it will satisfy the conditions since $f_n \to f$ in $\mathcal{X}_i$ norm implies that $f_n \to f$ in μ-measure. Consider the intersection $\mathcal{X}_1 \cap \mathcal{X}_2$ and the direct sum $\mathcal{X}_1 + \mathcal{X}_2 = \{f : f = f_1+f_2, f_i \in \mathcal{X}_i, i=1,2\}$. If we endow them with norms

$$\|f\| = \max(\|f\|_{\mathcal{X}_i}, i = 1, 2), \qquad f \in \mathcal{X}_1 \cap \mathcal{X}_2, \qquad (24)$$

and

$$\|f\|' = \inf\{\|f_1\|_{\mathcal{X}_1} + \|f_2\|_{\mathcal{X}_2} : f = f_1+f_2 \in \mathcal{X}_1+\mathcal{X}_2\}, \qquad (25)$$

then a straightforward but tedious computation shows that these are norms, and $(\mathcal{X}_1 \cap \mathcal{X}_2, \|\cdot\|)$ and $(\mathcal{X}_1 + \mathcal{X}_2, \|\cdot\|')$ are Banach spaces. Then for an interpolation couple $(\mathcal{X}_1, \mathcal{X}_2)$ a Banach space $\mathcal{X}$ is termed an *intermediate space* if $\mathcal{X}_1 \cap \mathcal{X}_2$ can be continuously embedded in $\mathcal{X}$ which in turn can be continuously embedded in $\mathcal{X}_1 + \mathcal{X}_2$. Thus, specializing Theorem 5 to the $L^p(\mu)$ case—the classical Riesz-Thorin form—it is seen that each $L^p(\mu)$, $1 < p < \infty$ is an intermediate space of the couple $(L^1(\mu), L^\infty(\mu))$. The corresponding result for Orlicz spaces is somewhat complicated, because Corollary 6 implies that there are non-Lebesgue (namely Orlicz) spaces which are intermediate to $L^1(\mathbb{R})$ and $L^\infty(\mathbb{R})$. Further, if $(\mathcal{X}_1, \mathcal{X}_2)$ is an interpolation couple and $\mathcal{X}$ is an intermediate space between $\mathcal{X}_1$ and $\mathcal{X}_2$, then $\mathcal{X}$ is called an *interpolation space* for $(\mathcal{X}_1, \mathcal{X}_2)$ whenever each bounded linear operator $T : \mathcal{X}_i \to \mathcal{X}_i$, $i = 1, 2$, is also defined on $\mathcal{X} \to \mathcal{X}$ and has a bound majorized by $(\|T\|_{\mathcal{X}_i}, i = 1, 2)$. With this concept, Calderón (1964) has shown that there are many "Lorentz spaces" between L^1 and L^∞, and characterized them. Thus an answer to our question is that there are several non-Orlicz spaces also between the given pair $L^{\Phi_i}(\mu)$, $i = 1, 2$, that qualify to be interpolation spaces. However, a closely related result, giving conditions for an Orlicz space $L^\Phi(\mu)$ to be an interpolation space between a pair of given spaces $L^{\Phi_i}(\mu)$, $i = 1, 2$, has been obtained by J. Gustavsson and J. Peetre (1977). This space need not be of the form $L^{\Phi_s}(\mu)$.

To see the general problem clearly, we note that $L^{\Phi_s}(\mu)$ can be defined alternatively as the set $A = \{f \in L^0(\mu) : |f| \leq \lambda |g|^{1-s} |h|^s$ for

some $\lambda > 0, N_{\Phi_1}(g) \le 1, N_{\Phi_2}(h) \le 1$. If we let $\|f\| = \inf\{\lambda > 0 : \lambda$ satis-
fies the above inequality$\}$, then Calderón (1964) has shown that $\{A, \|\cdot\|\}$
is a complete normed linear space, denoted by $(L^{\Phi_1}(\mu))^{1-s}(L^{\Phi_2}(\mu))^s$.
It is also shown that $A = L^{\Phi_s}(\mu)$ and the norms are equivalent in both
spaces. Observing that $\Phi_s^{-1} = (\Phi_1^{-1})^{1-s}(\Phi_2^{-1})^s = (\Phi_1^{-1})(\Phi_2^{-1}/\Phi_1^{-1})^s$,
$0 < s < 1$, let $\rho(x) = x^s$, $0 < s < 1$, so that $\rho(\cdot)$ is a concave function
on $I\!\!R^+$. Then we have $\Phi_\rho^{-1} = \Phi_1^{-1}\rho(\Phi_2^{-1}/\Phi_1^{-1})$, and this leads to more
general concave $\rho(\cdot)$ in this definition. In fact, if $\rho(\cdot) : I\!\!R^+ \to I\!\!R^+$ is continuous and is equivalent to a concave function (or "pseudoconcave"),
then Φ_ρ^{-1} given above leads to such a generalization. For instance, a
possible ρ is the following function where $t > 0$:

$$\rho(t) = t^s(\log(e+t))^\alpha(\log(e+\frac{1}{t}))^\beta, \qquad 0 < s < 1, \alpha, \beta \in I\!\!R. \quad (26)$$

With such a general ρ, define $A_\rho = \{f \in L^0(\mu) : |f| \le \lambda |g|\rho(|h|/|g|)$,
for some $\lambda > 0$, $N_{\Phi_1}(g) \le 1, N_{\Phi_2}(h) \le 1\}$ and let $\|f\|_\rho = \inf\{\lambda > 0$, for
which the above inequality holds$\}$. Then $(A_\rho, \|\cdot\|_\rho)$ is a Banach space
denoted by $L^{\Phi_1}(\mu)\odot\rho(L^{\Phi_2}(\mu)/L^{\Phi_1}(\mu))$ which in general is a subspace
of the Orlicz space $L^{\Phi_\rho}(\mu)$ where Φ_ρ is the inverse of Φ_ρ^{-1} given above.
Using a different argument (than that used for Theorem 5), the fol-
lowing general result has been established by Gustavsson and Peetre,
which we state for comparison.

Theorem 8. *Let Φ_i, $i = 1, 2$, be N-functions such that both Φ_i*
and their complementary functions Ψ_i be Δ_2-regular. Suppose that
the pseudoconcave ρ satisfies the condition that $\lim_{x\to\infty} \rho(\lambda x)/\rho(x) = 0$
for each $\lambda > 0$, and that $\rho(\cdot)$ is continuous. If Φ_ρ is the inverse
of $\Phi_1^{-1}\rho(\Phi_2^{-1}/\Phi_1^{-1})$ then $L^{\Phi_\rho}(\mu) = L^{\Phi_1}(\mu)\odot\rho(L^{\Phi_2}(\mu)/L^{\Phi_1}(\mu))$, and
$L^{\Phi_\rho}(\mu)$ is an interpolation space between $L^{\Phi_1}(\mu)$ and $L^{\Phi_2}(\mu)$.

The preceding somewhat detailed discussion shows that the answer
to our "natural question" is not a simple one. It is essentially positive
if the constants in (21) are not specified. With this understanding of
the subject we turn to other aspects of interpolation theory.

The result of Theorem 5 admits an easy extension to multilinear
operators. Recall that, if $\mathcal{X}_i, \mathcal{Y}$ are vector spaces and $\mathcal{X}_1 \times \cdots \times \mathcal{X}_n$ is
their product, then a mapping $T : \mathcal{X}_1 \times \cdots \times \mathcal{X}_n \to \mathcal{Y}$ is *multilinear* if
$T(x_1, \ldots, x_n)$ is linear in each x_i when the other components are fixed.
The desired extension can be given in the following form:

Theorem 9. *Let* $\mathcal{X}_i^j = L^{\Phi_{ij}}(\mu_i), \mathcal{Y}^j = L^{Q_j}(\nu)$ *be a family of Orlicz spaces on* $(\Omega_i, \Sigma_i, \mu_i)$ *and* $(S, \mathcal{S}, \nu)$, $i = 1, \ldots, n, j = 1, 2$. *Suppose that each* Φ_{ij} *is* Δ_2-*regular, and* Q_j *are continuous Young functions. If* T *is a multilinear operator simultaneously defined on the simple functions*

$$T : \underset{i=1}{\overset{n}{\times}} L^{\Phi_{ij}}(\mu_i) \to L^{Q_j}(\nu), \qquad j = 1, 2, \tag{27}$$

and Φ_{is} *and* Q_s *are the intermediate Young functions given by (2) such that (using the Orlicz norms in* $L^{\Phi_{ij}}(\mu_i)$ *and* $L^{Q_i}(\nu)$ *spaces)*

$$\|T(f_1, \ldots, f_n)\|_{Q_j} \le M_j \prod_{i=1}^{n} \|f_i\|_{\Phi_{ij}}, \quad j = 1, 2, \tag{28}$$

for f_i *in* $L^{\Phi_{i1}}(\mu_i)$ *or in* $L^{\Phi_{i2}}(\mu_i)$, *simple, then* T *is also defined on* $\underset{i=1}{\overset{n}{\times}} L^{\Phi_{is}}(\mu_i)$ *into* $L^{Q_s}(\nu)$ *and one has*

$$\|T(f_1, \ldots, f_n)\|_{Q_s} \le M_1^{1-s} M_2^{s} \prod_{i=1}^{n} \|f_i\|_{\Phi_{is}}, \tag{29}$$

for all $f_i \in L^{\Phi_{is}}(\mu)$, $i = 1, \ldots, n$.

Sketch of Proof: The fact that (28) holds for simple functions f_i, is proved with a repetition of the argument employed in (12)–(19) and the Three-Line theorem. This detail can be left to the reader. We now indicate how (28) can be extended to obtain (29).

Since Φ_{ij}'s are Δ_2-regular, simple functions are dense in each $L^{\Phi_{ij}}(\mu_i)$ as well as in $L^{\Phi_s}(\mu)$. Consequently for any f_i in $L^{\Phi_s}(\mu_i)$ there exist simple functions f_{im} such that $\|f_i - f_{im}\|_{\Phi_s} \to 0$ as $m \to \infty$, for each $1 \le i \le n$. But (29) is valid if $(f_1, \ldots, f_n)$ is replaced by $(f_{1m}, \ldots, f_{nm})$, for all $m \ge 1$. Since $T(f_{1m}, \ldots, f_{nm}) \in L^{Q_s}(\nu)$, $m \ge 1$, and the latter space is complete, it is sufficient to show that $\{T(f_{1m}, \ldots, f_{nm}), m \ge 1\}$ is a Cauchy sequence in $L^{Q_s}(\nu)$ so that when $f_{im} \to f_i$ in $L^{\Phi_s}(\mu)$ for $0 < s < 1$, the above limit defines $T(f_1, \ldots, f_m)$ in $L^{Q_s}(\nu)$ uniquely and (29) holds. Thus it is enough to verify that

$$\|T(f_{1m}, \ldots, f_{nm}) - T(f_{1k}, \ldots, f_{nk})\|_{Q_s} \to 0, \text{ as } m, k \to \infty. \tag{30}$$

This is deduced, with (29) for simple functions, as in the classical case:

$$\|T(f_{1m}, \ldots, f_{nm}) - T(f_{1k}, \ldots, f_{nk})\|_{Q_s} \le$$

$$\leq \sum_{i=1}^{n} \|T(f_{1m}, \ldots, f_{im}, 0, \ldots, 0) - T(f_{1k}, \ldots, f_{ik}, 0, \ldots, 0)\|_{Q_s}$$

$$\leq \sum_{i=1}^{n} \|T(f_{1m} - f_{1k}, \ldots, f_{im} - f_{1k}, 0, \ldots, 0)\|_{Q_s}$$

$$\leq M_1^{1-s} M_2^s \sum_{i=1}^{n} \prod_{j=1}^{i} \|f_{jm} - f_{jk}\|_{\Phi_j} \to 0, \text{ as } m, k \to \infty,$$

since $f_{jm} \to f_j$ in $\|\cdot\|_{Q_{js}}$ norm. This establishes (30) and hence (29) in the general case. ∎

In the above result, the product space $\overset{n}{\underset{i=1}{\times}} L^{\Phi_i}(\mu)$ was not endowed with a topology. For instance, it may be normed if we take $\|\cdot\| = \sum_{i=1}^{n} \|\cdot\|_{\Phi_i}$, or an L^p-type or an Orlicz norm instead of the displayed L^1-norm on a finite number of spaces. This motivates the following concept.

If $(A, \mathcal{A}, \nu)$ is a measure space, for each $\alpha \in A$ let $L^{\Phi_\alpha}(\mu_\alpha)$ be an Orlicz space on $(\Omega_\alpha, \Sigma_\alpha, \mu_\alpha)$ with norm $\|\cdot\|_\alpha$ (an Orlicz or gauge norm). Then $\mathcal{F} = \{\mathbf{f} : \mathbf{f} = [f_\alpha, \alpha \in A], f_\alpha \in L^{\Phi_\alpha}(\mu_\alpha), \text{ i.e., } \mathbf{f} \in \underset{\alpha \in A}{\times} L^{\Phi_\alpha}(\mu_\alpha)\}$ is a vector space under componentwise addition and scalar multiplication. The elements $\mathbf{f}$ are also called *vector fields*. If A is countable so that $\mathcal{A}$ is its power set, and $\nu(\{\alpha\}) \geq 0$ is a weight at 'α,' one may endow $\mathcal{F}$ with a norm $N_\theta^{\vec{\Phi}}(\cdot)$ defined for each Young function θ as:

$$N_\theta^{\vec{\Phi}}(\mathbf{f}) = \inf\{k > 0 : \sum_{\alpha \in A} \theta\left(\frac{\|f_\alpha\|_{\Phi_\alpha}}{k}\right) \nu(\{\alpha\}) \leq 1\}. \tag{31}$$

Then $\{\mathcal{F}, N_\theta^{\vec{\Phi}}(\cdot)\}$ becomes a normed vector space which is also complete. This fact and the relevant theory when A is a locally compact space and $\nu(\cdot)$ is a regular (or Radon) measure can be established, and this set is called an Orlicz space of vector fields. The structure theory of the latter which parallels that given in Chapter III is detailed in the standard work by N. Dinculeanu [(1974), Chapter VII]. If the $(A, \mathcal{A}, \nu)$ is an abstract measure space, the corresponding theory and some applications are given in Rao (1980,1981$_a$). Some related results and probabilistic applications of such a space will be considered in Chapter VIII. But here we discuss first a special case when A is a *finite* set and $\theta(t) = |t|^r$, $r \geq 1$, for the interpolation statement. This case was formulated by C. E. Cleaver (1972).

Thus let us denote the finite set A by $\{1, 2, \ldots, n\}$, and $\nu(\{i\}) = \lambda_i \geq 0$. If $\theta(t) = |t|^p$, then (31) becomes the classical p-norm

$$N_p^{\overrightarrow{\Phi}}(\mathbf{f}) = \left[\sum_{j=1}^{n} [N_{\Phi_j}(f_j)]^p \lambda_j\right]^{\frac{1}{p}}, \qquad 1 \leq p < \infty, \qquad (32)$$

and the corresponding essential supremum norm if $p = +\infty$.

Let $\mathcal{F} = \underset{\alpha \in A}{\times} \mathcal{X}_\alpha$, and $\mathcal{F}^* = \underset{\alpha \in A}{\times} \mathcal{X}_\alpha^*$ where $\mathcal{X}_\alpha^*$ is the adjoint space of $\mathcal{X}_\alpha$ ($= L^{\Phi_\alpha}(\mu_\alpha)$). Now $\mathcal{F}^*$ becomes a Banach space under the adjoint norm $N_{\bar{\theta}}^{\overrightarrow{\Psi}}(\cdot)$ to (31), and if $\mathbf{f} \in \mathcal{F}, \mathbf{g} \in \mathcal{F}^*$, then $\langle \mathbf{f}, \mathbf{g} \rangle = \{\langle f_\alpha, g_\alpha \rangle, \alpha \in A\} \in \mathbb{R}^A$ (or $\mathbb{C}^A$ in the complex case), where $\langle \cdot, \cdot \rangle$ is the duality pairing. It can be shown that (even in the general case) the Orlicz norm $\|\cdot\|_\theta^{\overrightarrow{\Phi}}$ equivalent to $N_\theta^{\overrightarrow{\Phi}}(\cdot)$ of (31) is given by

$$\|\mathbf{f}\|_\theta^{\overrightarrow{\Phi}} = \sup\left\{\sum_{\alpha \in A} |\langle \mathbf{f}, \mathbf{g} \rangle(\alpha)| \nu(\{\alpha\}) : N_{\bar{\theta}}^{\overrightarrow{\Psi}}(\mathbf{g}) \leq 1, \mathbf{g} \in \mathcal{F}^*\right\}, \qquad (33)$$

where $\bar{\theta}(\Psi)$ is complementary to $\theta(\Phi)$. [A proof of this under an inessential normalization of θ and $\bar{\theta}$, namely, $\theta(1) + \bar{\theta}(1) = 1$, is in Rao (1981a).] In the present case, the duality pairing may be written explicitly as:

$$\|\mathbf{f}\|_p^{\overrightarrow{\Phi}} = \sup\left\{\sum_{j=1}^{n} \lambda_j \left|\int_{\Omega_j} f_j dG_j\right| : N_{p'}^{\overrightarrow{\Psi}}(\mathbf{G}) \leq 1, \mathbf{G} \in \mathcal{F}^*\right\}, \qquad (34)$$

where $p' = p/(p-1)$, $1 \leq p \leq \infty$, and Ψ_j is complementary to Φ_j. Thus if Φ_{ij}, Q_{kj} are Young function pairs, $j = 1, 2$, let Φ_{is}, Q_{ks} be their intermediate functions, $1 \leq i \leq n, 1 \leq k \leq m$. Let $L^{\Phi_{ij}}(\mu_i)$ and $L^{Q_{kj}}(\mu_k')$ be the corresponding Orlicz spaces on $(\Omega_i, \Sigma_i, \mu_i)$ and $(\Omega_k', \Sigma_k', \mu_k')$, respectively. If $B = \{1, \ldots, n\}, \widetilde{B} = \{1, \ldots, m\}$, let $\nu(\{\alpha\}) = \lambda_i \geq 0$ and $\nu'(\{k\}) = \lambda_k' \geq 0$, to use (32), (33), and (34).

We may now formulate the desired interpolation result as:

Theorem 10. *Let* (Φ_{ij}, Φ_{is}) *and* (Q_{kj}, Q_{ks}), $0 \leq s \leq 1$, *be as defined above. Suppose that the* Φ_{ij} *are* Δ_2-*regular, and let* $r_j, t_j \geq 1$ *be reals with* r_s, t_s *defined by*

$$\frac{1}{r_s} = \frac{1-s}{r_1} + \frac{s}{r_2}, \qquad \frac{1}{t_s} = \frac{1-s}{t_1} + \frac{s}{t_2}. \qquad (35)$$

If T is a bounded linear operator simultaneously defined on $\big(\overset{n}{\underset{i=1}{\times}} L^{\Phi_{ij}}(\mu_i),$

$\| \cdot \|_{r_j}^{\overrightarrow{\Phi_j}} \big)$ *into* $\big(\overset{m}{\underset{k=1}{\times}} L^{Q_{kj}}(\mu'_k), \| \cdot \|_{t_j}^{\overrightarrow{Q_j}} \big),$ *$j = 1, 2,$ with bounds M_1*

and M_2 respectively, then T is also defined on $\big(\overset{n}{\underset{i=1}{\times}} L^{\Phi_{is}}(\mu_i), \| \cdot \|_{r_s}^{\overrightarrow{\Phi_s}} \big)$

into $\big(\overset{m}{\underset{k=1}{\times}} L^{Q_{ks}}(\mu'_k), \| \cdot \|_{t_s}^{\overrightarrow{Q_s}} \big),$ *$0 \le s \le 1,$ and*

$$\|T(\mathbf{f})\|_{t_s}^{\overrightarrow{Q_s}} \le M_1^{1-s} M_2^s \|\mathbf{f}\|_{r_s}^{\overrightarrow{\Phi_s}}, \mathbf{f} \in \overset{n}{\underset{i=1}{\times}} L^{\Phi_{is}}(\mu_i). \tag{36}$$

Proof: Although the argument follows the basic lines of the proof given for Theorem 5 above, there is a new element here since the range space of T is a product space and not a single Orlicz space based on a single measure space. So the essential detail will be given to illustrate the new conceptual problem.

Let Ψ_{ij}, R_{kj} be respectively complementary to Φ_{ij} and Q_{kj}, and let R_{ks} be intermediate to R_{k1} and R_{k2}. If $\mathbf{f} = (f_1, \ldots, f_n)$ and $\mathbf{g} = (g_1, \ldots, g_m)$ are elements from the $\overset{n}{\underset{i=1}{\times}} L^{\Phi_{ij}}(\mu_i)$ and $\overset{m}{\underset{k=1}{\times}} L^{R_{kj}}(\mu'_k)$ spaces in which each f_i and g_k are simple functions, then consider

$$I = \int_{\overline{B}} \langle T\mathbf{f}, \mathbf{g} \rangle d\nu' = \sum_{k=1}^m \langle T\mathbf{f}, \mathbf{g} \rangle (k) \lambda'_k. \tag{37}$$

Leaving the extension to the discontinuous Young function case aside which can be treated as before, we concentrate on the cases that Φ_{ij} and Q_{kj} are N-functions. Then using Theorem III.3.13, one can find constants $\alpha_{is} > 1$ such that

$$\frac{1}{\alpha_{is}} \Big[1 + \int_{\Omega_i} \Phi_{is} \Big(\frac{\alpha_{is}|f_i|}{\|f_i\|_{\Phi_{is}}} \Big) d\mu_i \Big] = \Big\| \frac{f_i}{\|f_i\|_{\Phi_{is}}} \Big\|_{\Phi_{is}} = 1, \tag{38}$$

where $\| \cdot \|_{\Phi_{is}}$ is the Orlicz norm and $i = 1, \ldots, n$. We intend to extend the integral I onto the strip $S = \{z \in \mathbb{C} : 0 \le \text{Re}(z) \le 1\}$. For this consider the inverses Φ_{is}^{-1} and R_{ks}^{-1}. Letting $f_j = |f_j|e^{iu_j}, g_k = |g_k|e^{iv_k}$, define the simple functions F_{jz} and G_{kz} depending on the complex parameter z as follows:

$$F_{jz} = \frac{1}{\alpha_{js}}[\|f_{js}\|_{\Phi_{js}}]^{\gamma(z)}\Phi_{jz}^{-1}\left(\Phi_{js}(\alpha_{js}|f_j|/\|f_j\|_{\Phi_{js}})\right)e^{iu_j},$$

and   (39)

$$G_{kz} = [N_{R_{ks}}(g_k)]^{\tau(z)}R_{kz}^{-1}\left(R_{ks}(|g_k|/N_{R_{ks}}(g_k))\right)e^{iv_k},$$

for $j = 1, \ldots, n$, $k = 1, \ldots, m$, where

$$\frac{\gamma(z)}{r_s} = \frac{1-z}{r_1} + \frac{z}{r_2} \quad \text{and} \quad \frac{\tau(z)}{t'_s} = \frac{1-z}{t'_1} + \frac{z}{t'_2}$$

$$(\, t'_\ell = t_\ell/(t_\ell - 1)), \ \ell = 1, 2), \qquad (40)$$

so that $\gamma(s) = 1$ and $\tau(s) = 1$. Let $\mathbf{F}_z = (F_{1z}, \ldots, F_{nz})$ and $\mathbf{G}_z = (G_{1z}, \ldots, G_{mz})$. Thus $\mathbf{F}_s = \mathbf{f}$ and $\mathbf{G}_s = \mathbf{g}$.

Now replacing $\mathbf{f}$ and $\mathbf{g}$ by $\mathbf{F}_z$ and $\mathbf{G}_z$ in (37), denote the resulting integral by $I(z)$. Next, if $f_j = \sum_{\ell=1}^{n_j} a_{j\ell}\chi_{E_{j\ell}}$ and $g_k = \sum_{\ell=1}^{m_k} b_{k\ell}\chi_{H_{k\ell}}$ are typical representations ($a_{j\ell}, b_{k\ell}$ being complex numbers), $E_{j\ell} \in \Sigma_j$, $H_{k\ell} \in \Sigma'_k$, then $I(z)$ is a linear combination of exponentials d^z, $d > 0$ with coefficients involving the integrals

$$\int_{\widetilde{B}} \langle T(\chi_{E_{1\ell_1}}, \ldots, \chi_{E_{n\ell_n}}), (\chi_{H_{1p_1}}, \ldots, \chi_{H_{mp_m}}) \rangle dv'. \qquad (41)$$

In general (41) may *not* be written as a sum of terms on individual measure spaces $(\Omega'_j, \Sigma'_j, \mu'_j)$. Since (41) is a scalar, $I(z)$ is continuous in z on the strip S, and bounded and regular in its interior. Furthermore $I(s) = I$. Thus we need to calculate the bounds of (37), i.e., for $I(z)$, on the boundary lines $\mathrm{Re}(z) = 0$ and $\mathrm{Re}(z) = 1$ in this case (of simple $\mathbf{f}$ and $\mathbf{g}$).

Using the norm inequalities for the duality pairings and then the Hölder inequality for the $(L^{t_s}(\nu'), L^{t'_s}(\nu'))$-spaces we get

$$|I(z)| \le \|T(\mathbf{F}_z)\|_{t_1}\|\mathbf{G}_z\|_{t'_1}, \ t_1^{-1} + (t'_1)^{-1} = 1,$$

$$= \|T(\mathbf{F}_z)\|_{t_1}^{\vec{Q}_1}\|\mathbf{G}_z\|_{t'_1}^{\vec{R}_1},$$

$$\le M_1\|\mathbf{F}_z\|_{r_1}^{\vec{\Phi}_1}\|\mathbf{G}_z\|_{t'_1}^{\vec{R}_1}. \qquad (42)$$

We need to simplify the right side norms of (42) for the cases that $\mathrm{Re}(z) = 0$ and $\mathrm{Re}(z) = 1$.

Thus let $\mathrm{Re}(z) = 0$, and consider, using (32) in this case,

$$\left(\|\mathbf{F}_z\|_{r_1}^{\vec{\Phi}_1}\right)^{r_1} = \sum_{j=1}^{n} \left(\|(F_{(iy)})_j\|_{\Phi_{j1}}\right)^{r_1} \lambda_j \qquad (i = \sqrt{-1}),$$

$$= \sum_{j=1}^{n} \left[\||f_j|\|_{\Phi_{js}}^{r_1 \gamma(iy)}| \cdot \left\|\frac{1}{\alpha_{js}}\Phi_{j(iy)}^{-1}(\Phi_{js}(\alpha_{js}|f_j|/\|f_j\|_{\Phi_{js}}))\right\|_{\Phi_{js}}\right]^{r_1} \lambda_j$$

$$\leq \sum_{j=1}^{n} \left[\|f_j\|_{\Phi_{js}}^{r_s}\right][1] \cdot \lambda_j, \qquad \text{by (38)},$$

$$= \left(\|\mathbf{f}\|_{r_s}^{\vec{\Phi}_s}\right)^{r_s}. \tag{43}$$

Regarding the second norm we get by a similar computation, on noting that (using, as usual, the gauge norms in adjoint spaces when Orlicz norms are employed in the original domains)

$$\left(\|\mathbf{G}_z\|_{t_1'}^{\vec{R}_1}\right)^{t_1'} \leq \sum_{j=1}^{m} \left[N_{R_{js}}(g_j)\right]^{t_s'} \lambda_j' = \left[\|\mathbf{g}\|_{t_s'}^{\vec{R}_s}\right]^{t_s'}. \tag{44}$$

Hence for simple $\mathbf{f}, \mathbf{g}$ satisfying $\|\mathbf{f}\|_{r_s}^{\vec{\Phi}_s} \leq 1, \|\mathbf{g}\|_{t_s'}^{\vec{R}_s} \leq 1$, we have $|I(iy)| \leq M_1$. Using a similar computation one gets $|I(1 + iy)| \leq M_2$ so that by the Three-Line theorem (Proposition 4), $|I(s)| \leq M_1^{1-s}M_2^s$. Taking the supremum over all such $\mathbf{g}$ of norms at most one, we have (36) for all simple functions.

For the general case, one uses the fact that the Φ_{ij} are Δ_2-regular so that simple functions are again dense in $\mathcal{F}$ as long as r and t are finite. Here we need to use the argument of the proof of Theorem 9 to extend (36) to all of $\mathcal{F}$. Finally the boundary cases can be separately treated, as in Theorem 5, and the latter argument need not be repeated here. ∎

Remark. One may interchange the gauge and Orlicz norms throughout in the above and obtain a corresponding inequality with the same procedure. Even if all the measure spaces are the same, the integral in (41) cannot always be expressed as a sum (when $m > 1$) of integrals since $T(\chi_{E_1}, \ldots, \chi_{E_n})$ need not be decomposable into m factors whatever $n \geq 1$ is. Even if $m = 2$, examples and simple forms of T exist to demonstrate this pathology.

Using the concepts and results of Orlicz spaces of vector fields, as given e.g., in Dinculeanu (1974) and Rao (1980), a further extension of Theorem 10 in which the Lebesgue spaces ℓ^p on a finite number of points is replaced by a general Orlicz space is possible. We shall omit this aspect here and give a specialization for applications of a packing problem in the next section.

For the desired simplification of the above result, consider the weighted sequence spaces $\ell^p_{\mathcal{X}}(\Lambda)$ based on a Banach space $\mathcal{X}$:

$$\ell^p_{\mathcal{X}}(\Lambda) = \{\mathbf{x}=(x_1,x_2,\ldots) : \|\mathbf{x}\|^p_p = \sum_{i=1}^{\infty} |x_i|^p_{\mathcal{X}} \lambda_i < \infty, x_i \in \mathcal{X}, \lambda_i \geq 0\}. \quad (45)$$

Then $\|\cdot\|_p$ is a norm and $\{\ell^p_{\mathcal{X}}(\Lambda), \|\cdot\|_p\}$ is again a Banach space. If the weights $\lambda_i = 0$ for $i > n$ [or $x_i = 0$ for $i > n$], denote the resulting finite dimensional space by the same symbol $\ell^p_{\mathcal{X}}(\Lambda)$. Taking, in particular $\mathcal{X} = L^{\Phi}(\mu)$, an Orlicz space, we have the following consequence of Theorem 10, essentially formulated by Cleaver (1976) [see also Hayden and Wells (1971) when $\mathcal{X} = L^p(\mu)$], where the above distinction for (41) was not observed.

Corollary 11. *Let Φ be a Δ_2-regular Young function and Φ_0 be given by $\Phi_0(x) = x^2$, $x \in \mathbb{R}$. If Φ_s, $0 \leq s \leq 1$ is the inverse of $(\Phi^{-1})^{1-s}(\Phi_0^{-1})^s$, then for any collection $\{f_i, 1 \leq i \leq n\} \subset L^{\Phi_s}(\mu)$ on a measure space (Ω, Σ, μ), and any $p_i > 0$, with $\sum_{i=1}^{n} p_i = 1$, one has*

$$\sum_{i=1}^{n}\sum_{j=1}^{n} p_i p_j \left(\|f_i - f_j\|_{\Phi_s}\right)^{\frac{2}{2-s}} \leq 2\alpha^{\frac{2(1-s)}{2-s}} \sum_{i=1}^{n} p_i \left(\|f_i\|_{\Phi_s}\right)^{\frac{2}{2-s}}, \quad (46)$$

where $0 < \alpha = \max_{1 \leq i \leq n} (1 - p_i) < 1$.

Proof: Let $\mathcal{X}_1 = L^{\Phi}(\mu), \mathcal{X}_2 = L^{\Phi_0}(\mu)$, $\lambda_i = p_i$, and $\lambda'_k = p_i p_j$. We let Λ be an n-vector and Λ' be an n^2-vector. Consider the spaces $\ell^1_{\mathcal{X}_1}(\Lambda)$ and $\ell^2_{\mathcal{X}_2}(\Lambda')$ and the operator $T : \ell^i_{\mathcal{X}_i}(\Lambda) \to \ell^i_{\mathcal{X}_i}(\Lambda')$, defined simultaneously on the spaces for $i = 1, 2$, by the equation

$$T\mathbf{f} = \mathbf{g}, \quad \mathbf{f} = (f_1, \ldots, f_n) \in \ell^i_{\mathcal{X}_i}(\Lambda). \quad (47)$$

Here $\mathbf{g} = (g_1, \ldots, g_{n^2})$ is taken as the vector of differences $(f_i - f_j)$, $1 \leq i, j \leq n$, with a fixed order of indices, say, the lexicographic ordering for definiteness. That such a $\mathbf{g}$ is in $\ell^i_{\mathcal{X}_i}(\Lambda')$ is clear.

It is also noted that T is well defined and linear. To see that in fact T is bounded on both spaces, consider

$$\|T\mathbf{f}\|_1^{\vec{\Phi}} = \sum_{i=1}^{n}\sum_{j=1}^{n} p_i p_j \|f_i - f_j\|_\Phi \le \sum_{\substack{i=1\\i\ne j}}^{n}\sum_{j=1}^{n} p_i p_j (\|f_i\|_\Phi + \|f_j\|_\Phi)$$

$$= \sum_{i=1}^{n}\sum_{j=1}^{n} p_i p_j (\|f_i\|_\Phi + \|f_j\|_\Phi) - 2\sum_{i=1}^{n} p_i^2 \|f_i\|_\Phi$$

$$= 2\sum_{i=1}^{n} \|f_i\|_\Phi (1 - p_i) p_i \le 2\alpha \sum_{i=1}^{n} p_i \|f_i\|_\Phi$$

$$= 2\alpha \|\mathbf{f}\|_1^{\vec{\Phi}}.$$

Since $\mathcal{X}_2$ is a Hilbert space, we have

$$\|T f\|_2^{\vec{\Phi}_0} = \left[\sum_{i=1}^{n}\sum_{j=1}^{n} p_i p_j \|f_i - f_j\|_{\Phi_0}^2\right]^{\frac{1}{2}}$$

$$\le \sqrt{2}\left[\sum_{i=1}^{n} p_i \|f_i\|_{\Phi_0}^2\right]^{\frac{1}{2}} = \sqrt{2}\|\mathbf{f}\|_2^{\vec{\Phi}_0}.$$

In this situation we can invoke Theorem 10, and conclude that $T : \ell_{\mathcal{X}_s}^{r_s}(\Lambda) \to \ell_{\mathcal{X}_s}^{t_s}(\Lambda')$ and

$$\|T\mathbf{f}\|_{t_s}^{\vec{\Phi}_s} \le (2\alpha)^{1-s} 2^{\frac{s}{2}} \|\mathbf{f}\|_{r_s}^{\vec{\Phi}_s}, \qquad 0 \le s \le 1, \tag{48}$$

where $\mathcal{X}_s = L^{\Phi_s}(\mu)$, $\frac{1}{r_s} = \frac{1-s}{1} + \frac{s}{2} = \frac{1}{t_s}$ so that $r_s = t_s = \frac{2}{2-s}$. But then raising both sides of (48) to the power r_s, one obtains (47). ∎

Remark. The proof shows that $\ell_{\mathcal{X}}^p(\Lambda)$ can be generalized to $\ell_{\mathcal{F}}^\theta(P)$ where θ is a Young function, (Ω, Σ, P) is a probability space, and $\mathcal{F} = \{\mathcal{X}_\omega, \omega \in \Omega\}$ is a family of Banach spaces satisfying some natural conditions as in, e.g., Rao (1980 or 1981$_a$). This will not be discussed further, as it leads to specialized studies.

In all the preceding considerations the basic measure space on which the domains of the operator T come into play is fixed. Similarly the ranges of T are also based on a fixed measure space. The slight changes coming in Theorem 10, also involve the same measures. If these are allowed to change, then several new problems arise, and one has to restrict the Δ_2-class further. We illustrate this by presenting a key result, with Young functions of Δ'-class. The following is adapted from Rao (1966), and it extends the L^p-case originally considered by E. M. Stein and G. Weiss (1958).

Let $(\Omega, \Sigma, \tilde{\mu}_i)$ and $(\Omega', \Sigma', \tilde{\mu}_i')$, $i = 1, 2$ be measure spaces and $L^{\Phi_i}(\tilde{\mu}_i)$ and $L^{Q_i}(\tilde{\mu}_i')$ be the Orlicz spaces based on them. Suppose that Φ_i, Q_i satisfy the Δ'-condition globally so that $\Phi_i(xy) \leq C_i \Phi_i(x) \Phi_i(y)$ and $Q_i(xy) \leq D_i Q_i(x) Q_i(y)$ for $x, y \geq 0$, $C_i > 0, D_i > 0$, $i = 1, 2$. Multiplying by C_i and replacing Φ_i by $\tilde{\Phi}_i = C_i \Phi_i$, we may assume that $C_i = 1$ and similarly $D_i = 1$ for convenience. Also suppose that $\Phi_i(1) = 1$ so that $\Phi_s(1) = 1$, $0 \leq s \leq 1$, and that $\Phi_i(x) > 0, Q_i(x) > 0$ for $x > 0$. Let μ and ν be measures on Σ and Σ', and $a_i > 0, b_i > 0$ be measurable functions such that for $i = 1, 2$,

$$\tilde{\mu}_i(E) = \int_E a_i d\mu, \quad \tilde{\mu}_i'(F) = \int_F b_i d\nu, \qquad E \in \Sigma, F \in \Sigma'. \qquad (49)$$

Note that if $\tilde{\mu}_i$ and $\tilde{\mu}_i'$ are both σ-finite, then $\mu = \tilde{\mu}_1 + \tilde{\mu}_2, \nu = \tilde{\mu}_1' + \tilde{\mu}_2'$ will suffice in (49) and then a_i and b_i are the Radon-Nikodým derivatives of $\tilde{\mu}_i$ and $\tilde{\mu}_i'$ relative to μ and ν. Thus (49) is merely an abstraction and relaxation of the σ-finiteness assumption there.

We now introduce a family of measures and the corresponding Orlicz spaces to formulate the desired result. Let $\bar{\Phi}_i = \max(1, \Phi_i)$ and $\overline{Q}_i = \max(1, Q_i)$ so that $\bar{\Phi}_i$ and $\overline{Q}_i$ are the principal parts of Φ_i and Q_i. If Φ_s and Q_s are the intermediate Young functions of (Φ_1, Φ_2) and (Q_1, Q_2) respectively, define measures μ_s, ζ on Σ and ν_s, η on Σ' by the following formulas with $\alpha_i = \Phi_i^{-1}$, $i = 1, 2$:

$$\mu_s(E) = \int_E \frac{\Phi_s[\alpha_1(a_1)] \cdot \Phi_s[\alpha_2(a_2)]}{\bar{\Phi}_1[\alpha_2(a_2)] \cdot \bar{\Phi}_2[\alpha_1(a_1)]} d\mu,$$

$$E \in \Sigma, \qquad (50)$$

$$\zeta(E) = \int_E [\bar{\Phi}_1(\alpha_2(a_2)) \cdot \bar{\Phi}_2(\alpha_1(a_1))]^{-1} d\mu,$$

$$\nu_s(F) = \int_F \frac{Q_s[\beta_1(b_1)] \cdot Q_s[\beta_2(b_2)]}{\overline{Q}_1[\beta_2(b_2)] \overline{Q}_2[\beta_1(b_1)]} d\nu,$$

$$F \in \Sigma', \qquad (51)$$

$$\eta(F) = \int_F [\overline{Q}_1(\beta_2(b_2)) \overline{Q}_2(\beta_1(b_1))]^{-1} d\nu,$$

where $\beta_i = Q_i^{-1}$, $i = 1, 2$. From definition it follows that $\mu_s \leq \tilde{\mu}_{s+1}$ and $\nu_s \leq \nu_{s+1}$ for $s = 0, 1$, and, moreover, these are equivalent measures.

Let $L^{\Phi_s}(\mu_s)$ and $L^{Q_s}(\nu_s)$ be the corresponding Orlicz spaces for $0 \le s \le 1$, with the changing measures. We denote the Orlicz (or gauge) norms in these spaces by $\|\cdot\|_{\Phi_s,\mu_s}[N_{\Phi_s,\mu_s}(\cdot)]$ and $\|\cdot\|_{Q_s,\nu_s}[N_{Q_s,\nu_s}(\cdot)]$, respectively. Then the following assertion holds.

Theorem 12. *With the notations introduced above, suppose that* $T : L^{\Phi_i}(\mu_i) \to L^{Q_i}(\nu_i)$, $i = 1, 2$, *is simultaneously defined on the simple functions, is linear, and satisfies*

$$\|Tf\|_{Q_i,\nu_i} \le K_i \|f\|_{\Phi_i,\nu_i}, \qquad i = 1, 2. \tag{52}$$

If Φ_i, Q_i *verify the (global)* Δ'-*condition, then for each* $0 \le s \le 1$, T *is also defined on* $L^{\Phi_s}(\mu_s)$ *into* $L^{Q_s}(\nu_s)$, *and*

$$\|Tf\|_{Q_s,\nu_s} \le D K_1^{1-s} K_2^s \|f\|_{\Phi_s,\mu_s}, \qquad f \in L^{\Phi_s}(\mu_s), \tag{53}$$

where $0 < D < \infty$ *is a constant depending only on the* μ_i, ν_i *and the* Φ_i, Q_i, $i = 1, 2$.

Proof: We present the details in steps for convenience.

Step 1. By construction $\tilde{\mu}_{s+1}$ and μ_s, $\tilde{\mu}'_{s+1}$ and ν_s are equivalent measures, for $s = 0, 1$. Since $\{\Phi_i, Q_i\} \subset \Delta' \subset \Delta_2$, simple functions are dense in the Orlicz spaces $L^{\Phi_i}(\tilde{\mu}_i)$, $L^{Q_i}(\tilde{\mu}'_i)$, $L^{\Phi_s}(\mu_s)$, and $L^{Q_s}(\nu_s)$, $0 \le s \le 1$.

Step 2. If $\mathcal{M}(\Omega, \Sigma)$ is the set of scalar measurable functions on Ω, then we assert that there exists a linear mapping $A : \mathcal{M}(\Omega, \Sigma) \to \mathcal{M}(\Omega, \Sigma)$ such that the restriction $A \mid L^{\Phi_s}(\mu_s)$ is a bounded invertible map into $L^{\Phi_s}(\zeta)$ satisfying

$$k_s \|f\|_{\Phi_s,\mu_s} \le \|Af\|_{\Phi_s,\zeta} \le \|f\|_{\Phi_s,\mu_s}, \qquad f \in L^{\Phi_s}(\mu_s), \tag{54}$$

for some absolute constant $0 < k_s < \infty$, depending continuously on $0 \le s \le 1$.

For, let $A : f \mapsto \alpha_1(a_1)\alpha_2(a_2)f$, $f \in \mathcal{M}(\Omega, \Sigma)$. Then A is an invertible linear operator. If $f \in L^{\Phi_s}(\mu_s)$, using the fact that $\Phi_s \in \Delta'$ (because the $\Phi_i \in \Delta'$, see Lemma 2), we have

$$\int\limits_{\Omega} \Phi_s(Af)d\zeta = \int\limits_{\Omega} \Phi_s(f)d\mu_s, \quad \text{by (50).}$$

This implies the right side inequality of (54). It is the other half that needs a more elaborate analysis and we present an indirect argument.

Suppose the inequality does not hold. Then there exists a sequence $\{f_n, n \geq 1\}$, $f_n \geq 0$ a.e., such that $\|f_n\|_{\Phi_s, \mu_s} = 1$ but $\lim_{n \to \infty} \|Af_n\|_{\Phi_s, \zeta} = 0$. Since $\Phi_i(x) = 0$ iff $x = 0$, each f_n vanishes outside of a set of σ-finite measure. Hence the sequence f_n, $n \geq 1$, vanishes outside a fixed σ-finite set of μ_s-measure, since the support of all these functions is a countable union of σ-finite sets. Thus for this step we may and do assume that μ_s itself is σ-finite. By the equivalence of ζ and μ_s, the Radon-Nikodým theorem asserts the existence of a positive $g_s = d\zeta/d\mu_s$. Hence we have, using the Δ'-condition of Φ_s once again and the fact that A is a multiplication so that $\Phi_s(f_n) = \Phi_s(A^{-1}Af_n) \leq \Phi_s(A^{-1})\Phi_s(Af_n)$,

$$
\begin{aligned}
0 \leq &\int_\Omega \frac{\Phi_s(f_n)d\mu_s}{\Phi_s(A^{-1})\Phi_s(\alpha_1(a_1))\Phi_s(\alpha_2(a_2))} \\
&\leq \int_\Omega \frac{\Phi_s(Af_n)d\mu_s}{\Phi_s(\alpha_1(a_1))\Phi_s(\alpha_2(a_2))} \\
&= \int_\Omega \Phi_s(Af_n) \cdot g_s d\mu_s = \int_\Omega \Phi_s(Af_n)d\zeta \to 0, \quad \text{as } n \to \infty, \quad (55)
\end{aligned}
$$

since $\|Af_n\|_{\Phi_s, \zeta} \to 0$. Next consider the set $B = \{\omega : V_s(\omega) \leq 1\}$ where $V_s = \Phi_s(A^{-1})\Phi_s(\alpha_1(a_1)\Phi_s(\alpha_2(a_2))$. Then

$$
\int_B \Phi_s(f_n)d\mu_s \leq \int_B \frac{1}{V_s}\Phi_s(f_n)d\mu_s \leq \int_\Omega \Phi_s(Af_n)d\zeta \to 0, \quad (56)
$$

as $n \to \infty$, by (55). Since on B^c, $V_s > 1$, we have again, by (55) and the convexity of Φ_s,

$$
\int_{B^c} \Phi_s\left(\frac{f_n}{V_s}\right)d\mu_s \leq \int_\Omega \frac{1}{V_s}\Phi_s(f_n)d\mu_s \to 0, \quad \text{as } n \to \infty. \quad (57)
$$

If X is the closed subspace of $L^{\Phi_s}(\mu_s)$ spanned by $\{f_n, n \geq 1\}$ restricted to B^c and Y that by $\{f_n/V_s, n \geq 1\}$, then the bounded map $U : f_n \mapsto f_n/V_s$ is linear and takes X onto Y. Since it is invertible, by the classical inverse boundedness theorem U^{-1} is bounded. Hence there is a constant $0 < C < \infty$ such that

$$
\|f_n\|_Y = \|U^{-1}\left(\frac{f_n}{V_s}\right)\|_Y \leq C\|\frac{f_n}{V_s}\|_X \to 0, \quad (58)
$$

as $n \to \infty$ by (57). Thus (56), (58) and the fact that Δ' implies Δ_2 yield that $\|f_n\|_{\Phi_s,\mu_s} \to 0$ as $n \to \infty$. But this contradicts the choice of f_n satisfying the condition $\|f_n\|_{\Phi_s,\mu_s} = 1$. Hence (54) must be true. A similar result holds if Φ_s, μ_s and ζ are replaced by Q_s, ν_s and η. Note also that this obtains for all $0 \le s \le 1$ and $k_s > 0$.

Step 3. Since simple functions are dense in $L^{\Phi_s}(\mu_s)$, $0 \le s \le 1$, the given operator T is densely defined on these spaces. Let A and $\tilde{A}$ be the mappings on $L^{\Phi_s}(\mu_s)$ and $L^{Q_s}(\nu_s)$, given by Step 2, and denote by $\mathcal{N}_{\zeta,s}$ and $\mathcal{N}_{\eta,s}$ the closures of $A[L^{\Phi_s}(\mu_s)]$ and $\tilde{A}[L^{Q_s}(\nu_s)]$ in $L^{\Phi_s}(\zeta)$ and $L^{Q_s}(\eta)$, respectively. Then $W = \tilde{A}TA^{-1} : \mathcal{N}_{\zeta,s} \to \mathcal{N}_{\eta,s}$ is linear and by the properties of Step 2, there are $0 < k_i < \infty$ such that

$$k_i \|A^{-1}f\|_{\Phi_i,\mu_i} \le \|f\|_{\Phi_i,\zeta} \le \|A^{-1}f\|_{\Phi_i,\mu_i}, \quad i = 1,2.$$

Consequently for $f \in \mathcal{N}_{\zeta,i}$, $i = 1,2$,

$$\|Wf\|_{Q_i,\eta} \le \|TA^{-1}f\|_{Q_i,\nu_i} \le K_i \|A^{-1}f\|_{\Phi_i,\mu_i,} \le \frac{K_i}{k_i}\|f\|_{\Phi_i,\zeta}.$$

Hence by Theorem 5 and Lemma 3, for each $0 \le s \le 1$,

$$\|Wf\|_{Q_s,\eta} \le 2\left(\frac{K_1}{k_1}\right)^{1-s}\left(\frac{K_2}{k_2}\right)^{s}\|f\|_{\Phi_s,\zeta}, \quad f \in \mathcal{N}_{\zeta,s}.$$

Once again using Step 2, there is a $d_s > 0$ such that

$$d_s\|Tf\|_{Q_s,\nu_s} \le \|\tilde{A}Tf\|_{Q_s,\eta} = \|WAf\|_{Q_s,\eta}$$
$$\le 2\left(\frac{K_1}{k_1}\right)^{1-s}\left(\frac{K_2}{k_2}\right)^{s}\|f\|_{\Phi_s,\mu_s,} \qquad (59)$$

using the preceding estimate. As noted at the end of Step 2, the $d_s > 0$ on $[0,1]$ and the hypothesis implies that it is continuous on $[0,1]$. Thus $2(d_s k_1^{1-s} k_2^s)^{-1}$ is bounded above by some $D > 0$. Using this in (59) we get (53). ∎

Remark. The preceding proof uses the Δ'-condition crucially, and it is not clear whether the result extends to other Orlicz spaces when Φ does not satisfy this restriction, to obtain a result similar to Corollary 6. The theorem reduces to the original L^p-version due to E. M. Stein and G. Weiss (1958), when $\Phi_i(x) = |x|^{p_i}$.

All the preceding considerations are based on the theory of complex functions and the Riesz-Thorin framework. Following the original real

variable approach of M. Riesz, a fundamental interpolation theorem
with weaker hypotheses (and a somewhat weaker conclusion) but also
useful in applications, where the former result is not applicable, has
been obtained by J. Marcinkiewicz (1939). An Orlicz space extension
of the latter was given by W. J. Riordan (1956) in his unpublished
thesis. We therefore include a version of this result as a final item of
this section.

Let us first give a motivation for such a new interpolation result.
The classical Riesz-Thorin theorem in its simplest form says that if T
is a bounded linear map on $L^{p_i}(\mu) \to L^{q_i}(\nu)$, $i = 1, 2$, $1 \le p_i, q_i \le \infty$,
then it is also bounded on $L^p(\mu) \to L^q(\nu)$ where $\frac{1}{p} = \frac{1-s}{p_1} + \frac{s}{p_2}$ and
$\frac{1}{q} = \frac{1-s}{q_1} + \frac{s}{q_2}$. Stated differently, if $\alpha_i = \frac{1}{p_i}$, $\beta_i = \frac{1}{q_i}$ and $\alpha = \frac{1}{p}$, $\beta = \frac{1}{q}$,
then the boundedness of T on the Lebesgue spaces with exponent pairs
(α_1, α_2) and (β_1, β_2) then it is also bounded on the intermediate pair
(α, β). Since $0 \le \alpha_i, \beta_i, \alpha, \beta \le 1$, this may be interpreted geometrically
as: the boundedness on the two end points of the line joining (α_1, α_2)
and (β_1, β_2) implies it is also bounded at each intermediate point (α, β).
However, there are important problems in which the operator is not
necessarily bounded at one of the end points of the line but has the
boundedness property at all the intermediate points. For instance if
$f \in L^p(0, 1)$, on the Lebesgue unit interval, then the operator T :
$(Tf)(t) = \frac{1}{t} \int_0^t f(x)dx$ is well defined for all $p \ge 1$, and is well known
to be bounded if $p > 1$ and unbounded for $p = 1$. Thus we have the
situation noted above. Many such problems arise in Fourier series.

To describe the desired result, for each measurable $f : \Omega \to \mathbb{R}$, let
$m_f(y) = \mu(\{\omega : |f(\omega)| > y\})$ where (Ω, Σ, μ) is the given measure space.
Then $m_f(\cdot)$ will be termed the *measure function* of f, determined by
μ. (Unfortunately the word "distribution function" is used for $m_f(\cdot)$ in
such standard works as Zygmund (1959), whereas in probability theory,
if $F_f(y) = \mu(\Omega) - m_f(y) = 1 - m_f(y)$, then $F_f(\cdot)$ is always called the
distribution function of f, determined by μ. So we shall avoid this
unnecessary conflict here.) If $T : L^p(\mu) \to L^q(\nu)$ is a bounded linear
operator, and $h = Tf \in L^q(\nu)$, then we have

$$\|f\|_p \|T\| \ge \|Tf\|_q = \left[\int_\Omega |h|^q d\nu \right]^{\frac{1}{q}} \ge \left(\int_{[|h| \ge y \ge 0]} |h|^q d\nu \right)^{\frac{1}{q}} \ge y(m_h(y))^{\frac{1}{q}}.$$

Hence

$$m_h(y) \le \left(\frac{\|T\|}{y}\|f\|_p\right)^q, \qquad y > 0. \tag{60}$$

If this inequality holds with a constant $\alpha > 0$ instead of $\|T\|$, for all $f \in L^p(\mu)$, and $h = Tf$, then T will be called an operator of *weak type* (p, q). In contrast, if $\|T\| = \alpha < \infty$, then T is of *strong* type (p, q) so that the strong type implies the weak type (but not conversely).

We now establish Riordan's extension of Marcinkiewicz's interpolation for certain Orlicz spaces. The result to be given is not the most general, but the basic ideas and difficulties will be clear. An application will be included in the next section. The result and proof hold for some non-linear T also. One says T is *quasi-linear* if $T : \mathcal{M}(\Omega_1, \Sigma_1) \to \mathcal{M}(\Omega_2, \Sigma_2)$ is defined, and

$$|T(f_1 + f_2)| \le C\left(|T(f_1)| + |T(f_2)|\right), \qquad 0 < C < \infty.$$

If $C = 1$ here, then T is called *sublinear*. (C depends only on T.) We then have the following:

Theorem 13. *Let* $T : \mathcal{M}(\Omega_1, \Sigma_1, \mu_1) \to \mathcal{M}(\Omega_2, \Sigma_2, \mu_2)$ *be a quasi-linear operator which is simultaneously of weak types* (p_1, q_1) *and* (p_2, q_2), *where* $1 \le p_i \le q_i < \infty$, $i = 1, 2$, $p_1 \ne p_2, q_1 \ne q_2$. *Let* $\alpha_0 = [(q_2/p_2) - (q_1/p_1)](q_2 - q_1)^{-1}$ *and* $s = (q_1^{-1} - q_2^{-1})(p_1^{-1} - p_2^{-1})^{-1}$, $q_1 > q_2$. *Suppose that* Φ_1 *is a continuous Young function such that for all* $u \ge 0$,

$$(i) \qquad \int_u^\infty t^{-q_1} d\Phi_1(t) \le C_1 u^{-q_1} \Phi_1(u),$$

$$\tag{61}$$

$$(ii) \qquad \int_0^u t^{-q_2} d\Phi_1(t) \le C_2 u^{-q_2} \Phi_1(u),$$

where $0 < C_i < \infty$, $i = 1, 2$, *are absolute constants.*

Then for any Young function Φ_2 *satisfying*

$$D_1\left[\Phi_1\left(\frac{x}{\Phi_2(x)^{\alpha_0}}\right)\right]^s \le \Phi_2(x) \le D_2\left[\Phi_1\left(\frac{x}{\Phi_2(x)^{\alpha_0}}\right)\right]^s, \quad x \ge 0, \tag{62}$$

for some constants $0 < D_1 \le D_2 < \infty$, *we have* $Tf \in L^{\Phi_1}(\mu_2)$ *for each* $f \in L^{\Phi_2}(\mu_1)$, *and in fact*

$$\|Tf\|_{\Phi_1} \le K\|f\|_{\Phi_2}, \qquad f \in L^{\Phi_2}(\mu_1), \tag{63}$$

for some $0 < K < \infty$ depending on T but not on f, $\| \cdot \|_{\Phi_i}$ being the usual norms in the Orlicz spaces $L^{\Phi_i}(\mu_i)$, $i = 1, 2$.

Remark. If $\Phi_2(x) = x^p$, $\Phi_1(x) = x^q$, $x \geq 0$, where $\frac{1}{p} = \frac{1-t}{p_1} + \frac{t}{p_2}$ and $\frac{1}{q} = \frac{1-t}{q_1} + \frac{t}{q_2}$, $0 < t < 1$, then the theorem reduces to the original Marcinkiewicz's interpolation theorem. We shall show later in applications that there exist several Φ_i's which are not mere power functions. After the proof, a relaxation of convexity of Φ_2 for (63) will be indicated. Considering various cases (and a tedious computation) it may be verified that $\Phi_1(x)$ is "between" x^{q_1} and x^{q_2}, and that $\Phi_2(x)$ is "between" x^{p_1} and x^{p_2}. [The inequalities need be shown, using (61) and (62), only for large x and for small x.] We proceed to establish (63).

Proof: Let us outline the essential points of the argument, leaving routine computations. Let $f \in L^{\Phi_2}(\mu_1)$ and by definition $L^{Q_i}(\mu_i) \subset M(\mu_i) = M(\Omega_i, \Sigma_i, \mu_i)$, $i = 1, 2$, $Q_1 = \Phi_2, Q_2 = \Phi_1$. Hence $h = Tf \in M(\mu_2)$ so that h is defined. Using (61) and (62), we show directly that (63) holds. Since no elegant result such as the Three-Line theorem is available, one has to estimate the norms $\|h\|_{\Phi_1}$ and $\|f\|_{\Phi_2}$ to obtain (63), and this is done here. For convenience we use the gauge norms.

Let $\lambda > 0$ be a number to be chosen suitably later, and let $y \mapsto n_h(\lambda y) = \mu_2\{\omega : |h(\omega)| > \lambda y\}$, be the measure function of h/λ. Then

$$\int_{\Omega_2} \Phi_1\left(\frac{h}{\lambda}\right) d\mu_2 = \int_0^{\infty} n_h(\lambda y) d\Phi_1(y), \tag{64}$$

by the standard image law (cf., e.g., Rao (1987), p.155 for finite measures but the same result holds in general). Let $p_1 > p_2$ so that $s > 0$, since $q_1 > q_2$. The case that $p_1 < p_2$ is similar. (See the remark after the proof.)

To use (61) and (62), let $f = f_1 + f_2$ where $f_1 = f\chi_{A_\alpha} + \alpha\chi_{A_\alpha^c}$, and $f_2 = f - f_1$ with $A_\alpha = \{\omega : |f(\omega)| \leq \alpha\}$. Then $f_i \in L^{p_i}(\mu_1)$, $i = 1, 2$. Indeed, for $u \geq u_0 > 0$, by the monotonicity of Φ_1 and (61)(i),

$$C_1 u_0^{-q_1} \Phi_1(u_0) \geq \int_u^{\infty} t^{-q_1-1} t\varphi_1(t) dt, \qquad \varphi_1 \text{ being the left derivative}$$

$$\text{of } \Phi_1,$$

$$\geq \int_u^{\infty} t^{-q_1-1} \Phi_1(t) dt, \qquad \text{since } t\varphi_1(t) \geq \Phi_1(t),$$

$$\geq \Phi_1(u) \cdot \frac{u^{-q_1}}{q_1},$$

so that $\Phi_1(u) \leq \tilde{C}_1 u^{q_1}$, $u \geq u_0$. Also from (61)(ii), for $u \geq u_0$,

$$K_0 = \int\limits_0^{u_0} t^{-q_2} d\Phi_1(t) \leq \int\limits_0^{u} t^{-q_2} d\Phi_1(t) \leq C_2 u^{-q_2} \Phi_1(u),$$

and hence $\Phi_1(u) \geq \tilde{C}_2 u^{q_2}$. Similarly for a fixed $u_1 > 0$, we get for some constants $K_i > 0$, $K_2 u^{q_2} \leq \Phi_1(u) \leq K_1 u^{q_1}$, $0 < u < u_1$, and it was shown that for $u \geq u_0$, $\tilde{C}_2 u^{q_1} \leq \Phi_1(u) \leq \tilde{C}_1 u^{q_2}$. Using these relations and (62), ($s > 0$), one deduces that for $u \geq u_0$, there are constants $\tilde{D}_1, \tilde{D}_2$ satisfying

$$\tilde{D}_2 u^{p_2} \leq \Phi_2(u) \leq \tilde{D}_2 u^{p_1} \qquad \text{(i.e., for large } u\text{)},$$

and

$$\tilde{D}_3 u^{p_1} \leq \Phi_2(u) \leq \tilde{D}_4 u^{p_2}, \qquad \text{for } 0 < u \leq u_1 \text{ (i.e., for small } u\text{)}.$$

These last pairs of inequalities imply that for $f \in L^{\Phi_2}(\mu_1)$, [so $\int\limits_{\Omega_1} \Phi_2(rf)d\mu_1$ $< \infty$ for some $r > 0$], $f_i \in L^{p_i}(\mu_1)$ as asserted.

If $h_i = Tf_i$, let $n_h(\cdot), n_i(\cdot), i = 1, 2$, be the corresponding measure functions of h and h_i. Now $|h| \leq |h_1| + |h_2|$, and hence $[|h| > y] \subset [|h_1| > \frac{y}{2}] \cup [|h_2| > \frac{y}{2}]$. Let $K_i > 0$ be constants that are involved in the weak types (p_i, q_i) of T. Since $f_i \in L^{p_i}(\mu_1)$, $i = 1, 2$, the following inequalities are obtained on using (60):

$$n_h(y) \leq n_1\left(\frac{y}{2}\right) + n_2\left(\frac{y}{2}\right)$$

$$\leq \left(\frac{2K_1}{y}\|f_1\|_{p_1}\right)^{q_1} + \left(\frac{2K_2}{y}\|f_2\|_{p_2}\right)^{q_2}, \qquad y > 0. \qquad (65)$$

Let $m_i(\cdot)$ be the measure function of f_i. Now using the same formula for $\int\limits_{\Omega_1} |f_i|^{p_i} d\mu_1$ given in (64), and substituting (65) in (64), we get the expression:

$$\int\limits_{\Omega_2} \Phi_1\left(\frac{h}{\lambda}\right) d\mu_2 \leq \left(\frac{2K_1}{\lambda}\right)^{q_1} \int\limits_0^\infty y^{-q_1}\left(\int\limits_0^\infty p_1 m_1(t) t^{p_1-1} dt\right)^{\frac{q_1}{p_1}} d\Phi_1(y)$$

$$+ \left(\frac{2K_2}{\lambda}\right)^{q_2} \int\limits_0^\infty y^{-q_2}\left(\int\limits_0^\infty p_2 m_2(t) t^{p_2-1} dt\right)^{\frac{q_2}{p_2}} d\Phi_1(y). \qquad (66)$$

To simplify the integrals on the right, observe that if $m_f(\cdot)$ is the measure function of f, then $m_1(t) = m_f(t)\chi_{[t \leq \alpha]}$ and $m_2(t) = m_f(t + \alpha)$, $t > 0$. Substituting $m_f(\cdot)$ for m_i in (66) we get the inside integrals as:

$$\int_0^\alpha m_f(t)t^{p_1-1}dt, \text{ and } \int_\alpha^\infty m_f(t)(t-\alpha)^{p_2-1}dt.$$

To eliminate the constant $\alpha > 0$, we choose it to satisfy the equation

$$A\alpha = y\Phi_2(A\alpha)^{\alpha_0} \qquad (A > 0 \text{ will be determined later.}). \qquad (67)$$

Since with our restrictions on the parameters, and the fact that for any $u_0 > 0$, there is an $x_0 > 0$ such that for $x \geq x_0$ we have

$$x \geq u_0\Phi_2(x)^{\alpha_0}, \qquad (68)$$

which is needed in the relations between $\Phi_2(u)$ and $|u|^{p_i}$, given in the third paragraph of this proof, we can conclude that y of (67) is an increasing function of α, for each fixed $\alpha_0 > 0$. With this information one can estimate the right side of (66).

Consider the first integral on the right side of (66). We have

$$P = \sup\left\{ \int_0^\infty y^{-q_1}\left(\int_0^\alpha m_f(t)t^{p_1-1}dt\right)g(y)d\Phi_2(y) : \|g\|_{k_1,\Phi_1,q_1} \leq 1\right\},$$

where $\|g\|_{k_1,\Phi_1,q_1} = \int_0^\infty |g(y)|^{k_1}y^{-q_1}d\Phi_1(y)$, with $\frac{1}{k_1} + \frac{p_1}{q_1} = 1$. Changing the order of integration and using the Hölder inequality, this gives

$$P \leq \sup\left\{ \int_0^\infty m_f(t)t^{p_1-1}dt \int_{t(\alpha)}^\infty y^{-q_1}g(y)d\Phi_1(y) : \|g\|_{k_1,\Phi_1,q_1} \leq 1\right\},$$

$$\text{with } t(\alpha) = A\alpha/\Phi_2(A\alpha)^{\alpha_0},$$

$$\leq \int_0^\infty m_f(t)t^{p_1-1}\left[\int_{t(\alpha)}^\infty y^{-q_1}d\Phi_1(y)\right]^{\frac{p_1}{q_1}}dt.$$

First use (61)(i) in the inner integral, and then (62) for the resulting expression. So, simplifying after substituting $t(\alpha)$ above, we get

$$P \leq C_1(p_1,q_1,s) \cdot A^{1-p_1}\int_0^\infty m_f(t) \cdot \frac{\Phi_2(At)}{At}dt$$

$$\leq C_1(p_1, q_1, s) \cdot A^{1-p_1} \int_0^\infty m_f\left(\frac{t}{A}\right)\varphi_2(t)\frac{dt}{A}, \text{ since}$$

$$t\varphi_2(t) \geq \Phi_2(t),$$

$$= C_1(p_1, q_1, s)A^{-p_1} \int_{\Omega_1} \Phi_2(Af)d\mu_1, \tag{69}$$

with the image law identity (64) again. Here $C_1(p, q, s) > 0$ is a constant.

Now we proceed with a similar estimation of the second integral in (66). Thus we let Q be the expression

$$Q = \sup\left\{\int_0^\infty y^{-q_2}\left(\int_\alpha^\infty m_f(t)(t-\alpha)^{p_2-1}dt\right)g(y)d\Phi_1(y) : \|g\|_{k_2,\Phi_1,q_2} \leq 1\right\},$$

where $\|g\|_{k_2,\Phi_1,q_2}$ is defined exactly as before (for P) with $\frac{1}{k_2} + \frac{p_2}{q_2} = 1$. Inverting the order of integration, setting $t(\alpha)$ as in the preceding computation and applying Hölder's inequality, we get, on simplifying exactly as above,

$$Q \leq C_2(p_2, q_2, s)A^{-p_2} \int_{\Omega_1} \Phi_2(Af)d\mu_1, \tag{70}$$

where $C_2(p_2, q_2, s) > 0$ is a constant.

Substituting (69) and (70) in (66) we get

$$\int_{\Omega_2} \Phi_1\left(\frac{h}{\lambda}\right)d\mu_2$$

$$\leq \left(\frac{2K_1}{\lambda}\right)^{p_1} p_1^{\frac{q_1}{p_1}}\tilde{C}_1(p_1, q_1, s) \cdot A^{-q_1}\left(\int_{\Omega_1}\Phi_2(Af)d\mu_1\right)^{\frac{q_1}{p_1}}$$

$$+ \left(\frac{2K_2}{\lambda}\right)^{p_2} p_2^{\frac{q_2}{p_2}}\tilde{C}_2(p_2, q_2, s)A^{-q_2}\left(\int_{\Omega_1}\Phi_2(Af)d\mu_1\right)^{\frac{q_2}{p_2}}. \tag{71}$$

We now take $A = N_{\Phi_2}(f)$ here, and then let the second free variable $\lambda > 0$ as $\lambda \geq KN_{\Phi_2}(f)$ where

$$K = \max\{2K_1p_1^{\frac{1}{p_1}}\tilde{C}_1(p_1, q_1, s)2^{-\frac{1}{q_1}}, 2K_2p_2^{\frac{1}{p_2}}\tilde{C}_2(p_2, q_2, s)2^{-\frac{1}{q_2}}\}.$$

Then $\int\limits_{\Omega_2} \Phi_1\left(\frac{h}{\lambda}\right)d\mu_2 \leq 1$, and since $\lambda \geq KN_{\Phi_2}(f)$, (71) reduces to (63) as desired. ∎

Remark. Although there are some omitted details in the above proof, we have included all the essential elements. The constant K above depends on p_i, q_i, K_i and s only. The adjustment for $s < 0$ proceeds the same way as above ($p_1 < p_2$ now) except that $f_1 \in L^{p_2}(\mu_1)$ and $f_2 \in L^{p_1}(\mu_1)$. Thus an interchange of these exponents will show the same computations and hence the end result (63) is unaltered.

An inspection of the proof shows that the convexity of Φ_2 is not crucial and only the Young function properties of Φ_1 are essential in estimating P and Q. We give the corresponding statement (also due to Riordan (1956)) leaving the modifications to the reader.

Theorem 14. *Let Φ_1 be a Young function satisfying (61), and Φ_2 be an increasing function on $\mathbb{R}^+$ such that $\Phi_2(0) = 0$, $\lim\limits_{x\to\infty} \Phi_2(x) = +\infty$, and moreover $\Phi_2(x) = [\Phi_1(x/\Phi_2(x)^{\alpha_0})]^s$ where α_0, s are as in Theorem 13, and where $T : \mathcal{M}(\Omega_1, \Sigma_1, \mu_1) \to \mathcal{M}(\Omega_2, \Sigma_2, \mu_2)$ is a quasilinear operator of types (p_1, q_1) and (p_2, q_2) simultaneously. Assume that the (p_i, q_i), $i = 1, 2$, satisfy the relations there. Then for each $f : \Omega_1 \to \mathbb{R}$ such that $\int\limits_{\Omega_1} \Phi_2(|f|)d\mu_1 \leq 1$, there is a constant $0 < K < \infty$, depending only on (p_i, q_i) as before, but not on f, such that $\int\limits_{\Omega_2} \Phi_1(Tf)d\mu_2 \leq K$.*

We present a few applications of the preceding interpolation theory in the next section to indicate how these results enter the analysis naturally. This will exhibit the power of these inequalities.

6.4. *Applications to fractional integration and packing problems*

Continuing the last part of the preceding section, we now give an application of the generalized Marcinkiewicz interpolation theorem to fractional integration, and later consider the other interpolation application to a sphere packing problem. Many other types of uses of both these results can be given, following Zygmund's classical work.

Let us introduce Φ_1 and Φ_2, which satisfy the hypothesis of Theorem 3.13 and then give the desired inequalities for them. Let $x_0 > x_1$ be a pair of numbers and ℓ_i be iterated logarithms, i.e., $\ell_0(x) = \log x$, and for $i \geq 0$, $\ell_{i+1}(x) = \log \ell_i(x)$ for $x > x_0$. If $\gamma_i \geq 0, \delta_i \geq 0$, define

$$
L(x) = \begin{cases} \prod_{i=1}^{n} \ell_i^{\gamma_i}(x), & \text{for } x > x_0, \\ \prod_{i=1}^{n} \ell_i^{\delta_i}(\frac{1}{x}), & \text{for } 0 < x < x_1. \end{cases}
$$

If x_0 is taken large enough and x_1 small enough, then $L(\cdot)$ is well defined, and for $1 < p < \infty$, using the notation of Theorem 3.13, let

$$
\Phi_2(x) = x^p L(x), \qquad x > 0, p \geq 1. \tag{1}
$$

Complete the definition of Φ_2 on $[0, \infty)$ by setting $\Phi(0) = 0$, and on $(0, \infty)$ take $x_1 < x_0$ so that Φ_2 is convex. [E.g. if x_1 is small and x_0 is large then $L(\cdot)$ is convex on $(0, x_1)$ and (x_0, ∞), and make Φ_2 convex by joining the two convex parts by a straight line.] Next let Φ_1 be defined, if $q \geq p$ and $L_1(x) = [L(x)]^{\frac{q}{p}}$, by the formula

$$
\Phi_1(x) = x^q L_1(x), \tag{2}
$$

for $0 < x < x_1' < x_0'$, and alsofor $x > x_0'$, i.e., for small enough x_1' and large enough x_0', and then make Φ_1 convex as before. If α_0 and s are as given in Theorem 3.13, we can and will take q to be

$$
q = \frac{1}{s} \frac{p}{1 - \alpha_0 p}, \qquad s > 0.
$$

To see that (Φ_1, Φ_2) satisfy the conditions of Theorem 3.13, note that

$$
\left[\Phi_1 \left(\frac{x}{\Phi_2(x)^{\alpha_0}} \right) \right]^s = x^{qs(1-\alpha_0 p)} L^{-\alpha_0 qs}(x) L_1^s \left(\frac{x^{1-\alpha_0 p}}{L(x)^{\alpha_0}} \right)
$$
$$
= \Phi_2(x) \left[L_1 \left(\frac{x^{1-\alpha_0 p}}{L^{\alpha_0}(x)} \right) \Big/ L^{\frac{q}{p}}(x) \right]^s. \tag{3}
$$

By considering the x near '0' and x large, one may verify that there exist $D_i > 0$, $i = 1, 2$ satisfying

$$
D_1 (L^{\frac{q}{p}}(x))^s < \left(L_1 \left(\frac{x^{1-\alpha_0 p}}{L(x)^{\alpha_0}} \right) \right)^s < D_2 (L^{\frac{q}{p}}(x))^s, \tag{4}
$$

which then shows that this implies (62) of the last section. In case $s < 0$, then L_1 is taken as $L_1(x) = L^{\frac{q}{p}}(\frac{1}{x})$, and the same inequalities

(4) hold. One can also verify that the Φ_1 thus defined satisfies (61) of Theorem 3.13. We now consider these Φ_i and apply the last theorem. The Φ_i satisfy the local Δ' condition. Some of their properties have been studied by R. O'Neil ((1968), Section 12).

Indeed, let $f : [0,1] \rightarrow \mathbb{R}$ be a periodic function of period 1, such that $\int_0^1 f(x)dx = 0$. Next let $f_1(x) = \int_0^x f(t)dt$ and for $i \geq 1$, $f_{i+1}(x) = \int_0^x f_i(t)dt$. Then the f_i have the same properties, and if

$$f(x) \sim \sum_{n=-\infty}^{\infty} c_n e^{2\pi i n x} \qquad (c_0 = 0),$$

define for any $\alpha > 0$,

$$f_\alpha(x) = \sum_{n=-\infty}^{\infty} c_n \frac{e^{2\pi i n x}}{(2\pi i n)^\alpha} \qquad (c_0 = 0). \tag{5}$$

This formula, which holds for integers α, defines a function f_α, for all $\alpha > 0$, called the αth *fractional integral* of f. The classical theory of trigonometric series shows that the series (5) converges a.e. Let $T_\alpha : f \mapsto f_\alpha$. Again the classical theory (cf. Zygmund (1959), Chapter 12, Section 8, p. 134) implies that T_α is of strong type (p,q) if $p > 1$, $0 < \alpha < \frac{1}{p}$ and $\frac{1}{p} - \frac{1}{q} = \alpha$. Using this and the generalized Marcinkiewicz theorem, we get the following:

Proposition 1. *Let Φ_1, Φ_2 be defined by (1) and (2) with $\frac{1}{p} - \frac{1}{q} = \alpha$. If $L(\cdot)$ and $L_1(\cdot)$ are taken with a single logarithm there [i.e., $L(x) = \log x$, $x > x_0$, and $L_1(x) = (\log x)^{\frac{q}{p}}$, for large x] and $|f|^p \log^+ |f| \in L^1(0,1)$, then $|f_\alpha|^q (\log^+ |f_\alpha|)^{\frac{q}{p}} \in L^1(0,1)$, and one has*

$$N_{\Phi_1}(T_\alpha f) = N_{\Phi_1}(f_\alpha) \leq K N_{\Phi_2}(f), \tag{6}$$

for a constant $K = K(p,q,s,\alpha) > 0$, being independent of f.

The following is an extension of a classical result on the boundedness of Fourier transforms from L^p into L^q. More precisely, we have:

Proposition 2. *Let $1 \leq p \leq 2$ and $q = p/(p-1)$, $f \in L^1(\mathbb{R}) \cap L^2(\mathbb{R})$ and $\hat{f}$ be its Fourier transform where $L^p(\mathbb{R})$ is the standard Lebesgue space on $\mathbb{R}$ with the Lebesgue measure. If $\Phi_2 : x \mapsto x^p L(x)$*

and $\Phi_1 : x \mapsto x^q L^{\frac{q}{p}}(\frac{1}{x})$, *where* $L(x) = \log x$, $x > 0$, *then* $T : f \mapsto \hat{f}$ *is linear and*

$$N_{\Phi_1}(\hat{f}) = N_{\Phi_1}(Tf) \leq K N_{\Phi_2}(f), \qquad f \in L^{\Phi_2}(\mathbb{R}), \tag{7}$$

for some constant $K = K(p,q) > 0$.

Proof: Since $\hat{f}(t) = \int_{\mathbb{R}} e^{itx} f(x) dx$, it follows that $\|\hat{f}\|_\infty \leq \|f\|_1$, and by Plancherel's theorem we also have $\|\hat{f}\|_2 = \|f\|_2$. Thus by the classical Riesz interpolation theorem (i.e., Theorem 3.5 if $\Phi_i(x) = |x|^{p_i}$, and $Q_i(x) = |x|^{q_i}$, $i = 1, 2$) with $p_1 = 1$, $p_2 = 2$, and $q_1 = \infty$ and $q_2 = 2$, so that for $0 < s < 1$, $p = p_s = \frac{2}{2-s}$ and $q = q_s = \frac{2}{s}$ or $1 < p < 2$ and $2 < q < \infty$, we get

$$\|Tf\|_q \leq \|f\|_p, \qquad f \in L^p(\mathbb{R}), \tag{8}$$

where $Tf = \hat{f}$. Thus T is of strong type (p,q). [This elegant method of proof using the interpolation theorem follows Weil (1938), p. 54, and the traditional argument for (8) is much more involved.]

Now let $\alpha_0 = 1$, $s = -1$ and $\Phi_1(x) = x^q L^{\frac{q}{p}}(\frac{1}{x})$, for small and large x and $\Phi_2(x) = x^p L(x)$ where $L(x) = \log x$ and where Φ_1 and Φ_2 are convex, with the same $1 \leq p \leq 2 \leq q \leq \infty$ as in (8). Then Theorem 3.13 applies to the Fourier transform operator T, and shows that for each $f \in L^{\Phi_2}(\mathbb{R})$, $\hat{f} \in L^{\Phi_1}(\mathbb{R})$ and that there is a constant K ($= K(p,q) > 0$) with which (7) holds. ∎

Our next application is to the packing problems of spheres in Orlicz spaces. We first recall the concept and a packing constant associated with a Banach space.

Definition 3. If $\mathcal{X}$ is a Banach space and B_1 is its (closed) unit ball of center 0, a collection of balls $\{B_r(x_i), i \in I\}$, $B_r(x_i) = \{y \in \mathcal{X} : \|y - x_i\| \leq r\}$, is a *packing* of B_1 by $B_r(x_i)$ when the following two conditions are met:

(a) $B_r(x_i) \subset B_1$, $i \in I$,

(b) the interiors of $B_r(x_i)$ are pairwise disjoint.

The *packing number* $\Lambda_{\mathcal{X}} = \sup\{r : B_r(x_i) \subset B_1$, for infinitely many i in $I\}$, with the convention that $\sup\{\emptyset\} = 0$.

It is seen that $\Lambda_{\mathcal{X}} = 0$ for finite dimensional $\mathcal{X}$ and $\Lambda_{\mathcal{X}} \leq \frac{1}{2}$ for the general case since by the triangle inequality one notes that balls with radius greater than $\frac{1}{2}$ cannot be packed in B_1. By a further analysis, C. A. Kottman (1970) has shown that if $\mathcal{X}$ is infinite dimensional, $\Lambda_{\mathcal{X}} \geq \frac{1}{3}$. In case $\mathcal{X}$ is an $L^p(0,1)$-space these bounds have been improved by special methods in the literature. Extending the latter to Orlicz spaces, and using Corollary 3.11 which is specially designed for this purpose, Cleaver (1976) has obtained the following bound for Λ_{Φ} ($= \Lambda_{L^{\Phi}(0,1)}$) using the Orlicz norm. We present details of the result with the gauge norm for convenience; and the computations with the Orlicz norm are similar and hence omitted. Some related work on these problems in the Orlicz space context was done by Ren (1985) which is used in the remark following the proof.

Theorem 4. *Let Φ be a Young function satisfying the Δ_2 condition, Φ_0 be given by $\Phi_0(x) = x^2$ and Φ_s, $0 \leq s \leq 1$ be the intermediate function between Φ and Φ_0. If $L^{\Phi}(0,1)$ is the Orlicz space on the Lebesgue unit interval, Λ_{Φ_s} is the packing number of $L^{\Phi_s}(0,1)$ with the gauge norm, then*

$$\max\left(\left[1 + \frac{\Phi_s^{-1}(2)}{\Phi_s^{-1}(1)}\right]^{-1}, [1 + 2\alpha_s]^{-1}\right) \leq \Lambda_{\Phi_s} \leq (1 + 2^{\frac{s}{2}})^{-1} \qquad (9)$$

where $\alpha_s = \liminf_{x \to \infty}[\Phi_s^{-1}(x)/\Phi_s^{-1}(2x)]$.

If, on the other hand, the Orlicz norm is used in $L^{\Phi_s}(0,1)$ and $\tilde{\Lambda}_{\Phi_s}$ is the corresponding number, then the right side bound of (9) is the same but the left side is to be $\max([1 + 2(\Psi_s^+)^{-1}(1)/(\Psi_s^+)^{-1}(2)]^{-1}, (\beta_s/(1 + \beta_s))$ where $\beta_s = \limsup_{x \to \infty}[(\Psi_s^+)^{-1}(x)/(\Psi_s^+)^{-1}(2x)]$ the Ψ_s^+ being the complementary Young function of Φ_s, and Ψ_s is intermediate to Ψ and Ψ_0.

Remark. As is evident from the definition, in general $\Lambda_{\Phi_s} \neq \tilde{\Lambda}_{\Phi_s}$ and thus the packing number depends on the particular, albeit equivalent, norm in a given Banach space, i.e., $L^{\Phi_s}(0,1)$ here. We shall establish (9), omitting the identical argument for the last part.

Proof: Let us establish the right side of (9) which uses the interpolation theorem referred to above, and the bound is sharper than $\frac{1}{2}$. Since the space is infinite dimensional, the work of Kottman shows that for some $0 < r \leq \frac{1}{2}$ there are infinitely many disjoint balls that can be packed in the unit ball B_1 of $L^{\Phi_s}(0,1)$. Let $f_1, \ldots, f_n$ be the centers of n balls of radius r that are packed in B_1. If $p_1, \ldots, p_n$ are positive, $\sum_{i=1}^{n} p_i = 1$, then by Corollary 3.11 we get

$$\sum_{i=1}^{n}\sum_{j=1}^{n} p_i p_j (\|f_i - f_j\|_{\Phi_s})^{\frac{2}{2-s}} \leq 2\alpha^{\frac{2(1-s)}{2-s}} \sum_{i=1}^{n} p_i (\|f_i\|_{\Phi_s})^{\frac{2}{2-s}}, \quad (10)$$

with $\alpha = \max_{1 \leq i \leq n} (1 - p_i) < 1$. The norm here could be either the gauge or Orlicz norm. Since the balls have disjoint interiors, $\|f_i - f_j\|_{\Phi_s} \geq 2r$ for $i \neq j$, and since we must have $r \leq \frac{1}{2}$, it follows that $\|f_i\|_{\Phi_s} \leq 1 - r$. Taking $p_i = \frac{1}{n}$ in (10), so that $\alpha = 1 - \frac{1}{n}$, one gets

$$\frac{1}{n^2} n(n-1)(2r)^{\frac{2}{2-s}} \leq 2(1 - \frac{1}{n})^{\frac{2(1-s)}{2-s}} \cdot \frac{1}{n} \cdot n(1-r)^{\frac{2}{2-s}}.$$

A simplification shows that

$$r \leq \left[1 + 2^{\frac{s}{2}}(1 - \frac{1}{n})^{\frac{s}{2}}\right]^{-1}.$$

Since n is arbitrary, it follows that $r \leq (1 + 2^{\frac{s}{2}})^{-1}$ and taking the supremum on r's here we have the right side of (9) for both norms. The measure space can be quite general for this part.

We now establish the left side inequality of (9) for the gauge norm. This computation does not depend on any interpolation results, but uses the diffuseness of the measure space. By definition of α_s, there exist $x_n < x_{n+1} \to \infty$ such that $\alpha_s = \lim_{x_n \to \infty} [\Phi_s^{-1}(x_n)/\Phi_s^{-1}(2x_n)]$. By considering a subsequence, we may assume that $\sum_{n=1}^{\infty} x_n^{-1} \leq 2$. Given $\varepsilon > 0$, choose n_0 such that $n \geq n_0$ implies

$$[\Phi_s^{-1}(x_n)/\Phi_s^{-1}(2x_n)] < \alpha_s + \varepsilon. \quad (11)$$

Choose a disjoint sequence of sets $\{E_n, n \geq 1\} \subset (0,1)$ such that $\mu(E_n) = (2x_n)^{-1}$, μ being the Lebesgue measure. Let $r = r_\varepsilon = (1 + 2(\alpha_s + \varepsilon))^{-1}$ and $f_n = 2r\Phi_s^{-1}(x_n) \cdot \chi_{E_n}$. Since $1 \leq 2\alpha_s \leq 2$ and $\Phi_s^{-1}(x) < \Phi_s^{-1}(2x) < 2\Phi^{-1}(x)$ (cf. Proposition II.1.1), we get

$$N_{\Phi_s}(f_n) = 2r\Phi_s^{-1}(x_n)/\Phi_s^{-1}(2x_n) < 2r(\alpha_s + \varepsilon) = 1 - r, \quad n \geq n_0.$$

But since the f_n have disjoint supports, one has, with $\Omega = [0,1]$,

$$\int_\Omega \Phi_s\left(\frac{f_n - f_m}{2r}\right) d\mu = \int_{E_n} \Phi_s\left(\frac{f_n}{2r}\right) d\mu + \int_{E_m} \Phi_s\left(\frac{f_m}{2r}\right) d\mu, \quad n \neq m,$$

$$= x_n\mu(E_n) + x_m\mu(E_m) = 1.$$

Since $\Phi_s \in \Delta_2$, this shows $N_{\Phi_s}(f_n - f_m) = 2r$ so that $\Lambda_{\Phi_s} \geq r$ being the supremum of all such r's. But $\varepsilon > 0$ is arbitrary, so $\Lambda_{\Phi_s} \geq (1 + 2\alpha_s)^{-1}$.

The second part of the maximum is similar. Let $E_{ni} = [(i-1)2^{-n}, i2^{-n})$ for $1 \leq i \leq 2^n, n \geq 1$. Define $r_0 = [1 + (\Phi_s^{-1}(2)/\Phi_s^{-1}(1))]^{-1}$ and let

$$f_n = (1 - r_0)\Phi_s^{-1}(1) \sum_{i=1}^{2^n} (-1)^{i+1} \chi_{E_{ni}}.$$

Then

$$\int_{\Omega} \Phi_s\left(\frac{f_n}{1 - r_0}\right) d\mu = \sum_{i=1}^{2^n} \mu(E_{ni}) = 1.$$

Hence Φ_s being in Δ_2, we get $N_{\Phi_s}(f_n) = 1 - r_0$, and

$$N_{\Phi_s}(f_n - f_m) = 2(1 - r_0)\Phi_s^{-1}(1)/\Phi_s^{-1}(2) = 2r_0.$$

As before this shows $\Lambda_{\Phi_s} \geq r_0 > \frac{1}{3}$, and implies that (9) holds as asserted. ∎

Using computations similar to those of the last part of the above theorem, Z. D. Ren has verified that for each nonreflexive $L^{\Phi}(0,1)$, with Φ as N-function, the packing value $\Lambda_{\Phi} = \frac{1}{2}$ under either of the norms. This was known for the $L^1(0,1)$ and $L^{\infty}(0,1)$ cases before. We omit the details.

Taking $\Phi(x) = |x|^p$, $p \geq 1$, and specializing the above theorem, one can obtain the following prior packing values for $L^p(0,1)$ due to Rankin (1955), and Wells and Williams (1975). This deduction is in Cleaver (1976).

Corollary 5. *If Λ_p denotes the packing value of $L^p(0,1)$, $p \geq 1$, then*

$$\Lambda_p = \begin{cases} (1 + 2^{1-\frac{1}{p}})^{-1}, & \text{for } 1 \leq p \leq 2, \\ (1 + 2^{\frac{1}{p}})^{-1}, & \text{for } 2 \leq p \leq \infty. \end{cases}$$

Further discussion of these applications will not be continued here.

6.5. Bases in M^{Φ}- and $L^{\Phi}(\mu)$-spaces

If $\mathcal{X}$ is a Banach space, then a sequence $\{x_n, n \geq 1\}$ is a *(Schauder) basis* for $\mathcal{X}$ provided each $x \in \mathcal{X}$ admits a *unique* representation $x = \sum_{i=1}^{\infty} a_i x_i$, the series converging in the norm of $\mathcal{X}$. Then the coefficient

$a_i = a_i(x)$ is termed the ith coordinate of x relative to the basis $\{x_n, n \geq 1\}$. If $\|x_n\| = 1$, $n \geq 1$, then the basis is said to be *normalized*, and it is seen that each Schauder basis can be normalized. Thus for the existence of such a basis, $\mathcal{X}$ must be separable. The standard Lebesgue spaces ℓ^p and the null convergent sequence space c_0 have such bases ($e_i(j) = \delta_{ij}$, $\delta_{ij} = 0$, $i \neq j$, $= 1$ for $i = j$, $i = 1, 2, \ldots$, $j \geq 1$ is such), $1 \leq p < \infty$. Moreover, if the expansion for each x is unconditionally convergent (i.e., $\sum\limits_{i=1}^{\infty} a_{\sigma(i)} x_{\sigma(i)}$ converges to x for all permutations $\sigma(\cdot)$ of the indices), then the $\{x_i, i \geq 1\}$ is an *unconditional basis*. In the early 1930s, S. Banach asked whether each separable Banach space has a Schauder basis. It was shown by S. Karlin (1948) that the answer is in the negative if the basis is to be unconditional. Finally in 1972, P. Enflo has shown that the answer is also negative for the original Banach problem by constructing a separable reflexive Banach space which does not admit a Schauder basis (cf. Enflo (1973)). This is a deep result, and hence it became important to know which concrete spaces appearing in applications have such bases.

Actually J. Schauder (1927) himself has shown that the space of continuous functions $C[0, 1]$ has a base and that $L^p[0, 1]$, $1 \leq p < \infty$ has Haar orthogonal system forming such a base. Moreover, in $L^2[0, 1]$ the trigonometric or Legendre polynomials and in $L^2(\mathbb{R}^+)$ the Laguerre and in $L^2(\mathbb{R})$ the Hermite functions form such bases. Thus a natural question here is to pose the same problem for separable Orlicz spaces. The existence of a basis allows one to extend a number of results from the finite dimensions to infinite dimensions. We thus include a brief account here. An extensive treatment of the subject in general Banach spaces is given by I. Singer (1970).

Since the Haar system plays an interesting part in these problems, let us recall them here for definiteness. These functions are defined on the unit interval $[0, 1]$: $H_0(\cdot) = 1$, $H_{2^k}(\cdot) = 2^{\frac{k}{2}}(\chi_{[0, 2^{-k-1})} - \chi_{[2^{-k-1}, 2^{-k})})$ for $k = 0, 1, \ldots$, and then if $1 \leq j < 2^k$, $k \geq 1$, let

$$H_{2^k+j}(x) = H_{2^k}(x - j2^{-k}) \chi_{[j2^{-k}, (j+1)2^{-k})}(x), \qquad x \in [0, 1]. \qquad (1)$$

Then it can be verified that the bounded functions $\{H_n, n \geq 1\}$ form a complete orthonormal system in $L^2[0, 1]$, and $H_k \in L^p[0, 1]$, $1 \leq p < \infty$. This means $\int\limits_0^1 f(x) H_k(x) dx = 0$, all k, implies $f = 0$, a.e. We then have the following result:

Theorem 1. *Let* Φ *be a continuous Young function with* $\Phi(x) > 0$ *for* $x > 0$. *Then the Haar system* $\{H_k, k \geq 0\} \subset M^\Phi \subset L^\Phi[0,1]$ *forms a Schauder basis for* M^Φ.

Sketch of Proof: A direct proof of this result, which is somewhat long, may be found in Krasnosel'skii and Rutickii's monograph (1961), p. 106. Here we sketch an alternative and "modern" argument based on the martingale convergence theorem. This concept was already discussed following Corollary V.2.6.

Thus let $f \in M^\Phi \subset L^1[0,1]$, and if $a_k(f) = \int\limits_0^1 f(x)H_k(x)dx$, consider

$$S_n(f) = \sum_{k=0}^{n} a_k(f)H_k, \qquad n \geq 1. \tag{2}$$

If $\mathcal{B}_n$ is the σ-algebra generated by $\{H_k, 1 \leq k \leq n\}$, so that it is also generated by the partitions of $[0,1]$ at the nth stage (each interval being equally divided into two parts of the preceding one) and since $\int\limits_0^1 H_k(x)dx = 0$, $k \geq 1$, it is verified that $\{S_n(f), \mathcal{B}_n, n \geq 1\}$ is a martingale, i.e., $\int_A S_{n+1}(f)(x)dx = \int_A S_n(f)(x)dx$, $A \in \mathcal{B}_n$. Also $S_n(f)$ is uniformly integrable. Hence from a classical martingale convergence theorem (since $\bigcup\limits_{n=1}^{\infty} \mathcal{B}_n$ generates the Borel σ-algebra of $[0,1)$), cf., e.g., Rao (1984), p. 177, $S_n(f) \to f$ a.e. and in L^1- and then, with a small computation, in $N_\Phi(\cdot)$-norm in M^Φ. This means $f = \lim\limits_{n \to \infty} S_n(f) = \sum\limits_{k=0}^{\infty} a_k(f)H_k$ in M^Φ, so that the $\{H_k, k \geq 0\}$ is a basis in M^Φ. ∎

Remark. Actually the preceding type of argument is included in standard texts on Probability Theory as a novel application of martingale convergence and several generalizations are available. Again we refer to Rao (1984), p. 213, for these results (given as exercises) with necessary references. The work shows that there are other systems that qualify to be a Schauder base in these spaces.

Since $L^\Phi[0,1]$ is separable iff it coincides with M^Φ or Φ is Δ_2-regular, by Theorem III.5.1, we have the following consequence.

Corollary 2. *Let* Φ *be a continuous Young function with* $\Phi(x) = 0$ *iff* $x = 0$. *Then* $L^\Phi[0,1]$ *has a Schauder base iff* $L^\Phi[0,1] = M^\Phi$, *or equivalently (since the measure is diffuse) iff* $\Phi \in \Delta_2$.

For the last part we invoke Proposition III.5.2. It is now natural to ask about the unconditional bases in M^Φ. This problem has been solved in Gaposkin (1968). His proof is quite long and depends on several other estimates. So we state it here for comparison, without proof, since we do not use it in any applications later.

Theorem 3. *Let Φ be an N-function. Then $L^\Phi[0,1]$ admits an unconditional basis iff it is reflexive, i.e., iff $\Phi \in \Delta_2 \cap \nabla_2$.*

We remark that the Fourier expansion of f by Haar functions, given by (2), defines $P_n : f \mapsto S_n(f)$ as a positive contractive projection, for each $n \geq 1$, on L^Φ into M^Φ. In fact $P_n = E^{\mathcal{B}_n}(\cdot)$ the conditional expectation operator. What the above theorem shows is that $P_n \to I$, the identity in the strong operator topology only when we restrict it to M^Φ, and not otherwise. Such an identification immediately leads to other properties in these Banach spaces. An intimately related question here is the approximation property. Does the identity operator admit the uniform approximation on compact sets of the space by linear operators of finite rank? Again a negative answer is obtained by Enflo's results quoted above. But for concrete Banach spaces, this problem has interest. It can be proved that the $L^\Phi(\mu)$-spaces have the approximation property and even all Banach function spaces and their adjoint spaces have this (cf., e.g., Rao (1975)). We shall not consider these, which constitute a digression, and move on to other geometrical aspects of Orlicz spaces in the next chapter.

Bibliographical Notes: Studies of subspaces of operators from $B(\mathcal{X}, \mathcal{Y})$ extend in many directions. The simplest of these is the space of integral operators when $\mathcal{X}$ and $\mathcal{Y}$ are Banach function spaces. They start with the convolution operators when $\mathcal{X}$ and $\mathcal{Y}$ are based on the Lebesgue spaces $(\mathbb{R}^n, \mathcal{B}, \mu)$, and lead to tensor products and those that can be represented as products of some other operators with a nicer structure. Sections 1 and 2 treat with these aspects. A major study on integral operators on Orlicz spaces is that of R. O'Neil's (1968). Several properties of integral operators were also given more recently by Gretsky and Uhl (1981), and Schep (1979). The book by Halmos and Sunder (1978) should also be consulted in this context.

We have given a small portion of factorization of the operators in Section 2. A general result on characterization of a bounded linear operator to be an integral operator through factorization properties is due to A. Grothendieck (1955) when $\mathcal{X} = L^1(\mu)$. We have not included this aspect, but Theorem 2.3 is a sample result.

The theory of interpolation of linear as well as sub- or quasi-linear operators is quite vast. In the context of L^p-spaces, the classical theory in many of its ramifications is given in a recent book by C. Bennett and R. Sharpley (1988). There are numerous contributors for this theory, starting with the basic Riesz-Thorin theorem. Many generalizations are found in the works of Aronszajn and Gagliardo (1965), Calderón (1964) as well as Lions (1961) and a new method in Gustavsson-Peetre (1977). We have presented the results that treat explicitly the Orlicz spaces, exact references being given in the text. Some new ideas appear in interpolation with change of measures and when the range space is a (tensor) product of Banach spaces. The conceptual change in the latter, first occurring in the L^p-case in Hayden and Wells (1971) and later in Cleaver (1976) for the Orlicz spaces, has not been recognized explicitly. It leads to interpolation with spaces of vector fields. This aspect has been briefly but carefully discussed in the text since the distinction is important for the subject. But we did not go deep into the subject here since one has to deal with Orlicz spaces of vector fields and this would have taken us far from the main theme. A treatment of the latter subject is available in Dinculeanu (1974) and further progress is given in Rao (1980,1981$_a$). However, we included another useful interpolation theorem for weak types which unfortunately was unpublished. It is due to W. J. Riordan (1956) and is an Orlicz space generalization of the classical Marcinkiewicz interpolation theorem. Thus we have discussed virtually all the theory of this subject that is available for the $L^\Phi(\mu)$ spaces.

A brief and immediate illustration of the interpolation theory to fractional integration and packing problems is presented in Section 4. This will already show not only the great potential for many applications, but also indicate a limitation of the interpolation for general Banach function spaces. The basic Riesz interpolation is very much tied to Lebesgue spaces in that the manipulation of the exponents $1 \leq p \leq \infty$, covers all L^p-spaces, but it includes only a subclass of Orlicz spaces. If Φ_1 and Φ_2 are two Young functions, then the "intermediate functions" Φ_s between Φ_1 and Φ_2 do *not* contain all Φ satisfying $\Phi_1 \prec \Phi \prec \Phi_2$ in whatever reasonable way ' $\prec$ ' is defined. A new computation to include the "middle" L^Φ-spaces is generally needed. This is illustrated by the left side bounds of Theorem 4.4. The same is true in other contexts such as the Hausdorff-Young theorem (cf. Rao (1968$_b$)). There are further extensions of interpolation theory for Lorentz' spaces, which are given in the book by Bennett and Sharpley noted above. We have not treated this, as it does not fit into our scheme. Finally, an indication of the basis problem for Orlicz spaces is included for completeness, but it is really a separate subject. However, we hope that the account has a fair view of the general analysis and shows further possibilities of these problems on Orlicz spaces.

VII
Geometry and Smoothness

Unlike the Lebesgue spaces, many geometric and smoothness, or norm differentiability properties can be distinguished for Orlicz spaces. These include the rotundity, reflexivity and uniform convexity/smoothness properties. This chapter is devoted to an account of these results that are specific to Orlicz spaces. We also include a treatment of the modulus of convexity for these spaces extending a part of the Clarkson type study for the classical Lebesgue space. A few applications to some of the results with the "double norm" spaces are given and these are utilized to derive another interpolation theorem for linear operators on Orlicz spaces with weights.

7.1. *Strict convexity and reflexivity*

In recent years, studies of geometric properties of general Banach spaces have been very visible. These include convexity or rotundity, reflexivity, uniform convexity and several subclasses of these properties. Banach spaces have been classified and are being studied for these properties. Orlicz spaces can be given as examples of many of these classes. But that will not be the main focus here. We shall be mainly concerned with those aspects that distinguish themselves for Orlicz spaces. Let us start therefore with the strict convexity properties and introduce the necessary concepts.

If $\{\mathcal{X}, \|\cdot\|\}$ is a Banach space, it is said to be *strictly convex* (or *rotund*) whenever $\|\alpha x + (1-\alpha)y\| < \alpha\|x\| + (1-\alpha)\|y\|$, for each pair of distinct elements x, y of $\mathcal{X}$, $\|x\| = \|y\| = 1$, and $0 < \alpha < 1$. Thus the concept is a property of the norm and in fact if there are a pair of equivalent norms on $\mathcal{X}$, rotundity can be present for one but not for the other norm. Since Orlicz spaces have two natural (gauge and Orlicz) norms, it is often necessary to consider conditions separately, as already noted following Theorem V.1.1. Let us start with a preliminary result which serves as a motivation for others that follow.

Proposition 1. *Let* (Φ, Ψ) *be a complementary pair of Young functions. If* $L^\Phi(\mu)$ *is an Orlicz space on* (Ω, Σ, μ) *and* Φ' *is the (left) derivative of* Φ, *let* $T : f \mapsto \Phi'(f)$ *be a mapping for* $f \in L^\Phi(\mu)$. *Then one has:*

(i) $T(M^\Phi) \subset L^\Psi(\mu)$ *with either norm, where* $M^\Phi = \{f \in L^\Phi(\mu) : \int_\Omega \Phi(\alpha f) d\mu < \infty$, *all* $\alpha > 0\}$;

(ii) *if* $U^\Phi = \{f \in L^\Phi(\mu), \|f\|_\Phi \leq 1\}$, $f \in U^\Phi \Rightarrow N_\Psi(Tf) \leq 1$; *but this inclusion can fail when in the definition of* U^Φ, *the gauge norm is used.*

Proof: (i) Since Φ' is nondecreasing and $\Phi'(0) = 0$ (by definition), supose first that Φ' is bounded. [$\Phi'(x) \equiv 0$ is inadmissible.] If $\alpha = \sup_x \Phi'(x)$, then $0 < \alpha < \infty$, and the function Ψ' satisfies $\Psi'(x) = +\infty$ for $x > \alpha$ so that $\Psi(x) = +\infty$ for $x > \alpha$. In this case $L^\Psi(\mu) \subset L^\infty(\mu)$ (cf. Proposition III.4.3). If $\Psi(x) = 0$ for $0 \leq x \leq \alpha$, then there is equality in the last inclusion. Since $\Phi'(f)$ is bounded for each $f \in L^\Phi(\mu)$, it follows that $\Phi'(f) \in L^\Psi(\mu)$. If, on the other hand, $\Psi(x) = 0$ for $0 \leq x < x_1 < \alpha$, $0 < \Psi'(x) < \infty$ for $x_1 \leq x < \alpha$, and $\Psi(x) = +\infty$ for $x > \alpha$, then it should be shown that $\Phi'(|f|)\chi_{[x_1 \leq |f| < \alpha]}$ is in $L^\Psi(\mu)$ for $f \in M^\Phi$. We assert that the result holds without further restriction.

If $f \in M^\Phi$, then $kf \in M^\Phi$ for all $k > 0$. Taking $k = 2$ and using the equality condition in Young's inequality we have

$$\Psi(\Phi'(|f|)) = |f|\Phi'(|f|) - \Phi(f) \leq \Phi(2|f|) - \Phi(f), \quad \text{a.e.} \quad (1)$$

Integrating this and using the fact that $\int_\Omega \Phi(kf)d\mu < \infty$, we get $\Phi'(f) \in L^\Psi(\mu)$. This implies (i) in a more general form.

(ii) We establish this part using an argument due to Krasnosel'skii and Rutickii (1961). So, if Φ' is bounded and $\Psi(x) = 0$ for $0 < x \leq \alpha$, $= +\infty$ for $x > \alpha$, then by (i) the result is true. Suppose then $0 < \Psi(x) < \infty$

for $x_1 \leq x < \alpha$, where $0 < \alpha < \infty$ as in (i), and $\Psi(x) = +\infty$ for $x > \alpha$. Thus $\Psi'(x) > 0$ on an interval, and the argument of the above authors can be applied. For this, suppose that the desired inclusion is false. Then there is an $0 \leq f \in U^{\Phi}$ such that $N_{\Psi}(Tf) > 1$. By the structure theorem there exist simple functions $0 \leq f_n \uparrow f$ pointwise so that by the Lebesgue monotone convergence criterion

$$\lim_{n \to \infty} \int_{\Omega} \Psi(\Phi'(f_n)) d\mu = \int_{\Omega} \Psi(\Phi'(f)) d\mu > 1. \tag{2}$$

Consequently there is an $n_0 > 1$ such that for all $n \geq n_0$, $\int_{\Omega} \Psi(\Phi'(f_n)) d\mu > 1$. But then, using the equality conditions in Young's inequality we get

$$\Psi(\Phi'(f_n)) < \Psi(\Phi'(f_n)) + \Phi(f_n) = f_n \Phi'(f_n), \tag{3}$$

holding on a set of positive μ-measure. Integrating (3) and using the first inequality in the proof of Proposition III.3.3, we get for $n \geq n_0$,

$$\int_{\Omega} \Psi(\Phi'(f_n)) d\mu < \int_{\Omega} f_n \Phi'(f_n) d\mu$$

$$\leq \|f_n\|_{\Phi} \int_{\Omega} \Psi(\Phi'(f_n)) d\mu \leq \int_{\Omega} \Psi(\Phi'(f_n)) d\mu, \tag{4}$$

since $\|f_n\|_{\Phi} \leq \|f\|_{\Phi} \leq 1$. The contradiction in (4) implies that $N_{\Psi}(Tf) \leq 1$ is true.

To see that $T(L^{\Phi}(\mu)) \subset L^{\Psi}(\mu)$ need not hold, suppose that μ is diffuse and $L^{\Phi}(\mu) \neq M^{\Phi}$. We now use the example constructed in the converse part of Proposition III.4.8. Thus $f_0 = \sum_{n=1}^{\infty} x_n \chi_{\Omega_n}$ is the element with the property that $N_{\Phi}(f_0) = 1$. If $Tf_0 \in L^{\Psi}(\mu)$, then for some $k_0 > 0$ we must have

$$\infty > \int_{\Omega} \Psi\left(\frac{\Phi'(f_0)}{k_0}\right) d\mu = \sum_{n=1}^{\infty} \Psi\left(\frac{\Phi'(x_n)}{k_0}\right) \mu(\Omega_n)$$

$$\geq \sum_{n=1}^{\infty} \left[x_n \frac{\Phi'(x_n)}{k_0} - \Phi(x_n)\right] \mu(\Omega_n), \text{ by Young's inequality,}$$

$$\geq \frac{1}{k_0} \sum_{n=1}^{\infty} \Phi(x_n)[2^n + 1 - k_0] \mu(\Omega_n)$$

$$= \frac{1}{k_0} \sum_{n=1}^{\infty} \left(1 + \frac{1 - k_0}{2^n}\right) = +\infty. \tag{5}$$

This contradiction shows that when $\Phi \notin \Delta_2$, $T(U^\Phi, N_\Phi(\cdot)) \not\subset L^\Psi(\mu)$ can obtain. Thus all parts of the result are established. ∎

Motivated by (ii) of the above proposition, we first consider the rotundity of $L^\Phi(\mu)$ for the normalized complementary Young pair (Φ, Ψ), so that $\Phi(1) + \Psi(1) = 1$. Although every Young complementary pair can be normalized, the following result will indicate various extensions. The proof is adapted from Rao (1965, p. 681). The statement there was not quite as explicit as desirable. The precise enunciation is as follows. As usual $\rho_\Phi(f) = \int_\Omega \Phi(f)d\mu$.

Theorem 2. *Let (Φ, Ψ) be a normalized complementary Young pair, and Φ be continuous. If $(L^\Phi(\mu), N_\Phi(\cdot))$ and $(L^\Psi(\mu), N_\Psi(\cdot))$ are the corresponding Orlicz spaces, suppose that $T(S^\Phi \cap M^\Phi) \subset S^\Psi$ where $S^\Phi = \{f : N_\Phi(f) = 1\}$ and $S^\Psi = \{g : N_\Psi(g) = 1\}$. Then for each $f \in S^\Phi$, $\rho_\Phi(f) = \Phi(1)$ and if Φ is also strictly convex, we have $(L^\Phi(\mu), N_\Phi(\cdot))$ to be a rotund Banach space.*

Proof: Let $0 \le f \in S^\Phi$. Then there exist simple $0 \le f_n \uparrow f$ pointwise, so that $N_\Phi(f_n) \uparrow N_\Phi(f) = 1$. Since $f_n \in M^\Phi$, $\Phi'(f_n) \in L^\Psi(\mu)$ by Lemma 1(i). Now by the fact that $f_n/N_\Phi(f_n) \in S^\Phi \cap M^\Phi$, we have $h_n = \Phi'(f_n/N_\Phi(f_n)) \in S^\Psi$ because of the supposition. Hence both f_n and h_n are simple and by the equality in Young's inequality,

$$\int_\Omega \frac{f_n}{N_\Phi(f_n)} h_n d\mu = \int_\Omega \Phi\left(\frac{f_n}{N_\Phi(f_n)}\right) d\mu + \int_\Omega \Psi(h_n) d\mu = \Phi(1) + \Psi(1) = 1.$$

From this one gets

$$N_\Phi(f_n) = \int_\Omega f_n h_n d\mu \le \int_\Omega f h_n d\mu \le 1. \tag{6}$$

Letting $n \to \infty$, and noting that $N_\Phi(f_n) \uparrow 1$, (6) implies

$$\lim_{n \to \infty} \int_\Omega f h_n d\mu = 1. \tag{7}$$

Thus for given $\varepsilon > 0$, there is an $n_0(\varepsilon)$ such that if $n \ge n_0(\varepsilon)$, (7) implies

$$1 - \varepsilon \le \int_\Omega f h_n d\mu \le \int_\Omega \Phi(f)d\mu + \int_\Omega \Psi(h_n)d\mu,$$

$$\le \int_\Omega \Phi(f)d\mu + \Psi(1), \text{ since } h_n \in S^\Psi.$$

This means

$$\Phi(1) - \varepsilon \le \int_\Omega \Phi(f)d\mu \le \Phi(1). \tag{8}$$

Since $\varepsilon > 0$ is arbitrary, (8) implies that $\rho_\Phi(f) = \Phi(1)$.

Finally if $f_i \in S^\Phi$ and $f_1 \ne f_2$, then by the strict convexity of Φ (used for the first time), we have $\Phi\left(\frac{f_1+f_2}{2}\right) < \frac{1}{2}[\Phi(f_1) + \Phi(f_2)]$, on a set of positive measure. Hence integration of this inequality gives

$$\int_\Omega \Phi\left(\frac{f_1 + f_2}{2}\right)d\mu < \frac{1}{2}\int_\Omega \Phi(f_1)d\mu + \frac{1}{2}\int_\Omega \Phi(f_2)d\mu = \Phi(1).$$

But then, by the first part $N_\Phi\left(\frac{f_1+f_2}{2}\right) < 1$ must hold. Consequently $(L^\Phi(\mu), N_\Phi(\cdot))$ is rotund. ∎

It should be observed that the above conditions on (Φ, Ψ) are always satisfied for the case that $\Phi(x) = \frac{|x|^p}{p}$, $p \ge 1$. On the other hand, one can verify by examples that as in the last part of Proposition 1(ii), $Tf \in S^\Psi$ for each $f \in S^\Phi$ need not hold for all normalized (Φ, Ψ). Thus the impact of this condition is that Φ' should not grow fast, and the $L^p(\mu)$-case suggests that $\Phi \in \Delta_2$ should be considered. Clearly $\rho_\Phi(f) = \Phi(1)$ is all that is needed for the rotundity part. For brevity and precision, we call (Φ, Ψ) *strongly normalized* if it is a normalized complementary Young pair and if $Tf \in S^\Psi$ for each $f \in S^\Phi \cap M^\Phi$, where gauge norms are used in both spaces $L^\Phi(\mu)$ and $L^\Psi(\mu)$. From (6), since $h_n \to h = \Phi'(f)$, pointwise, as $n \to \infty$, one has (by Fatou's lemma):

$$\int_\Omega |f|\Phi'(|f|)d\mu = 1, \qquad f \in S^\Phi, \tag{9}$$

which will be of interest in some computations.

Since the verification of (strong) normalization is not simple for applications, it is useful to analyze the argument leading to $\rho_\Phi(f/N_\Phi(f)) = \Phi(1)$. This has been done by Turett (1976) who obtained the following result for not necessarily normalized complementary pairs of Young functions.

Theorem 3. *Let $(L^\Phi(\mu), N_\Phi(\cdot))$ be an Orlicz space on a measure triple (Ω, Σ, μ). Then $L^\Phi(\mu)$ is rotund (for the gauge norm) provided (i) Φ is strictly convex and (ii) for each f with $N_\Phi(f) = 1, \rho_\Phi(f) = 1$.*

On the other hand, if μ is diffuse on a set of positive measure, then the rotundity of $L^\Phi(\mu)$ implies the conditions (i) and (ii) above.

Proof: The direct part is the same as the last paragraph of the preceding proof with $\Phi(1)$ replaced by 1 since (Φ, Ψ) is not normalized. Instead (ii) is assumed. So the same argument applies verbatim, and we need to consider only the converse direction under an additional hypothesis on μ.

As in the preceding chapters, we present an indirect proof. Thus suppose $L^{\Phi}(\mu)$ is rotund, but (i) does not hold. We assert: (a) $\Phi(I\!\!R) \subset I\!\!R^{+}$, and (b) there exists a pair of distinct functions f_1, f_2 such that $N_{\Phi}(f_i) = 1$, $i = 1, 2$, with $N_{\Phi}\left(\frac{f_1 + f_2}{2}\right) = 1$ to contradict the rotundity of $L^{\Phi}(\mu)$; similarly with (ii).

Thus if (a) is false, there exists a point $u_0 \in I\!\!R^{+}$ such that $\Phi(u) = +\infty$ for $u > u_0$ while $\Phi(u) < \infty$ for $0 \leq u < u_0$. By diffuseness and the finite subset property of μ, we can find disjoint measurable sets A_n of positive measure such that if $n_0 > \frac{1}{u_0}$ and $u_n = u_0 - (n + n_0)^{-1}$, then

$$0 < \mu(A_n) < 2^{-n}\left(1 \wedge \Phi(u_n)^{-1}\right) \ (\wedge = \min).$$

Let $f = \sum\limits_{n=1}^{\infty} u_n \chi_{A_n}$. Then

$$\int_{\Omega} \Phi(f)d\mu = \sum_{n=1}^{\infty} \Phi(u_n)\mu(A_n) \leq 1$$

so that $N_{\Phi}(f) \leq 1$. On the other hand if $0 < \alpha < 1$, and n_1 is chosen such that $(u_{n_1}/\alpha) > u_0 + 1$, then $\int_{\Omega} \Phi(f/\alpha)d\mu \geq \Phi(u_0 + 1)\mu(A_{n_1}) = +\infty$, implying that $N_{\Phi}(f) = 1$. Using a similar construction we can find a g (even having a disjoint support with f) such that $N_{\Phi}(g) = 1$, and then $N_{\Phi}(f + g) = 2$. This contradicts the rotundity of $L^{\Phi}(\mu)$ so that we must have $\Phi(I\!\!R) \subset I\!\!R^{+}$. We next show that (b) holds unless Φ is strictly convex, which then shows that (i) is necessary.

Now if Φ is not strictly convex, then there exist $0 < x_1 < x_2$ such that

$$\Phi\left(\frac{x_1 + x_2}{2}\right) = \frac{1}{2}\left[\Phi(x_1) + \Phi(x_2)\right], \quad \Phi(x_i) > 0, \tag{10}$$

so that it is linear between x_1 and x_2, its (left) derivative Φ' being a constant there. By the diffuseness of μ we can find disjoint measurable sets A_1, A_2, A_3 and a real number $\alpha > 1$ such that

$$0 < \mu(A_1) = \mu(A_2) = [\alpha(\Phi(x_1) + \Phi(x_2))]^{-1} \leq \mu(A_3) < \infty. \tag{11}$$

Since Φ is continuous, by the preceding paragraph, and $\Phi(x) \searrow 0$ as $x \searrow 0$, there is an $x_3 > 0$ such that

$$\Phi(x_3)\mu(A_3) = 1 - \tfrac{1}{\alpha} > 0. \tag{12}$$

If now we take $f_1 = x_1\chi_{A_1} + x_2\chi_{A_2} + x_3\chi_{A_3}$ and $f_2 = x_2\chi_{A_1} + x_1\chi_{A_2} + x_3\chi_{A_3}$, then $f_1 \neq f_2$ on a set of positive measure, and by (11)–(12) we get

$$\int_\Omega \Phi(f_1)d\mu = \sum_{i=1}^{3} \Phi(x_i)\mu(A_i) = \tfrac{1}{\alpha} + \Phi(x_3)\mu(A_3) = 1,$$

and similarly

$$\int_\Omega \Phi(f_2)d\mu = 1.$$

Since f_i are simple, these two facts imply that $N_\Phi(f_i) = 1$, $i = 1, 2$. But also by (10) and (12) we get $\int_\Omega \Phi\left(\frac{f_1+f_2}{2}\right)d\mu = 1$ so that $N_\Phi(f_1 + f_2) = 2$, and $L^\Phi(\mu)$ is not rotund. Thus Φ must be strictly convex.

Next, since $\Phi \in \Delta_2$ implies the equivalence of $N_\Phi(f) = 1$ and $\rho_\Phi(f) = 1$, it is enough to show that under diffuseness of μ and the rotundity of $L^\Phi(\mu)$, Φ must be Δ_2-regular to complete the proof. Suppose then $\Phi \notin \Delta_2$. To derive a contradiction, we may suppose that $\mu(\Omega) = 1$, and μ is diffuse on Ω. Then using the familiar argument, we find $0 < a_n \uparrow \infty$, $\Phi(a_1) \geq 1$, such that

$$\Phi\left((1 + \tfrac{1}{n})a_n\right) > 2^n\Phi(a_n). \tag{13}$$

Choose disjoint A_n ($\in \Sigma$) such that $\mu(A_n) = 2^{-n-1}(\Phi(a_n))^{-1}$. Thus, $\mu\left(\bigcup_{n=1}^{\infty} A_n\right) \leq \tfrac{1}{2}$. Let $A_0 \subset \left(\bigcup_n A_n\right)^c$ such that $0 < \mu(A_0) < \tfrac{1}{2\alpha}$ where $\alpha = 1 \vee \Phi(1)$. Let $g_1 = \sum_{n=1}^{\infty} a_n\chi_{A_n}$, $g_2 = g_1 + \chi_{A_0}$ so that $g_1 \neq g_2$ on A_0. It then follows that

$$\int_\Omega \Phi(g_1)d\mu \leq \int_\Omega \Phi(g_2)d\mu = \sum_{n=1}^{\infty} \Phi(a_n)\mu(A_n) + \Phi(1)\mu(A_0) < 1. \tag{14}$$

For each $0 < \lambda < 1$ let n_0 be chosen such that $\tfrac{1}{\lambda} \geq 1 + \tfrac{1}{n_0}$. Then using the choice of a_n in (13), we get

$$\int_\Omega \Phi\left(\frac{g_2}{\lambda}\right)d\mu \geq \int_\Omega \Phi\left(\frac{g_1}{\lambda}\right)d\mu \geq \sum_{n \geq n_0} \Phi\left((1 + \tfrac{1}{n})a_n\right)\mu(A_n) = +\infty.$$

It follows that $N_\Phi(g_i) = 1$, $i = 1, 2$. Also by (14), $\rho_\Phi\left(\frac{g_1+g_2}{2}\right) \leq \rho_\Phi(g_2) < 1$. If $0 < \lambda < 2$ and n_1 is chosen such that $\frac{2}{\lambda} \geq 1 + \frac{1}{n_1}$, then

$$\int_\Omega \Phi\left(\frac{g_1 + g_2}{\lambda}\right) d\mu \geq \int_\Omega \Phi\left(\frac{2g_1}{\lambda}\right) d\mu \geq \sum_{n \geq n_1} \Phi\left((1 + \tfrac{1}{n})a_n\right)\mu(A_n) = \infty,$$

by (13). Hence $N_\Phi(g_1 + g_2) = 2$ so that $L^\Phi(\mu)$ is not rotund. ∎

As a consequence of this proof, Proposition III.4.8, and Theorem IV.4.5, we have the following result.

Theorem 4. *Let Φ be a Young function and $L^\Phi(\mu)$ be the corresponding Orlicz space on (Ω, Σ, μ), μ being diffuse on a set of positive measure. Then the following are equivalent statements:*

(i) *Φ is Δ_2-regular;*

(ii) *for each $0 \neq f \in L^\Phi(\mu)$, one has $\rho_\Phi(f/N_\Phi(f)) = 1$;*

(iii) *for $f, g \in L^\Phi(\mu)$, $f \wedge g = 0$, a.e. (i.e., f, g have disjoint supports) and $N_\Phi(f) = 1 = N_\Phi(g) \Rightarrow N_\Phi(f + g) \neq 1$;*

(iv) *for all $f \in L^\Phi(\mu)$, $\sup\{\rho_\Psi(\varphi(|f|)) : N_\Phi(f) \leq 1\} < \infty$, where Ψ is complementary to Φ and φ is the (left) derivative of Φ;*

(v) *$L^\Phi(\mu)$ does not contain a closed subspace isomorphic to c_0;*

(vi) *$L^\Phi(\mu)$ does not contain a closed subspace isomorphic to ℓ^∞;*

(vii) *$0 \leq f_n \leq f_{n+1}$ in $L^\Phi(\mu)$, and $\sup_n N_\Phi(f_n) < \infty$ implies $\{f_n, n \geq 1\}$ is Cauchy in norm (and hence converges to an element in $L^\Phi(\mu)$).*

Proof: The relations (iv) ⇔ (i) ⇔ (v) ⇔ (vi) are consequences of Proposition III.4.8 and Theorem IV.4.5. That (i) ⇒ (ii) ⇒ (iii) are immediate without restrictions on measures, and the same holds for (i) ⇒ (vii). It is thus enough to show that (iii) ⇒ (i) and (vii) ⇒ (i) when μ is diffuse on a set of positive measure. This is done as follows.

(iii) ⇒ (i). If this is false, then by diffuseness we can find a pair of disjoint measurable sets A and B such that $0 < \mu(A) \leq 1$, $0 < \mu(B) \leq 1$, and μ is diffuse on both A and B. Next construct elementary functions f_1 on A and f_2 on B exactly as in (11)–(13) (see also the construction given in the next part) such that $N_\Phi(f_i) = 1$, $i = 1, 2$, and $\int_\Omega \Phi(f_i)d\mu < \frac{1}{2}$. It then follows that

$$\int_\Omega \Phi(f_1 + f_2)d\mu = \int_\Omega \Phi(f_1)d\mu + \int_\Omega \Phi(f_2)d\mu < 1,$$

since f_1 and f_2 have disjoint supports. Hence $N_\Phi(f_1 + f_2) \leq 1$. But one also has $N_\Phi(f_1 + f_2) = N_\Phi(|f_1| + |f_2|) \geq N_\Phi(|f_1|) = 1$ so that $N_\Phi(f_1 + f_2) = 1$ and (iii) does not hold. it should be remarked that in the construction, one needs to consider two cases that $\Phi(u) < \infty$ for $u < u_0$ and $\Phi(u) = +\infty$ for $u > u_0$; and $\Phi(\mathbb{R}) \subset \mathbb{R}^+$. However, the method is the same as that used in the above proof, and we leave it to the reader.

(vii) $\Rightarrow$ (i). If this is false, we again construct an elementary function f in $L^\Phi(\mu)$, supported by a set of positive finite diffuse μ-measure, such that $N_\Phi(f) = 1$ and $\int_\Omega \Phi(f)d\mu < \frac{1}{2}$ just as before. Indeed, let A be such a set, $0 < \mu(A) \leq \frac{1}{2}$, $0 < a_n \nearrow \infty$ satisfying (13), and disjoint sets $A_n \subset A$ such that $\mu(A_n) \leq 2^{-n}\mu(A)\,(1 \wedge (\Phi(A_n))^{-1})$. Let $f = \sum_{n=1}^\infty a_n \chi_{A_n}$. To see that this satisfies the conditions, let $g_n = \sum_{k=1}^n a_k \chi_{A_k}$. Thus $g_n \leq g_{n+1}$ and, $\sup_n N_\Phi(g_n) \leq N_\Phi(f) = 1$. But for $n > m$, we get

$$\int_\Omega \Phi(2(g_n - g_m))d\mu = \sum_{k=m+1}^n \Phi(2a_k)\mu(A_k) \geq \sum_{k=m+1}^n \mu(A) = (n - m)\mu(A).$$

So $\{g_n, n \geq 1\}$ is not Cauchy in norm and (vii) does not hold. ∎

An interesting deduction of the above two results is given by

Corollary 5. *An Orlicz space $(L^\Phi(\mu), N_\Phi(\cdot))$ on a measure space (Ω, Σ, μ), with μ diffuse on a set of positive measure, is rotund iff $\Phi(\mathbb{R}) \subset \mathbb{R}^+$, Φ is strictly convex and is Δ_2-regular. Consequently if $(L^\Phi(\mu), N_\Phi(\cdot))$ and $(L^\Psi(\mu), N_\Psi(\cdot))$ are both rotund where Ψ is complementary to Φ, then $(\Psi(\mathbb{R}) \subset \mathbb{R}^+$ and) $L^\Phi(\mu)$ is reflexive and $(L^\Phi(\mu), N_\Phi(\cdot))^* = (L^\Psi(\mu), \|\cdot\|_\Psi)$ and $(L^\Psi(\mu), N_\Psi(\cdot))^* = (L^\Phi(\mu), \|\cdot\|_\Phi)$.*

Using Theorem V.1.1, we can also conclude that a reflexive $L^\Phi(\mu)$ is isomorphic to a strictly convex reflexive $L^{\Phi_1}(\mu)$. However we cannot always assert that $\Phi_1 = \Phi$ itself as the following simple example shows.

Example: Let $\Phi(x) = |x|\chi_{[|x|\leq 1]} + x^2\chi_{[|x|>1]}$. Then $\Phi \in \Delta_2 \cap \nabla_2$. So $L^\Phi(\mu)$ is reflexive if (Ω, Σ, μ) is the Lebesgue unit interval, but it is not rotund since Φ is not strictly convex.

We can now complete the discussion following the proof of Proposition V.1.2. Consider the Lebesgue space $L^\infty([0,1])$ with a new norm

$\|f\| = \|f\|_\infty + \|f\|_2$. With this new norm, the space is rotund as is easily seen. But the new space is not an Orlicz space since otherwise, we will have $L^\Phi([0,1]) = L^\infty([0,1])$ so that Φ and $\Phi_0 : x \mapsto \begin{cases} 0, & |x| \le 1 \\ +\infty, & |x| > 1 \end{cases}$ are equivalent Young functions. Thus $\Phi(\mathbb{R}) \subset \overline{\mathbb{R}}^+$ and not $\mathbb{R}^+$. So by Theorem 3 (or Corollary 5) above, $(L^\Phi([0,1]), N_\Phi(\cdot)) = (L^\infty([0,1]), \|\cdot\|)$ and it cannot be rotund. This means $\|\cdot\|$ is not an equivalent gauge or Orlicz norm. But by Theorem V.1.1 and the above Corollary, if Φ is Δ_2-regular, then $L^\Phi(\mu)$ is isomorphic to a rotund Orlicz space $(L^{\Phi_1}(\mu), N_{\Phi_1}(\cdot))$ and $\Phi \sim \Phi_1$, μ being a measure.

Remark. We now discuss a point related to Theorems 2 and 3 regarding the (strongly) normalized and general Young complementary pairs. The crucial step in the proof of Theorem 2 with the normalized strictly convex Young pair is the equality in (6) and (7). These two relations also imply that the gauge and Orlicz norms give the same value for the $L^\Phi(\mu)$ spaces, which thus strengthens the relation in Remark 2 following Proposition III.3.4. It will be useful to find alternative conditions on (strongly) normalized complementary pairs (Φ, Ψ) in order that one can easily check for the equality of these two norms. From a practical point of view, Theorems 3 and 4 give simple conditions that can be used for any complementary pair. The points raised at the end of Section IV.2 should be understood in this light.

We now present the corresponding rotundity result with the Orlicz norms for general Young functions. This case was primarily investigated by H. W. Milnes (1957) and our presentation essentially follows his work.

Unlike the case of gauge norm, difficulties start with the very definition of the Orlicz norm since it involves both Φ and Ψ, in addition to the kind or type of the underlying measure space. For instance, if (Ω, Σ, μ) is a finite measure space, then the functional $I : \alpha \mapsto \rho_\Psi(\Phi'(\alpha|f|))$ is continuous for a bounded measurable f, Φ' being the left derivative of Φ. For general f in $L^\Phi(\mu)$, Proposition III.2.2 shows the type of difficulties that might arise.

Thus if $V_0 = \sup\{v : \Psi'(v) < \infty\}$, Ψ' being the left derivative of Ψ, then Ψ is called *essentially C^1-continuous* if both Ψ and Ψ' are continuous on the half-open interval $[0, V_0)$, and if $\lim_{v \nearrow V_0} \Psi'(v) = +\infty$, and $\lim_{v \nearrow V_0} \Psi(v) = +\infty$. If Ψ is essentially C^1-continuous, then we say that Φ is essentially strictly convex, i.e., strictly convex on $[0, V_0)$. The conditions for rotundity of $L^\Phi(\mu)$ are given by Milnes on the complementary function Ψ in contrast to the hypotheses of all the preceding results. This however should not be surprising since the calculation of $\|f\|_\Phi$ uses $\rho_\Psi(\cdot)$ at every stage (cf. e.g., the proof of Theorem III.3.13). Thus a characterization of rotundity which is a companion to Theorem 3 can be given for Orlicz normed spaces as follows:

Theorem 6. *Let* (Φ, Ψ) *be a complementary pair of Young functions and* (Ω, Σ, μ) *be a σ-finite diffuse measure space. Then* $(L^\Phi(\mu), \|\cdot\|_\Phi)$ *is rotund iff the complementary function Ψ is essentially C^1-continuous.*

Proof-in-outline: We sketch the *ideas* of the argument. The actual detail takes over ten pages, involving the behavior of $\rho_\Phi(\Phi'(\alpha|f|))$ for bounded and then for general f.

1. The simpler part is to show that $L^\Phi(\mu)$ is not rotund if Ψ is discontinuous, and a considerably more involved computation is needed to reach the same conclusion if Ψ' is discontinuous. The point of the proof is to show if $\|f_i\|_\Phi = 1$, $f_1 \neq f_2$, and $\|f_1 + f_2\|_\Phi = 2$, one still has $\|f_1 - f_2\|_\Phi = 2$ for a pair of functions. The construction uses the diffuseness of μ.

2. In the opposite direction, assuming that Φ is essentially strictly convex and μ is diffuse, one first shows that for each $f \in L^\Phi(\mu)$ with $\|f\|_\Phi > 0$, there exists a $0 < k < \infty$ such that $\|f\|_\Phi = \frac{1}{k}(1 + \rho_\Phi(kf))$. This depends on constructing a $g \geq 0$ such that $kf = \Psi'(g)$ and $\rho_\Psi(g) = 1$, satisfying $\|f\|_\Phi = \int_\Omega fg d\mu$. (This is a generalization of Theorem III.3.13 which was proved for N-functions that are not necessarily strictly convex and hence is weaker than what is needed above.) With these properties, Milnes shows that for any pair of elements $f_i, \|f_i\|_\Phi = 1$ and $\|f_1 + f_2\|_\Phi = 2$, one must have $\|f_1 - f_2\|_\Phi = 0$. This will imply the desired rotundity of $L^\Phi(\mu)$. We refer to the original paper for details. However, we include a direct argument for an important special case below, using Theorem III.3.13.

An interesting consequence of Theorem 6 and Corollary 5 is that if (Ω, Σ, μ) is a diffuse measure space, (Φ, Ψ) is a complementary pair of Young functions, then $(L^\Phi(\mu), N_\Phi(\cdot))$, $(L^\Psi(\mu), \|\cdot\|_\Psi)$ and $(L^\Phi(\mu), \|\cdot\|_\Phi)$ can all be rotund without being reflexive. In fact if Φ is an N-function which is strictly convex (in this case essential C^1-continuity of Ψ clearly coincides with strict convexity of Φ since $V_0 = +\infty$), then Ψ is C^1-continuous on all of $\mathbb{R}$. Thus Theorem 6 implies at once the following:

Corollary 7. *If (Ω, Σ, μ) is a diffuse σ-finite measure space and Φ an N-function, then $(L^{\Phi}(\mu), \|\cdot\|_{\Phi})$ is rotund iff Φ is strictly convex.*

Proof: This special case can be proved directly also, and we provide some details. In the sufficiency proof the measure μ can be arbitrary. Thus let Φ be a strictly convex N-function. Then by Theorem III.3.13, for any $f_i \in L^{\Phi}(\mu)$, there exist $0 < k_i \, (= k_{f_i}) < \infty$ such that (k_i need not be unique, but will be unique for our Φ)

$$\|f_i\|_{\Phi} = \frac{1}{k_i}\Big(1 + \rho_{\Phi}(k_i f_i)\Big), \qquad i = 1, 2.$$

Suppose that f_1 and f_2 differ on a set of positive measure and let $\|f_i\|_{\Phi} = 1$. Then for the set $\Omega_0 = \{\omega : k_1 f_1(\omega) = k_2 f_2(\omega)\}$, $\mu(\Omega_0^c) > 0$. For otherwise $k_1 f_1 = k_2 f_2$ a.e. and hence, using $\|f_i\|_{\Phi} = 1$, we get $k_1 = k_2$ so that $f_1 = f_2$ a.e., contrary to the choice of f_1, f_2. Hence on Ω_0^c, we have for any $0 < \alpha < 1$,

$$\Phi\big(\alpha k_1 f_1 + (1 - \alpha)k_2 f_2\big)(\omega) < \alpha \Phi(k_1 f_1)(\omega) + (1 - \alpha)\Phi(k_2 f_2)(\omega),$$

so that on integration one gets

$$\rho_{\Phi}(\alpha k_1 f_1 + (1 - \alpha)k_2 f_2) < \alpha \rho_{\Phi}(k_1 f_1) + (1 - \alpha)\rho_{\Phi}(k_2 f_2).$$

Taking $\alpha = k_2(k_1 + k_2)^{-1}$ in the above we have with the choice of the k_i,

$$
\begin{aligned}
2 = \|f_1\|_{\Phi} + \|f_2\|_{\Phi} &= \frac{k_1 + k_2}{k_1 k_2}\Big(1 + \alpha \rho_{\Phi}(k_1 f_1) + (1 - \alpha)\rho_{\Phi}(k_2 f_2)\Big) \\
&> \frac{k_1 + k_2}{k_1 k_2}\Big(1 + \rho_{\Phi}\big(\alpha k_1 f_1 + (1 - \alpha)k_2 f_2\big)\Big) \\
&= \frac{1}{k}\Big(1 + \rho_{\Phi}\big(k(f_1 + f_2)\big)\Big), \qquad \text{where } \frac{1}{k} = \frac{1}{k_1} + \frac{1}{k_2} > 0.
\end{aligned}
$$

Hence taking the infimum over all $k > 0$ and using Theorem III.3.13 again, one has

$$2 > \|f_1 + f_2\|_{\Phi}.$$

Thus $(L^{\Phi}(\mu), \|\cdot\|_{\Phi})$ is rotund.

For the necessity, we need the fact that μ is diffuse. The idea is to suppose that the N-function Φ is not strictly convex and construct a pair (f_1, f_2) of bounded functions (thus they are elements of M^Φ) such that $\|f_1+f_2\|_\Phi = \|f_1\|_\Phi + \|f_2\|_\Phi$ where $f_1 \neq f_2$ on a set of positive measure. This computation is similar to that used in Theorem 3. One needs to consider the case that Φ' is constant on an interval when Φ is not strictly convex. In fact the construction is essentially a modification of Milnes's Theorem 6. The latter may be simplified slightly using Theorem III.3.13 again, and we omit it here. ∎

Considering the example $\Phi(x) = (1 + |x|)\log(1 + |x|) - |x|$ and its complementary N-function Ψ given by $\Psi(x) = e^{|x|} - |x| - 1$, where Φ is Δ_2-regular (but Ψ is not), if (Ω, Σ, μ) is a finite diffuse measure space, then $(L^\Phi(\mu), N_\Phi(\cdot))$ is rotund but not reflexive, and $(L^\Phi(\mu), N_\Phi(\cdot))^* = (L^\Psi(\mu), \|\cdot\|_\Psi)$. By the preceding work $L^\Phi(\mu)$ is rotund in both norms while $(L^\Psi(\mu), \|\cdot\|_\Psi)$ is rotund in the Orlicz but not the gauge norm. By the classical Banach space theory, a space $\mathcal{X}$ is smooth if its dual $\mathcal{X}^*$ is rotund. [Here smoothness of $\mathcal{X}$ means that its unit ball has exactly one supporting hyperplane at each point of its boundary. This property will be examined later.] Moreover, for reflexive Banach spaces $\mathcal{X}$, the classical theory further implies (cf., M. M. Day (1962), p. 112) that $\mathcal{X}$ is rotund (smooth) iff $\mathcal{X}^*$ is smooth (rotund). Thus the Orlicz spaces exhibit some distinctions. This may be stated for reference as follows:

Proposition 8. *There exist rotund and smooth nonreflexive Orlicz spaces whose duals are rotund Orlicz spaces.*

This property again distinguishes Orlicz spaces from the Lebesgue classes, in addition to showing clear differences in the geometrical properties of the gauge and Orlicz norms. The smoothness questions are better studied through the differentiability properties of the norm functionals. We now turn to this property.

7.2. Norm differentiability, uniform convexity and smoothness

A Banach space $\mathcal{X}$ is termed *uniformly convex* if for each $0 < \varepsilon \leq 2$, there is a $0 < \delta(\varepsilon) < 1$ such that for any pair x_1, x_2, $\|x_i\| = 1$, $i = 1, 2$, satisfying $\|x_1 - x_2\| \geq \varepsilon$, one has $\|x_1 + x_2\| \leq 2(1 - \delta(\varepsilon))$, and $\delta(\varepsilon)$ is called the *modulus of convexity* of $\mathcal{X}$. The dual concepts are obtained if smoothness is substituted for convexity (= rotundity). They are better stated (more analytically) in terms of equivalent norm (= strong), or weak differentiability notions. We state the latter forms precisely and apply them to Orlicz spaces.

Definition 1. If $\mathcal{X}$ is a Banach space and $S = \{x \in \mathcal{X} : \|x\| = 1\}$ is its unit sphere, then the norm of $\mathcal{X}$ is said to be *weakly* or *Gâteaux differentiable* at a point x of S if the following limit exists:

$$G(x, h) = \lim_{t \to 0} \frac{\|x+th\| - \|x\|}{t}, \qquad h \in S. \tag{1}$$

If the limit in (1) exists uniformly in h for each $x \in S$, then the norm is *strongly* or *Fréchet differentiable* at each x. If moreover the limit exists in (1) uniformly in the pair $(x, h) \in S \times S$, then the norm is *uniformly Fréchet differentiable* in $\mathcal{X}$.

Thus $\mathcal{X}$ is said to be *smooth* if its norm is weakly differentiable at each point of its unit sphere. The space is *uniformly smooth* if for each $\varepsilon' > 0$, there is a $\delta'(\varepsilon') > 0$ such that $\|x-y\| \leq \delta' \Rightarrow \|x+y\| \geq \|x\| + \|y\| - \varepsilon'\|x - y\|$, for x, y in S. Some classical results of V. L. Šmulian and of M. M. Day imply that for a Banach space $\mathcal{X}$ the following are equivalent: (i) uniform rotundity of $\mathcal{X}$, (ii) uniform Fréchet differentiability of $\mathcal{X}^*$, and (iii) uniform smoothness of $\mathcal{X}^*$. [For references and proofs, see Day (1962), p. 114.] Here we intend to find conditions on the Young functions in order that $\mathcal{X} = L^\Phi(\mu)$ has [or has not] these properties and apply them later to some concrete problems. It will be seen that for the Lebesgue spaces $L^p(\mu)$, $1 < p < \infty$, both uniform rotundity and smoothness are present while this is not true for Orlicz spaces. If one only desires the isomorphism results, then several of these latter distinctions coalesce because a reflexive Orlicz space can be isomorphic to a rotund Orlicz space which has a Fréchet differentiable norm. With these remarks as motivation, we now present an analysis of these properties for general Orlicz spaces and give explicit conditions for use in applications.

As in the preceding section, we start with a strongly normalized complementary pair (Φ, Ψ) of Young functions, and later present the corresponding result for the general (i.e., nonnormalized) case. This gives a better insight for the later considerations, as well as a comparison with the Lebesgue spaces. The following result is again from Rao (1965) with a more complete and explicit statement. See also Krasnosel'skii and Rutickii (1961).

Theorem 2. *Let* (Φ, Ψ) *be a strongly normalized complementary pair of continuous Young functions with* $\Phi'(x) > 0$ *for* $x > 0$ *and be continuous. Then the norm functional* $N_\Phi(\cdot)$ *is weakly differentiable at each point of* $M^\Phi - \{0\}$ *in* $(L^\Phi(\mu), N_\Phi(\cdot))$ *on a general measure space* (Ω, Σ, μ). *For each* f_0, f *in* $S^\Phi \cap M^\Phi$ *one has*

$$G(f_0; f) = \frac{d}{dt} N_\Phi(f_0 + tf)\Big|_{t=0} = \int_\Omega f \Phi'(f_0) d\mu, \qquad (2)$$

where $\Phi'(f) = \Phi'(|f|)\mathrm{sgn}(f)$, *by definition. Further, the norm is strongly [or Fréchet] differentiable at* f_0 *if* Ψ *is* Δ_2-*regular and* (2) *will then be the strong derivative.*

Remark. One of the useful things of these differentiability results is that there are no restrictions on the underlying measure space, and this aspect is especially desirable in applications.

Proof: For each $0 \neq f \in M^\Phi$, the mapping $h_f : k \mapsto \int_\Omega \Phi(\frac{f}{k})d\mu$ is continuous, $h_f(\mathbb{R}^+) \subset \mathbb{R}^+$ and $\lim_{k \to 0^+} h_f(k) = \infty$, $\lim_{k \to \infty} h_f = 0$. Also for each f_0, f in M^Φ of unit N_Φ-norm, if $k = k_t = N_\Phi(f_0 + tf)$ then for each t,

$$\int_\Omega \Phi\left(\frac{f_0 + tf}{k}\right) d\mu = \Phi(1). \qquad (3)$$

If $F(k, t)$ denotes the integrand in (3), then the differential dF is given by

$$dF = -\Phi'\left(\frac{f_0 + tf}{k}\right)\frac{f_0 + tf}{k^2}dk + \Phi'\left(\frac{f_0 + tf}{k}\right)\frac{f}{k}dt, \quad \text{a.e.,} \qquad (4)$$

by the classical Lebesgue theory. For $|t| < 1, k_t \geq \alpha > 0$, and hence the right side of (4) is dominated by $2\Phi'(\frac{|f_0|+|f|}{\alpha})(\frac{|f_0|+|f|}{\alpha^2})$ which is integrable by Lemma 1.1(i) and Hölder's inequality. Hence (3) and (4) yield

$$0 = -\frac{1}{k^2}\left[\int_\Omega \Phi'(\frac{f_0 + tf}{k})(f_0 + tf)d\mu\right]dk + \frac{1}{k}\left[\int_\Omega \Phi'(\frac{f_0 + tf}{k})f d\mu\right]dt. \qquad (5)$$

Consequently $\frac{dk}{dt}$ exists and one has ($k_0 = 1$) on using the strong normalization:

$$\frac{dk_t}{dt}\Big|_{t=0} = \int_\Omega f \Phi'(f_0) d\mu, \qquad (6)$$

since by the equality conditions in the Young inequality, $\int\limits_\Omega f_0 \Phi'(f_0) d\mu$
$= 1$, see equation (1.1.9). This establishes (2). It remains to consider the strong differentiability.

First observe that $G(f_0, \cdot)$ of (2) is continuous in f_0 under the present hypothesis. Indeed, if f_0, f, h are elements of M^Φ with $N_\Phi(f_0) = 1$, then

$$\left| G\left(\frac{f_0 + h}{N_\Phi(f_0 + h)}, f\right) - G(f_0, f) \right| \leq \int\limits_\Omega \left| f\left(\Phi'\left(\frac{f_0 + h}{N_\Phi(f_0 + h)}\right) - \Phi'(f_0)\right) \right| d\mu$$

$$\leq N_\Phi(f) N_\Psi\left(\Phi'\left(\frac{f_0 + h}{N_\Phi(f_0 + h)}\right) - \Phi'(f_0)\right)$$

$$\longrightarrow 0 \qquad \text{as } N_\Phi(h) \to 0, \tag{7}$$

by the absolute continuity of norm $N_\Phi(\cdot)$ on M^Φ. Since this is uniform in f in (7), the assertion follows. Now using a classical result in integral calculus, the strong differentiability of norm is deduced as follows. For f_0, h in M^Φ, by (2)

$$\frac{d}{dt} N_\Phi(f_0 + th) = G(f_0 + th, h), \qquad -1 < t \leq 1, \tag{8}$$

where we use the fact that for any $g \in M^\Phi - \{0\}$, $G(g, \cdot) = G\left(\frac{g}{N_\Phi(g)}, \cdot\right)$
From this it follows that

$$N_\Phi(f_0 + h) - 1 = \int\limits_0^1 G(f_0 + th, h) dt, \quad \text{since } N_\Phi(f_0) = 1.$$

Consequently, with the linearity of $G(f_0, \cdot)$ in (2), one gets

$$\left| \frac{N_\Phi(f_0 + h) - 1 - G(f_0, h)}{N_\Phi(h)} \right| \leq \int\limits_0^1 |G(f_0 + th, h) - G(f_0, h)| \frac{dt}{N_\Phi(h)}$$

$$\leq \int\limits_0^1 \|G(f_0 + th, \cdot) - G(f_0, \cdot)\| dt$$

$$\longrightarrow 0, \qquad \text{as } N_\Phi(h) \to 0, \text{ by (7)}, \tag{9}$$

This shows that $N_\Phi(\cdot)$ is strongly differentiable at f_0 with derivative $G(f_0, \cdot)$. ∎

If (Φ, Ψ) is not a strongly normalized pair, then the useful Hölder equality formula and the result of Theorem 1.2 cannot be directly employed. However, the desired modifications are readily obtained as in Theorem 1.3. We now give this version which is easier for applications. [As before, $\Phi'(f) = \Phi'(|f|)\mathrm{sgn}(f)$ for real f and $\Phi'(x) > 0$ for $x > 0$.] In connection with this result see also Wang and Chen (1987).

Theorem 3. *Let (Φ, Ψ) be a complementary pair of continuous Young functions and $(L^\Phi(\mu), N_\Phi(\cdot))$ be the Orlicz space on a measure space (Ω, Σ, μ). If Φ' is continuous and $\Phi'(x) > 0$ for $x > 0$, then $N_\Phi(\cdot)$ is weakly differentiable at every $0 \neq f_0 \in M^\Phi \subset L^\Phi(\mu)$ and the weak derivative $G(f_0, \cdot)$ of $N_\Phi(\cdot)$ at f_0 is given by*

$$G(f_0, f) = \frac{\int\limits_\Omega f \Phi'\left(\frac{f_0}{N_\Phi(f_0)}\right) d\mu}{\int\limits_\Omega \frac{f_0}{N_\Phi(f_0)} \Phi'\left(\frac{f_0}{N_\Phi(f_0)}\right) d\mu}, \quad f \in M^\Phi. \tag{10}$$

On the other hand, if μ is diffuse on a set of positive measure, then $N_\Phi(\cdot)$ is differentiable at all $0 \neq f_0 \in M^\Phi$ implies the continuity of Φ' as above.

Proof: In the preceding proof the normalization hypothesis of (Φ, Ψ) was not used until (5), and hence (10) obtains without further work, except that $\Phi(1)$ is replaced by 1 in (3). We now consider the converse under the additional restriction.

Thus let $N_\Phi(\cdot)$ be weakly differentiable but $\Phi'(\cdot)$ not be continuous, and μ be diffuse on a set of positive measure. Then there is an $E \in \Sigma$, $0 < \mu(E) < \infty$, on which μ is diffuse. For simplicity, we assume $E = \Omega$ and derive a contradiction. We consider an equivalent condition of smoothness. Namely, a normed vector space is smooth if at each point x_0 with $\|x_0\| = 1$, there is a unique supporting hyperplane, i.e., there is a *unique* linear functional x_0^* such that $x_0^*(x_0) = 1$, $\|x_0^*\| = 1$, and the set $\{x : x_0^*(x) = 1\}$ lies on one side of the ball $\{x : \|x\| \leq 1\}$ with equality only at x_0. (See Day (1962), p. 111.) In (11) $x_0^* = G(f_0, \cdot)$. Since Φ' is not continuous and Ψ is continuous, it is not strictly convex so that $\Psi'(\cdot)$ is "flat" on an interval, i.e., there exist $0 < v_1 < v_2 < \infty$ with $\Psi'(v) = a > 0$ for $v_1 \leq v \leq v_2$. Using the diffuseness of μ, we can find $\Omega_1 \in \Sigma$ such that $0 < \Phi(a)\mu(\Omega_1) < 1$ and $\Omega_2 \in \Sigma$, $\Omega_1 \cap \Omega_2 = \emptyset$, $b > 0$ such that

$$\Phi(\Psi'(b))\mu(\Omega_2) = 1 - \Phi(a)\mu(\Omega_1). \tag{11}$$

Let A, B in $\Sigma(\Omega_1)$ be chosen such that $A \cap B = \emptyset$, and $\mu(A) = \mu(B) = \frac{1}{2}\mu(\Omega_1)$. This is possible by the diffuseness of μ. Finally define

$$f = a\chi_{\Omega_1} + \Psi'(b)\chi_{\Omega_2}, \quad h_1 = v_1\chi_A + v_2\chi_B + b\chi_{\Omega_2},$$
$$h_2 = v_2\chi_A + v_1\chi_B + b\chi_{\Omega_2}.$$

From construction (11) we get $\rho_\Phi(f) = 1$ so that $N_\Phi(f) = 1$ since $f \in M^\Phi$. Also we have

$$\rho_\Phi(\Psi'(h_i)) = \Phi(a)[\mu(A)+\mu(B)]+\Phi(\Psi'(b))\mu(\Omega_2) = 1, \quad i = 1, 2. \quad (12)$$

By Proposition III.3.14, one has

$$\|h_i\|_\Psi = \int_\Omega \Psi'(h_i)h_i d\mu, \quad i = 1, 2. \quad (13)$$

Let $g_i = h_i/\|h_i\|_\Psi$. Since clearly $\Psi'(h_i) = f$, $i = 1, 2$, (13) implies

$$1 = N_\Phi(f) = \int_\Omega \Psi'(h_i)g_i d\mu = \int_\Omega fg_i d\mu = x_i^*(f), \quad i = 1, 2. \quad (14)$$

Since $g_1 \neq g_2$ on Ω_1, we conclude that x_1^* and x_2^* are two distinct continuous functionals on M^Φ which yield different support planes at f of M^Φ so that the space is not smooth, i.e., the norm is not weakly differentiable at f. This contradiction shows that Φ' must be continuous as asserted. ∎

As a consequence of this result and Theorem 1.4, we have the following.

Corollary 4. *Let Φ be an N-function and (Ω, Σ, μ) be a measure space with μ being diffuse on a set of positive measure. Then the Orlicz space $(L^\Phi(\mu), N_\Phi(\cdot))$ is smooth iff Φ' is continuous and Φ is Δ_2-regular.*

An analogous result for the Orlicz norm can be given, generalizing a theorem of Krasnosel'skii and Rutickii [(1961), p. 192]:

Theorem 5. *Let (Φ, Ψ) be a complementary pair of continuous Young functions and (Ω, Σ, μ) be a measure space. If $M^\Phi \subset (L^\Phi(\mu), \|\cdot\|_\Phi)$ is the subspace determined by the simple functions and Φ' is continuous $\Phi'(x) > 0$ for $x > 0$, then the norm $\|\cdot\|_\Phi$ is weakly* *differentiable at each point of M^Φ except the origin, and the derivative $G(f_0, \cdot)$ at $f_0 \in M^\Phi$, $\|f_0\|_\Phi = 1$, is given by*

$$G(f_0, f) = \int_\Omega f\Phi'(kf_0)d\mu \qquad (14)$$

where $\Phi'(f)$ is $\Phi'(|f|)\text{sgn}(f)$ for real f, and $k > 1$ is a solution of the equation

$$k - 1 = \rho_\Phi(kf_0). \qquad (15)$$

On the other hand, if μ is diffuse on a set of positive measure, then the weak differentiability of $\|\cdot\|_\Phi$ at each f_0, $\|f_0\| = 1$ implies that Φ' is continuous as given.

Proof: As in the proof of the last part of Theorem 3, we employ the alternative form of smoothness definition with the unique support hyperplane concept and establish (14) and (15) which is simpler than using the differential calculus. Now by the classical Hahn-Banach theorem, for each f with $\|f\|_\Phi = 1$, there is an $x_f^* \in (M^\Phi)^*$ such that $1 = x_f^*(f)$, $\|x_f^*\| = 1$, and since by the continuity hypotheses on the pair (Φ, Ψ), f vanishes outside a set of σ-finite measure, we may assume for convenience that μ is σ-finite. Hence $(M^\Phi)^* = L^\Psi(\mu)$. Since a support hyperplane is uniquely determined by such a linear functional, it suffices to show, for each $f \in M^\Phi$, the corresponding x_f^* is unique. Using the fact that $x_f^*(f) = \int_\Omega fg_f d\mu$, $\|x_f^*\| = N_\Psi(g_f)$, by Theorem IV.1.6 $[(M_\Phi, \|\cdot\|_\Phi)^* \cong (L^\Psi(\mu), N_\Psi(\cdot))$ since the norms "alternate"], it is sufficient to show the uniqueness of x_f^* in this representation. We do this now.

Observe that for each $0 \neq f \in M^\Phi$, equation (15) has a solution by proposition III.1.2.

Suppose if possible there are two such functionals and hence g_f^i, $i = 1, 2$ in $L^\Psi(\mu)$, so that $N_\Psi(g_f^i) = $ norm of the functional $= 1$. Since $\|f\|_\Phi = g_f^{*i}(f)$, $i = 1, 2$ one has

$$1 = \|f\|_\Phi = \int_\Omega fg_f^1 d\mu = \int_\Omega fg_f^2 d\mu, \qquad (16)$$

by the representation theorem recalled above. This implies that there is equality in Hölder's inequality so that by Proposition III.4.9(a), there is a $0 < k < \infty$ satisfying (15) and

$$k|f||g_i| = \Phi(kf) + \Psi(g_i), \quad \text{a.e.,} \quad i = 1, 2 \ (g_i = g_f^i). \tag{17}$$

But this implies (due to the equality conditions in the Young inequality) that $g_1 = g_2 = \Phi'(kf) \ (= \Phi'(k|f|)\mathrm{sgn}(f))$ a.e. So the support functional is unique and hence M^Φ is smooth at each $f \neq 0$ in it.

Finally the support functional $x_{f_0}^*$ at f_0 is asserted to be $G(f_0, \cdot)$. When this is shown, the representation of $x_{f_0}^*$ and (17) imply $G(f_0, f)$ is given by (14). The assertion follows from an explicit computation. If f_0, f are two elements of M^Φ of unit norm and $t \in \mathbb{R}$, then by the Hahn-Banach theorem there are $x_{f_0}^*$ and $x_{f_0+tf}^*$ in $(M^\Phi)^*$ of unit norms such that $x_{f_0}^*(f_0) = 1$ and $x_{f_0+tf}^*(f_0 + tf) = \|f_0 + tf\|$. Hence for $t > 0$,

$$
\begin{aligned}
x_{f_0}^*(f) &= (x_{f_0}^*(tf) + x_{f_0}^*(f_0) - 1)/t = (x_{f_0}^*(f_0 + tf) - 1)/t \\
&\leq \frac{1}{t}\Big(\|f_0 + tf\| - 1\Big) \leq \frac{1}{t}\Big(x_{f_0+tf}^*(f_0 + tf) - |x_{f_0+tf}^*(f_0)|\Big) \\
&\qquad \text{since } |x_{f_0+tf}^*(f_0)| \leq \|f_0\| = 1, \\
&= x_{f_0+tf}^*(f) + \frac{1}{t}\Big(x_{f_0+tf}^*(f_0) - |x_{f_0+tf}^*(f_0)|\Big) \leq x_{f_0+tf}^*(f), \tag{18}
\end{aligned}
$$

since the omitted term on the right is nonpositive. Thus

$$x_{f_0}^*(f) \leq \frac{1}{t}\Big(\|f_0 + tf\| - 1\Big) \leq x_{f_0+tf}^*(f), \quad t > 0, \tag{19}$$

and taking $t < 0$ so that the inequalities reverse in (18), one has

$$x_{f_0}^*(f) \geq \frac{1}{t}\Big(\|f_0 + tf\| - 1\Big) \geq x_{f_0+tf}^*(f), \quad t < 0. \tag{20}$$

But by the first part, the space was shown to be smooth and hence, by an earlier observation, the norm $\|\cdot\|_\Phi$ is weakly differentiable. Thus from (19) and (20) one gets $x_{f_0}^*(f) = \lim_{t \to 0} \frac{1}{t}(\|f_0 + tf\|_\Phi - 1) = G(f_0, f)$, and (14) holds.

For the last part, if Φ' is not continuous, using the diffuseness of μ, we can again construct a pair of distinct functions g_i and an $f \in M^\Phi$ satisfying (16), such that $1 = \|f\|_\Phi = \int_\Omega f g_i d\mu = x_i^*(f)$, $i = 1, 2$. This will imply that M^Φ is not smooth, contradicting the norm differentiability hypothesis. The details can be left to the reader. ∎

Even though in the presence of diffuseness of μ a solution k of (15) need not exist for $f \in L^\Phi(\mu) - M^\Phi$ so that (14) does not hold,

one still has a precise analog of Corollary 4 for the Orlicz norm also.
(An independent proof of this result is also in Milnes (1957).) We omit
further discussion here. It should be noted, however, that rotundity and
smoothness results are very much the properties of particular norms
and hence one cannot attribute any of these statements from $N_\Phi(\cdot)$
to $\| \cdot \|_\Phi$ and vice versa. In other words, isomorphism results are too
insensitive for the deductions of these geometrical properties from one
space to its image. In fact we shall see later that all separable Orlicz
spaces are isomorphic to smooth Orlicz spaces which may also be chosen
to be rotund at the same time.

We next consider uniform rotundity and smoothness properties
of these spaces to complete our discussion on their geometry. The
following two results play a key role in this work. The next one in
its essentials is due to Milnes (1957) and its final form to Akimovič
(1972) while a useful deduction from it, given afterwards, belongs to
the former author.

Proposition 6. *Let* Φ *be a continuous Young function. Then
the following three statements are mutually, holding locally [globally],
equivalent:*

(i) *for each* $0 < a < 1$, *there is a* $0 < \delta_a < 1$ *such that*

$$\Phi\left(\frac{x + ax}{2}\right) \leq \frac{1}{2}(1-\delta_a)(\Phi(x)+\Phi(ax)), \quad x \geq x_0, [x_0{=}0]; \quad (21)$$

(ii) *for each* $\varepsilon > 0$, *there is a* $k_\varepsilon > 1$ *and* $x_1(\varepsilon) > 0$, $[x_1(\varepsilon) = 0]$
such that

$$\Phi'((1 + \varepsilon)x) \geq k_\varepsilon \Phi'(x), \qquad x \geq x_1(\varepsilon) \geq 0; \quad (22)$$

(iii) *for each* $\varepsilon > 0$, *there is an* $\eta_\varepsilon > 0$ *such that* $|x - y| \geq$
$\varepsilon \max(x, y) \geq \varepsilon x_3$, *for some (each)* $x_3 > 0$ *implies*

$$\Phi\left(\frac{x + y}{2}\right) \leq \frac{1}{2}(1 - \eta_\varepsilon)(\Phi(x) + \Phi(y)). \quad (23)$$

Moreover, if Φ *satisfies any of these equivalent conditions, then it
is* ∇_2-*regular (and not necessarily conversely).*

Proof: (i) $\Rightarrow$ (ii). For any $\varepsilon > 0$, let $a = \frac{1}{1+\varepsilon}$. Then by (21) there is a
$\delta_a > 0$ such that for some $x_0 \geq 0$,

$$2\Phi\left(\frac{x + ax}{2}\right) \leq (1 - \delta_a)(\Phi(x) + \Phi(ax))$$

which simplifies to

$$\Phi\left(\frac{(2+\varepsilon)x}{2(1+\varepsilon)}\right) - \Phi\left(\frac{x}{1+\varepsilon}\right)$$

$$\leq \Phi(x) - \Phi\left(\frac{(2+\varepsilon)x}{2(1+\varepsilon)}\right) - \delta_a\left[\Phi(x) + \Phi\left(\frac{x}{1+\varepsilon}\right)\right]. \quad (24)$$

But we also have from the convexity of Φ (Φ' being its left derivative),

$$\frac{\varepsilon x}{2(1+\varepsilon)}\Phi'\left(\frac{x}{1+\varepsilon}\right) \leq \int_{x(1+\varepsilon)^{-1}}^{(2+\varepsilon)x(2+2\varepsilon)^{-1}} \Phi'(t)dt \leq \Phi\left(\frac{(2+\varepsilon)x}{2(1+\varepsilon)}\right) - \Phi\left(\frac{x}{1+\varepsilon}\right)$$

$$+ \delta_a\left(\Phi(x) + \Phi\left(\frac{x}{x+\varepsilon}\right)\right), \quad (25)$$

since a nonnegative term is added,

$$\leq \Phi(x) - \Phi\left(\frac{(2+\varepsilon)x}{2(1+\varepsilon)}\right), \qquad \text{by (24)},$$

$$= \int_{(2+\varepsilon)x(2+2\varepsilon)^{-1}}^{x} \Phi(t)dt \leq \frac{\varepsilon x}{2(1+\varepsilon)}\Phi'(x). \quad (26)$$

Dividing (25) by the first term and using the last inequality of (26), we get after an elementary simplification,

$$\frac{\Phi'(x)}{\Phi'\left(\frac{x}{1+\varepsilon}\right)} \geq 1 + \delta_a\frac{\Phi(x) + \Phi\left(\frac{x}{1+\varepsilon}\right)}{\Phi\left(\frac{2+\varepsilon}{2(1+\varepsilon)}x\right) - \Phi\left(\frac{x}{1+\varepsilon}\right)} \geq 1 + \delta_a, \quad x \geq x_0,$$

because the coefficient of δ_a, is at least 1. Setting $k = 1 + \delta_a$, $x_1 = \frac{x_0}{1+\varepsilon}$, we have (22).

(ii) $\Rightarrow$ (iii). Let $0 < \varepsilon < 1$. Then there is a $k_\varepsilon > 1$, $x_2(\varepsilon) \geq 0$, such that by (22),

$$\Phi'(x) \geq k_\varepsilon \Phi'((1 - \tfrac{\varepsilon}{2})x), \quad x \geq x_2(\varepsilon). \quad (27)$$

If $x_3 = (1 + \varepsilon)x_2(\varepsilon)$, and x, y satisfy $|x - y| \geq \varepsilon \max(|x|, |y|) \geq \varepsilon x_3$, then we may assume $x - y \geq \varepsilon x > \varepsilon y > \varepsilon x_3$, or $(1 - \varepsilon)x > y > x_3$. Let $x_3 \leq s \leq x$, and

$$F(s) = \Phi(x) + \Phi(s) - 2\Phi\left(\frac{x+s}{2}\right) \qquad (\geq 0 \text{ by convexity}).$$

Since $F'(s) \leq 0$, so that $F(\cdot)$ is decreasing on (y, x), we have with $y < (1 - \varepsilon)x$,

$$F(y) \geq \Phi(x) + \Phi((1 - \varepsilon)x) - 2\Phi\left(\left(1 - \frac{\varepsilon}{2}\right)x\right)$$

$$= \int_{(1-\frac{1}{2}\varepsilon)x}^{x} \left(\Phi'(t) - \Phi'(t - \tfrac{\varepsilon}{2}x)\right)dt, \quad \text{by a change of variable,}$$

$$\geq \int_{(1-\frac{1}{2}\varepsilon)x}^{x} \left[\Phi'(t) - \Phi'((1 - \tfrac{\varepsilon}{2})t)\right]dt, \quad \text{since } \Phi'(\cdot) \text{ is nondecreasing,}$$

$$\geq \left(1 - \frac{1}{k_\varepsilon}\right)(\Phi(x) - \Phi((1 - \tfrac{\varepsilon}{2})x)), \quad \text{using (27)},$$

$$> \left(1 - \frac{1}{k_\varepsilon}\right)\left[\Phi(x) - (1 - \tfrac{\varepsilon}{2})\Phi(x)\right], \quad \text{by the convexity of } \Phi,$$

$$\geq \frac{\varepsilon}{4}\left(1 - \frac{1}{k_\varepsilon}\right)\left[\Phi(x) + \Phi(y)\right], \quad \text{since } \Phi(y) \leq \Phi(x). \tag{28}$$

Hence (28) implies (23), if we set $\eta_\varepsilon = \frac{\varepsilon}{4}(1 - \frac{1}{k_\varepsilon})$.

(iii) $\Rightarrow$ (i). For any $0 < \varepsilon < 1$, take $a = 1 - \varepsilon$, $\delta_a = \eta_\varepsilon$ and $x_0 = x_3(\varepsilon) \geq 0$. If $x - y > \varepsilon x$ and $x \geq x_0$, let $a = \frac{y}{x}$. Then (23) implies

$$\Phi\left(\frac{x + ax}{2}\right) = \Phi\left(\frac{x + y}{2}\right) \leq \frac{1}{2}(1 - \delta_a)(\Phi(x) + \Phi(y)), \quad x \geq x_0 \geq 0,$$

which is (21).

Regarding the last part, consider (ii). Taking $\varepsilon = \frac{1}{2}$ there we have, since Φ' is increasing, $\Phi'(2t) \geq \Phi'((1 + \frac{1}{2})t) \geq k_\varepsilon \Phi'(t)$, $t \geq x_2 \geq 0$ for some $k_\varepsilon > 1$. Hence

$$\Phi'(2t) \geq k_\varepsilon^2 \Phi'(t/2) \geq \cdots \geq k_\varepsilon^n \Phi'(t/2^{n-1}), \quad \text{for } t \geq 2^{n-1}x_2 \geq 0,$$

and so $\Phi'(2^n t) \geq k_\varepsilon^n \Phi'(t)$, $t \geq x_2 \geq 0$. But $\Phi(2x) \geq x\Phi'(x)$, for $x > 0$. Hence

$$\Phi(2^{n+1}x) \geq 2^n x\Phi'(2^n x) \geq 2^n k_\varepsilon^n x\Phi'(x)$$

$$\geq 2^n k_\varepsilon^n \Phi(x), \quad x \geq x_2 \geq 0. \tag{29}$$

This shows that $\Phi(x) \leq \frac{1}{2^n k_\varepsilon^n}\Phi(2^{n+1}x)$ which implies that Φ is ∇_2-regular if we set $\ell = 2^{n+1}$ after taking n such that $k_\varepsilon^n \geq 2^2$ so that $2^n k_\varepsilon^n \geq 2\ell$. It is possible to construct simple examples to show that

$\Phi \in \nabla_2$ does not imply (i). ∎

Remark. The last part is immediate if Φ' is strictly increasing so that it has Ψ' as its inverse; and $\Phi'(2t) \geq k_\varepsilon \Phi'(t)$ implies at once that $\Psi'(k_\varepsilon x) \leq 2\Psi'(x)$, in case $x_2 = 0$. This implies $\Psi(k_\varepsilon x) \leq 2k_\varepsilon \Psi(x)$ so that Ψ is Δ_2- or Φ is ∇_2-regular.

We next establish an important inequality for the complementary Young function Ψ if Φ satisfies condition (ii) of the preceding proposition.

Lemma 7. *Let (Φ, Ψ) be a complementary pair of Young functions, such that Φ' satisfies condition (ii) [hence any one of the equivalent three condition] of Proposition 6, then for each pair x_1, x_2 such that $|x_1 - x_2| \geq \delta_\varepsilon x_1$ ($\delta_\varepsilon > 0$ depends only on $\varepsilon > 0$ of (ii)) we have*

$$\Psi(y_2) \geq \Psi(y_1) + \Psi'(y_1)(y_2 - y_1)$$
$$+ L_\varepsilon \Phi(|x_1 - x_2|), \quad y_i = \Phi'(x_i), i = 1, 2, \quad (30)$$

where $L_\varepsilon > 0$ depends only on k_ε and δ_ε hence only on $\varepsilon > 0$.

Proof: By hypothesis, for each $0 < \varepsilon < \frac{1}{4}$ there are $k_\varepsilon > 1$ and $x_0(\varepsilon) \geq 0$ such that

$$\Phi'((1 - \varepsilon)x) \geq k_\varepsilon \Phi'((1 - \varepsilon)^2 x), \quad x \geq x_0(\varepsilon), \quad (31)$$

and hence $\Phi'(I\!R^+) \subset I\!R^+$. To establish (30), let $x_0 \leq x_i$ and suppose $x_1 < x_2$. Then taking $\delta_\varepsilon = \max\left(\frac{x_0(\varepsilon)}{x_1}, \frac{2\varepsilon - \varepsilon^2}{(1-\varepsilon)^2}\right)$, so that $|x_1 - x_2| = x_2 - x_1 \geq \delta_\varepsilon x_1$, we have for $(1 - \varepsilon)^2 x_2 \geq x_1 \geq x_0(\varepsilon)$, the following with $y_i = \Phi'(x_i), i = 1, 2$ ($\Psi'(I\!R^+) \subset I\!R^+$ by (31)):

$$\Psi(y_2) - \Psi(y_1) = \int_{y_1}^{y_2} \Psi'(t)dt$$

$$\geq \Psi'(y_1)(y_2 - y_1) + \text{ area of the rectangle of}$$

vertices $(x_1, \Phi'(x_1))$ and $\left((1-\varepsilon)x_2, \Phi'((1-\varepsilon)x_2)\right)$,

$$= \Psi'(y_1)(y_2 - y_1) + \left[\Phi'(x_2) - \Phi'((1-\varepsilon)x_2)\right][(1-\varepsilon)x_2 - x_1],$$

seen by drawing a picture,

$$\geq \Psi'(y_1)(y_2 - y_1) + (1 - \tfrac{1}{k_\varepsilon})\Phi'(x_2)[(1-\varepsilon) - (1-\varepsilon)^2]x_2,$$

since $x_2(1-\varepsilon)^{-1} > x_2 \geq x_1 \geq x_0$,

$$= \Psi'(y_1)(y_2 - y_1) + (1 - \tfrac{1}{k_\varepsilon})\varepsilon(1 - \varepsilon)\Phi'(x_2)x_2,$$

$$\geq \Psi'(y_1)(y_2 - y_1) + (1 - \tfrac{1}{k_\varepsilon})\varepsilon(1 - \varepsilon)\Phi(|x_2 - x_1|). \quad (32)$$

Taking $P_\varepsilon = (1 - \frac{1}{k_\varepsilon})\varepsilon(1 - \varepsilon) > 0$, (32) reduces to (30) in this case.

If $x_0 \le x_2 < x_1$, then we interchange y_1, y_2 (x_1, x_2 in the above) and the same procedure (with reversing the inequalities) gives an inequality with $Q_\varepsilon > 0$ as a corresponding constant in (32). If $L_\varepsilon = \min(P_\varepsilon, Q_\varepsilon) > 0$, then (32) holds in all cases. ∎

In what follows we need to use a couple of classical theorems from the Banach space theory. There is no special simplification. Hence we state them here without proof. The first fact was established independently [it is due to several authors] at about the same time, of which we give B. J. Pettis (1939) for ready reference. The second result is due to V. L. Šmulian (1939), which we already noted following Definition 1. In fact rotundity and smoothness are dual concepts for reflexive spaces.

Fact 1. Every uniformly convex Banach space is reflexive.

Fact 2. The unit sphere of a Banach space is uniformly convex iff the unit sphere of its adjoint space is uniformly smooth, i.e., iff the norm of the adjoint space is uniformly Fréchet differentiable at each point of its unit sphere.

The following result is due to Milnes (1957).

Theorem 8. *Let Φ be a Young function and (Ω, Σ, μ) be a measure space. Then the Orlicz space $(L^\Phi(\mu), \| \cdot \|_\Phi)$ is uniformly convex if for each $\varepsilon > 0$, there is a $k_\varepsilon > 1$ such that $\Phi'((1 + \varepsilon)x) \ge k_\varepsilon \Phi'(x)$, $x \ge x_0(\varepsilon) \ge 0$ [$x_0(\varepsilon) = 0$ if $\mu(\Omega) = \infty$], and Φ is Δ_2-regular and strictly convex. If μ is diffuse on a set of positive measure, then conversely the uniform convexity of the space implies the growth condition on Φ' given above, and also that Φ is Δ_2-regular as well as strictly convex.*

Proof: Suppose the two conditions on Φ and Φ' hold. Then not only is Φ Δ_2-regular but the hypothesis on Φ' implies that the complementary function Ψ is also Δ_2-regular. To see this, since $\Phi'((1+\varepsilon)x) \ge k_\varepsilon \Phi'(x)$, choose an integer $m > 0$ such that $(1 + k_\varepsilon)^m > 2^{m+1}$ so that

$$\Phi'((1 + \varepsilon)^m x) \ge \left(\frac{1 + k_\varepsilon}{2}\right)^m \Phi'(x) \ge 2\Phi'(x), \quad x \ge x_0(\varepsilon). \tag{33}$$

Considering the inverse Ψ' of Φ', (33) implies, writing $y = \Phi'(x)$,

$$(1 + \varepsilon)^m \Psi'(y) \ge \Psi'(2y), \quad \text{if } y \ge y_\varepsilon = \Phi'(x_0(\varepsilon)) \ge 0.$$

Hence integration gives for $y \geq y_\varepsilon \geq 0$,

$$(1+\varepsilon)^m \Psi(y) \geq \int_{y_\varepsilon}^{y} \Psi'(2v)dv + (1+\varepsilon)^m \Psi'(y_\varepsilon) > \frac{1}{2}[\Psi(2y) - \Psi(2y_\varepsilon)]. \quad (34)$$

This shows that Ψ is Δ_2-regular, so that $L^\Phi(\mu)$ is already reflexive.

Thus to prove uniform convexity of $(L^\Phi(\mu), \|\cdot\|_\Phi)$, we use the second (equivalent) definition of the concept: if $\{f_n, n \geq 1\}$ and $\{g_n, n \geq 1\}$ are any sequences in the unit ball of the space such that $\|f_n + g_n\|_\Phi \to 2$, then $\|f_n - g_n\|_\Phi \to 0$. First consider $f_n \geq 0, g_n \geq 0$ in this statement. Let $\eta = (2\varepsilon - \varepsilon^2)/(1-\varepsilon)^2$ for $0 < \varepsilon < 1$. Then for $|x - x'| \geq \eta x > 0$, one has by Lemma 8 that there is an $L_\eta > 0$ such that with $y = \Phi'(x)$

$$\Psi(y') \geq \Psi(y) + \Psi'(y)(y' - y) + L_\eta \Phi(|x - x'|). \quad (35)$$

Also since Φ is Δ_2-regular, the norm of f_n can be calculated using Proposition III.3.14 as follows. We have, for a constant $k_n > 0$ such that $\rho_\Psi(\Phi'(k_n f_n))d\mu = 1$, the equation

$$1 = \|f_n\|_\Phi = \int_\Omega \Phi'(k_n f_n) f_n d\mu. \quad (36)$$

Since $\|f_n\|_\Phi = 1 > N_\Phi(f_n)$, it follows that $1 < k_n < \infty$ (cf. Propositions III.3.17 and III.3.9). Let $h_n = \Phi'(k_n f_n)$. So $N_\Psi(h_n) = 1$. Similarly, considering $\|f_n + g_n\|_\Phi \leq 2$, there is $\tilde{h}_n = \Phi'\left(\tilde{k}_n\left(\frac{f_n + g_n}{2}\right)\right)$ satisfying $N_\Psi(\tilde{h}_n) = 1$, $1 < \tilde{k}_n < \infty$.

With these expressions for the norms, we deduce from the alternative definition of uniform convexity the desire conclusion. Thus consider the set $A_n(\eta)$ defined by

$$A_n(\eta) = \left\{\omega : k_n f_n(\omega) - \frac{\tilde{k}_n}{2}(f_n + g_n)(\omega) \geq k_n \eta f_n(\omega)\right\}.$$

Hence putting $y' = \tilde{h}_n(\omega)$ and $y = h_n(\omega)$ in (35) and integrating on $A_n(\eta)$ and adding it to the integrals of $\tilde{h}_n$ and h_n over $A_n^c(\eta)$, with the convexity inequality $\Psi(y') \geq \Psi(y) + \Psi'(y)(y' - y)$, we get

$$1 = \int_\Omega \Psi(\tilde{h}_n)d\mu \geq \int_\Omega \Psi(h_n)d\mu + k_n \int_\Omega f_n(\tilde{h}_n - h_n)d\mu$$

$$+ L_\eta \int_{A_n(\eta)} \Phi\left(k_n f_n - \frac{\tilde{k}_n}{2}(f_n + g_n)\right)d\mu.$$

Since $\int_\Omega \Psi(h_n)d\mu = 1$ also, one has on simplification

$$\int_\Omega f_n(h_n - \tilde{h}_n)d\mu \geq \frac{L_\eta}{k_n} \int_{A_n(\eta)} \Phi\left(k_n f_n - \frac{\tilde{k}_n}{2}(f_n + g_n)\right)d\mu$$

$$\geq L_\eta \int_{A_n(\eta)} \Phi\left(f_n - \frac{\tilde{k}_n}{2k_n}(f_n + g_n)\right)d\mu, \text{ since } k_n > 1. \ (37)$$

Using the analog of (36), we have

$$\|f_n + g_n\|_\Phi = \int_\Omega (f_n + g_n)\tilde{h}_n d\mu$$

$$= \int_\Omega f_n \tilde{h}_n d\mu + \int_\Omega g_n \tilde{h}_n d\mu \leq \|f_n\|_\Phi + \|g_n\|_\Phi \leq 2.$$

Letting $n \to \infty$ and using the fact that the left side tends to 2, the middle integrals tend to one. Since $0 \leq \int_\Omega f_n \tilde{h}_n d\mu \leq \|f_n\|_\Phi = \int_\Omega f_n h_n d\mu$, one has $\int_\Omega f_n(h_n - \tilde{h}_n)d\mu \to 0$ as $n \to \infty$. Hence (37) implies, since $0 < L_\eta < \infty$,

$$\int_{A_n(\eta)} \Phi\left(f_n - \frac{\tilde{k}_n}{2k_n}(f_n + g_n)\right)d\mu \to 0, \qquad \text{as } n \to \infty. \qquad (38)$$

Considering the complementary set $A_n^c(\eta)$, since $0 < \eta \leq 1$, and Φ is convex, we get

$$\int_{A_n^c(\eta)} \Phi\left(f_n - \frac{\tilde{k}_n}{2k_n}(f_n + g_n)\right)d\mu \leq \int_{A_n^c(\eta)} \Phi(\eta f_n)d\mu \leq \eta \int_\Omega \Phi(f_n)d\mu \leq \eta. \ (39)$$

Adding (38) and (39), and using the fact that $\eta > 0$ is arbitrary and Φ is Δ_2-regular (so that $\rho_\Phi(\cdot)$- and norm-convergences are equivalent), we deduce that

$$\left\|f_n - \frac{\tilde{k}_n}{2k_n}(f_n + g_n)\right\|_\Phi \to 0 \qquad \text{as } n \to \infty. \qquad (40)$$

Hence $\left|\frac{\tilde{k}_n}{2k_n}\|f_n + g_n\|_\Phi - \|f_n\|_\Phi\right| \to 0$, and so $\frac{\tilde{k}_n}{k_n} \to 1$ since $\|f_n\|_\Phi = 1$ and $\|f_n + g_n\|_\Phi \to 2$. But then (40) implies that $\|f_n - g_n\|_\Phi \to 0$ as

$n \to \infty$. So it only remains to consider the case that f_n and g_n are not necessarily positive elements of the unit ball.

Therefore, if f_n, g_n are general elements in the unit ball such that $\|f_n + g_n\|_\Phi \to 2$, and $\|f_n\|_\Phi = \|g_n\|_\Phi$, define F_n and G_n as

$$F_n = |f_n|, \text{ if } (f_n + g_n) \text{ has the sign of } f_n, \text{ and } = 0 \text{ otherwise,}$$
$$G_n = |g_n|, \text{ if } (f_n + g_n) \text{ has the sign of } g_n, \text{ and } = 0 \text{ otherwise.}$$

Then $F_n + G_n \geq |f_n + g_n|$, $0 \leq F_n \leq |f_n|$, $0 \leq G_n \leq |g_n|$, $n \geq 1$, and $2|F_n - G_n| \geq |f_n - g_n| \geq |F_n - G_n|$, $n \geq 1$. Hence $\|F_n\|_\Phi \leq 1$, $\|G_n\|_\Phi \leq 1$, and $\liminf_n \|F_n - G_n\|_\Phi \geq 2$, $\|f_n - g_n\| \leq 2\|F_n - G_n\|_\Phi$. By the special case of positive elements of the preceding paragraph applied to F_n and G_n we get $\|f_n - g_n\|_\Phi \to 0$, so that $(L^\Phi(\mu), \|\cdot\|_\Phi)$ is uniformly convex.

For the converse direction suppose that $(L^\Phi(\mu), \|\cdot\|_\Phi)$ is uniformly convex and the measure μ is diffuse on a set of positive measure. Since a uniformly convex space is both reflexive and strictly convex, it follows by the work of the preceding section and the diffuseness of μ that Φ may be taken Δ_2 regular and strictly convex (cf. Corollary IV.2.12 and Φ can be assumed to be an N-function). Thus we need to verify that Φ' satisfies also the growth condition of the theorem. We now produce a contradiction by supposing the opposite conclusion.

Suppose there is an $\varepsilon_0 > 0$ and a sequence $0 < x_n \nearrow \infty$ such that

$$\lim_{n\to\infty} [\Phi'((1 + \varepsilon_0)x_n)/\Phi'(x_n)] = 1. \tag{41}$$

For the present purposes (Ω, Σ, μ) can be taken to be a finite diffuse measure space. If $\lambda_n = [\Psi(\Phi'(x_n)) + \Psi(\Phi'((1 + \varepsilon_0)x_n))]^{-1}$, we may assume that $2\lambda_n < \mu(\Omega)$, $n \geq 1$. Clearly $\lambda_n \to 0$ as $n \to \infty$. Let $y_n = \Psi^{-1}(\frac{1}{2\lambda_n})$, and choose A_n, B_n in Σ, disjoint and satisfying $\mu(A_n) = \mu(B_n) = \lambda_n$. This is possible by the diffuseness of μ. Thus we have

$$\Phi'((1 + \varepsilon_0)x_n) \geq y_n \geq \Phi'(x_n). \tag{42}$$

Let $k_n = \lambda_n[(1 + \varepsilon_0)x_n\Phi'((1 + \varepsilon_0)x_n) + x_n\Phi'(x_n)]$, and define f_n and g_n as:

$$k_n f_n = [(1 + \varepsilon_0)x_n\chi_{A_n} + x_n\chi_{B_n}], \quad k_n g_n = [x_n\chi_{A_n} + (1 + \varepsilon_0)x_n\chi_{B_n}].$$

A simple computation shows that $\rho_\Psi(\Phi'(k_n f_n)) = 1 = \rho_\Psi(\Phi'(k_n g_n))$. Hence

$$\|f_n\|_\Phi = \int_\Omega \Phi'(k_n f_n) f_n d\mu = 1 = \int_\Omega \Phi'(k_n g_n) g_n d\mu = \|g_n\|_\Phi.$$

But by definition of the sequences $\{x_n, n \geq 1\}$ and $\{y_n, n \geq 1\}$, we have

$$\lim_{n\to\infty} \frac{\Phi'((1+\varepsilon_0)x_n)}{y_n} = 1 = \lim_{n\to\infty} \frac{\Phi'(x_n)}{y_n}.$$

Consequently

$$\begin{aligned}
\left\|\frac{f_n + g_n}{2}\right\|_\Phi &= \frac{(1+\frac{1}{2}\varepsilon_0)x_n}{k_n}\|\chi_{A_n \cup B_n}\|_\Phi \\
&= \frac{(2+\varepsilon_0)x_n}{k_n}\lambda_n \Psi^{-1}\left(\frac{1}{2\lambda_n}\right) \\
&= \frac{(2+\varepsilon_0)y_n}{(1+\varepsilon_0)\Phi'((1+\varepsilon_0)x_n) + \Phi'(x_n)} \to 1, \qquad \text{as } n \to \infty.
\end{aligned}$$

However, since $|f_n - g_n| = \frac{\varepsilon_0 x_n}{k_n}\chi_{A_n \cup B_n}$, we get

$$\lim_{n\to\infty}\|f_n - g_n\|_\Phi = \frac{2\varepsilon_0}{2+\varepsilon_0}\lim_{n\to\infty}\left\|\frac{f_n+g_n}{2}\right\|_\Phi = \frac{2\varepsilon_0}{2+\varepsilon_0} > 0.$$

Thus $(L^\Phi(\mu), \|\cdot\|_\Phi)$ is not uniformly convex. ∎

Using the classical fact (recalled prior to the theorem) that in any Banach space, uniform convexity and uniform smoothness (= uniform strong differentiability of norm) are dual concepts, we have the following form of the above result which we state for a reference.

Corollary 9. *Let (Φ, Ψ) be a pair of continuous complementary Young functions and $(L^\Phi(\mu), \|\cdot\|_\Phi)$ be an Orlicz space on (Ω, Σ, μ). If the derivative $\Phi'(\cdot)$ satisfies the growth condition (22) on $\mathbb{R}^+$ and $\Phi \in \Delta_2$, then $(L^\Psi(\mu), N_\Psi(\cdot))$ is uniformly smooth.*

Here the fact that $(L^\Phi(\mu), \|\cdot\|_\Phi)^* = (L^\Psi(\mu), N_\Psi(\cdot))$ is used. To obtain an analogous result with the norms $\|\cdot\|_\Phi$ and $N_\Phi(\cdot)$ interchanged will be of obvious interest. However, once a direct evaluation of norm $\|\cdot\|_\Phi$ given by (36) is known, no special properties of this functional are utilized in the proof. Consequently the same argument with simple modifications establishes the uniform convexity of $(L^\Phi(\mu), N_\Phi(\cdot))$ also, and hence $(L^\Psi(\mu)), \|\cdot\|_\Psi)$ is then uniformly smooth. Let us therefore state this result for a convenient reference. This equivalence was noted by Akimovič (1972).

Theorem 10. *Let (Φ, Ψ) be a pair of complementary Young functions. Then the Orlicz space $(L^{\Phi}(\mu), N_{\Phi}(\cdot))$ on a measure space (Ω, Σ, μ) is uniformly convex if for each $\varepsilon > 0$, there exists a $k_{\varepsilon} > 1$ such that $\Phi'((1 + \varepsilon)x) \geq k_{\varepsilon}\Phi'(x)$, $x \geq x_0(\varepsilon) \geq 0$ (and $x_0(\varepsilon) = 0$ if $\mu(\Omega) = +\infty$), with Φ being Δ_2-regular and strictly convex. The converse holds if, moreover, the measure μ is diffuse on a set of positive measure.*

Leaving the details for proving this result, which follow closely those of Theorem 9, we present essentials for the smoothness of $(L^{\Phi}(\mu), N_{\Phi}(\cdot))$, using the differentiability of its norm (with Theorem 4) for variety. The argument is adapted from Rao (1965, p. 678) omitting the unnecessary normalization and other hypotheses there.

Theorem 11. *Let (Ω, Σ, μ) be a measure space and Φ be a Young function such that $(L^{\Phi}(\mu), N_{\Phi}(\cdot))$ is reflexive. Then its norm is uniformly strongly differentiable (or the sphere is uniformly smooth) iff*

$$\lim_{N_{\Phi}(f) \to 0} \frac{1}{N_{\Phi}(f)} \int_{\Omega} f_0 \left[\Phi' \left(\frac{f_0 + f}{N_{\Phi}(f_0 + f)} \right) - \Phi'(f_0) \right] d\mu = 0, \qquad (43)$$

uniformly in $f_0 \in S^{\Phi} = \{h : N_{\Phi}(h) = 1\}$. This condition holds if the (left) derivative Ψ' of the complementary function Ψ of Φ satisfies: for each $\varepsilon > 0$ there is a $k_{\varepsilon} > 1$ such that

$$\Psi'((1 + \varepsilon)y) \geq k_{\varepsilon}\Psi'(y), \quad y \geq y_0(\varepsilon) \geq 0 \quad [= 0 \text{ if } \mu(\Omega) = \infty],$$

and Ψ is also strictly convex when $\mu(\Omega) < \infty$.

Proof: The first statement that (43) is equivalent to uniform strong differentiability of norm is immediate from definition. Regarding the second (main) part, since the mapping $t \mapsto N_{\Phi}(f_0 + tf)$ is convex and continuous on $[-1, 1]$, it can be expressed as an indefinite integral of its derivative. But now the norm is (weakly) differentiable by Theorem 3. Hence $\frac{d}{d\alpha}N_{\Phi}(f_0 + t\alpha f) = G(f_0 + t\alpha f; tf)$, outside of a set of (Lebesgue) measure zero by the standard theory (cf., e.g., Rao (1987), p. 242). So for all $f \in S^{\Phi}$, $|t| < 1$, one has

$$\left| \frac{N_{\Phi}(f_0 + tf) - 1}{t} - G(f_0; f) \right| \leq \int_0^1 |G(f_0 + t\alpha f; f) - G(f_0; f)| d\alpha$$

$$\leq \int_0^1 \|G(f_0 + t\alpha f; \cdot) - G(f_0; \cdot)\| d\alpha,$$

$$\text{since } N_\Phi(f) = 1,$$

$$= \int_0^1 \left\|G\left(\frac{f_0 + t\alpha f}{N_\Phi(f_0 + t\alpha f)}, \cdot\right) - G(f_0; \cdot)\right\| d\alpha,$$

$$\text{since } G(\beta f_0, \cdot) = G(f_0, \cdot), \quad \beta > 0,$$

$$\leq 2 \int_0^1 N_\Psi\left(\Phi'\left(\frac{f_0 + t\alpha f}{N_\Phi(f_0 + t\alpha f)}\right) - \Phi'(f_0)\right) d\alpha,$$

$$\text{since } G(f_0, \cdot) \in (L^\Phi(\mu))^*. \qquad (44)$$

Here we have used the representation of the space $(L^\Phi(\mu), N_\Phi(\cdot))^* = (L^\Psi(\mu), \|\cdot\|_\Psi)$, by Theorem IV.1.6, and also $\|\cdot\|_\Psi \leq 2N_\Psi(\cdot)$. Since Φ and Ψ are both Δ_2-regular, the right side integral tends to zero uniformly in (f_0, f) as $t \to 0$ iff the following integral equation holds uniformly in (f_0, f):

$$\lim_{t \to 0} \int_0^1 \int_\Omega \Psi\left[\Phi'\left(\frac{f_0 + t\alpha f}{N_\Phi(f_0 + t\alpha f)}\right) - \Phi'(f_0)\right] d\mu d\alpha = 0. \qquad (45)$$

Thus, by the bounded convergence theorem, it is enough to show that the inner integral tends to zero in the desired sense. For this we need to use the given conditions. The latter imply the hypothesis of Lemma 8. So for each $\eta \ (= \eta(\varepsilon) \searrow 0$ as $\varepsilon \searrow 0)$, there is an $L_\eta > 0$ satisfying

$$\Phi(u') - \Phi(u) \geq \Phi'(u)(u' - u) + L_\eta \Psi(|v' - v|)$$

whenever $|v' - v| > \eta v > 0$ with $v = \Psi'(u)$ and $v' = \Psi'(u') > 0$. Consider $f \geq 0$ a.e., and $0 < t < 1$. Taking $u = f_0(\omega)$, $u' = \frac{f_0 + t\alpha f}{N_\Phi(f_0 + t\alpha f)}(\omega)$, $\Omega_1 = \{\omega : |v' - v| \geq \eta v > 0\}$ in the above inequality and integrating we obtain

$$\int_{\Omega_1} \Psi(v' - v) d\mu \leq \frac{1}{L_\eta} \int_{\Omega_1} [\Phi(u') - \Phi(u) - \Phi'(u)(u' - u)] d\mu. \qquad (46)$$

Since for any convex function Φ one has $\Phi(x') - \Phi(x) \geq \Phi'(x)(x' - x)$ always, we can integrate this expression and add it to (46) to get the right side of (46) with Ω in place of Ω_1. Since $\int_\Omega \Phi(u')d\mu = 1 = \int_\Omega \Phi(u)d\mu$, this result becomes:

$$\int_{\Omega_1} \Psi(v' - v)d\mu \leq \frac{1}{L_\eta} \int_\Omega \Phi'(u)(u - u')d\mu$$

$$\leq \frac{2}{L_\eta} N_\Phi\left(f_0 - \frac{f_0 + t\alpha f}{N_\Phi(f_0 + t\alpha f)}\right), \qquad \Phi'(f_0) \in S^\Psi,$$

$$\text{and Hölder's inequality is used.} \qquad (47)$$

But for any f_0, f in S^Φ, we also have for $0 \leq \alpha \leq 1$,

$$N_\Phi\left(f_0 - \frac{f_0 + t\alpha f}{N_\Phi(f_0 + t\alpha f)}\right) \leq N_\Phi(f_0 - (f_0 + t\alpha f))$$

$$+ N_\Phi\left(f_0 + t\alpha f - \frac{f_0 + t\alpha f}{N_\Phi(f_0 + t\alpha f)}\right)$$

$$\leq |t\alpha| + [N_\Phi(f_0 + t\alpha f) - 1]$$

$$\leq 2|t|\alpha. \qquad (48)$$

Substituting (48) in (47) and letting $t \to 0^+$, the right side tends to zero uniformly in f_0 and f in S^Φ. Since $|v' - v| \leq \eta v$ for $0 < \eta < 1$ on Ω_1^c, we have

$$\int_{\Omega_1^c} \Psi(v' - v)d\mu \leq \int_\Omega \Psi(\eta v)d\mu \leq \eta \rho_\Psi(v) \leq \eta. \qquad (49)$$

From (46)–(49) we get (45) in this case. If now f_0 and f are arbitrary in S^Φ, we may reduce this to the positive case using the same trick as in the proof (of Theorem 9) following (40). This completes the argument. ∎

It should be noted from these three results that for uniform convexity and smoothness of Orlicz spaces the conditions on (Φ, Ψ) are the *same* for both gauge and Orlicz norms. This is of interest since the same statement is not true for strict convexity! Thus without mention of the particular measure, we may state the following result as a consequence of the above work.

Corollary 12. *Let $L^\Phi(\mu)$ be a reflexive Orlicz space on a triple (Ω, Σ, μ). If Φ' satisfies the Milnes growth condition and Φ is strictly convex, then $L^\Phi(\mu)$ [$L^\Psi(\mu)$] is uniformly convex [smooth].*

As noted in Theorem 9, if μ is restricted to be diffuse on a set of positive measure, then the conditions appearing in Theorems 8 and 11 and hence Corollary 12 are the best possible in the sense of the former result. We shall next prove that isomorphism condition is too weak a property to distinguish between uniform convexity and reflexivity for Orlicz spaces.

7.3. Remarks on reflexivity and convexity properties

It was already seen in Theorem V.1.1, in conjunction with Theorems 1.1 and 1.2 of this chapter, that an Orlicz space can be isomorphic to a strictly convex Orlicz space without itself being strictly convex (or rotund). Here we show that a similar property holds between uniform convexity and reflexivity. The following result, due to Akimovič (1972), illuminates this point.

Proposition 1. Let Φ be an N-function which is Δ_2- and ∇_2-regular, i.e., $\Phi \in \Delta_2 \cap \nabla_2$. Then there is an equivalent N-function Φ_1 which is Δ_2-regular and whose (left) derivative Φ_1' satisfies the Milnes growth condition (i.e., (22) of Section 2).

Proof: For simplicity we consider only the global relation of Δ_2 and ∇_2. Thus $\Phi \in \Delta_2 \cap \nabla_2$. Then by Theorem II.3.3, there exists $\delta > 0$ and $k > 2$ such that

$$(2 + \delta)\Phi(x) \leq \Phi(2x) \leq k\Phi(x), \qquad x \geq 0. \tag{1}$$

Let $\tilde{\Phi}_1 : x \mapsto \int\limits_0^{|x|} \frac{\Phi(t)}{t} dt$ and $\Phi_1 : x \mapsto \int\limits_0^{|x|} \frac{\Phi_1(t)}{t} dt$. Then $\Phi \sim \tilde{\Phi}_1 \sim \Phi_1$ where '$\sim$' denotes equivalence in the sense of Chapter II. From this we get

$$\frac{1}{k}\Phi(x) \leq \Phi(\frac{x}{2}) = \frac{2}{x} \int\limits_0^{x/2} \Phi(\frac{x}{2}) dt \leq \int\limits_{x/2}^{x} \frac{\Phi(t)}{t} dt = \tilde{\Phi}_1(x) - \tilde{\Phi}_1(\frac{x}{2})$$

$$< \tilde{\Phi}_1(x) = \int\limits_0^{x/2} \frac{\Phi(t)}{t} dt + \int\limits_{x/2}^{x} \frac{\Phi(t)}{t} dt$$

$$\leq \Phi(\frac{x}{2}) + \frac{1}{2}\Phi(x) \leq \left(\frac{1}{2 + \delta} + \frac{1}{2}\right)\Phi(x) = \frac{1}{c}\Phi(x) \text{ (say)}.$$

Thus we have on noting that $\tilde{\Phi}_1 \in \Delta_2$ (so the last inequality of (1) holds),

$$1 < c = \frac{4 + 2\delta}{4 + \delta} \leq \frac{\Phi(t)}{\tilde{\Phi}_1(t)} = \frac{t\tilde{\Phi}_1'(t)}{\tilde{\Phi}_1(t)} \leq k, \quad t > 0. \tag{2}$$

Integrating this we have for $\theta > 1$,

$$c \log \theta = c \int_x^{\theta x} \frac{dt}{t} \leq \int_x^{\theta x} \frac{\tilde{\Phi}_1'(t)}{\tilde{\Phi}_1(t)} dt \leq k \int_x^{\theta x} \frac{dt}{t} = k \log \theta,$$

so that $\theta^c \tilde{\Phi}_1(x) \leq \tilde{\Phi}_1(\theta x) \leq \theta^k \tilde{\Phi}_1(x)$, $x > 0$. Hence for $\varepsilon > 0$,

$$\Phi_1'((1+\varepsilon)x) = \frac{\tilde{\Phi}_1((1+\varepsilon)x)}{(1+\varepsilon)x} \geq \frac{(1+\varepsilon)^c \tilde{\Phi}_1(x)}{(1+\varepsilon)x} = (1+\varepsilon)^{c-1} \Phi_1'(x), \quad x > 0. \tag{3}$$

Since $(1 + \varepsilon)^{c-1} > 1$, Φ_1' satisfies the Milnes growth condition. This shows that Φ_1 is Δ_2-regular, and has the desired properties. ∎

The preceding construction of Φ_1 has an additional property. Thus let Ψ_1 be the complementary function of Φ_1. Then the fact that $\Phi_1 \in \Delta_2 \cap \nabla_2$ implies that $\Psi_1 \in \Delta_2 \cap \nabla_2$ by Theorem II.3.3. Moreover, since Ψ_1' is the inverse of Φ_1' by definition, consider $\alpha(\cdot)$ given by the following equation:

$$\Psi_1'((1+\varepsilon)y) = \alpha(y)\Psi_1'(y), \quad y \geq 0. \tag{4}$$

Clearly $\alpha(y) > 1$ since Ψ_1' is strictly increasing. Writing $x = \Psi_1'(y)$ we get from (4) and the earlier relations for (3),

$$(1+\varepsilon)\Phi_1'(x) = \Phi_1'(\alpha(y)x) = \frac{\tilde{\Phi}_1(\alpha(y)x)}{\alpha(y)x} \leq \frac{\alpha(y)^k \tilde{\Phi}_1(x)}{\alpha(y)x} = \alpha(y)^{k-1}\Phi_1'(x).$$

This shows that $\alpha(y) \geq (1 + \varepsilon)^{\frac{1}{k-1}} > 1$, for all $y \geq 0$. This and (4) imply that Ψ_1' also satisfies the Milnes growth condition. Thus we have the following consequence:

Theorem 2. *Let Φ be an N-function satisfying $\Phi \in \Delta_2 \cap \nabla_2$ and $L^\Phi(\mu)$ be an Orlicz space on (Ω, Σ, μ) with either norm. Then there exists an equivalent N-function Φ_1 (i.e., $\Phi \sim \Phi_1$) such that the corresponding space $L^{\Phi_1}(\mu)$ on the same (Ω, Σ, μ) satisfies $L^\Phi(\mu) = L^{\Phi_1}(\mu)$ as sets; and their gauge (respectively Orlicz) norms are equivalent with the following additional structure:*

(i) *both spaces are reflexive,*

(ii) $L^{\Phi}(\mu)$ *need not be strictly convex but* $L^{\Phi_1}(\mu)$ *is both uniformly convex and uniformly smooth, and*

(iii) *the spaces are isomorphic.*

All the assertions are implied by the preceding work. The point of this result is that isomorphism does not distinguish several sharper aspects of the spaces. On the other hand for the Lebesgue spaces $L^p(\mu)$, $1 < p < \infty$, we have both the uniform convexity and smoothness simultaneously. While these are not present for a given reflexive Orlicz space, they are restored under isomorphism, and thus several important geometric properties of the $L^p(\mu)$-spaces are inherited in this weaker form. We next turn to some analytical properties related to convexity and smoothness in the general case.

7.4. *Modulus of convexity and Clarkson type inequalities*

In the preceding sections we discussed the (uniform) convexity properties of an Orlicz space on a measure space (Ω, Σ, μ). Also at the beginning of Section 2 above, we defined the modulus of uniform convexity associated with a uniformly convex Banach space $\mathcal{X}$; namely, for each $0 < \varepsilon \le 2$ there is a $0 < \delta(\varepsilon) < 1$ ($\delta(\varepsilon) \to 0$ as $\varepsilon \searrow 0$) such that $\|x_i\| \le 1$ and $\|x_1 - x_2\| > \varepsilon \Rightarrow \|x_1 + x_2\| \le 2(1 - \delta(\varepsilon))$. The function $\delta(\cdot)$ plays an interesting role in the geometry of Banach spaces, and its special structure has long been analyzed for the $L^p(\mu)$-spaces, $1 < p < \infty$. In this section we consider the corresponding functions $\delta_\Phi(\cdot)$ for Orlicz spaces $L^\Phi(\mu)$ and illustrate their utility.

To motivate the general case, we recall the classical Lebesgue case. The following inequalities are due to J. A. Clarkson (1936). Let $L^p(\mu)$ be the Lebesgue space on (Ω, Σ, μ).

(i) If $2 \le p < \infty$ and $f, g \in L^p(\mu)$, one has

$$\left\|\frac{f+g}{2}\right\|_p^p + \left\|\frac{f-g}{2}\right\|_p^p \le \frac{1}{2}[\|f\|_p^p + \|g\|_p^p]; \tag{1}$$

(ii) and if $1 < p < 2$, $f, g \in L^p(\mu)$, then letting $p' = \frac{p}{p-1}$,

$$\left\|\frac{f+g}{2}\right\|_p^{p'} + \left\|\frac{f-g}{2}\right\|_p^{p'} \le \left[\frac{1}{2}(\|f\|_p^p + \|g\|_p^p)\right]^{\frac{1}{p-1}}. \tag{2}$$

Because of the presence of both p and (its conjugate exponent) $p' = \frac{p}{p-1}$ in (2), it is considerably more difficult to establish it than (1). Note that if $p = 2$ $(= p')$ then equality holds in both (1) and (2), so that they reduce to the classical parallelogram identity. The latter actually characterizes a Hilbert space. Here we shall consider some analogs of these inequalities for Orlicz spaces. We start by proving an easy extension of the simple inequality (1).

The *modulus of convexity*, $\delta(\varepsilon) = \delta_{\mathcal{X}}(\varepsilon)$ of $(\mathcal{X}, \|\cdot\|)$ introduced above, can be rewritten as:

$$\delta_{\mathcal{X}}(\varepsilon) = \inf\left\{1 - \tfrac{1}{2}\|x_1 + x_2\| : \|x_i\| \leq 1, \right.$$
$$\left. \|x_1 - x_2\| \geq \varepsilon, x_i \in \mathcal{X}\right\}, \qquad 0 < \varepsilon \leq 2. \tag{3}$$

Thus $\mathcal{X}$ is uniformly convex if $\delta_{\mathcal{X}}(\varepsilon) > 0$, for each $0 < \varepsilon \leq 2$. The following special result for certain $\mathcal{X} = L^{\Phi}(\mu)$ was given in Rao ((1987), pp. 307–308, Exercises 3–6), and independently also by Hudzik (1987) where a lower bound for $\delta_{\mathcal{X}}(\varepsilon)$ was noted.

Proposition 1. *Let Φ be a continuous Young function so that $\Phi(x) = \theta(x^2)$ where θ is a Δ_2-regular Young function. Then $(L^{\Phi}(\mu), N_{\Phi}(\cdot))$ is uniformly convex and one has*

$$\delta_{\mathcal{X}}(\varepsilon) = \delta_{\Phi}(\varepsilon) \geq p(q(\tfrac{\varepsilon}{2})), \qquad 0 < \varepsilon \leq 2, \mathcal{X} = L^{\Phi}(\mu), \tag{4}$$

where for $0 < \lambda < 1$,

$$p(\lambda) = \sup\left\{0 < \sigma < 1 : \sup\left(\Phi(\tfrac{u}{1-\sigma})/\Phi(u) : u > 0\right) \leq \tfrac{1}{1-\lambda}\right\}, \; and$$

$$q(\varepsilon) = \inf\left\{\Phi(u)/\Phi(\tfrac{u}{\varepsilon}) : u > 0\right\}.$$

Proof: Let f_1, f_2 be a pair of elements in $L^{\Phi}(\mu)$, and set $a = \tfrac{1}{2}(f_1 + f_2)$, $b = \tfrac{1}{2}(f_1 - f_2)$. Then

$$\Phi\left(\frac{f_1 + f_2}{2}\right) + \Phi\left(\frac{f_1 - f_2}{2}\right) = \theta(a^2) + \theta(b^2)$$
$$\leq \theta(a^2 + b^2), \text{ since } \theta \text{ is a Young function,}$$
$$= \theta(\tfrac{1}{2}(f_1^2 + f_2^2)) \leq \tfrac{1}{2}[\theta(f_1^2) + \theta(f_2^2)],$$
$$\theta \text{ being convex.} \tag{5}$$

Suppose now that $N_{\Phi}(f_i) \leq 1$, $i = 1, 2$. Then by Theorem III.2.3, this is equivalent to the condition that $\rho_{\Phi}(f_i) = \int_{\Omega} \Phi(f_i) d\mu \leq 1$. Hence integrating (5) and using this assumption, we get

$$\rho_{\Phi}\left((f_1 + f_2)/2\right) \leq 1 - \rho_{\Phi}\left((f_1 - f_2)/2\right). \tag{6}$$

But θ and hence Φ are Δ_2-regular. Consequently, $N_\Phi(f) \to 0$ iff $\rho_\Phi(f) \to 0$ (cf. Lemma V.2.1) so that $N_\Phi((f_1+f_2)/2) \geq \varepsilon > 0$ iff $\rho_\Phi((f_1+f_2)/2) \geq \eta_\varepsilon$ for some $\eta_\varepsilon > 0$ which tends to zero as $\varepsilon \to 0$. From this we get

$$\tilde{\delta}_\Phi(\varepsilon) = \inf\{1 - \rho_\Phi((f_1+f_2)/2) : \rho_\Phi(f_i) \leq 1, \rho_\Phi((f_1-f_2)/2) \geq \eta_\varepsilon\} \geq \eta_\varepsilon,$$

and hence

$$\delta_\Phi(\varepsilon) = \inf\left\{1 - N_\Phi((f_1+f_2)/2) : N_\Phi(f_i) \leq 1, N_\Phi((f_1-f_2)/2) \geq \varepsilon\right\}$$
$$\geq \tilde{\eta}_\varepsilon. \tag{7}$$

These two equivalent conditions imply that $(L^\Phi(\mu), N_\Phi(\cdot))$ is uniformly convex. However, it is useful to have an explicit expression for $\tilde{\eta}_\varepsilon$. We shall show that the right side of (4) is a possible value.

Let $0 < \sigma < 1$ and note that the convexity of Φ implies

$$\Phi(u) \leq (1-\sigma)\Phi\left(\frac{u}{1-\sigma}\right), \quad \text{or} \quad \sup\left\{\Phi\left(\frac{u}{1-\sigma}\right)/\Phi(u) : u > 0\right\} \geq \frac{1}{1-\sigma}.$$

Given $\varepsilon > 0$, by the Δ_2-regularity of Φ one has $\Phi(\frac{u}{\varepsilon}) \leq k_\varepsilon \Phi(u)$, so that $q(\varepsilon) \geq k_\varepsilon^{-1} > 0$ and

$$\Phi(u) \geq q(\varepsilon)\Phi(\tfrac{u}{\varepsilon}), \qquad u \geq 0. \tag{8}$$

If $N_\Phi(f_1 - f_2) \geq \varepsilon$, then taking $2u = |f_1 - f_2|$ in (8), and observing that if $N_\Phi((f_1 - f_2)/\varepsilon) \geq 1$ or equivalently $\rho_\Phi((f_1-f_2)/\varepsilon) \geq 1$, we have

$$\rho_\Phi\left(\frac{f_1-f_2}{2}\right) \geq q\left(\frac{\varepsilon}{2}\right)\rho_\Phi\left(\frac{f_1-f_2}{\varepsilon}\right) \geq q\left(\frac{\varepsilon}{2}\right). \tag{9}$$

Hence (9) and (6) imply

$$\rho_\Phi\left(\frac{f_1+f_2}{2}\right) \leq 1 - q\left(\frac{\varepsilon}{2}\right). \tag{10}$$

But by definition of $p(\lambda)$, $0 < \lambda < 1$, we get

$$\Phi\left(\frac{u}{1-p(\lambda)}\right) \leq \frac{1}{1-\lambda}\Phi(u), \qquad u \geq 0. \tag{11}$$

Let $\lambda = q(\frac{\varepsilon}{2})$ and $2u = |f_1 - f_2|$ in (11). Then using (10) we get

$$\rho_\Phi\left(\frac{f_1+f_2}{2(1-p(\lambda))}\right) \leq \frac{1}{1-\lambda}\rho_\Phi\left(\frac{f_1+f_2}{2}\right) \leq 1,$$

so that

$$N_\Phi\left(\frac{f_1 + f_2}{2}\right) \le 1 - p(q(\tfrac{\varepsilon}{2})),$$

and this establishes (4); or $\tilde{\eta}_\varepsilon = p(q(\tfrac{\varepsilon}{2}))$ in (7). ∎

It should be noted that the lower bound given for $\delta_\chi(\varepsilon)$ is not an exact expression. If $\Phi(x) = |x|^{p_1}$, $2 \le p_1 < \infty$, then $\delta_\chi(\varepsilon) = \delta_{p_1}(\varepsilon)$ and (5) is the original (easy) part of J. A. Clarkson's inequality. But even here, $\delta_p(\varepsilon)$ is not simple if $p_1 \ne 2$. To have some idea of the form of $\delta_\chi(\varepsilon)$, we consider an example from Hudzik (1987).

Example 2. Let $\Phi(x) = \max(|x|^{p_1}, |x|^{p_2})$, $2 \le p_1 \le p_2 < \infty$. This function evidently is convex, and satisfies the assumption of the above result (with $p_1 = p_2$ giving the Clarkson's simpler case). We need to calculate $p(q(\tfrac{\varepsilon}{2}))$ of (4). Thus, for $0 < \varepsilon < 1$,

$$\frac{\Phi(x)}{\Phi\left(\frac{x}{\varepsilon}\right)} = \begin{cases} \varepsilon^{p_1}, & 0 < x \le \varepsilon \\ \varepsilon^{p_2} x^{p_1 - p_2}, & \varepsilon < x \le 1 \\ \varepsilon^{p_2}, & x > 1 \end{cases}$$

and

$$= \begin{cases} \varepsilon^{p_1}, & 0 < x \le 1, \ \varepsilon = 1, \\ \varepsilon^{p_1} x^{p_2 - p_1}, & 1 < x \le \varepsilon, \ \varepsilon > 1, \\ \varepsilon^{p_2}, & x > \varepsilon > 1. \end{cases}$$

Consequently, $q(\varepsilon) = \min(\varepsilon^{p_1}, \varepsilon^{p_2})$, $\varepsilon > 0$.

Regarding $p(\lambda)$, consider for $0 < \sigma < 1$,

$$\frac{\Phi\left(\frac{x}{1-\sigma}\right)}{\Phi(x)} = \begin{cases} (1 - \sigma)^{-p_1}, & 0 < x \le 1 - \sigma \\ (1 - \sigma)^{-p_2} x^{p_1 - p_2}, & 1 - \sigma < x \le 1 \\ (1 - \sigma)^{-p_2}, & x > 1. \end{cases}$$

Hence taking the supremum of this over $u > 0$, and noting that this is $(1 - \sigma)^{-p_2}$, we get

$$p(\lambda) = \sup\{0 < \sigma < 1 : (1 - \sigma)^{-p_2} \le (1 - \lambda)^{-1}\} = 1 - (1 - \lambda)^{1/p_2}.$$

This gives $p(q(\tfrac{\varepsilon}{2})) = 1 - [1 - (\tfrac{\varepsilon}{2})^{p_2}]^{1/p_2} = 1 - \frac{1}{2}(2^{p_2} - \varepsilon^{p_2})^{1/p_2}$, $0 < \varepsilon \le 2$, which is the asserted lower bound for the modulus of convexity of $(L^\Phi(\mu), N_\Phi(\cdot))$.

It is known that for $1 < p < 2$, the classical Clarkson inequality is much more difficult than the case $p \ge 2$. Consequently it is not surprising that the general case of $(L^\Phi(\mu), N_\Phi(\cdot))$ spaces is even more involved. We now present this result which is essentially taken from R. P. Maleev and S. L. Troyanski (1975).

Theorem 3. *Let Φ be a continuous Young function of class Δ_2 such that for each $\varepsilon > 0$ there is a $K_\varepsilon > 1$ such that its (left) derivative Φ' satisfies*

$$\Phi'((1+\varepsilon)x) \geq K_\varepsilon \Phi'(x), \qquad x \geq 0. \tag{12}$$

Then there exists a Young function Φ_1 equivalent to Φ such that the modulus of convexity δ_{Φ_1} of $\mathcal{X} = L^{\Phi_1}(\mu)$ satisfies the inequality

$$\delta_{\Phi_1}(\varepsilon) \geq C p_\Phi(\varepsilon), \quad 0 < \varepsilon \leq 1 \,(0 < C < \infty \text{ a constant}), \tag{13}$$

where

$$p_\Phi(\varepsilon) = \varepsilon^2 \inf\{\Phi(xy)/x^2\Phi(y) : \varepsilon \leq x \leq 1, y > 0\}.$$

As we have seen in Theorems 2.8 and 2.10, the inequality (12) implies that $(L^\Phi(\mu), N_\Phi(\cdot))$ as well as $(L^\Phi(\mu), \|\cdot\|_\Phi)$ are uniformly convex. The introduction of Φ_1 here allows one to find conditions on Φ so that the lower bound in (13) can be evaluated. One can take Φ_1 as:

$$\Phi_1(x) = \int_0^x \left[\int_0^v \frac{\Phi(t)}{t} dt\right] \frac{dv}{v}.$$

The details of proof of (13) will not be included here since they are long and unrevealing. The point of the result is that the analog of the Lebesgue case cannot be expected since an explicit form of Φ is not given. A weakness of the result is that one has the bound (13) only for an equivalent norm with Φ_1. Since Akimovič's (1972) result states that a reflexive Orlicz space $L^\Phi(\mu)$ is isomorphic to both a uniformly convex and uniformly smooth Orlicz space $L^{\Phi_1}(\mu)$, one can also present an analogous inequality for uniform smoothness of the same space. [Without isomorphism $L^\Phi(\mu)$ is uniformly convex iff $L^\Psi(\mu)$ is uniformly smooth! See Corollary 2.12. Thus the equivalence implied here is not sharp.] The *modulus of smoothness* is defined as:

$$\eta_\mathcal{X}(\varepsilon) = \tfrac{1}{2}\sup\left(\|x_1+\varepsilon x_2\| + \|x_1-\varepsilon x_2\| - 2\,:\right.$$
$$\left.\|x_i\| = 1, x_i \in \mathcal{X}, i=1,2\right), \qquad \varepsilon > 0. \tag{14}$$

$\mathcal{X}$ is uniformly smooth if $\lim_{\varepsilon \searrow 0} \eta_\mathcal{X}(\varepsilon)/\varepsilon = 0$. Thus when Φ satisfies (12) we have:

$$\eta_{\Phi_1}(\varepsilon) \le C_1 q_\Phi(\varepsilon), \qquad 0<C_1<\infty, \qquad (15)$$

where $q_\Phi(\varepsilon) = \varepsilon^2 \sup \{\Phi(xy)/x^2\Phi(y) : \varepsilon \le x \le 1, y>0\}$.

The details of both (13) and (15) can be found in the Maleev-Troyanski paper. More work and better conditions seem necessary before one can consider the result to be in a final form. Consequently we omit further discussion. For a comparison we note that the (more difficult) Clarkson inequality for $1 < p < 2$ given by (2) involves both p and its conjugate exponent p', and thus it depends on complicated computations. This also explains why (4) is much less difficult than (13) (and (15)). For a general study of these problems, one may consult Day (1944) and a survey is found in Cudia (1963). See also Figiel (1976). We conclude this chapter with an application.

7.5. *The $L^{\Phi_1\Phi_2}$-space and applications*

Using the ideas and results of the preceding sections, we first present a characterization of the adjoint space of $L_{\mathcal{X}}^{\Phi}$, an Orlicz space of $\mathcal{X}$-valued functions where $\mathcal{X}$ is a Banach space. Then we specialize the result when $\mathcal{X}$ is an Orlicz space on a (possibly different) measure space, and use it for a couple of applications.

The next result is a companion to Theorem VI.1.10, and is given in terms of norm differentiability hypotheses as announced prior to the statement there. We recall that if $\mathcal{X}$ is a (real) Banach space with norm $\|\cdot\|$, and $f : \Omega \to \mathcal{X}$ is a strongly measurable function (i.e., $f(\Omega)$ is essentially a separable set and x^*f is a (real) measurable function for each $x^* \in \mathcal{X}^*$, the adjoint space), then $L^\Phi(\mu, \mathcal{X})$ is the space of all (equivalence classes of) strongly measurable $f : \Omega \to \mathcal{X}$ on (Ω, Σ, μ) such that $\int_\Omega \Phi(\alpha\|f\|)d\mu < \infty$ for some $\alpha > 0$. The corresponding gauge norm $N_\Phi(\cdot)$ is given by:

$$N_\Phi(f) = \inf\{k > 0 : \int_\Omega \Phi\Big(\frac{\|f\|}{k}(\omega)\Big)d\mu(\omega) \le 1\}. \qquad (1)$$

And the Orlicz norm $\|\cdot\|_\Phi$ is similarly defined. These are equivalent norms, and $L^\Phi(\mu, \mathcal{X})$ is a complete normed linear (i.e., a Banach) space, with simple functions dense in case Φ is Δ_2-regular. The following result will be useful for $L^{\Phi_1\Phi_2}$-spaces. It is an extension of the Lebesgue case due to E. J. McShane (1950). A reason for its inclusion here is that it illuminates the structure of the double norms spaces and their application to be discussed below. As usual (Φ, Ψ) is a complementary Young pair.

Theorem 1. *Let (Φ, Ψ) be Δ_2-regular and also be strictly convex. Suppose $\mathcal{X}$ is a real Banach space whose norm is weakly differentiable. Suppose also that its adjoint space $\mathcal{X}^*$ has a similarly differentiable norm, except at the origin. If $L^\Phi(\mu, \mathcal{X})$ is the Orlicz space introduced above, then for each $x^* \in [L^\Phi(\mu, \mathcal{X})]^*$, there is a unique g_{x^*} $[\in L^\Phi(\mu, \mathcal{X})]$ of unit norm such that for all f in $L^\Phi(\mu, \mathcal{X})$,*

$$x^*(f) = \|x^*\| \frac{\int_\Omega \Phi'(\|g_{x^*}\|) \frac{d}{dt}[\|g_{x^*} + tf\|]_{t=0} d\mu}{\int_\Omega \|g_{x^*}\| \Phi'(\|g_{x^*}\|) d\mu}. \tag{2}$$

Proof: By hypothesis, via Theorem IV.1.10, $L^\Phi(\mu)$ is reflexive, $(L^\Phi(\mu))^* = L^\Psi(\mu)$, and Φ is strictly convex; hence by Theorem 1.3, $(L^\Phi(\mu), N_\Phi(\cdot))$ is strictly convex. [The same will be true with the Orlicz norm also by the direct part of Corollary 1.7.] We use the gauge norm in one space, say the L^Φ, and the Orlicz norm in the adjoint L^Ψ, so that the duality pair can be used without further comment. We now specialize a classical result on vector valued function spaces due to I. Halperin (1954, p. 205) to conclude that $L^\Phi(\mu, \mathcal{X})$ and $L^\Psi(\mu, \mathcal{X}^*)$ are both reflexive and each can be identified as the adjoint space of the other, since $\mathcal{X}$ is reflexive by hypothesis. Moreover, in a reflexive Banach space strict convexity (= rotundity) and weak differentiability of norm are dual concepts (cf., e.g., Day (1962, p. 114). It also follows that $L^\Phi(\mu)$, $L^\Psi(\mu)$, $\mathcal{X}$ and $\mathcal{X}^*$ are all rotund. We now assert that $L^\Phi(\mu, \mathcal{X})$ [and $L^\Psi(\mu, \mathcal{X}^*)$] are rotund and then establish (2).

The rotundity assertion is simple. Indeed, if $L^\Phi(\mu)$ and $\mathcal{X}$ are rotund but $L^\Phi(\mu, \mathcal{X})$ is not, then there exist $0 \notin f_i \in L^\Phi(\mu, \mathcal{X})$, $i = 1, 2$, such that $f_1 \neq t f_2$ a.e. (t a scalar $\neq 1$) and $\lambda(\|f_1 + f_2\|) = \lambda(\|f_1\|) + \lambda(\|f_2\|)$ where $\lambda(\cdot)$ denotes the norm of $L^\Phi(\mu, \mathcal{X})$. Since $\lambda(\|f_i\|) > 0$, for simplicity we can take $\lambda(\|f_1 + f_2\|) = 1$ here. The rotundity of $\mathcal{X}$ implies $\|f_1 + f_2\| < \|f_1\| + \|f_2\|$ on a set of positive μ-measure. Thus $u = \|f_1\| + \|f_2\| - \|f_1 + f_2\| \geq 0$ satisfies $\lambda(u) > 0$. Hence

$$1 = \lambda(\|f_1 + f_2\|) = \lambda(\|f_1\|) + \lambda(\|f_2\|)$$

$$\geq \lambda(\|f_1\| + \|f_2\|)$$
$$= \lambda(\|f_1 + f_2\| + \alpha u), \qquad 0 \leq \alpha \leq 1,$$
$$\geq \lambda(\|f_1 + f_2\|) = 1, \text{ since } \lambda \text{ is monotone.} \qquad (3)$$

Thus if $g_1 = \|f_1 + f_2\|$ and $g_2 = \|f_1 + f_2\| + \alpha u$, then $\lambda(g_i) = 1$, $i = 1, 2$, and $\lambda((g_1 + g_2)/2) = \lambda(\|f_1 + f_2\| + \frac{\alpha}{2}u) = 1$, by (3). But the rotundity of $L^{\Phi}(\mu)$ then implies $\lambda(g_1 - g_2) = 0$ so that $u = 0$ a.e., a contradiction. Thus $L^{\Phi}(\mu, \mathcal{X})$ is rotund and similarly $L^{\Psi}(\mu, \mathcal{X}^*)$ is rotund since $L^{\Psi}(\mu)$ and $\mathcal{X}^*$ have the same property (recall that $\mathcal{X}$ has a weakly differentiable norm). [Note that there is no conflict in using the same norm symbol for $L^{\Phi}(\mu, \mathcal{X})$ and $L^{\Phi}(\mu)$ since $u \in L^{\Phi}(\mu), x \in \mathcal{X}$ with $\|x\| = 1$ implies $ux \in L^{\Phi}(\mu, \mathcal{X})$, $\lambda(\|ux\|) = \lambda(u)$, and the set of such functions is norm determining for $L^{\Phi}(\mu, \mathcal{X})$.] Hence $L^{\Phi}(\mu, \mathcal{X})$ and $L^{\Psi}(\mu, \mathcal{X}^*)$ have weakly differentiable norms and these should be calculated. An auxiliary result will be used for completing the argument.

We first motivate a reason for this effort. It is a consequence of a classical result of R. C. James (1957) and others that $0 \neq \ell \in \mathcal{Y}^*$ implies the existence of a unique $g_\ell \in \mathcal{Y}$ (a reflexive rotund Banach space), $\|g_\ell\| = 1$ satisfying $\ell(g) = \|\ell\|$. However, we need to find an expression for $\ell(y)$, for all $y \in \mathcal{Y}$. In the present case, it is possible to apply an extension of the procedure of Theorem IV.1.6 using several properties of Bochner's integration theory. This constitutes a great digression and cannot be included here. Fortunately, we can present an alternative method when $\mathcal{Y}$ also has a differentiable norm. This is plausible from the fact that the (weak) derivative of the norm functional is linear and continuous, as seen from Theorems 2.2 and 2.3. We make this remark precise by establishing the following result essentially due to McShane (1950):

Lemma 2. *Let $\mathcal{Y}$ be a reflexive Banach space with norm $\|\cdot\|$, which is rotund and which is also smooth, i.e., has a (weakly) differentiable norm. Then we have for each $\ell \in (\mathcal{Y}^*, \|\cdot\|^*)$,*

$$\ell(g) = \|\ell\|^* \frac{d}{dt}(\|g_\ell + tg\| + i\|g_\ell - itg\|)_{t=0}, \quad g \in \mathcal{Y}, \qquad (4)$$

for a unique unit vector g_ℓ of $\mathcal{Y}$.

Proof: Since $\mathcal{Y}$ is given to be reflexive, rotundity and smoothness are dual properties in such a space as noted before. So it follows that

$\mathcal{Y}^*$ is also rotund with a (weakly) differentiable norm. Let us assume that $\ell \neq 0$ since otherwise (4) is trivial. Also as noted above, there exists a $g \in \mathcal{Y}$, $\|g\| = 1$ such that $\ell(g) = \|\ell\|^* > 0$. For convenience we take $\|\ell\|^* = 1$ in the proof. The key idea now is to observe that $\ell(g + z[f - \ell(f)g]) = \ell(g) = 1 = \|\ell\|^*$, for all $z \in \mathbb{C}$ and all $f \in \mathcal{Y}$, so that

$$1 = \ell(g + z[f - \ell(f)g]) \leq \|g + z(f - \ell(f)g)\|. \tag{5}$$

But for any real t, $|t| < 1$, we have

$$\|g + tf\| - \|g\| = |(1 + t\ell(f))|\|g + t[1 + t\ell(f)]^{-1}(f - \ell(f)g)\| - 1$$
$$\geq |1 + t\ell(f)| \geq t\,\mathrm{Re}(\ell(f)), \qquad \text{by (5)}. \tag{6}$$

Dividing by $t > 0$ and letting $t \downarrow 0$, and similarly for $t < 0$ and $t \uparrow 0$, we get (since the norm is (weakly) differentiable as noted above):

$$\mathrm{Re}(\ell(f)) = \frac{d}{dt}(\|g + tf\|)_{t=0}. \tag{7}$$

Replacing f by $-if$ in (7), we get

$$\mathrm{Im}(\ell(f)) = \frac{d}{dt}(\|g - itf\|)_{t=0}. \tag{8}$$

Adding (7) and (8) gives (4), except that we need to show that $g = g_\ell$ is unique. This however is immediate since, if there are two such elements g_1, g_2, then $g_\alpha = \alpha g_1 + (1-\alpha)g_2$ will also give $\ell(g_\alpha) = \|\ell\|^* = 1$, for $0 < \alpha < 1$. Since this implies $\|g_\alpha\| = 1$, the unit ball of $\mathcal{Y}$ contains the segment g_α, $0 < \alpha < 1$, so that $\mathcal{Y}$ is not rotund, contrary to hypothesis. ∎

We are now ready to complete the proof of Theorem 1.

Proof of Theorem 1 (continued): If f_0, f are on the unit sphere of $L^\Phi(\mu, \mathcal{X})$ and $k(t) = N_\Phi(\|f_0 + tf\|)$, then $\frac{dk}{dt}$ can be calculated as in Theorem 2.3. Thus

$$\int_\Omega \Phi\left(\frac{\|f_0 + tf\|}{k(t)}\right) d\mu = 1$$

can be differentiated and take the derivative under the integral sign using the dominated convergence theorem, since for $|t| < 1$ there is

$0 < \alpha \le k(t)$ such that the integrand is dominated by $\Phi\left(\frac{\|f_0\|+\|f\|}{\alpha}\right)$ which is integrable. Since $k(0) = 1$, we get

$$G(f_0; f) = \left.\frac{dk(t)}{dt}\right|_{t=0} = \frac{\int_\Omega \Phi'(\|f_0\|)\frac{d}{dt}[\|f_0 + tf\|]_{t=0}d\mu}{\int_\Omega \|f_0\|\Phi'(\|f_0\|)d\mu}, \qquad (9)$$

as in equation (2.10). But we already noted that $L^\Phi(\mu, \mathcal{X})$ is reflexive, rotund, and smooth. hence by Lemma 2, there is a unique unit vector g_ℓ in $L^\Phi(\mu, \mathcal{X})$ such that $\ell(f) = \|\ell\|^*G(g_\ell, f)$ where $G(g_\ell, f)$ is given by (9) with $f_0 = g_\ell$. Note that our space being real, the imaginary part in (4) (or (8)) disappears. This gives (2) in all detail. ∎

An immediate consequence of the above result is the useful

Corollary 3. *Let* (Φ, Ψ) *be a complementary Young pair, both* Δ_2-*regular, and strictly convex. If* $\mathcal{X}$ *is a uniformly convex Banach space with a (weakly) differentiable norm, in particular, if* $\mathcal{X}$ *is a Hilbert space, then for each* $\ell \in [L^\Phi(\mu, \mathcal{X})]^*$ *there is a unique unit vector* g_ℓ *such that* $\ell(f)$ *is given by* (2).

This result when $\Phi(x) = |x|^p$, $1 < p < \infty$ is due to McShane (1950), and Theorem 1 is an extension of it, given in Rao (1965) with an unnecessarily stronger restriction that $\mathcal{X}^*$ shall have a Fréchet differentiable norm. It should be noted that the same result of Theorem 1 holds if the Orlicz norm is used and the spaces are complex valued with simple modifications. These will be omitted here.

The above considerations will be specialized if $\mathcal{X} = L^{\Phi_2}(\mu_2)$ denoting the resulting $L^{\Phi_1}(\mu_1, \mathcal{X})$ as $L^{\Phi_1\Phi_2}(\mu)$ which includes the case $\Phi_1 = \Phi$, $\Phi_2 = \Psi$, the complementary Young function of Φ in the original treatment in Zaanen (1953). We now present the precise result and then discuss some consequences as applications. Here is the main representation.

Theorem 4. *Let* $(\Omega_i, \Sigma_i, \mu_i)$ *be measure spaces,* (Φ_i, Ψ_i) *be complementary Young pairs which are* Δ_2-*regular and strictly convex,* $i = 1, 2$. *If* $L^{\Phi_1}(\mu_1), L^{\Psi_1}(\mu_1)$ *and* $L^{\Phi_2}(\mu_2), L^{\Psi_2}(\mu_2)$ *are the corresponding (real) Orlicz spaces, let* $L^{\Phi_1\Phi_2}$ *(* $= L^{\Phi_1}(\mu_1, L^{\Phi_2}(\mu_2))$ *) be the Orlicz space of functions* $f : \Omega_1 \to L^{\Phi_2}(\mu_2)$ *[i.e.,* $f(\omega_1) = f(\cdot, \omega_1) \in L^{\Phi_2}(\mu_2)$ *and* $N_{\Phi_2}(f(\cdot, \cdot)) \in L^{\Phi_1}(\mu_1)$ *with norm* $N_{\Phi_1\Phi_2}(\cdot) = N_{\Phi_1}(N_{\Phi_2}(\cdot))$ *] and similarly* $L^{\Psi_1\Psi_2}$ *be defined. Then for each* ℓ *in* $[L^{\Phi_1\Phi_2}]^*$ *there is a unique* $g \in L^{\Psi_1\Psi_2}$ *such that*

$$\ell(f) = \int\limits_{\Omega_1} [\int\limits_{\Omega_2} g(\omega_2, \omega_1) f(\omega_2, \omega_1) d\mu_2] d\mu_1, \quad f \in L^{\Phi_1 \Phi_2} \qquad (10)$$

and

$$\|\ell\| = \big\| \|g\|_{\Psi_2} \big\|_{\Psi_1} = \|g\|_{\Psi_1 \Psi_2}. \qquad (11)$$

Proof: The result is a simple identification of various parts of Theorem 1, and we indicate the main points. The hypothesis here implies that $L^{\Phi_i}(\mu_i)$, $i=1,2$, are reflexive and rotund. Analogous statement holds for $L^{\Psi_i}(\mu_i)$ also. Hence taking $\mathcal{X} = L^{\Phi_2}(\mu_2)$ in Theorem 1, $L^{\Phi_1 \Phi_2}$ is $L^{\Phi_1}(\mu_1, \mathcal{X})$, and we get for each ℓ in $(L^{\Phi_1 \Phi_2})^*$, a unique g_ℓ in $L^{\Phi_1 \Phi_2}$ such that for all $f \in L^{\Phi_1 \Phi_2}$

$$\ell(f) = \|\ell\| \frac{\int\limits_{\Omega_1} \Phi_1'(N_{\Phi_2}(g_\ell)) \frac{d}{dt}[N_{\Phi_2}(g_\ell + tf)]_{t=0} d\mu}{\int\limits_{\Omega_1} N_{\Phi_2}(g_\ell) \Phi_2'(N_{\Phi_2}(g_\ell)) d\mu_1}. \qquad (12)$$

Substituting the value of the derivative from Theorem 2.3 here, we get

$$\ell(f) = \int\limits_{\Omega_1} \int\limits_{\Omega_2} \left[\frac{\|\ell\|}{\alpha} \Phi_1'\left(N_{\Phi_2}(g_\ell(\cdot, \omega_1))\right) \Phi_2'\left(\frac{g_\ell(\omega_2, \omega_1)}{N_{\Phi_2}(g_\ell(\cdot, \omega_1))}\right) \times \right.$$
$$\left. \times \frac{1}{\int\limits_{\Omega_2} N_{\Phi_2}(g_\ell(\cdot, \omega_1)) \Phi_2'(\omega_2, \omega_1) d\mu_2} \right] f(\omega_2, \omega_1) d\mu_2 d\mu_1, \quad (13)$$

where $\alpha = \int\limits_{\Omega_1} N_{\Phi_2}(g_\ell(\cdot, \omega_1)) \Phi_2'\left(N_{\Phi_2}(g_\ell(\cdot, \omega_1))\right) d\mu_1(\omega_1) > 0$. Calling the coefficient of f in (13) by g, we arrive at (10). Since the Gateâux derivative satisfies the norm conditions, we must have (11) for g also. It should be observed that because of the assumed conditions on Φ_1, Φ_2, the g_ℓ and f are supported by a σ-finite $\mu_1 \otimes \mu_2$ set, although this set may vary with f. If the pairs (Φ_i, Ψ_i) are (strongly) normalized, then $\alpha = 1$, and the integrals in the denominators will be unity a.e.$[\mu_1]$, and the formulas simplify slightly. Thus (10) holds as stated. ∎

In the preceding theorems we also demanded strict convexity of the Young functions (Φ_i, Ψ_i), $i = 1, 2$ (and $\mathcal{X}$) to use the work of the preceding sections. However, for the representation of continuous linear functionals, this (additional) restriction can be disbanded. The reason for this is as follows. We discuss the detail for Theorem 4. If (Φ_i, Ψ_i), $i = 1, 2$ are Δ_2-regular, then the classical results recalled at the beginning of the proof of Theorem 1 imply that $L^{\Phi_i}(\mu_i)$, $i = 1, 2$, are reflexive and also $L^{\Phi_1 \Phi_2}$ is a reflexive Banach space. Now excluding the trivial case that this space is finite dimensional, so that $L^{\Phi_i}(\mu_i)$, $i = 1, 2$, may be taken infinite dimensional, we may also assume that $\Phi_i(x) \nearrow \infty$ as $x \nearrow \infty$, and continuous at $x = 0$, so that these are N-functions. In this case, each $L^{\Phi_i}(\mu_i)$ is isomorphic (= topologically equivalent) to a uniformly convex and smooth Orlicz space by Theorem 3.2 on the same measure spaces. Hence $L^{\Phi_1 \Phi_2}$ is also isomorphic to $L^{\tilde{\Phi}_1 \tilde{\Phi}_2}$ on the same measure space. We may now apply the preceding theorem to the letter space, and can go back to the given space by the method of "diagram chasing," since all the maps are topological equivalences with norm topologies.

$$
\begin{array}{ccc}
L^{\Phi_1 \Phi_2} & \overset{\tau}{\longrightarrow} & L^{\tilde{\Phi}_1 \tilde{\Phi}_2} \\
\downarrow T_1 & & \downarrow \tilde{T}_1 \\
(L^{\Phi_1 \Phi_2})^* = \quad L^{\Psi_1 \Psi_2} & \overset{\tau^*}{\longleftarrow} & (L^{\tilde{\Phi}_1 \tilde{\Phi}_2})^* \quad = L^{\tilde{\Psi}_1 \tilde{\Psi}_2}
\end{array}
$$

We state this result for reference as:

Theorem 5. Let $(\Phi_i, \Psi_i) \in \Delta_2$, $i = 1, 2$, and $(\Omega_i, \Sigma_i, \mu_i)$, $i = 1, 2$, be measure spaces. Then $(L^{\Phi_1 \Phi_2})^* = L^{\Psi_1 \Psi_2}$ and the representation (10) holds. Using the gauge norms in the $L^{\Phi_i}(\mu_i)$ and the Orlicz norms in $L^{\Psi_i}(\mu_i)$, one may then conclude (11) for the equivalence, in the usual way.

Note that by demanding that (Φ_i, Ψ_i) obey global Δ_2-conditions, we need not mention the dimensionality of the spaces in question. This is sufficient for most applications, and in particular for the following ones.

As a simple consequence we present the following.

Proposition 6. Let (Φ, Ψ) be a complementary Young pair such that Φ obeys the Δ_2-condition as well as the Milnes's condition: $\Phi'((1+\varepsilon)x) \geq k_\varepsilon \Phi'(x)$, $x \geq 0$, for each $\varepsilon > 0$ and a $k_\varepsilon > 1$. Let $L^\Phi(\mu)$ be the corresponding Orlicz space on (Ω, Σ, μ), and $\{f_n, n \geq 1\} \subset L^\Phi(\mu)$ be a sequence such that $N_\Phi(f_n) \leq 1$. If there is a g in $L^\Psi(\mu)$, with $a_n = \int_\Omega f_n g d\mu \to a = \|g\|_\Psi$, then $f_n \to f$, for some f, in norm. In particular, if $\Phi(x) = \frac{x^2}{2}$, then $f = g[\int_\Omega f^2 d\mu / \int_\Omega g^2 d\mu]^{\frac{1}{2}}$, a.e. in the last assertion.

Proof: Under the hypothesis on Φ, $L^\Phi(\mu)$ is uniformly convex, and it has also a (weakly) differentiable norm. [This statement holds for both the gauge and Orlicz norms.] If $\ell_g(f_n) = \int_\Omega f_n g d\mu = a_n$, then the hypothesis and the representation of such functionals imply (cf. Theorem IV.1.6) $\|\ell_g\| = \|g\|_\Psi$. Since $N_\Phi(f_n) \leq 1$, we have for $\varepsilon > 0$, $m, n \geq n(\varepsilon)$,

$$\ell_g\left(\frac{[f_m + f_n]}{2}\right) = \frac{1}{2}[\ell_g(f_m) + \ell_g(f_n)] > \|\ell_g\|(1 - \delta_\varepsilon) \qquad (14)$$

for some $\delta_\varepsilon > 0$ by the uniform convexity of $L^\Phi(\mu)$. Hence $N_\Phi(f_n + f_m) > 2(1 - \delta_\varepsilon)$, and this implies $N_\Phi(f_m - f_n) < \varepsilon$. So $f_n \to f$ strongly for some f, and $N_\Phi(f) = 1$ since

$$\|\ell_g\| = \lim_n \ell_g(f_n) = \ell_g(f) \leq \|\ell_g\| N_\Phi(f) \leq \|\ell_g\|. \qquad (15)$$

Finally, if $\Phi(x) = \frac{x^2}{2}$, then (Φ, Ψ) is strongly normalized, and the equality conditions in the Hölder inequality gives the last part. ∎

This application with (Ω, Σ, μ) as a probability space and $\{f_n, n \geq 1\}$ as a sequence of second order, but not necessarily independent, random variables and $f_n = E^{\mathcal{B}_n}(h)$ for a sequence $\mathcal{B}_n \subset \mathcal{B}_{n+1} \subset \Sigma$, σ-algebras, then a standard martingale convergence theorem shows that $f_n \to f = E^{\mathcal{B}_\infty}(h)$, a.e. and in norm where $\mathcal{B}_\infty = \sigma(\bigcup_n \mathcal{B}_n)$. Thus, taking $\|h\|_2 = 1$, one sees that $g = f$ a.e. in the proposition. However, the martingale condition is not assumed here, and this shows how other types of limit results can be obtained in this context with suitable alternative conditions. We turn to a more substantive application.

In fact, using the representation of Theorem 5 (or 4) we wish to present an analog of the interpolation result (Theorem VI.3.12) with weights under somewhat different restrictions. This has interest in certain aspects of harmonic analysis, especially the multiplier theory. We start with an auxiliary result.

Proposition 7. *Let Φ_i, $i = 0, 1$, be continuous Young functions and Φ_s be the intermediate function to them (cf., Definition VI.3.1). If $(L^{\Phi_s}(\mu), \|\cdot\|_{\Phi_s})$, $0 \leq s \leq 1$, is an Orlicz space on (Ω, Σ, μ), u, v are measurable functions and $a = |u|^{1-s}|v|^s$, then for each f such that $uf \in L^{\Phi_0}(\mu)$, $vf \in L^{\Phi_1}(\mu)$ one has $af \in L^{\Phi_s}(\mu)$, $0 < s < 1$. Moreover,*

$$\|af\|_{\Phi_s} \leq (\|uf\|_{\Phi_0})^{1-s}(\|vf\|_{\Phi_1})^s. \qquad (16)$$

Proof: We first briefly sketch the alternative representation of $L^{\Phi_s}(\mu)$, noted in the discussion preceding Theorem VI.3.8, due to Calderón. It is used to establish (16) and this will prove the proposition. Thus let A_{Φ_s} be defined as:

$$A_{\Phi_s} =$$
$$\{f \in L^0(\mu) : |f| \le \lambda |g|^{1-s}|h|^s, a.e., \lambda > 0, N_{\Phi_0}(g) \le 1, N_{\Phi_1}(h) \le 1\} \quad (17)$$

then A_{Φ_s} is also representable as:

$$A_{\Phi_s} =$$
$$\{f \in L^0(\mu) : |f| \le \lambda (\Phi_0^{-1}(|g|))^{1-s}(\Phi_1^{-1}(|h|))^s, a.e., \lambda > 0, g, h \in U\}, \quad (18)$$

where U is the unit ball of $L^1(\mu)$. For A_{Φ_s} the norm is given by: $\|f\| = \inf\{\lambda > 0 : \lambda$ satisfies the inequality in (18)$\}$. Now for each g, h in $L^1(\mu)$, and $f \in L^0(\mu)$ such that

$$|f| \le \lambda (\Phi_0^{-1}(|g|))^{1-s}(\Phi_1^{-1}(|h|))^s$$
$$\le 2\lambda \left(\Phi_0^{-1}\left(\frac{|g|+|h|}{2}\right)\right)^{1-s}\left(\Phi_1^{-1}\left(\frac{|g|+|h|}{2}\right)\right)^s, \quad \text{since } \Phi_i^{-1}$$

$$\text{are concave, nonnegative, and increasing,}$$

$$= 2\lambda \Phi_s^{-1}\left(\frac{|g|+|h|}{2}\right), \quad \text{by Definition VI.3.1.}$$

Hence $\int_\Omega \Phi_s\left(\frac{|f|}{2\lambda}\right)d\mu \le 1$ and $N_{\Phi_s}(f) \le 2\lambda$ so that $N_{\Phi_s}(f) \le 2\|f\|$. Thus $A_{\Phi_s} \subset L^{\Phi_s}(\mu)$. But from definition of $L^{\Phi_s}(\mu)$, it follows that $L^{\Phi_s}(\mu) \subset A_{\Phi_s}$ and then the equivalence of the norms is a consequence of the closed graph theorem. We therefore utilize the form of $A_{\Phi_s} = L^{\Phi_s}(\mu)$ as given in (17) with $\|f\| \le N_{\Phi_s}(f)$.

To use the equivalent Orlicz norm and establish (16), consider

$$I = \int_\Omega |af||g|d\mu, \quad g \in L^{\Psi_s}(\mu) \qquad (af \text{ is as in the statement}). \quad (19)$$

With g in A_{Ψ_s}, where Ψ_s is the intermediate function to Ψ_0 and Ψ_1, the complementary Young functions of Φ_0 and Φ_1, we can find $g_i^n \in L^{\Psi_i}(\mu)$, $i = 0, 1$, $\lambda_n > 0$ such that $\lambda_n \to \|g\|$ where

$$|g| \le \lambda_n |g_0^n|^{1-s}|g_1^n|^s, \quad N_{\Psi_i}(g_i^n) \le 1, i = 0, 1. \quad (20)$$

Substituting (20) in (19) and using the Hölder inequality twice, one gets

$$I \le \lambda_n \left(\int_\Omega |uf||g_0^n|d\mu\right)^{1-s}\left(\int_\Omega |vf||g_1^n|d\mu\right)^s$$

$$\leq \lambda_n(\|uf\|_{\Phi_0})^{1-s}(\|vf\|_{\Phi_1})^s$$
$$\rightarrow \|g\|(\|uf\|_{\Phi_0})^{1-s}(\|vf\|_{\Phi_1})^s, \quad \text{as } n \rightarrow \infty,$$
$$\leq N_{\Psi_s}(g)(\|uf\|_{\Phi_0})^{1-s}(\|vf\|_{\Phi_1})^s. \tag{21}$$

But now taking the supremum in (21) as g's vary in the unit ball of $L^{\Psi_s}(\mu)$, one gets the Orlicz norm of af in (19), and so (16) holds. ∎

We can now present the desired convexity theorem with weights, and the method of proof does not involve the complex method [hence no need for the maximum modulus theorem] in contrast to the work of Section 6.3. Instead, Theorem 5 [or 4] above will be employed. The treatment follows W. T. Kraynek (1970).

Theorem 8. *Let (Φ_i, Ψ_i) and (Q_i, R_i), $i = 1, 2$, be pairs of complementary Young functions all of which are in Δ_2. If $(\Omega_i, \Sigma_i, \mu_i)$ are measure spaces, $L^{\Phi_i}(\mu_i), L^{Q_i}(\mu_i)$ are the Orlicz spaces on them, let $u_i, v_i \geq 0$ be locally in $L^{\Phi_i}(\mu_i), L^{Q_i}(\mu_i)$ in the sense that they are in these spaces when restricted to sets of finite measures, $i = 1, 2$. Suppose that $\ell \in (L^{\Phi_1\Phi_2})^* \cap (L^{Q_1Q_2})^*$ satisfying*

$$|\ell_1(f)| = \left|\ell\left(\frac{f}{u_1 v_1}\right)\right| \leq A_1 N_{\Phi_1\Phi_2}(f), \quad f \in L^{\Phi_1\Phi_2},$$

and $\tag{22}$

$$|\ell_2(f)| = \left|\ell\left(\frac{f}{u_2 v_2}\right)\right| \leq A_2 N_{Q_1 Q_2}(f), \quad f \in L^{Q_1 Q_2}.$$

If $a = u_1^{1-s} v_1^s$, $b = u_2^{1-s} v_2^s$, and Φ_s, Q_s are the intermediate Young functions of (Φ_1, Φ_2) and (Q_1, Q_2), then for each $f = f_1 f_2$, $f_1 \in L^{\Phi_s}(\mu_1), f_2 \in L^{Q_s}(\mu_2)$, such that $f \in L^{\Phi_s Q_s}$, we have

$$\left|\ell\left(\frac{f}{ab}\right)\right| \leq C A_1^{1-s} A_2^s N_{\Phi_s}(f_1) N_{Q_s}(f_2), \quad 0 < s < 1. \tag{23}$$

Thus for each f_1 with $af_1 \in L^{\Phi_s}(\mu_1)$ and f_2 with $bf_2 \in L^{Q_s}(\mu_2)$ the bound (23) holds, where $0 < A_i < \infty$, are constants and $0 < C < \infty$ is another constant (≤ 16 can be taken).

Proof: By (22), ℓ_1, ℓ_2 are continuous linear functionals on $L^{\Phi_1\Phi_2}$ and $L^{Q_1Q_2}$, which are reflexive by the conditions on Φ_i's and Q_i's. Hence by Theorem 5, we have

$$\ell_i(f) = \int_{\Omega_1} \left[\int_{\Omega_2} g_i(\omega_1, \omega_2) f(\omega_1, \omega_2) d\mu_2\right] d\mu_1, \tag{24}$$

which, with Orlicz norms in the original spaces and gauge norms in the duals, show $\|\ell_1\| = N_{\Psi_1\Psi_2}(g_1) \leq A_1$, $\|\ell_2\| = N_{R_1R_2}(g_2) \leq A_2$. Now let $A \in \Sigma_1, B \in \Sigma_2$ of finite measures and put $h = \chi_A\chi_B$. Then (22) and (24) give for $f_i = u_iv_ih$

$$\ell_1(f_1) = \ell_2(f_2) = \ell(h),\tag{25}$$

and hence

$$\iint_{AB} g_1(\omega_1,\omega_2)u_1(\omega_1)v_1(\omega_2)d\mu_2d\mu_1 = \iint_{AB} g_2(\omega_1,\omega_2)u_2(\omega_1)v_2(\omega_2)d\mu_2d\mu_1.$$

Since A, B are arbitrary and of finite measure, this shows that the integrands can be identified $a.e.(\mu_1\otimes\mu_2)$ using the standard arguments. Let the common value be denoted by g. Let f be a measurable function on $\Omega_1 \times \Omega_2$ such that $u_1v_1f \in L^{\Phi_1\Phi_2}$ and $u_2v_2f \in L^{Q_1Q_2}$. Then (25) implies

$$\ell(f) = \int_{\Omega_1}\int_{\Omega_2} g(\omega_1,\omega_2)f(\omega_1,\omega_2)d\mu_2d\mu_1.\tag{26}$$

Moreover, the definition g gives $(u_1v_1)^{-1}g \in L^{\Psi_1\Psi_2}$, $(u_2v_2)^{-1}g \in L^{R_1R_2}$, so that $v_1^{-1}g(\omega_1,\cdot) \in L^{\Psi_2}(\mu_2)$, $v_2^{-1}g(\omega_1,\cdot) \in L^{R_2}(\mu_2), a.a.(\omega_1)$. Hence by Proposition 7, $b^{-1}g(\omega_1,\cdot) \in L^{\Psi_s}(\mu_2)$, and then for $a.a.(\omega_1)$,

$$N_{\Psi_s}(b^{-1}g(\omega_1,\cdot)) \leq 2(\|v_1^{-1}g(\omega_1,\cdot)\|_{\Psi_1})^{1-s}(\|v_2^{-1}g(\omega_1,\cdot)\|_{\Psi_2})^s.\tag{27}$$

Let $f(\omega_1,\omega_2) = \tilde{f}_1(\omega_1)\tilde{f}_2(\omega_2)$ with $\tilde{f}_i$ as simple functions. Then (26) becomes

$$|\ell(f)| \leq \int_{\Omega_1}\int_{\Omega_2} |g(\omega_1,\omega_2)||\tilde{f}_1(\omega_1)||\tilde{f}_2(\omega_2)|d\mu_2d\mu_1$$

$$= \int_{\Omega_1}\Big[\int_{\Omega_2} |b^{-1}(\omega_2)g(\omega_1,\omega_2)||b(\omega_2)\tilde{f}_2(\omega_2)|d\mu_2\Big]|\tilde{f}_1(\omega_1)|d\mu_1$$

$$\leq \int_{\Omega_1}|\tilde{f}_1(\omega_1)|N_{\Psi_s}(b^{-1}g(\omega_1,\cdot))\|b\tilde{f}_2\|_{\Phi_s}d\mu_1,\text{ by Hölder's inequality,}$$

$$\leq 2\int_{\Omega_1}|\tilde{f}_1|(\|v_1^{-1}g(\omega_1,\cdot)\|_{\Psi_1})^{1-s}(\|v_2^{-1}g(\omega_1,\cdot)\|_{\Psi_2})^s d\mu_1 \cdot \|b\tilde{f}_2\|_{\Phi_s},$$

$$\text{by (27),}$$

$$= 2\int_{\Omega_1}(b|\tilde{f}_1|)(\|v_1^{-1}u_1(\omega_1)^{-1}g(\omega_1,\cdot)\|_{\Psi_1})^{1-s}\times$$

$$(\|v_2^{-1}u_2(\omega_1)^{-1}g(\omega_1,\cdot)\|_{\Psi_2})^s d\mu_1 \cdot \|b\tilde{f}_2\|_{\Phi_s}.\tag{28}$$

But by hypothesis on u_i, v_i, we get $b\tilde{f}_1 \in L^{\Phi_s}(\mu_1)$, by Proposition 7, and there exist $\lambda_n > 0$, $h_i^n \in L^{\Phi_i}(\mu_1)$, $N_{\Phi_i}(h_i^n) \leq 1$, such that

$$|b\tilde{f}_1| \leq \lambda_n |h_1^n|^{1-s} |h_2^n|^s, \qquad a.e., \ n \geq 1, \tag{29}$$

and $\lim_{n \to \infty} \lambda_n = N_{\Phi_s}(af_1)$. Hence (28) and (29) imply with the Hölder inequality

$$|\ell(f)| \leq 2\lambda_n \left(\int_{\Omega_1} \|v_1^{-1} u_1(\omega_1)^{-1} g(\omega_1, \cdot)\|_{\Psi_1} |h_1^n(\omega_1)| d\mu_1 \right)^{1-s} \times$$

$$\left(\int_{\Omega_1} \|v_2^{-1} u_2(\omega_1)^{-1} g(\omega_1, \cdot)\|_{\Psi_2} |h_2^n(\omega_1)| d\mu_1 \right)^s \|b\tilde{f}_2\|_{\Phi_s},$$

$$\leq 8\lambda_n (\|v_1^{-1} u_1^{-1} g\|_{\Psi_1, R_1})^{1-s} (\|v_2^{-1} u_2^{-1} g\|_{\Psi_2, R_2})^s \|bf_2\|_{\Phi_s}$$

$$\leq 8\lambda_n A_1^{1-s} A_2^s \|bf_2\|_{\Phi_s}, \quad \text{by (22)}. \tag{30}$$

Letting $n \to \infty$ in (30), this reduces to (23) if f is a simple function. But then the general case is an immediate consequence. It remains only to replace the Orlicz norms with twice the gauge norms here. ∎

As an immediate deduction of this result, we can obtain the following interpolation for bilinear (and hence multilinear) forms.

Corollary 9. *Let Φ_i, Q_i, $(\Omega_i, \Sigma_i, \mu_i)$, $u_i, v_i \geq 0$, $i = 1, 2$ be as in the above theorem so that $L^{\Phi_i}(\mu_i), L^{Q_i}(\mu_i)$ are reflexive. Let $\ell :$ $(L^{\Phi_1}(\mu_1) \times L^{\Phi_2}(\mu_2)) \cap (L^{Q_1}(\mu_1) \times L^{Q_2}(\mu_2)) \to I\!R$ be a linear functional such that (analogous to (22))*

$$\left| \ell \left(\frac{f_1}{u_1}, \frac{f_2}{v_1} \right) \right| \leq A_1 N_{\Phi_1}(f_1) N_{\Phi_2}(f_2), \quad (f_1, f_2) \in L^{\Phi_1}(\mu_1) \times L^{\Phi_2}(\mu_2),$$

and

$$\left| \ell \left(\frac{\tilde{f}_1}{u_2}, \frac{\tilde{f}_2}{v_2} \right) \right| \leq A_2 N_{Q_1}(\tilde{f}_1) N_{Q_2}(\tilde{f}_2), \quad (\tilde{f}_1, \tilde{f}_2) \in L^{Q_1}(\mu_1) \times L^{Q_2}(\mu_2).$$

Then

$$\left| \ell \left(\frac{f_1}{a}, \frac{f_2}{b} \right) \right| \leq C A_1^{1-s} A_2^s N_{\Phi_s}(f_1) N_{Q_s}(f_2)$$

for $(f_1, f_2) \in L^{\Phi_s}(\mu_1) \times L^{Q_s}(\mu_2)$, where $a = u_1^{1-s} v_1^s$, $b = u_2^{1-s} v_2^s$, $0 < s < 1$ and Φ_s, Q_s are intermediate to $(\Phi_1, \Phi_2), (Q_1, Q_2)$, and A_1, A_2 and C are absolute constants.

This simple modification yields the useful form with weights for operators. If $T : L^{\Phi_1}(\mu_1) \to L^{\Phi_2}(\mu_2)$ and also on $L^{Q_1}(\mu_1) \to L^{Q_2}(\mu_2)$, then define a bilinear form as:

$$\ell\left(\frac{f_1}{u_i}, \frac{f_2}{v_i}\right) = \int_{\Omega_2} (u_i^{-1} f_1)(\omega_2) T(v_i^{-1} f_2)(\omega_2) d\mu_2,$$

$$i=1,2, \, f_i \in L^{\Psi_2}(\mu_2) \cap L^{R_2}(\mu_2).$$

Under appropriate hypotheses, so that the integral exists, one can obtain the multilinear interpolation theorem from the above result, using the same argument as for Theorem VI.3.9. We state the result precisely and leave the details to the reader.

Proposition 10. Let $\mathcal{X}_i^j = L^{\Phi_{ij}}(\mu_i), \mathcal{Y}^j = L^{Q_j}(\nu)$ be reflexive Orlicz spaces on measure spaces $(\Omega_i, \Sigma_i, \mu_i)$ and $(S, \mathcal{S}, \nu)$, $i = 1, \ldots, n, j = 1, 2$. Let T be a multilinear operator simultaneously defined for simple functions on $\underset{i=1}{\overset{n}{\times}} \mathcal{X}_i^j \to \mathcal{Y}^j$, $j = 1, 2$ such that if $u_i \geq 0$, $i = 1, \ldots, n$, $v_j \geq 0$, $j = 1, 2$, are measurable functions satisfying

$$\|v_j^{-1} T(u_1^{-1} f_1, \ldots, u_n^{-1} f_n)\|_{\mathcal{Y}^j} \leq A_j \prod_{i=1}^{n} \|f_i\|_{\mathcal{X}_i^j}, \quad j=1,2, \, f_i \in \mathcal{X}_i^1 \cap \mathcal{X}_i^2,$$

then

$$\|b_s^{-1} T(a_{1s}^{-1} f_1, \ldots, a_{ns}^{-1} f_n)\|_{Q_s} \leq C A_1^{1-s} A_2^s \prod_{i=1}^{n} \|f_i\|_{\Phi_{is}}, \quad f_i \in L^{\Phi_{is}}(\mu_i),$$

where $b_s = v_1^{1-s} v_2^s$, $a_{is} = (u_i^1)^{1-s}(u_i^2)^s$, $i = 1, \ldots, n$, and $A_j > 0, C > 0$, $j = 1, 2$, are some absolute constants.

The difference between this and the corresponding theorems in Section 6.3 is that here we are always assuming the reflexivity of the spaces whereas in the former case, nonreflexive spaces are admitted. On the other hand, when the weight functions $u_i, v_i \geq 0$ are considered before (cf. Theorem VI.3.12), the (possibly nonreflexive) spaces admitted are $L^{\Phi}(\mu)$ in which Φ is of Δ'-class, and the $L^{\Phi}(\mu)$ can be more general here although they are reflexive. Thus both cases complement each other while they simultaneously extend the results of the Lebesgue classes.

The treatment here shows how the geometry of the underlying spaces contribute to further extend and clarify the structure of problems considered. They can also be used in other contexts, notably in harmonic analysis. This line of development and general considerations will be concluded, as the general viewpoint is sufficiently exemplified.

Bibliographical Notes: Several authors have investigated the geometric properties of general Banach spaces, and such studies are still quite popular in relating them with the analytical aspects. A summary of the work may be found in Day (1962, Chapter VII) and Cudia (1963). Also a detailed analysis and classifications of various convexity properties are available in Cudia (1964). There are two aspects of these studies as they relate to Orlicz spaces. One is to specialize the general properties to these spaces, and the other is to investigate the basic structure of Orlicz spaces towards the geometric aspects motivated by the Lebesgue theory. We have touched on the former and preferred to go into more detail for the latter. The basic study on convexity properties is due to H. W. Milnes (1957) and several of the results and ideas in Sections 1–3 are based on his work. A conversion of some of these to the smoothness aspects was given by Rao (1965). However, the statements of the results in the latter are not as explicit as desirable, although the proofs indicate clearly what is meant. There, an attempt is made to use only the gauge norm by using the normalized Young pairs and certain continuities of the derivatives. The matter is clarified here by using the concept of *strongly normalized* complementary pairs. For the study of strict convexity, this condition admits a relaxation. Analyzing the proof of the results, Turett (1976) has obtained Theorem 1.3 and its consequences. The corresponding work has been established, with the Orlicz norm given as Theorems 1.6 and 2.8, by Milnes (1957) earlier. The isomorphism between the (uniform) convexity and reflexivity, included as Theorem 3.2, is due to Akimovič (1972). This clarifies the relation between the exact Lebesgue theory and the (weaker) Orlicz space isomorphism properties. Similarly, the results on the modulus of convexity are more complicated (and less satisfactory) as seen from Theorem 4.3 which is due to Maleev and Troyanski (1975). Further research seems necessary for this problem, and the current work appears not to be in its final form. In this connection, see also T. Figiel [(1976), Section II].

There are several other articles on the geometric aspects of Orlicz spaces, specialized from the general Banach space theory, due to Wu (1978), Wang, Wang and Li (1986), Kaminska and Kurc (1986),

Cui and Wang (1987), and many others. Most of these papers analyze the problems assuming the measure spaces to be restricted (nonatomic finite or at most σ-finite, or purely atomic, etc.) and using the N-functions with further conditions. For related extensions and an extensive bibliography to the literature on Orlicz and modular spaces, one may refer to Musielak (1983). We did not spend much space for these works, since we intended mainly to present results that lead to substantive applications.

To indicate the last possibility we have treated the interpolation theory and other results briefly in Section 5. Propositions 5.7–5.10 are taken from Kraynek (1970) with simple modifications. As noted in the text, Theorem 5.8 complements the work of Section 6.3 in some respects. In the Lebesgue context, this result was formulated by G. O. Okikiolu (1966), but the treatment there is much more involved and is quite different. It is possible to apply these results to multiplicity theory in harmonic analysis. But that is an independent subject for study. We refer the relevant aspect of it to Edwards and Gaudry (1977) for the interested reader to pursue.

VIII
Orlicz Spaces Based on Sets of Measures

An extension of the preceding theory of Orlicz spaces based on a measure space is considered if the underlying space has an (possibly) uncountable set of measures. Such spaces arise in some important applications in statistical theory, and related subjects. We therefore treat again compactness, reflexivity and (uniform) convexity properties. Also an application to "sufficiency" is given and a corresponding martingale concept is formulated. The existence, uniqueness and construction problems of optimal estimators of functions of the measures are also included. The material in this chapter is slightly technical as it depends heavily on measure theoretical considerations.

8.1. Orlicz spaces $L^\Phi(\theta, G)$

Let us motivate the subject of this chapter by noting a typical problem of importance in applications. Suppose that X is a random variable (= a real measurable function) on a probability space (Ω, Σ, P) having a Gaussian distribution with mean α and variance 1, so that

$$P[X < x] = (2\pi)^{-\frac{1}{2}} \int_{-\infty}^{x} e^{-\frac{1}{2}(u-\alpha)^2} du, \qquad -\infty < \alpha < \infty. \qquad (1)$$

If the right side is denoted $F(x, \alpha)$, then $\{F(\cdot, \alpha), \alpha \in \mathbb{R}\}$ defines a family of Lebesgue-Stieltjes measures. A natural problem here is to estimate a function of α, say $G(\alpha)$ by a function $H(X)$ of X so that the "loss" or the error $|H(X) - G(\alpha)|$ is a minimum. Based on several other important properties of an optimum choice of H, depending on the form of this error, one considers a nonnegative convex function Φ of the errors and minimizes the expected loss. This means one has to find an $H(\cdot)$ such that

$$\int_\Omega \Phi(|H(X) - G(\alpha)|)dP_\alpha = \int_{\mathbb{R}} \Phi(|H(u) - G(\alpha)|)dF(u, \alpha), \quad (2)$$

is a minimum if $\alpha \in I$, a prescribed subset of $\mathbb{R}$. The above equation from the Ω-space to the $(\mathbb{R}, \mathcal{B}, F)$ is just the image law, $(P \circ X^{-1})(-\infty, x)$ $= F(x, \alpha)$. Stated differently, we have a set of probability measures P_α (inducing $F(\cdot, \alpha)$), $\alpha \in I$, and functions $H(\cdot)$ such that $\rho_\Phi^\alpha(H - G(\alpha)) = \int_\Omega \Phi(|H(u) - G(\alpha)|)dP_\alpha < \infty$ and one must find an $H(\cdot)$ having certain extremal properties in a space $\bigcap_{\alpha \in I} L^\Phi(P_\alpha)$. Thus when Φ is a Young function, we need to consider the structure of the Orlicz space $L^\Phi(M)$ where M is a family of probability measures on (Ω, Σ). We therefore introduce the spaces which include the above special case so that the results are of interest in other situations as well.

Since $f \in \bigcap_\alpha L^\Phi(P_\alpha)$ holds when $\sup_\alpha N_{\Phi, \alpha}(f) < \infty$, and since 'sup' is a special convex function, one can consider the set

$$L^\Phi(\theta, G) = \left\{ f \in \mathcal{M}(\Omega, \Sigma) : \sum_{\alpha \in M} \theta\left(\frac{N_{\Phi, \alpha}(f)}{k}\right) G(\alpha) < \infty, \text{ some } k > 0 \right\}, \quad (3)$$

where $\mathcal{M}(\Omega, \Sigma)$ is the space of scalar measurable functions, θ is a convex function on the set M of measures, and $G(\cdot) : M \to (0, \infty)$ is a function. Thus $L^\Phi(\theta, G)$ corresponds to the set of "optimal estimators" for (2), and one needs to know its structure which depends on the convex functions θ and Φ. In fact, when $\Phi = |x|^p$, $p \geq 1$, and $\theta(x) = 0$ for $|x| \leq 1$, $= +\infty$ for $|x| > 1$, it was originally considered by T. S. Pitcher (1965). Most of the results in this case were formulated by him. If M is a dominated set of probability measures (i.e., $P_\alpha \ll \mu$ for a fixed σ-finite measure μ) and $G(\alpha) = 1$, then the above stated statistical optimization problem was extensively treated in the literature. The latter generalization was motivated by a desire to solve the same type of problems when M is not dominated. This is the situation that is

at the back of the work in this chapter, and we present an outline of
the elements of the general theory with a few applications. The Orlicz
space extensions of Pitcher's results are due to R. L. Rosenberg (1968,
cf. also 1970) whose work will be used for the most part.

To analyze the set $L^\Phi(\theta, G)$, we assume for convenience that the
measure space (Ω, Σ, M) is complete, i.e., each set $A \subset \Omega$ such that
$\mu^*(A) = 0$ for the outer measure μ^* of each μ in M then $A \in \Sigma$. So $f_i :$
$\Omega \to I\!R$, $i = 1, 2$, and $f_1 = f_2$ a.e.$[\mu]$ for each μ in M, then one says that
$f_1 = f_2[M]$. With this understanding, for each $f \in \bigcap\limits_{\mu \in M} L^\Phi(\Omega, \Sigma, \mu)$,
let $F_f : M \to [0, \infty]$ be defined as $F_f(\mu) = \|f\|_{\Phi,\mu}$ where $\|\cdot\|_{\Phi,\mu}$ is the
(Orlicz) norm of $L^\Phi(\mu)$ $[= L^\Phi(\Omega, \Sigma, \mu)]$. Then we set

$$N_{\Phi,\theta}(f) = N_\theta(F_f) = \inf \Big\{ k > 0 : \sum_{\mu \in M} \theta\Big(\frac{F_f(\mu)}{k}\Big) G(\mu) \leq 1 \Big\}, \quad (4)$$

the gauge norm of $\ell^\theta(M, \mathcal{P}(M), G)$, the weighted sequence Orlicz space.
We then have:

Theorem 1. *The set $\{L^\Phi(\theta, G), N_{\Phi,\theta}(\cdot)\}$ is a Banach space over
the scalar field.*

Remark. Although the space given by (1) involves two norms, it is
different from the "double norm" spaces considered in Section 7.5 and
there exist both conceptual as well as technical distinctions between
these objects. Therefore this result needs a separate proof. Also, to
minimize the already complex notations, we keep M to be the set of
probability measures here and throughout the chapter.

Proof of Theorem: The fact that $N_{\Phi,\theta}(\cdot)$ is a norm and hence $L^\Phi(\theta, G)$
is a normed linear space is easily inferred from the computation: $f, g \in$
$L^\Phi(\theta, G)$, α a scalar, implies with (4),

$$N_{\Phi,\theta}(\alpha f + g) = N_\theta(F_{\alpha f + g}) \leq N_\theta(|\alpha| F_f + F_g)$$
$$\leq |\alpha| N_\theta(F_f) + N_\theta(F_g) = |\alpha| N_{\Phi,\theta}(f) + N_{\Phi,\theta}(g);$$

and $N_{\Phi,\theta}(f) = 0$ yields $F_f(\mu) = \|f\|_{\Phi,\mu} = 0$, $\mu \in M$ so that $f = 0$,
a.e.$[M]$.

For the completeness, which is not that easy, let $\{f_n, n \geq 1\}$ be a
Cauchy sequence in $\{L^\Phi(\theta, G), N_{\Phi,\theta}(\cdot)\}$ so that $N_{\Phi,\theta}(f_n - f_m) \to 0$

as $n, m \to \infty$. Next choose a subsequence $n_1 < n_2 < \cdots$ such that $N_{\Phi, \theta}(f_{n_{k+1}} - f_n) < \frac{1}{2^k}$ for $n > n_{k+1}$. Then

$$\sum_{k=1}^{\infty} N_{\Phi, \theta}(f_{n_k} - f_{n_{k+1}}) < N_{\Phi, \theta}(f_{n_1}) + 1 < \infty. \qquad (5)$$

But by definition of $N_{\Phi, \theta}$ in (4), for each $\varepsilon > 0$, and any $f \in L^{\Phi}(\theta, G)$,

$$\theta\left(\frac{F_f(\mu)}{N_\theta(F_f) + \varepsilon}\right) \le G(\mu)^{-1}, \qquad \mu \in M. \qquad (6)$$

This implies $\|f\|_{\Phi, \mu} = F_f(\mu) \le (N_\theta(F_f) + \varepsilon)\theta^{-1}(1/G(\mu))$. Since $\varepsilon > 0$ is arbitrary, this estimate substituted in (5) gives

$$\sum_{k=1}^{\infty} \|f_{n_k} - f_{n_{k+1}}\|_{\Phi, \mu} \le \theta^{-1}(1/G(\mu))(N_{\Phi, \theta}(f_{n_1}) + 1) < \infty, \qquad \mu \in M. \quad (7)$$

But $L^{\Phi}(\mu) \subset L^1(\mu)$ when μ is a finite measure so that $\|f_n - f_m\|_{1, \mu} \le C\|f_n - f_m\|_{\Phi, \mu}$ where $0 < C = [\Psi^{-1}(1)]^{-1} < \infty$. Choose $k_0(\mu)$ such that the tail part of the series in (7) is bounded by $\frac{1}{C}4^{-k_0}$. Thus we get

$$\sum_{k \ge k_0(\mu)} \|f_{n_k} - f_{n_{k+1}}\|_{1, \mu} < 4^{-k_0}. \qquad (8)$$

This implies for $m > k \ge k_0(\mu)$

$$\mu\{\omega : [|f_{n_k} - f_{n_{k+1}}| + \cdots + |f_{n_{m-1}} - f_{n_m}|](\omega) \ge 2^{-k}\} < 2^{-k_0}.$$

Let $E_{k,m}$ be the set in braces and $E_\mu = \bigcap_{k \ge k_0} \bigcup_{m \ge k} E_{k,m}$. Then $\mu(E_\mu) = 0$ and $f_{n_k} \to f$ pointwise on $\Omega - E_\mu$. Since Σ is M-complete, f is M-measurable and $\|f_{n_k} - f_\mu\|_{\Phi, \mu} \to 0$, so that $\|f_{n_k} - f_\mu\|_{1, \mu} \to 0$ as $k \to \infty$, $\mu \in M$. Hence $f = f_\mu$, a.e. Thus $f \in \bigcap_{\mu \in M} L^{\Phi}(\mu)$, and $\|f_{n_k} - f_\mu\|_{\Phi, \mu} \to 0$. Also $F_{(f_{n_k} - f)}(\mu) = \lim_{m \to \infty} F_{(f_{n_k} - f_{n_m})}(\mu)$. Consequently we have

$$N_{\Phi, \theta}(f_{n_k} - f) = N_\theta(F_{(f_{n_k} - f)}) \le N_\theta\left(\sum_{m \ge k} F_{(f_{n_m} - f_{n_{m+1}})}\right)$$

$$= \lim_{m \to \infty} N_\theta\left(\sum_{\ell = k}^{m} F_{(f_{n_\ell} - f_{n_{\ell+1}})}\right)$$

$$\le \lim_{m \to \infty} \sum_{\ell = k}^{m} N_\theta\left(F_{(f_{n_\ell} - f_{n_{\ell+1}})}\right)$$

$$= \sum_{\ell \ge k} N_{\Phi, \theta}(f_{n_\ell} - f_{n_{\ell+1}}) \to 0, \qquad \text{as } k \to \infty.$$

Hence $f \in L^{\Phi}(\theta, G)$. From this the standard argument shows that this space is complete. ∎

Remark. The $L^{\Phi}(\theta, G)$ is solid in the sense that if $f \in L^{\Phi}(\theta, G)$, g is M-measurable, and $|g| \leq |f|$ a.e.$[\mu]$, then $g \in L^{\Phi}(\theta, G)$; and in fact $N_{\Phi,\theta}(g) \leq N_{\Phi,\theta}(f)$. This follows immediately from (3) and (4).

A simple extension of Theorem III.3.13, given for N-functions, will be useful and is included in the following result. The second assertion is of interest in the structural study of the space $L^{\Phi}(\theta, G)$.

Proposition 2. *Let Φ be a Young function and $\mu \in M$, the set of probability measures on (Ω, Σ). Then there is an equivalent Young function Φ_0, so that $L^{\Phi}(\mu) = L^{\Phi_0}(\mu)$ and for each $f \in L^{\Phi}(\mu)$ one has*

$$\|f\|_{\Phi_0,\mu} = \inf\left\{\frac{1}{k}\left(1 + \int_{\Omega} \Phi(kf)d\mu\right) : k > 0\right\}. \tag{9}$$

Further, if M is a convex set but not a singleton, then it has infinitely many distinct members and

$$N_{\Phi_0,\theta}(f) = \inf\{k > 0 : \sup_{\mu \in M} \|f\|_{\Phi_0,\mu} \leq k\theta^{-1}(0)\}. \tag{10}$$

Hence if $\theta^{-1}(0) = 0$ then $L^{\Phi}(\theta, G) = \{0\}$ for any weight function G, and if θ is discontinuous (so that θ^{-1} is bounded for the Young function θ), $L^{\Phi}(\theta, G)$ does not depend on G and hence can be denoted by L_{θ}^{Φ}.

Proof: Formula (9) was proved in Theorem III.3.13 for all Young functions. It is immediate if $\Phi(x) = |x|$; or $\Phi(x) = 0$ for $0 \leq |x| \leq 1$ and $= +\infty$ for $|x| > 1$. In the former case $L^{\Phi}(\mu) = L^1(\mu)$ and in the latter case $L^{\Phi}(\mu) = L^{\infty}(\mu)$ since μ is a finite measure. If Φ' is bounded, then $L^{\Phi}(\mu) = L^1(\mu)$ so that $\Phi = \Phi_0$ with $\Phi_0(x) = |x|$, and if Φ is discontinuous $L^{\Phi}(\mu) = L^{\infty}(\mu)$ with $\Phi = \tilde{\Phi}_0$ of the (normalized) discontinuous type above. Similarly if $\Phi'(x) \nearrow \infty$, as $|x| \to \infty$, but $\Phi'(x) \to \alpha > 0$ as $|x| \to 0$, then Φ is a principal part of an N-function Φ_0 so that $L^{\Phi}(\mu) = L^{\Phi_0}(\mu)$. In all these cases the norms are equivalent (by the closed graph theorem), i.e.,

$$C_1\|f\|_{\Phi_0,\mu} \leq \|f\|_{\Phi,\mu} \leq C_2\|f\|_{\Phi_0,\mu}, \quad f \in L^{\Phi}(\mu). \tag{11}$$

Here the constants $C_i > 0$ depend on Φ (or Φ_0), and since $\mu(\Omega) = 1$, they are independent of μ for all $\mu \in M$. But (9) is true for all Lebesgue spaces, and hence for any $L^{\Phi_0}(\mu)$. In view of (11) we can (and will) use (9) for $L^{\Phi}(\theta, G)$ in its structural analysis.

By the convexity, if $\mu_1, \mu_2 \in M$, then $\alpha\mu_1 + (1-\alpha)\mu_2 \in M$ for $0 < \alpha < 1$ and these are distinct probability measures. Thus M has infinitely many elements. But if μ_1, μ_2 are two distinct elements of M and $\nu = \alpha\mu_1 + (1-\alpha)\mu_2$, $0 < \alpha < 1$, then $f \in L^{\Phi}(\mu_1) \cap L^{\Phi}(\mu_2) \Rightarrow f \in L^{\Phi}(\nu)$ and $F_f(\cdot)$ is a concave function on the convex set M. To see this, by definition of the Orlicz norm,

$$\|f\|_{\Phi, \nu} = \sup\{\int_{\Omega} |fg| d\nu : \rho_{\Psi, \nu}(g) \leq 1\}$$

$$\leq \sup\{\int_{\Omega} |f(\alpha g)| d\mu_1 : \int_{\Omega} \Psi(\alpha g) d\mu_1 \leq 1\}$$

$$+ \sup\{\int_{\Omega} |f(1-\alpha)g| d\mu_2 : \int_{\Omega} \Psi((1-\alpha)g) d\mu_2 \leq 1\}$$

$$= \|f\|_{\Phi, \mu_1} + \|f\|_{\Phi, \mu_2} < \infty,$$

and hence $F_f(\nu) < \infty$. Also by (9) if Φ is equivalent to Φ_0, then

$$\|f\|_{\Phi_0, \nu} = \inf_{k>0} \{\frac{1}{k}(1 + \int_{\Omega} \Phi_0(kf) d\nu)\}$$

$$\geq \alpha \inf_{k>0} \{\frac{1}{k}(1 + \int_{\Omega} \Phi_0(kf) d\mu_1)\}$$

$$+ (1-\alpha) \inf_{k>0} \{\frac{1}{k}(1 + \int_{\Omega} \Phi_0(kf) d\mu_2)\}$$

$$= \alpha\|f\|_{\Phi_0, \mu_1} + (1-\alpha)\|f\|_{\Phi_0, \mu_2}.$$

Finally, if $\sup_{\mu \in M} \|f\|_{\Phi, \mu} \leq k\theta^{-1}(0)$ so that $\theta(\|f\|_{\Phi, \mu}/k) = 0$, and thus if $\theta(F_f(\mu)/k) = 0$, $\mu \in M$, one gets

$$\sum_{\mu \in M} \theta(F_f(\mu)/k)G(\mu) = 0 \leq 1.$$

Hence $k \geq N_{\Phi, \theta}(f)$. If $\theta(F_f(\mu_0)/k) > 0$ for some $\mu_0 \in M$, let $\mu_1 \in M$ be such that $\mu_0 \neq \mu_1$. Then for any $0 < \beta < \alpha < 1$, the concavity of $F_f(\cdot)$ and the monotonicity of θ imply

$$\theta\Big(\frac{F_f(\alpha\mu_0 + (1-\alpha)\mu_1)}{\beta k}\Big) \geq \theta\Big(\frac{F_f(\alpha\mu_0)}{\alpha k}\Big) \geq \theta\Big(\frac{F_f(\mu_0)}{k}\Big) > 0.$$

Hence

$$\sum_{\nu \in M} \theta(F_f(\nu)/\beta k) G(\nu) = \infty$$

since there are uncountably many nonzero terms in this sum. Hence $\beta k < N_{\Phi,\theta}(f)$. Letting $\beta \nearrow 1$, this shows that (10) must hold for $N_{\Phi,\theta}(f)$. If $\theta^{-1}(0) > 0$, then clearly $N_{\Phi,\theta}(f) = \sup_{\mu \in M}(\|f\|_{\Phi,\mu}/\theta^{-1}(0))$ is independent of G, and $= 0$ if $\theta^{-1}(0) = 0$. This establishes all the assertions. ∎

Remark. It may be noted that if θ is discontinuous and G is bounded, then $N_{\Phi,\theta}(f) = \sup\{F_f(\mu) : \mu \in M\}$ is unaltered when M is replaced by its convex hull $M_c = \{\nu = \sum_{i=1}^{n} \alpha_i \mu_i : \mu_i \in M, \alpha_i > 0, \sum_{i=1}^{n} \alpha_i = 1, n \geq 1\}$. However, this need not hold if θ is continuous or G is unrestricted. Thus for (10) above, the convexity assumption of M cannot be dropped in general.

8.2. Compact subsets of $L^\Phi(\theta, G)$

Although $L^\Phi(\theta, G)$ is a Banach space, its strong (or norm) topology is not appropriate for a study of the structure of its bounded subsets. Hence we introduce a new condition on the family M of measures, following Pitcher (1965). This is more general than the assumption of domination by a fixed (finite) measure. Then we discuss the structure of the unit ball of $L^\Phi(\theta, G)$ under a variety of conditions on θ, G and Φ.

For a complementary pair (Φ, Ψ) and $\mu \in M$, consider the continuous linear functional $\ell_{h,\mu}(\cdot)$ on $L^\Phi(\mu)$ defined by $\ell_{h,\mu}(f) = \int_\Omega f h d\mu$, $h \in L^\Psi(\mu)$. If h is bounded, then the fact that $L^\Phi(\theta, G) \subset L^\Phi(\mu)$ implies $\ell_{h,\mu} \in (L^\Phi(\theta, G))^*$. Let $\Gamma_\Psi(M)$ denote the linear span of $\{\ell_{h,\mu} : \mu \in M, \text{ and } h \text{ bounded measurable } (\Sigma)\}$. Define the $\Gamma_\Psi(M)$-topology of $L^\Phi(\theta, G)$ by the neighborhood system:

$$N(f; A, \varepsilon) = \{g \in L^\Phi(\theta, G) : |\ell(f) - \ell(g)| < \varepsilon, \ell \in A\}, \qquad (1)$$

where $f \in L^\Phi(\theta, G)$, A is a finite subset of $\Gamma_\Psi(M)$, and $\varepsilon > 0$. Since $\Gamma_\Psi(M)$ is evidently total on $L^\Phi(\theta, G)$, i.e., $\ell_{h,\mu}(f) = 0$ for all $\ell_{h,\mu} \in \Gamma_\Psi(M) \Rightarrow f = 0$, a.e.$[M]$, it follows that the $\Gamma_\Psi(M)$-topology is Hausdorff on $L^\Phi(\theta, G)$ under which it becomes a locally convex linear topological space. The fact that $\Gamma_\Psi(M)$ is total on $L^\Phi(\theta, G)$ implies, by some elementary results in abstract analysis, that the continuous linear functionals on $L^\Phi(\theta, G)$ for the $\Gamma_\Psi(M)$-topology are precisely those in $\Gamma_\Psi(M)$ (cf., e.g., Dunford-Schwartz (1958), V.3.9). Hence if $U^\Phi(\theta, G) = \{f \in L^\Phi(\theta, G) : N_{\Phi,\theta}(f) \leq 1\}$, the closed unit ball, it is also closed in the $\Gamma_\Psi(M)$-topology. This is a consequence of the separation theorem for closed disjoint convex sets in such spaces (cf., again the above reference, V.2.14). We use this fact in the following work.

The desired generalization of domination noted above is given by:

Definition 1. The family of measures (Ω, Σ, M) is said to obey the *compactness condition relative to θ and G* [or *(θ, G)-compact, for short*] if for some Young function Φ, the unit ball $U^\Phi(\theta, G)$ of $L^\Phi(\theta, G)$ is compact in the $\Gamma_\Psi(M)$-topology.

Recall that M is dominated if there is a (σ-) finite measure ν on (Ω, Σ) such that each μ in M is ν-continuous, i.e., $\mu \ll \nu$. In case θ is discontinuous and $G : M \to [\alpha, \beta]$, $0 < \alpha < \beta < \infty$, then it will follow from our work below that M is (θ, G)-compact for each Φ whose complementary function Ψ is continuous. The preoccupation with such a nonintuitive condition is that the structural analysis of $L^\Phi(\theta, G)$ is complicated when M is not dominated. This (θ, G)-compactness is also strictly weaker than the dominated hypothesis in most cases, and allows one to extend important special results based on dominated families to the general case formulated in Definition 1.

For simplicity set $\mathcal{X} = L^\Phi(\theta, G)$ and let $\mathcal{Y} = I\!R^{\mathcal{X}}$ ($= \underset{x \in \mathcal{X}}{\times} I\!R^x$, $I\!R^x = I\!R$) and give $\mathcal{Y}$ the product topology, i.e., a neighborhood of the origin of $\mathcal{Y}$ is taken as $\underset{x \in \mathcal{X}}{\times} N_x$ where N_x is a neighborhood of '0' in $I\!R$ for a finite set of x's and $N_x = I\!R$ for all the rest of the elements x. Then $\mathcal{X}$ can be embedded in $\mathcal{Y}$ since the topology given by (1) for $\mathcal{X}$ is the relativized product topology of $\mathcal{Y}$ and the mapping $\tau : \mathcal{X} \to \mathcal{Y}$ defined by $\tau(f) = (\ell(f), \ell \in \Gamma)$ is one-to-one by the fact that Γ is total on $\mathcal{X}$ (i.e., $\ell(x) = 0$ for all $\ell \in \Gamma \Rightarrow x = 0$). Note that τ is just the diagonal map, i.e., $\mathcal{X}$ is identified with the diagonal in $\mathcal{Y}$, in the sense that $f \mapsto (f_\mu, \mu \in M)$ where $f_\mu = f$ a.e.$[\mu]$, $\mu \in M$. Since τ is continuous, we also conclude that if $C^\Phi(\theta, G)$ is $\Gamma_\Psi(M)$-compact, then $\tau(C^\Phi(\theta, G)) \subset \mathcal{Y}$ is a closed set. The converse implication that τ^{-1} (closed bounded set of $\mathcal{Y}$) is $\Gamma_\Psi(M)$-compact in $\mathcal{X}$ is valid only if the complementary function Ψ of Φ is continuous. We sketch the reasoning here. If $\mathcal{K}$ is the set of all bounded measurable functions on (Ω, Σ), then $\mathcal{K} \subset L^\Psi(\mu), \mu \in M$, and if Ψ is continuous, then the closed subspace determined by $\mathcal{K}$ denoted $\mathcal{M}^\Psi(\mu)$, satisfies $(\mathcal{M}^\Psi)^* \cong L^\Phi(\mu)$ by Theorem IV.1.7, and this does not hold if Ψ is not continuous. It follows from this that the unit ball $U^\Phi(\mu)$ of $L^\Phi(\mu)$ is weak*-compact, $\mu \in M$. Since then by the Tychonov product theorem, $\underset{\mu \in M}{\times} U^\Phi(\mu)$ is compact in the product topology. But the latter is the $\Gamma_\Psi(M)$-topology and hence $\tau(U^\Phi(\theta, G))$ is bounded and closed. However τ^{-1} is continuous, so that $U^\Phi(\theta, G)$ is $\Gamma_\Psi(M)$-compact.

Summarizing the above, we have established the following.

Proposition 2. *Let* (Φ, Ψ) *be a complementary pair of Young functions and* (Ω, Σ, M) *be a family of probability measure spaces. If* $U^\Phi(\theta, G)$ *is the closed unit ball of* $L^\Phi(\mu)$ *on* (Ω, Σ, M) *and* $\Gamma(M) \subset (L^\Phi(\theta, G))^*$ *is as in Definition 1, then (i)* $U^\Phi(\theta, G)$ *is* $\Gamma(M)$-*closed, and (ii)* $\tau(U^\Phi(\theta, G))$ *is closed in the product topology of* $\underset{\mu \in M}{\times}(\theta^{-1}(\frac{1}{G(\mu)})U^\Phi(\mu))$ *provided* $U^\Phi(\theta, G)$ *is* $\Gamma(M)$-*compact. Moreover, if* Ψ *is continuous, then* $\tau(U^\Phi(\theta, G))$ *is closed in the product topology iff* $U^\Phi(\theta, G)$ *is* $\Gamma(M)$-*compact.*

We now present different sets of conditions for the $\Gamma(M)$-compactness of $U^\Phi(\theta, G)$ and then use these in applications of the compactness of (Ω, Σ, M). All these results are helpful in finding a Radon-Nikodým theorem for classes M_1 and M_2 on (Ω, Σ). Such a result is not possible in general when the M_i are not dominated families. With the help of the compactness condition here (and its properties), we will be able to establish later in the next section the desired Radon-Nikodým type result for certain σ-algebras contained in Σ, which are termed "sufficient." Thus the rest of this section is devoted to examining the hereditary and permanence properties of $\Gamma(M)$- [or (θ, G)-]compactness of (Ω, Σ, M).

Proposition 3. *Let* $f_\mu \in L^\Phi(\mu), \mu \in M$. *If for each finite subset* $\{\mu_i, 1 \le i \le n\} \subset M$, *there is an* f *such that* $N_{\Phi, \theta}(f) \le 1$ *and* $f = f_{\mu_i}$ *a.e.*$[\mu_i]$, $i = 1, \ldots, n$, *then* $\{f_\mu, \mu \in M\}$ *is in the closure of* $\tau(U^\Phi(\theta, G))$. *On the other hand, for each such element in this closure, and each countable set* $\{\mu_n, n \ge 1\} \subset M$, *there is an* $f \in U^\Phi(\theta, G)$, *possibly depending on this countable set, such that* $f = f_{\mu_n}$, *a.e.*$[\mu_n]$.

In particular, if the complementary function Ψ of Φ is continuous, and M is countable, then $U^\Phi(\theta, G)$ is $\Gamma_\Psi(M)$-compact. [τ and U^Φ are as in Proposition 2.]

Proof: The first statement is immediate from the definition of the $\Gamma_\Psi(M)$-topology. In fact, each $\Gamma_\Psi(M)$ neighborhood of $\{f_\mu, \mu \in M\}$ is of the form (1) which by hypothesis has a nonempty intersection with $U^\Phi(\theta, G)$. This means each such neighborhood intersects $\tau(U^\Phi(\theta, G))$ so that $\{f_\mu, \mu \in M\} \in \bar{\tau}(U^\Phi(\theta, G))$, the closure in $\Gamma_\Psi(M)$-topology.

In the opposite direction, suppose that the last condition holds, i.e., $\{f_\mu, \mu \in M\} \in \bar{\tau}(U^\Phi(\theta, G))$. Let $\mu_n \in M, n \geq 1$. Then there is always a dominating measure μ (i.e., $\mu_n \ll \mu$). For instance $\mu = \sum_{n=1} 2^{-n} \mu_n \in M$ and $\mu_n \ll \mu$. We now construct an $f \in U^\Phi(\theta, G)$ such that $f = f_{\mu_n}$, a.e.$[\mu_n]$. Consider the sets:

$$A_{m,n} = \{\omega : \frac{d\mu_m}{d\mu}(\omega) > 0, \frac{d\mu_n}{d\mu}(\omega) > 0\},$$
$$C_{m,n} = \{\omega \in A_{m,n} : f_{\mu_m}(\omega) > f_{\mu_n}(\omega)\}. \tag{2}$$

Note that $0 < \frac{d\mu_m}{d\mu_n}(\omega) < \infty$ for a.a. $\omega \in A_{m,n}$, by the chain rule. We claim that $\mu_m(C_{m,n}) = 0$. Indeed, in the opposite case, if $C_{m,n}(k) = \{\omega \in C_{m,n} : 0 < \frac{d\mu_n}{d\mu_m}(\omega) \leq k\}$, then by the monotone convergence theorem $\mu_m(C_{m,n}(k_0)) > 0$ for some $k_0 > 0$. Thus $\chi_{C_{m,n}(k_0)} \in L^\Psi(\mu), \mu \in M$. Since $\{f_\mu, \mu \in M\} \in \bar{\tau}(U^\Phi(\theta, G))$, there is an $f_\varepsilon \in U^\Phi(\theta, G)$ such that a neighborhood of f_ε contains f_{μ_n} so that $|\int_\Omega (f_{\mu_n} - f_\varepsilon) d\mu_n| < \varepsilon/2$, and

$$|\int_\Omega (f_{\mu_m} - f_\varepsilon) d\mu_n| = |\int_\Omega (f_{\mu_m} - f_\varepsilon) \chi_{C_{m,n}(k_0)} \frac{d\mu_n}{d\mu_m} d\mu_m| < \varepsilon/2.$$

It follows that

$$|\int_{C_{m,n}(k_0)} (f_{\mu_n} - f_{\mu_m}) \frac{d\mu_n}{d\mu_m} d\mu_m| = |\int_\Omega (f_n - f_{\mu_m}) \chi_{C_{m,n}(k_0)} d\mu_n| < \varepsilon. \tag{3}$$

Since the left side of (3) does not depend on $\varepsilon > 0$, it follows that $f_{\mu_n} = f_{\mu_m}$, a.e.$[\mu_m]$ on $C_{m,n}(k_0)$ contradicting the assumption that $\mu_m(C_{m,n}(k_0)) > 0$. Thus $f_{\mu_m} > f_{\mu_n}$ on $A_{m,m}$ is impossible. Interchanging m and n here, we find that $f_{\mu_m} < f_{\mu_n}$ on $A_{m,n}$ is also not possible. Thus $f_{\mu_m} = f_{\mu_n}$ a.e. on $A_{m,n}$ for both μ_m and μ_n. Also $\mu_n(\{\omega : \frac{d\mu_n}{d\mu}(\omega) > 0\}) = 1$, by the choice of μ above.

Let D_n be the set on which $\frac{d\mu_n}{d\mu} > 0$ for the first time, i.e., $D_n = \{\frac{d\mu_n}{d\mu} > 0, \frac{d\mu_k}{d\mu} = 0,$ for $1 \leq k \leq n-1\}$. The D_n are measurable for Σ, and disjoint. Let $f = \sum_n f_{\mu_n} \chi_{D_n}$. Then f is measurable for Σ, and

$$\{\omega : f(\omega) \neq f_{\mu_n}(\omega), n \geq 1\}$$

$$\subset \left\{\omega : \frac{d\mu_n}{d\mu}(\omega) = 0\right\} \cup \bigcup_{j=1}^{n-1} \left\{\omega \in D_j : \frac{d\mu_n}{d\mu}(\omega) > 0, f_{\mu_j}(\omega) \neq f_{\mu_n}(\omega)\right\}$$

$$\subset \left\{\omega : \frac{d\mu_n}{d\mu}(\omega) = 0\right\} \cup \bigcup_{j=1}^{n-1} \left\{\omega \in A_{j,n} : f_{\mu_j}(\omega) \neq f_{\mu_n}(\omega)\right\}. \quad (4)$$

Each of the right side sets of (4) has μ_n-measure zero. Thus $f = f_{\mu_n}$ a.e.$[\mu_n], n \geq 1$, and this f satisfies the conditions of the proposition.

Note that $f \in L^\Phi(\mu_n)$, $n \geq 1$, by construction. Also $(f_\mu, \mu \in M) \in \bar{\tau}(U^\Phi(\theta, G))$ so that there is a $g \in U^\Phi(\theta, G)$ such that f_μ is in a $\Gamma_\Psi(M)$-neighborhood, so that f constructed above ($f = f_{\mu_n}$ a.e.$[\mu_n], n \geq 1$) satisfies that it is also in such a neighborhood. But by Proposition 2, $U^\Phi(\theta, G)$ is $\Gamma_\Psi(M)$-closed. Hence $f \in U^\Phi(\theta, G)$, where f can be taken as zero a.e.$[\mu]$ for $\mu \in M - \{\mu_n, n \geq 1\}$.

If the complementary function Ψ of Φ is continuous, then $L^\Phi(\mu)$ is an adjoint space for each $\mu \in M$, and hence the $\Gamma_\Psi(M)$-topology of $U^\Phi(\theta, G)$ is the weak*-topology. In this case, since it is $\Gamma_\Psi(M)$-closed by Proposition 2, it follows by the classical Alaoglu-Birkhoff-Kakutani theorem that it is $\Gamma_\Psi(M)$-compact. ∎

As noted before, if Φ is discontinuous, $L^\Phi(\mu) = L^\infty(\mu)$ and if Φ' is bounded (but Φ is continuous), then $L^\Phi(\mu) = L^1(\mu)$. We denote, in these cases, $U^\Phi(\theta, G)$ as $U^\infty(\theta, G)$ and $U^1(\theta, G)$. Similarly $\Gamma_\Psi(M)$ in these cases may be expressed as Γ_∞- and Γ_1-topologies, the general case being simply abbreviated to Γ_Ψ. With this notation, we can establish the following result which connects the Γ_Ψ-topological properties with the compactness condition of (Ω, Σ, M) given by Definition 1 on $U^\Phi(\theta, G)$.

Theorem 4. *Consider the $\Gamma_\Psi(M)$-topology restricted to the ball $U^\Phi(\theta, G)$ of $L^\Phi(\theta, G)$. Then for any Φ, the Γ_∞- and Γ_Ψ-topologies agree on $U^\infty(\theta, G)$. In case (Ω, Σ, M) is (θ, G)-compact, then $U^\infty(\theta, G)$ is Γ_1-compact. In the opposite direction, if Ψ is continuous, $U^\infty(\theta, G)$ contains a strictly positive element, and is Γ_1-compact, then $U^\Phi(\theta, G)$ is Γ_Ψ-compact for each such Φ.*

Proof: Since $L^\infty(\mu) \subset L^\Phi(\mu)$ (μ being a finite measure) Γ_Ψ-neighborhoods contain each a Γ_∞-neighborhood and hence Γ_Ψ-topology is stronger than that determined by Γ_∞. On the other hand, consider an element $\ell_{h,\mu}(\cdot) \in \Gamma_\Psi$. If $h^{(n)} = h\chi_{[|h|\leq n]}$, then $\ell_{h^{(n)},\mu}(\cdot) \in \Gamma_\infty$, $n \geq 1$, and if $f \in U^\infty(\theta, G)$ then the computation for (6) of Section 1 shows that (recall that for discontinuous Φ, $L^\Phi(\mu) = L^\infty(\mu)$)

$$|f| \leq \theta^{-1}\left(\frac{1}{G(\mu)}\right)\|f\|_1 \leq \theta^{-1}\left(\frac{1}{G(\mu)}\right), \quad a.e.[\mu].$$

Hence

$$|(\ell_{h,\mu} - \ell_{h^{(n)},\mu})(f)| = |\int_\Omega f(h - h^{(n)})d\mu|$$

$$\leq \theta^{-1}(G(\mu)^{-1}) \int_{(|h|>n)} |h|d\mu \to 0, \qquad (5)$$

uniformly in f as $n \to \infty$ since $L^\Psi(\mu) \subset L^1(\mu)$. Thus $\ell_{h,\mu}$ of Γ_Ψ can be approximated on $U^\infty(\theta, G)$ uniformly by elements of Γ_∞, so that the Γ_∞- and Γ_Ψ-topologies agree on the ball $U^\infty(\theta, G)$.

Suppose that (Ω, Σ, M) is (θ, G)-compact so that by Definition 1 this says, $U^\Phi(\theta, G)$ is Γ_Ψ-compact for some Φ. Since $L^\infty(\mu) \subset L^\Phi(\mu)$, embedding being continuous, we have $U^\infty(\theta, G) \subset U^\Phi(\theta, G)$, and since $U^\infty(\theta, G)$ is Γ_1-closed by Proposition 2(i), the preceding paragraph implies that $U^\infty(\theta, G)$ is Γ_Ψ- (and hence Γ_1-)compact.

For the opposite direction suppose that Ψ is continuous, and $U^\infty(\theta, G)$ is Γ_1-compact, containing a positive element, say f_0. To show that $U^\Phi(\theta, G)$ is Γ_Ψ-compact, it suffices to prove that $\tau(U^\Phi(\theta, G))$ is Γ_Ψ-closed because of Proposition 2(ii). Thus let $\{f_\mu, \mu \in M\}$ be in the closure of $\tau(U^\Phi(\theta, G))$. But by Proposition 3 above, for each $\{\mu_n, n\geq 1\} \subset M$, there is an f' such that $f' = f_{\mu_n}$ a.e.$[\mu_n], n \geq 1$. If $N > 0$ is an integer and $f^{(N)} = f\chi_{[|h|\leq N]}$, then $\frac{1}{N}f_0(f')^{(N)} = \frac{1}{N}f_0 f_{\mu_n}^{(N)}$ a.e.$[\mu_n], n \geq 1$. But clearly $\frac{1}{N}f_0(f')^{(N)} \in U^\infty(\theta, G)$ by the monotonicity of norm $N_{\Phi,\theta}$. Hence $\{\frac{1}{N}f_0 f_\mu^{(N)}, \mu \in M\}$ is also in $\bar{\tau}(U^\infty(\theta, G)$, $N > 0$. Since $B_\infty(\theta, G)$ is Γ_1-closed, it follows that there is a $g_N \in U^\infty(\theta, G)$ such that $g_N = \frac{1}{N}f_0 f_\mu^{(N)}$ a.e.$[\mu]$, $\mu \in M$. But $f_\mu^{(N)} \to f_\mu$ a.e.$[\mu]$, so that $(Ng_N/f_0) = f_\mu^{(N)} \to f$ a.e.$[\mu]$, as $N \to \infty$, and $f = f_\mu$ a.e.$[\mu], \mu \in M$. So $f \in U^\Phi(\theta, G)$ by the preceding proposition. Hence $\tau(U^\Phi(\theta, G))$ is closed, so that it is Γ_Ψ-compact. $\blacksquare$

Remark. The condition that $U^\infty(\theta, G)$ has a positive element is essentially the demand that $\theta(x_0) = 0$ for some $x_0 > 0$. Indeed, this implies that $f_0 = x_0$ and so $\sum_{\mu \in M} \theta(\|f_0\|_{\infty,\mu}) G(\mu) = 0 \leq 1$, and $N_{\Phi,\theta}(f_0) \leq 1$ with $0 < f_0 \in U^\infty(\theta, G)$. On the other hand, if M is uncountable and $0 < f_0 \in U^\infty(\theta, G)$, then $N_\infty(f_0) \leq 1 \Rightarrow \sum_{\mu \in M} \theta(\|f_0\|_{\infty,\mu}) G(\mu) \leq 1$ so that for some μ_0, $\theta(\|f_0\|_{\infty,\mu_0}) = 0$ whence $\theta(x_0) = 0$ for some $0 < x_0 = \|f_0\|_{\infty,\mu_0}$ since $G(\mu) > 0$. Thus the existence of a positive element in $U^\infty(\theta, G)$ forces θ to be vanishing on $[0, x_0]$ for some $x_0 > 0$.

We present another characterization of (θ, G)-compactness of (Ω, Σ, M) and then show that this is a more general condition then domination, in preparation to an application of sufficiency (or a new Radon-Nikodým theorem) in the next section.

If $N \subset M$ then we denote the unit ball in $L^\Phi(\theta, G)$ based on N with $G|N$, by $U^\Phi(\theta, G, N)$ and if $N = M$ by $U^\Phi(\theta, G)$ as before. The statement that $f = g$ a.e.$[\mu]$ for $\mu \in N$ will be abbreviated to $f = g[N]$. With this notation we have:

Proposition 5. *Let $U^\Phi(\theta, G)$ be $\Gamma_\Psi(M)$-compact and $N \subset M$. If for each element g of $U^\Phi(\theta, G, N)$ there is an $f \in U^\Phi(\theta, G)$ such that $f = g[N]$, then the ball $U^\Phi(\theta, G, N)$ is $\Gamma_\Psi(N)$-compact.*

Proof: The identity map $I : U^\Phi(\theta, G) \to U^\Phi(\theta, G, N)$ taking $f \to f[N]$ is continuous, since the inverse image of a neighborhood $\{f : |\ell_{h_i,\mu_i}(f)| < \varepsilon, i=1,2,\ldots,n\}$, $\mu_i \in N, h_i \in L^\Psi(\mu_i)$ (h_i is bounded) is just $\{f : |\ell_{h_i,\mu_i}(f)| < \varepsilon, i=1,\ldots,n\}$ and the map is, by hypothesis, onto. Thus the continuous image of $U^\Phi(\theta, G)$ being $U^\Phi(\theta, G, N)$ is compact in $\Gamma_\Psi(N)$-topology. ∎

We now establish the result which gives conditions on enlarging a given set of measures preserving the compactness property.

Theorem 6. *Suppose $N \subset M$ is such that*
  (i) *(Ω, Σ, N) is (θ, G) compact, i.e., $U^\Phi(\theta, G, N)$ is $\Gamma_\Psi(N)$ compact for some Φ,*
  (ii) *for each $\mu \in M$ there exist $\{\mu_n, n \geq 1\} \subset N$ such that $\mu \ll \sum_{n=1}^\infty 2^{-n} \mu_n$,*
  (iii) *$f \in U^\infty(\theta, G, N) \Rightarrow |f| \leq K[N]$, for some $K > 0$, and*
  (iv) *$f \in U^\infty(\theta, G, N) \Rightarrow f \in U^\infty(\theta, G)$.*

Then (Ω, Σ, M) *is* (θ, G) *compact, i.e.,* N *can be replaced by* M *in (i) above. In particular, if* M *is dominated by a* σ-*finite measure* ν *on* Σ, Ψ *is continuous and* $\theta^{-1}\left(\frac{1}{G(\cdot)}\right)$ *is bounded on* M, *then* (Ω, Σ, M) *is compact, i.e.,* $U^{\Phi}(\theta, G)$ *is* $\Gamma_{\Psi}(N)$ *compact.*

Proof: Observe that the class of M-null sets is contained in the null sets of N ($\subset M$), and the reverse inclusion follows by condition (ii). Thus the classes of null sets for both M and N are the same. This and condition (iv) imply that $U^{\infty}(\theta, G, N) = U^{\infty}(\theta, G)$. Hence (i) and Theorem 4 yield that $U^{\infty}(\theta, G)$ is $\Gamma_{\Psi}(N)$-compact. We now assert that the $\Gamma_{\infty}(N)$- and $\Gamma_{\infty}(M)$-topologies coincide on $U^{\infty}(\theta, G)$, so that the compactness of (Ω, Σ, M) will follow. Note that on $U^{\infty}(\theta, G)$ the Γ_{∞}- and Γ_{Ψ}-topologies coincide for any Ψ. Thus it suffices to show that each element of $\Gamma_{\infty}(M)$ is a uniform limit of a sequence from $\Gamma_{\infty}(N)$ on $U^{\infty}(\theta, G)$.

Indeed, let $\mu \in M$ and $\{\mu_n, n \geq 1\} \subset N$ be such that (ii) holds. If $\nu = \sum_{n=1}^{\infty} 2^{-n} \mu_n$, so that $\mu \ll \nu$, let $f_0 = \frac{d\mu}{d\nu}$. Then

$$\mu(A) = \int_A f_0 d\nu = \sum_{n=1}^{\infty} \int_A f_n d\mu_n, \quad f_n = 2^{-n} f_0, \quad A \in \Sigma. \quad (6)$$

Let $h \in L^{\infty}(\mu)$, so that there is some n_0 satisfying $|h| \leq n_0$ a.e.[N], and hence

$$\mu_n \{\omega : \min[(f_n, k)|h|](\omega) > k n_0 \}$$
$$\leq \mu_n \{\omega : |h(\omega)| > n_0, f_0(\omega) > 0 \} = 0, \quad n \geq 1, k \geq 1,$$

we deduce that $\min(f_n, k)h \in L^{\infty}(\mu)$. Thus if $\ell_{h,\mu} \in \Gamma_{\infty}(M)$, then

$$\sum_{n=1}^{k} \ell_{\min(f_n, k)h, \mu_n} \in \Gamma_{\infty}(N). \quad (7)$$

Let $f \in U^{\infty}(\theta, G)$ so that $|f| \leq 1[M]$, and then

$$\left| \ell_{h,\mu}(f) - \sum_{n=1}^{k} \ell_{\min(f_n, k)h, \mu_n}(f) \right|$$
$$= \left| \sum_{n=1}^{\infty} \int_{\Omega} f h f_n d\mu_n - \sum_{n=1}^{k} \int_{\Omega} f h \min(f_n, k) d\mu_n \right|$$

$$\leq \sum_{n=1}^{k} \int_{\Omega} |fh|(f_n - \min(f_n, k))d\mu_n + \sum_{n=k+1}^{\infty} \int_{\Omega} |fh|f_n d\mu_n$$

$$\leq Kn_0 \Big[\sum_{n=1}^{k} \int_{\Omega} (f_n - \min(f_n, k))d\mu_n + \sum_{n=k+1}^{\infty} \int_{\Omega} f_n d\mu_n \Big],$$

$$\text{since } |fh| \leq Kn_0[M], \text{ by (iii).} \tag{8}$$

But $\sum_{n=1}^{\infty} \int_{\Omega} f_n d\mu_n = \mu(\Omega) = 1$, by (6). So given $\varepsilon > 0$, choose n_ε such that the tail part of this series satisfies $\sum_{n \geq n_\varepsilon + 1} \int_{\Omega} f_n d\mu_n < \frac{\varepsilon}{2Kn_0}$, and

then choose k_ε such that $\sum_{n=1}^{n_\varepsilon} \int_{(f_n > k_\varepsilon)} f_n d\mu_n < \frac{\varepsilon}{2Kn_0}$. Then (8) becomes with $k \geq k_\varepsilon$, so that $[f_n > \min(f_n, k)] \subseteq [f_n > k_\varepsilon]$,

the left side $\leq Kn_0 \Big[\sum_{n=1}^{\infty} \int_{(f_n > k_\varepsilon)} f_n d\mu_n + \sum_{n=k+1}^{\infty} \int_{\Omega} f_n d\mu_n \Big]$

$$= Kn_0 \Big[\sum_{n=1}^{n_\varepsilon} \int_{(f_n > k_\varepsilon)} f_n d\mu_n + \sum_{n>n_\varepsilon} \int_{\Omega} f_n d\mu_n + \sum_{n>k} \int_{\Omega} f_n d\mu_n \Big]$$

$$< \frac{\varepsilon}{2} + \frac{\varepsilon}{2} + \sum_{n>k} \int_{\Omega} f_n d\mu_n.$$

Now letting $k \to \infty$, the last term $\to 0$ for any $\varepsilon > 0$. From this we get the functionals in (7) tending to an $\ell_{h,\mu}$ uniformly for f in $U^\infty(\theta, G)$. This gives the main part of the theorem, and we turn to the last assertion.

Let M be the given countable set, and let $\widetilde{M} = M \cup \{\nu\}$. So $\widetilde{M}$ and M play the roles of M and N in the preceding part. If θ_1 is the discontinuous Young function such that $\theta_1(x) = 0$ for $|x| \leq 1$, $= +\infty$ for $|x| > 1$, then $N_{\infty,\theta_1}^M(f) = \sup_{\mu \in M} \|f\|_{\infty,\mu}$. Since the complementary function of the discontinuous Φ is Ψ giving $L^\Psi(\mu) = L^1(\mu)$, Proposition 3 implies that $U^\infty(\theta_1, G)$ is $\Gamma_1(\nu)$-compact. Also every element of $\widetilde{M}$ is dominated by ν, an element of the "subset" $\{\nu\}$. If $|f| \leq 1[\nu]$, then $|f| \leq 1[\widetilde{M}]$, so that by the preceding part $U^\infty(\theta_1, G, \widetilde{M})$ is $\Gamma_\Psi(\widetilde{M}) = \Gamma_1(\widetilde{M})$-compact. Since $M \subset \widetilde{M}$, for each $g \in U^\infty(\theta_1, G, M)$, so that $|g| \leq 1[M]$, we get $f = g\chi_{[|g| \leq 1]}$ to be an element of $U^\infty(\theta_1, G, \widetilde{M})$. Hence by Proposition 5, $U^\infty(\theta_1, G, M)$ is $\Gamma_1(M)$-compact. Since $1 \in U^\infty(\theta_1, G, M)$, by

Theorem 4 we conclude that $U^\infty(\theta_1, G, M)$ is $\Gamma_\Psi(M)$-compact. When $\theta^{-1}(1/G(\cdot))$ is bounded then $U^\Phi(\theta, G, M) \subset cU^\Phi(\theta_1, G, M)$ for some $c > 0$. So $U^\Phi(\theta, G, M)$ is $\Gamma_\Psi(M)$-closed. Since $cU^\Phi(\theta_1, G, M)$ is $\Gamma_\Psi(M)$-compact, it follows that $U^\Phi(\theta, G, M)$ is also $\Gamma_\Psi(M)$-compact, proving the last part. ∎

This result implies that the compactness condition of Definition 1 is an extension of domination. The following examples show that the concept is a proper extension, but that there also exist noncompact families M on (Ω, Σ), as noted by Pitcher (1965).

Examples. A. Let $(\Omega_\alpha, \Sigma_\alpha, M_\alpha)$, $\alpha \in I$, be a family of measure spaces with M_α as a compact set of probability measures on Σ_α (in the sense of Definition 1) and I an uncountable index set. Consider the direct sum: $\Omega = \cup\{\{\alpha\} \times \Omega_\alpha : \alpha \in I\}$, called the (disjoint) sum, and one identifies Ω_α with $\{\alpha\} \times \Omega_\alpha$; $\Sigma = \{A \subset \Omega : A \cap (\{\alpha\} \times \Omega_\alpha) = A \cap \Omega_\alpha \in \Sigma_\alpha, \alpha \in I\}$ and extend $\mu \in M_\alpha$ to Σ by defining $\mu(A) = \mu(A \cap \Omega_\alpha)$. If we let $M = \bigcup_{\alpha \in I} M_\alpha$, then (Ω, Σ, M) is a family of (an uncountable set of) probability measures. This M is not dominated. But it is (θ, G) compact if we take $\theta(x) = 0$, $0 \leq |x| \leq 1$ and $= +\infty$ for $|x| > 1$, and $G(\mu) = 1$ for all $\mu \in M$. Indeed, take Φ to be discontinuous so that Ψ is continuous and if $\{f_\mu, \mu \in M\}$ is in the closure of $\tau(U^\infty(\theta, G))$, then there is a function f_α such that $f_\alpha = f_\mu[M_\alpha]$, for each α by compactness of M_α. Let $f = f_\alpha$ on Ω_α, $\alpha \in I$. Then by construction of the σ-algebra Σ, f is measurable for Σ, and $\tau(f) = \{f_\mu, \mu \in M\}$ so that by Theorem 4, M is compact although cannot be dominated. Thus Definition 1 is more general than domination.

B. Let $\Omega = [0, 1]$, $\Sigma =$ the Borel σ-algebra, and $M = \{\delta_x, x \in \Omega\}$, the uncountable set of Dirac measures on (Ω, Σ). Every subset $A \subset [0, 1]$, gives rise to an element of $\tau(U^\infty(\theta, G))$ with θ, G as in the above example. However, if A is not in Σ, then the set $\{f_\mu, \mu \in A\}$ where $f_\mu = 1$ if $\mu = \delta_x$, and $= 0$ otherwise, is not a $\tau(f)$ for any f since any such function should be of the form χ_A and it is not Σ-measurable. Thus (Ω, Σ, μ) is not (θ, G)-compact.

Because of these examples, compactness condition of Definition 1 is of interest and we shall present some alternative forms for its fulfillment later. In the following section we consider an important application which in fact motivated the study, and indicate how it opens up new parts of analysis for future work in the subject.

8.3. Sufficiency for σ-subalgebras of (Ω, Σ, M)

To introduce the problem let $\mathcal{B} \subset \Sigma$ be a σ-subalgebra and $\mu \in M$ with $\mu_1 = \mu | \mathcal{B}$, the restriction. Then for each bounded Σ-measurable $f : \Omega \to I\!R$, there exists a μ-unique $\tilde{f}_\mu$ such that

$$\int_A f d\mu = \int_A \tilde{f}_\mu d\mu_1, \qquad A \in \mathcal{B}.$$

This is just the Radon-Nikodým theorem applied to $\nu : A \mapsto \int_A f d\mu$, which is σ-additive satisfying $\nu \ll \mu_1$ on $\mathcal{B}$. We denote $\tilde{f}_\mu$ as $E^{\mathcal{B}}_\mu(f)$ and call it the conditional expectation of f given (or relative to) $\mathcal{B}$ and μ on Σ. Evidently $\tilde{f}_\mu$ depends both on $\mathcal{B}$ and μ, and if we fix these two quantities, then $E^{\mathcal{B}}_\mu(\cdot)$ is a linear positivity preserving operator on $L^1(\mu)$. It is easy to verify that $E^{\mathcal{B}}_\mu$ is a contractive projection operator on $L^1(\mu)$. Our problem here is to find σ-algebras $\mathcal{B} \; (\subset \Sigma)$ such that there is a single element $E^{\mathcal{B}}(f) = E^{\mathcal{B}}_\mu(f)$, a.e.$[\mu]$ for all μ in M. It is clear that this is a more stringent requirement than when M is a finite set. Let us introduce the concept precisely as follows.

Definition 1. Let (Ω, Σ, M) be a complete space and Φ be a Young function. If $\mathcal{B}$ is a σ-subalgebra of Σ, then it is called *sufficient* for the family M of probability measures if there is a $\mathcal{B}$-measurable element, denoted $E^{\mathcal{B}}(f)$ (or $E(f|\mathcal{B}, M)$ in more detail), such that

$$\int_\Omega E^{\mathcal{B}}(f) g d\mu_1 = \int_\Omega f g d\mu, \quad f \in L^\Phi(\Omega, \Sigma, \mu), g \in L^\Psi(\Omega, \mathcal{B}, \mu), \mu \in M. \quad (1)$$

It is clear that $\mathcal{B} = \Sigma$ is always sufficient for each M since $E^{\mathcal{B}}(f) = f$ a.e.$[M]$ will do. On the other hand, if $\mathcal{B} = \{\phi, \Omega\}$, it is observed that $E^{\mathcal{B}}(f)$ exists iff $M = \{\mu\}$, a singleton. To see this, if μ, ν are two measures in M, then (1) implies with $f = \chi_A$, $A \in \Sigma$, and $g = 1$ (all $\mathcal{B}$-measurable g are constants!),

$$\mu(A) = \int_\Omega f d\mu = \int_\Omega E^{\mathcal{B}}_\mu(f) d\mu_1 = \int_\Omega E^{\mathcal{B}}_\nu(f) d\nu_1 = \int_\Omega f d\nu = \nu(A).$$

Thus $\mu = \nu$. This shows that the existence of nontrivial (or minimal) σ-subalgebras $\mathcal{B}$ satisfying (1) is not simple. In general such $\mathcal{B}$ need not exist. We shall first discuss how this problem can be considered as a question in vector measure theory and show that there are difficulties even in the abstract point of view, giving us no hint for a solution; and then present an answer using the concept given in Definition 2.1. This will indicate a need for further study in the subject.

Let $B(I)$ be the Banach space of real bounded functions on the set I where I indexes the elements of M. Then the collection of all measures $M = \{\mu_i, i \in I\}$ can be regarded as a vector from Σ into $B(I)$, i.e., $M : \Sigma \to B(I)$ with $M(A) = \{\mu_i(A), i \in I\}$. But since $0 \leq \mu_i(A) \leq 1$, the set $M(\Sigma)$ is contained in the positive part of the unit ball of $B(I)$, the latter being a Banach lattice under the uniform norm. It is easily seen that $M(\cdot)$ is additive on Σ. In general, however, it is not σ-additive in the norm topology of $B(I)$. For otherwise, there exists a probability measure ν that dominates M. This would imply that the set M of measures is dominated and, as noted in Example B above, this need not happen. Moreover, the vector space $B(I)$ is neither reflexive nor separable. All the known vector Radon-Nikodým results demand that $M(\cdot)$ be σ-additive and have finite variation with values in reflexive or separable adjoint Banach spaces (or those that have modifications related to these spaces, cf., e.g., N. Dinculeanu (1967), or J. J. Uhl and J. Diestel (1977)). None of these conditions is satisfied in our particular case. Thus these abstract considerations have not been able to provide a solution for our case.

Using the concrete structure of the problem at hand, we shall present a solution here. Let M_c be the convex hull of the set of probability measures M. For each σ-subalgebra $\mathcal{B}$, let $\widehat{\mathcal{B}}$ be the M_c completion in that

$$\widehat{\mathcal{B}} = \{A \subset \Omega : \mu \in M_c \Rightarrow \mu^*(A \Delta A_\mu) = 0 \text{ for some } A_\mu \in \mathcal{B}\}, \quad (2)$$

where μ^* is the outer measure generated by $(\mu, \mathcal{B})$ with the standard Carathéodory procedure. (Cf., e.g., Rao (1987), Ch. 2.) This collection $\widehat{\mathcal{B}}$ has some desirable properties, as seen from the proposition below.

Proposition 2. Let $(\Omega, \Sigma, M, \mathcal{B}, \widehat{\mathcal{B}})$ be as above. Then we have: (i) $\mathcal{B} \subset \widehat{\mathcal{B}}$, (ii) $\widehat{\widehat{\mathcal{B}}} = \widehat{\mathcal{B}}$, (iii) $\widehat{\mathcal{B}}$ is a σ-algebra, and (iv) $f : \Omega \to \mathbb{R}$ is $\widehat{\mathcal{B}}$-measurable iff there is a $\mathcal{B}$-measurable f_μ such that $f = f_\mu$, a.e.$[\mu]$, for each $\mu \in M_c$. Further, if (Ω, Σ, M) is (θ, G) compact and

$U^\infty(\theta, G)$ *contains a positive element, then* $\widehat{\Sigma} = \Sigma$ *and* $(\Omega, \widehat{\mathcal{B}}, M)$ *is* (θ, G)-*compact.*

Proof: (i) is clear since $A = A_\mu$, $\mu \in M_c$ will imply the result for any $A \in \mathcal{B}$. Regarding (ii), let $A \in \widehat{\widehat{\mathcal{B}}}$. Then by definition there exist $A_\mu \in \widehat{\mathcal{B}}$ and $\tilde{A}_\mu \in \mathcal{B}$ such that $\mu^*(A \Delta A_\mu) = 0$ and $\mu^*(A_\mu \Delta \tilde{A}_\mu) = 0$, so that $\mu^*(A \Delta \tilde{A}_\mu) = 0$. Hence $A \in \widehat{\mathcal{B}}$, i.e., $\widehat{\mathcal{B}} \subset \widehat{\widehat{\mathcal{B}}} \subset \widehat{\mathcal{B}}$. As for (iii), if $A_n \in \widehat{\mathcal{B}}$ so that $A_{n\mu} \in \mathcal{B}, n \geq 1 \Rightarrow$

$$\mu^*((\bigcup_n A_n) \Delta (\bigcup_n A_{n\mu})) \leq \sum_{n=1}^\infty \mu^*(A_n \Delta A_{n\mu}) = 0, \quad \mu \in M_c,$$

by the σ-subadditivity of μ^*. So $\bigcup_n A_n \in \widehat{\mathcal{B}}$, and $0 = \mu^*(A \Delta A_\mu) = \mu^*(A^c \Delta A_\mu^c) \Rightarrow A \in \widehat{\mathcal{B}} \Leftrightarrow A^c \in \widehat{\mathcal{B}}$ so that it is a σ-algebra.

Regarding (iv), the result is true if f is a simple function and the general case follows by the structure theorem for measurable functions. In the opposite direction, the result follows from the fact that $\{\omega : f(\omega) < x\} \Delta \{\omega : f_\mu(\omega) < x\}$ is μ-null for all $\mu \in M_c$ and all $x \in \mathbb{R}$. Thus (i)–(iv) hold.

Suppose that (Ω, Σ, M) is (θ, G)-compact and $0 < f_0 \in U^\infty(\theta, G)$. Since $\Sigma \subset \widehat{\Sigma}$ by (i), let $A \in \widehat{\Sigma}$. Then for $f = \chi_A$ there is $f_\mu = f$ a.e.$[\mu]$, measurable for Σ by (iv). So if $\mu_i \in M$, $1 \leq i \leq n$ and $\mu = \frac{1}{n} \sum_{i=1}^n \mu_i \in M_c$ we get $f = f_\mu$ a.e.$[\mu_i]$, so that $f_\mu = f_{\mu_i}$, a.e.$[\mu_i]$. By Theorem 2.4, there is an f_1, measurable for Σ, such that $f_1 = f_\mu$, a.e., $\mu \in M$, since $U^\infty(\theta, G)$ also contains a positive element. Hence $f_1 = f$ a.e.$[M]$, and f is measurable for Σ which is complete by assumption. Thus $A \in \Sigma$ or $\widehat{\Sigma} \subset \Sigma \subset \widehat{\Sigma}$, as desired.

For the last part, by Proposition 2.2, it suffice to show that $\tau(U^\infty(\theta, G, \widehat{\mathcal{B}}))$ is closed where $U^\infty(\theta, G, \widehat{\mathcal{B}})$ contains the $\widehat{\mathcal{B}}$-measurable elements of $U^\infty(\theta, G)$. Thus let $\{f_\mu, \mu \in M\}$ be an element of the $\Gamma_1(M)$-closure of $\tau(U^\infty(\theta, G, \widehat{\mathcal{B}}))$. Then by the earlier Proposition 2.2, for each finite subset $\{\mu_1, \dots, \mu_n\}$ of M, there is a $\widehat{\mathcal{B}}$-measurable $\tilde{f}$ such that $\tilde{f} = f_{\mu_i}$ a.e.$[\mu_i]$, $i = 1, \dots, n$, and since $\widehat{\mathcal{B}} \subset \widehat{\Sigma} = \Sigma$, we deduce that $\tilde{f}$ and f_{μ_i} are also Σ-measurable. So there is a Σ-measurable f such that $f = f_\mu$, a.e.$[\mu]$, $\mu \in M$, and $f \in U^\infty(\theta, G)$. Since $\mu = \sum_{i=1}^n \alpha_i \mu_i \in M_c$ for each convex combination, the $\widehat{\mathcal{B}}$-measurable $\tilde{f}$ satisfies $f = f_{\mu_i} = \tilde{f}$, a.e.$[\mu_i]$, so that $f = \tilde{f}$, a.e.$[\mu]$, and f is $\widehat{\mathcal{B}}$-measurable. Consequently

$f \in U^\infty(\theta, G, \widehat{\mathcal{B}})$, and hence $\{f_\mu, \mu \in M\} \in \tau(U^\infty(\theta, G, \widehat{\mathcal{B}}))$. So the latter is weakly closed as asserted. ∎

We now connect these relations with sufficient σ-subalgebras in Σ.

Proposition 3. *Let $\mathcal{B}$ be a sufficient σ-subalgebra for $(\Omega, \widehat{\Sigma}, M)$. Then $\widehat{\mathcal{B}} = \mathcal{B}$.*

Proof: Since $\mathcal{B} \subseteq \widehat{\mathcal{B}}$, let $A \in \widehat{\mathcal{B}}$ so that $f = \chi_A$ is bounded and $\widehat{\mathcal{B}}$-(hence $\widehat{\Sigma}$-)measurable and $\tilde{f} = E(f|\mathcal{B}, M)$ exists and is bounded. If $\mu \in M$, then $f - \tilde{f} = g_\mu$ a.e.$[\mu]$ is bounded and $\mathcal{B}$-measurable so that

$$\int_\Omega (f - \tilde{f})^2 d\mu = \int_\Omega (f - \tilde{f}) g_\mu d\mu = \int_\Omega f g_\mu d\mu - \int_\Omega E(f|\mathcal{B}, M) g_\mu d\mu = 0,$$

by (1). Hence $f = \tilde{f}$, a.e.$[\mu]$. Since $\tilde{f}$ is $\mathcal{B}$-measurable and the latter σ-algebra (in the definition of sufficiency) is complete, f is $\mathcal{B}$-measurable also, so that $A \in \mathcal{B}$. Thus $\widehat{\mathcal{B}} \subset \mathcal{B} \subset \widehat{\mathcal{B}}$ as desired. ∎

We can now present a solution of the existence problem.

Theorem 4. *If (Ω, Σ, M) is (θ, G) compact and $U^\infty(\theta, G)$ has a positive element, then there is a unique minimal sufficient σ-algebra $\mathcal{B} \subset \Sigma$, so that if $\mathcal{B}_1$ is any other sufficient σ-subalgebra of Σ, then $\mathcal{B} \subset \mathcal{B}_1$.*

Proof: Let $\{\mu_n, n \geq 1\}$ be a finite (or countable) subset of M. Let λ_0 be a finite measure that dominates all μ_n. (For instance, $\lambda_0 = \sum_i \alpha_i \mu_i$, $\alpha_i > 0$, $\sum_i \alpha_i = 1$ will give $\mu_i \ll \lambda_0$.) If $f_n = \frac{d\mu_n}{d\lambda_0}$ and $\mathcal{B}_0$ is the smallest σ-algebra relative to which each f_n is measurable (and completed for $\mu_n, n \geq 1$), then it is asserted that $\mathcal{B}_0$ is sufficient for $\mu_n, n \geq 1$. This follows from an easy computation: Let $A \in \Sigma$, and consider for $B \in \mathcal{B}_n$,

$$\int_B E_{\lambda_0}^{\mathcal{B}_0}(\chi_A) d\mu_n = \int_B E_{\lambda_0}^{\mathcal{B}_0}(\chi_A) f_n d\lambda_0$$

$$= \int_B E_{\lambda_0}^{\mathcal{B}_0}(f_n \chi_A) d\lambda_0, \text{ since } f_n \text{ is } \mathcal{B}_0\text{-measurable,}$$

$$= \int_B f_n \chi_A d\lambda_0, \text{ by definition of } E_{\lambda_0}^{\mathcal{B}_0}(\cdot)$$

$$= \int_B \chi_A d\mu_n. \tag{3}$$

Hence by linearity of $E_{\lambda_0}^{\mathcal{B}_0}$ and of the integral, we get

$$\int_\Omega fg d\mu_n = \int_\Omega E_{\lambda_0}^{\mathcal{B}_0}(f)g d\mu_n = \int_\Omega \tilde{f}g d\mu_n, \tag{4}$$

for any bounded Σ-measurable f and bounded $\mathcal{B}_0$-measurable g, $[\mu_n, n \geq 1]$, and a unique $\mathcal{B}_0$-measurable $\tilde{f}$. This implies $\mathcal{B}_0$ is sufficient for the set $\{\mu_n, n \geq 1\}$. Moreover, (4) also shows that $E_{\mu_n}^{\mathcal{B}_0}(f) = \tilde{f}$ a.e.$[\mu_n], n \geq 1$. This is now extended as follows.

For every finite subset of M, let $\mathcal{B}_{fin}$ be the σ-algebra obtained as above, and let $\mathcal{B}$ be the smallest σ-algebra containing all these $\mathcal{B}_{fin}$ as we consider all the finite subsets of M. Then we use the fact that $\mathcal{B}_{fin} \subset \mathcal{B}$, and by a commutativity property of $E^{\mathcal{B}}$, $E_{\lambda_0}^{\mathcal{B}} E_{\lambda_0}^{\mathcal{B}_{fin}} = E_{\lambda_0}^{\mathcal{B}_{fin}}$ for any λ_0 which dominates the finite subset. Hence (4) implies that for each $(\mu_1, \ldots, \mu_n)$ and a bounded (Σ) measurable f, there is a $\mathcal{B}$-measurable $\tilde{f}$ such that $E_\mu^{\mathcal{B}}(f) = \tilde{f}$ a.e.$[\mu_k], 1 \leq k \leq n, n \geq 1$. Next we utilize further the hypothesis that (Ω, Σ, M) is (θ, G)-compact and that $U^\infty(\theta, G)$ has a positive element (so $\theta^{-1}(0) > 0$). Thus $\{E_\mu^{\mathcal{B}}(f), \mu \in M\}$ is in the closure of $\tau(U^\infty(\theta, G))$. But the latter set is closed and Γ_1-compact since the complementary function Ψ of the discontinuous Φ is continuous, and the last part of Theorem 2.12 (actually the construction of f from f_0 in that proof) can be invoked. Hence there exists a $g \in U^\infty(\theta, G)$ such that $E_\mu^{\mathcal{B}}(f) = g$ a.e.$[\mu], \mu \in M$. We claim that g is $\widehat{\mathcal{B}}$-measurable and $g = E^{\widehat{\mathcal{B}}}(f)$.

To prove the claim, note that for each $(\mu_1, \ldots, \mu_n)$ from M, there is an $\tilde{f}$ such that $\tilde{f} = E_{\mu_i}^{\mathcal{B}}(f)$, a.e.$[\mu_i], i=1, \ldots, n$, and hence $g = \tilde{f}$, a.e.$[\mu_i], i=1, \ldots, n$. This implies that g is $\widehat{\mathcal{B}}$-measurable by Proposition 2. Further if $A \in \widehat{\mathcal{B}}$ and $A_\mu \in \mathcal{B}$ satisfying $\mu(A \Delta A_\mu) = 0$, then we get

$$\int_A g d\mu = \int_{A_\mu} E_\mu^{\mathcal{B}}(f) d\mu = \int_{A_\mu} f d\mu, \text{ by definition of } E_\mu^{\mathcal{B}}(\cdot),$$

$$= \int_A f d\mu = \int_A E^{\widehat{\mathcal{B}}}(f) d\mu, \text{ since } A \in \widehat{\mathcal{B}}.$$

But A is arbitrary in $\widehat{\mathcal{B}}$, so that $g = E^{\widehat{\mathcal{B}}}(f)$. This implies that $\widehat{\mathcal{B}}$ is sufficient for (Ω, Σ, M). [Thus if $\theta^{-1}(\frac{1}{G})$ is bounded and M is dominated (cf., Theorem 2.6), then $\widehat{\mathcal{B}}$ is sufficient if (and obviously only if) it is sufficient for each finite (in fact two) family of subsets of M.]

Since $\Sigma = \widehat{\Sigma}$, if $\mathcal{B}_1$ is any sufficient σ-subalgebra of Σ (for M), then $\mathcal{B}_{fin} \subset \mathcal{B}_1$ so that $\mathcal{B} = \sigma(\cup \mathcal{B}_{fin}) \subset \mathcal{B}_1$ and then by Proposition 3, $\mathcal{B}_1 = \widehat{\mathcal{B}}_1$. Thus $\widehat{\mathcal{B}} \subset \widehat{\mathcal{B}}_1 = \mathcal{B}_1$, and $\widehat{\mathcal{B}}$ constructed here is also minimal sufficient. ∎

As a consequence of the above result and its proof, one has

Corollary 5. *Let (Ω, Σ, M) be (θ, G)-compact and $U^{\infty}(\theta, G)$ contain a positive element. If $\Sigma_0 \subset \Sigma$ is any sufficient σ-subalgebra of Σ and $\Sigma_0 \subset \Sigma_1 \subset \Sigma$ is any other σ-subalgebra, then $\widehat{\Sigma}_1$ is also sufficient.*

We can use the last result and if $\{\mathcal{F}_n, n \geq 1\}$ is an increasing sequence of σ-subalgebras of (Ω, Σ, M) which satisfies the hypothesis of Corollary 5, and $\mathcal{F}_1$ is sufficient for M, then (since $\widehat{\mathcal{F}}_1 = \mathcal{F}_1$ and $\widehat{\Sigma} = \Sigma$) $\{\widehat{\mathcal{F}}_n, n \geq 1\}$ is an increasing sequence of sufficient σ-algebras contained in Σ. This fact can be used in defining a (sub-)martingale on (Ω, Σ, M) as follows.

Definition 6. If $\{f_n, \mathcal{F}_n, n \geq 1\} \subset L^{\Phi}(\theta, G)$ with f_n measurable for $\mathcal{F}_n$ and each $\mathcal{F}_n$ ($\subset \mathcal{F}_{n+1}$) sufficient for (Ω, Σ, M) then the sequence is said to be an M- *(sub-)martingale* if for each $n \geq 1$ one has

$$E^{\mathcal{F}_n}(f_{n+1}) = (\geq)f_n, \qquad a.e.[M]. \tag{5}$$

It will be interesting to develop martingale theory in this general context. One should note that there are subtle difficulties. For instance given a pair of sufficient σ-algebras (Σ_1, Σ_2) from Σ, it is not generally true that $\Sigma_1 \cap \Sigma_2$ or $\sigma(\Sigma_1 \cup \Sigma_2)$ is sufficient. A number of counter-examples illustrating some pathological aspects is given by Burkholder (1961). We include a simple result here exemplifying the positive aspect of the concepts.

Proposition 7. *Let $\{f_n, \mathcal{F}_n, n \geq 1\}$ be in a ball of $L^{\Phi}(\theta, G)$. If each $\mathcal{F}_n$ is sufficient for M, and the f_n sequence is an M-submartingale, $G(M) \subset [\alpha, \infty)$ for an $\alpha > 0$, then $f_n \to f$ a.e.[M], and $N_{\Phi,\theta}(f) \leq \liminf_n N_{\Phi,\theta}(f_n)$.*

Proof: Let $\mu \in M$. Then (5) shows $f_n \leq E_\mu^{\mathcal{F}_n}(f_{n+1})$, a.e., and the f_n sequence is a submartingale on (Ω, Σ, μ). Thus the hypothesis implies (cf. definition of $N_{\Phi,\theta}$ and Theorem 1.1) that $f_n \to f$, a.e.[μ], for each $\mu \in M$, by the standard Doob submartingale convergence theorem (see, e.g., Rao (1984), p. 184). We set $f = 0$ where this limit fails to exist.

Then $\mathcal{F}_\infty = \sigma(\bigcup_n \mathcal{F}_n) \subset \Sigma$ and f is $\mathcal{F}_\infty$-measurable. If $\mathcal{F}_\infty$ is completed for M, then $\widehat{\mathcal{F}}_\infty$ is sufficient for M, and $f_n \to f$, a.e.$[M]$. Since $N_{\Phi,\theta}(\cdot)$ is a function norm and since $N_\Phi(f) \leq \liminf_n N_\Phi(f_n)$, one gets the desired result from this and the definition of the $N_{\Phi,\theta}(\cdot)$ norm. ∎

The hypothesis of this proposition is satisfied if (Ω, Σ, μ) is (θ, G)-compact, $U^\infty(\theta, G)$ contains a positive element, and $\mathcal{F}_1$ is sufficient, in addition to the condition on G. In particular if M is dominated, all these things reduce to the classical case. In view of the fact that compactness of (Ω, Σ, M) plays a key role in the above considerations, we now present some conditions where it is automatically satisfied.

8.4. Reflexivity and uniform convexity of $L^\Phi(\theta, G)$

The importance of the Γ_Ψ-compactness of (Ω, Σ, M) is illustrated in the sufficiency study in the preceding section. Since Γ_Ψ is by definition the weak*-topology of $L^\Phi(\theta, G)$, and since in a reflexive Banach space the weak and weak*-topologies coincide, we have the following useful sufficient condition.

Proposition 1. *If $L^\Phi(\theta, G)$ is a reflexive (Banach) space under the norm $N_{\Phi,\theta}(\cdot)$ (or any equivalent norm), then (Ω, Σ, M) is (θ, G) compact.*

Motivated by this observation, it is instructive to present easily verifiable conditions to verify the reflexivity of $L^\Phi(\theta, G)$. A detailed analysis of the dual of this space is somewhat involved, but some good sufficient conditions can be inferred by embedding $L^\Phi(\theta, G)$ in the "substitution space" $L^\Phi(\mu, \ell^\theta(G))$ and using some classical results of the general Banach space theory. We can present the following result.

Theorem 2. *Let $L^\Phi(\theta, G)$ be the space introduced before on (Ω, Σ, M) and weight function $G : \Omega \to (0, \infty)$, θ being another Young function. Suppose that Ψ and ζ are the complementary Young functions of Φ and θ. If Φ is Δ_2-regular and θ satisfies (a) $\sup\limits_{0 < x < \infty} (\theta(2x)/\theta(x)) < \infty$, or (b) $\sup\limits_{0 < x \leq a} (\theta(2x)/\theta(x)) < \infty$ for each $0 < a < \infty$ and $G(M) \subset [\alpha, \infty)$ for some $\alpha > 0$, then $(L^\Phi(\theta, G))^* \cong L^\Psi(\zeta, G)$, i.e., these are topologically equivalent. Hence, if both (Φ, Ψ) are Δ_2-regular and (θ, ζ) both satisfy either (a) or (b), then $L^\Phi(\theta, G)$ is reflexive.*

The conditions here imply that $(L^\Phi(\mu))^* \cong L^\Psi(\mu)$ and $(\ell^\theta(G))^* \cong \ell^\varsigma(G)$ and the general theory of substitutions and the work of Chapter IV imply the result. The last part is then a consequence since under the given conditions both $L^\Phi(\mu)$ and $\ell^\theta(G)$ are reflexive so that $L^\Phi(\mu, \ell^\theta(G))$ is also reflexive. However, Rosenberg (1970) has given complete details of the case at hand. We shall omit the formal proof since the result will not be needed.

It should be noted that the conditions given in Theorem 2 are not necessary. The reason is that $L^\Phi(\theta, G)$ is a "sparse" subspace of $L^\Phi(\mu, \ell^\theta(G))$. Without going into a precise discussion of the last statement we give an example of an infinite dimensional space $L^\Phi(\theta, G)$ which is reflexive without the corresponding $L^\Phi(\mu, \ell^\theta(G))$ being reflexive. It is adapted from Pitcher (1965). If $\theta(x) = 0$ for $|x| \le 1, = +\infty$ for $|x| > 1$, then $N_{\Phi,\theta}(f) = \sup\{\|f\|_{\Phi,\mu} : \mu \in M\}$ and does not depend on G. So we take it as $G(\mu) \equiv 1$ for simplicity and in this case $L^\Phi(\theta, G)$ will be written simply as $E^\Phi(\Omega, \Sigma, M) = \{f : N_{\Phi,\theta}(f) = \sup_\mu \|f\|_{\Phi,\mu} < \infty\}$ which is a Banach space.

Example 3. Let (Ω, Σ, μ) be the Lebesgue unit interval (with $\Omega = [0,1]$), $\Phi(x) = |x|^p$, and $\omega_0 \in \Omega$ be an arbitrarily fixed point. If $1 \le p < q < \infty$, $L^p(\mu)$ $(\supset L^q(\mu))$ be the Lebesgue space. If $0 \ne g \in L^q(\mu)$, define a measure μ_g by the equation:

$$\int_\Omega f d\mu_g = \left[\int_\Omega f|g|^{q-p} d\mu \Big/ \left(\int_\Omega |g|^q d\mu\right)^{(q-p)/q}\right] + k_g f(\omega_0), \quad f \in L^{\frac{q}{p}}(\mu), \quad (6)$$

with

$$k_g = 1 - (\|g\|_{q-p}/\|g\|_q)^{q-p} \quad (\ge 0). \tag{7}$$

It follows, on using the Hölder inequality, that μ_g is well defined and is a probability measure for each such g. If $g = 0$, let μ_0 be the point measure concentrating at ω_0. If $M = \{\mu_g, g \in L^q(\mu)\}$, then (Ω, Σ, M) is our candidate. We now claim that for each $f \in E^p(\Omega, \Sigma, M)$,

$$C_1\|f\|_{q,\mu+\mu_0} \le N_{p,\theta}(f) = \|f\|_{p,M} \le C_2\|f\|_{q,\mu+\mu_0}. \tag{8}$$

Since $L^q(\mu+\mu_0)$ is reflexive and since, by (8), $E^p(\Omega, \Sigma, M)$ and $L^q(\mu+\mu_0)$ are topologically equivalent, it follows that $E^p(\Omega, \Sigma, M)$ is reflexive for all $1 \leq p < q$, establishing our initial assertion. Thus it remains to establish (8).

To this end consider (6) with $|f|^p$ in place of f:

$$\int_\Omega |f|^p d\mu_g \leq \left[\left(\int_\Omega |f|^q d\mu \right)^{\frac{p}{q}} \left(\int_\Omega |g|^q d\mu \right)^{\frac{(q-p)}{q}} \right] \left[\int_\Omega |g|^q d\mu \right]^{\frac{(p-q)}{q}} + k_g |f(\omega_0)|^p,$$

$$\text{by Hölder's inequality with } \left(\frac{q}{p}, \frac{q}{q-p} \right);$$

$$\leq (\|f\|_q)^p + |f(\omega_0)|^p, \quad \text{since } 0 \leq k_g \leq 1,$$

$$\leq [\|f\|_q + |f(\omega_0)|]^p \leq [2\|f\|_{q,\mu+\mu_0}]^p.$$

Hence $\|f\|_{p,M} = \sup_g \|f\|_{p,\mu_g} \leq 2\|f\|_{q,\mu+\mu_0} < \infty$, giving half of (8) with $C_2 = 2$. On the other hand, if $f \in L^p(\Omega, \Sigma, M)$, let $g_n = \inf(|f|, n)$. Then

$$\infty > \|f\|_{p,M}^p \geq \int_\Omega |f|^p d\mu_g \geq \|g_n\|_q, \qquad n \geq 1.$$

So $f \in L^q(\mu)$. Consequently,

$$\|f\|_{p,M}^p \geq \frac{1}{2} \int_\Omega |f|^p d(\mu_f + \mu_0), \quad \text{since } \frac{1}{2}(\mu_f + \mu_0) \in M,$$

$$= \frac{1}{2}(\|f\|_q^p + c_f |f(\omega_0)|^p + |f(\omega_0)|^p), \qquad c_f \geq 0,$$

$$\geq \frac{1}{2^p}(\|f\|_q + |f(\omega_0)|)^p, \quad \text{since } (u+v)^p \leq 2^{p-1}(u^p + v^p),$$

$$\geq \frac{1}{2^p} \left(\int_\Omega |f|^q d(\mu + \mu_0) \right)^{\frac{p}{q}}, \quad \text{since } (u+v)^{\frac{1}{q}} \geq u^{\frac{1}{q}} + v^{\frac{1}{q}} \text{ for } q \geq 1.$$

Hence $\|f\|_{p,M} \geq \frac{1}{2}\|f\|_{q,\mu+\mu_0}$ which proves the left side of (8) with $C_1 = \frac{1}{2}$. ∎

This example shows that there are (infinite dimensional) $L^\Phi(\theta, G)$ spaces which are reflexive and hence (Ω, Σ, M) is (θ, G) compact without M being dominated. [There is no σ-finite measure dominating M here.] Taking $p = 1$, we see that $L^1(\theta, G)$ is a reflexive subspace of $L^q(\theta, G)$, $q > 1$. Thus finding necessary and sufficient conditions for the reflexivity of $L^\Phi(\theta, G)$ spaces is not entirely a consequence of the general theory of Banach spaces.

Using the work of Chapter VII, we can present conditions for strict and uniform convexity of the spaces $L^\Phi(\theta, G)$ to give a feeling for the geometry of these "subspaces of substitution spaces." Remembering the definitions of strict and uniform convexity from Sections 7.1 and 7.2, an argument analogous to that employed for Theorem VII.5.1 (so the substitution spaces have the convexity properties if the component spaces have the same property), we can obtain the following result. We omit the proof here. But it can be constructed using the work of Sections 7.1 and 7.2.

Theorem 4. *(a) If θ' is continuous and strictly increasing and Φ is in Δ_2 and also strictly convex, then $\left(L^\Phi(\theta, G), N_{\Phi,\theta}(\cdot)\right)$ is rotund.*

(b) If both θ and Φ are moreover in Δ_2, and for each $\varepsilon > 0$, there is a $k_\varepsilon > 1$ such that

(i) $\theta'((1+\varepsilon)x) \geq k_\varepsilon \theta'(x)$, and (ii) $\Phi'((1+\varepsilon)x) \geq k_\varepsilon \Phi'(x)$, $x \geq x_\varepsilon \geq 0$,

then $\left(L^\Phi(\theta, G), N_{\Phi,\theta}(\cdot)\right)$ is uniformly convex.

The details have also been essentially given by Rosenberg (1968, 1970).

8.5. *Further remarks and applications*

In this section we complement the discussion on considering $L^\Phi(\theta, G)$ as a space of functions relative to a vector valued measure, considered in Section 3 above. Further, a solution of the optimization problem mentioned as a motivation at the beginning of Section 1 will be outlined.

Let $(\mathcal{X}, \|\cdot\|)$ be a Banach space and if $f : \Omega \to \mathcal{X}$ is strongly measurable (i.e., $f(\omega) = \lim_n f_n(\omega)$ for a sequence of simple functions $f_n = \sum_{i=1}^{n} a_i^n \chi_{A_i^n}$, $a_i^n \in \mathcal{X}$, $\omega \in \Omega$, $A_i \in \Sigma$) so that $\|f\| : \Omega \to \overline{\mathbb{R}}^+$ is measurable for Σ, suppose that if $F_f(\mu) = \|f\|_{\Phi,\mu}$ we have

$$\inf\left\{k > 0 : \sum_{\mu \in M} \theta\left(\frac{F_f(\mu)}{k}\right) G(\mu) \leq 1\right\} = N_{\Phi,\theta}(\|f\|) < \infty. \quad (1)$$

Here, as usual

$$F_f(\mu) = \sup\left\{\int_\Omega \|f(\omega)\| |g(\omega)| d\mu : \rho_\Psi(g) \leq 1\right\}, \quad (2)$$

is the Orlicz norm. The equivalent gauge norm is given by

$$N_\Phi(f) = \inf\left\{k > 0 : \int_\Omega \Phi\left(\frac{\|f\|}{k}\right) d\mu \leq 1\right\}. \quad (3)$$

If $L_{\mathcal{X}}^{\Phi}(\theta, G)$ is the set of strongly measurable $f : \Omega \to \mathcal{X}$ such that $N_{\Phi,\theta}(\|f\|) < \infty$, then a simple modification of the argument of Theorem 1.1 shows that $(L_{\mathcal{X}}^{\Phi}(\theta, G), N_{\Phi,\theta}(\cdot))$ is a Banach space, i.e., is complete so that it may be termed an Orlicz space of $\mathcal{X}$-valued functions on (Ω, Σ, M) relative to (θ, G). The special case that θ is two-valued, i.e., $\theta(x) = 0$, $0 \leq |x| \leq 1$, $= +\infty$ for $|x| > 1$, and that $G(\mu) \equiv 1$ for all $\mu \in M$, was recently considered by S. K. Roy and N. D. Chakraborty (1986) to obtain the same conclusion. Also one can extend several properties of the Orlicz spaces given in Chapter IV. It is, however, of interest to generalize the work of Sections 1 and 2 to the vector case. For the work that uses the Radon-Nikodým theorem one has to restrict the spaces $\mathcal{X}$ to be either reflexive or separable adjoint spaces. [This substantiates the assumption that $\mathcal{X}$ has the Radon-Nikodým property, i.e., each vector measure on Σ into $\mathcal{X}$, of finite variation admits the Radon-Nikodým derivative.] We do not have this property.

In considering (Ω, Σ, M) we have $M \subset ca(\Omega, \Sigma)$, the space of σ-additive bounded set functions on the σ-algebra Σ so that $M(\Sigma) \subset I\!\!R^+$, as a bounded set. If I indexes the measures in M, then we may consider M as a vector valued function, $M : \Sigma \to B(I)$, which is finitely additive. Since weak and strong σ-additivity concepts are equivalent (according to a classical result by B. J. Pettis), and since if M is σ-additive it necessarily will have a dominating (also called a "control") measure; it is not σ-additive in general in the norm topology of the Banach lattice $B(I)$. Moreover if $\{f_n, n \geq 1\}$ is a bounded sequence in $B(I)$, then it has a least upper bound or equivalently each monotone increasing bounded sequence has a supremum in $B(I)$, so that the space is boundedly σ-complete. Using this property, we may consider the vector measure $M : \Sigma \to B(I)$ and note the following properties: (i) $M(A) \geq 0$, (ii) $M(\emptyset) = 0$ and (iii) for a disjoint family $A_n \in \Sigma$,

$$M\left(\bigcup_{n=1}^{\infty} A_n \right) = \sup_{n \geq 1} \sum_{k=1}^{n} M(A_k). \tag{4}$$

This property holds for $M(\cdot)$ so that it is σ-additive in the *order* topology of $B(I)$, which is strictly weaker than the norm topology. For such a vector measure, an integration, and the consequent Lebesgue-type limit theory, has been developed by J. D. M. Wright (1969$_a$) who also obtained a Radon-Nikodým theorem. We now present this result

which complements Theorem 3.4 and helps to understand why many
σ-subalgebras of Σ (completed) are not sufficient.

To discuss the Radon-Nikodým theorem, we need to state the concept of the integral of a scalar function relative to the vector valued measure $M(\cdot)$ which satisfies (4). As usual, if $f_n = \sum\limits_{i=1}^{n} a_i \chi_{A_i}$, then one defines

$$\int\limits_{\Omega} f_n d\mu = \sum_{i=1}^{n} a_i M(A_i), \tag{5}$$

and if $f : \Sigma \to \overline{I\!\!R}^+$ is measurable for (Σ), let $0 \leq f_n \uparrow f$ so $f(\omega) = \sup\limits_n f_n(\omega)$, $\omega \in \Omega$. We then set $\int\limits_{\Omega} f d\mu = \sup\limits_n \int\limits_{\Omega} f_n dM$, or $= +\infty$ if $M(\{\omega : f(\omega) > c\}) = +\infty$, for some $c > 0$. Finally for $f : \Omega \to \overline{I\!\!R}$, one lets $\int\limits_{\Omega} f dM = \int\limits_{\Omega} f^+ dM - \int\limits_{\Omega} f^- dM$ if both terms are finite. It can be verified that the integral is well defined and the mapping $f \mapsto \int\limits_{\Omega} f dM$ is linear. Also the monotone and dominated convergence theorems are valid. However, the integral is somewhat weaker and cannot be deduced from the Daniell (or Lebesgue) theory. We also can introduce the $L^{\Phi}(M)$ spaces as follows.

Definition 1. Let Φ be a Young function and (Ω, Σ, M) be as before. Then consider $L^{\Phi}(M)$ as the set of measurable $f : \Omega \to I\!\!R$ such that $\Phi(|f|/\alpha)$ is integrable relative to M in the sense of the M-integral given above.

It can be shown as an extension of the work of Wright (1969_b) that $L^{\Phi}(M)$ is a vector lattice. We may introduce a topology on $L^{\Phi}(M)$ by defining a semi-norm on it.

Definition 2. Let θ be another Young function and $G : I \to (0, \infty)$ be a weight function where I is the index set of M. For each $f \in L^{\Phi}(M)$, let

$$N_{\Phi,i}(f) = \inf \left\{ \alpha > 0 : \left(\int\limits_{\Omega} \Phi(\tfrac{f}{\alpha}) dM \right)(i) \leq 1 \right\}$$

and

$$N_{\Phi,\theta}^{G}(f) = \inf \left\{ k > 0 : \sum_{i \in I} \theta\left(\frac{N_{\Phi,i}(f)}{k} \right) G(\{i\}) \leq 1 \right\}. \tag{6}$$

Set $L^{\Phi}(\theta, G) = \{ f \in L^{\Phi}(M) : N_{\Phi,\theta}^{G}(f) < \infty \}$.

It may be verified that $N_{\Phi,\theta}^G(\cdot)$ is a (semi-norm) and that the equivalence of $N_{\Phi,i}(\cdot)$ and $\|\cdot\|_{\Phi,i}$ implies the same property between $N_{\Phi,\theta}^G(\cdot)$ of (6) and $N_{\Phi,\theta}^G(\cdot)$ of (1). Indeed, $\frac{1}{2}N_{\Phi,\theta}^G(f) \le N_{\Phi,\theta}(f) \le N_{\Phi,\theta}^G(f)$. Thus Theorem 1.1 implies the following assertion.

Theorem 3. *The set* $\{L^\Phi(\theta, G), N_{\Phi,\theta}^G(\cdot)\}$ *is a Banach lattice over the reals.*

We now proceed to introduce certain concepts that allow a presentation of the desired Radon-Nikodým theorem. Although for a pair of vector valued measures $M_j : \Sigma \to B(I)$, $j = 1, 2$, the absolute continuity of M_1 relative to M_2 is straightforward, written as $M_1 \ll M_2$ to denote $M_2(A) = 0 \Rightarrow M_1(A) = 0$, this is too weak a concept to admit a unique Radon-Nikodým derivative if the M_i are not scalar valued. The following example illustrates a possible pathology.

Example. Let I be a two point set, $\Omega = \{\omega_1, \omega_2\}$, $\Sigma =$ power set of Ω, and $M_i : \Sigma \to B(I) = \mathbb{R}^2$, $i = 1, 2$. Suppose that $M_1(\{\omega_i\}) = a_i = M_2(\{\omega_{3-i}\})$, $i = 1, 2$, where a_1, a_2 are a pair of linearly independent vectors. ($M_1 \ll M_2 \ll M_1$.) But there cannot exist a scalar function $\alpha : \Omega \to \mathbb{R}$ such that $M_1(\{\omega_1\}) = \int_{\{\omega_1\}} \alpha(\omega) dM_2(\omega)$ since then we must have $a_1 = \alpha(\omega_1) a_2$ and hence a_1, a_2 are linearly dependent. In this case there exists a 2×2-matrix valued $\alpha(\cdot)$ and then it will not be unique. Thus a unique scalar $\alpha(\cdot)$ does not exist.

This example shows that the absolute continuity between the vector or $B(I)$-valued M_1 and M_2 should be strengthened. The precise condition that eliminates this difficulty and yields a Radon-Nikodým theorem is not known at this time. If the M_i are "modular measures" [i.e., if $\pi : B(I) \to L^\infty(M_i)$ is an algebra homomorphism satisfying $\int_\Omega \pi(a)f dM_i = a \int_\Omega f dM_i$, $i = 1, 2$, $a \in B(I)$] and if they satisfy a further condition called "amply absolutely continuous" (modulo π), then Wright (1969$_b$) shows that $M_1(E) = \int_E f dM_2$ holds, $E \in \Sigma$ for a unique (Σ-) measurable f. However, the very first modular condition is not satisfied for our M_i and so this result cannot be applied. Thus the only available Radon-Nikodým theorem in our context is that given by Theorem 3.4 for a sufficient σ-subalgebra and in fact the subject awaits a detailed study.

We now turn to the optimization problem noted before. It is as follows. If $F : M \to I\!R$ is a function, it is desired to estimate $F(\mu)$ subject to the condition that the "mean error" $H_f(\mu) = \|f - F(\mu)\|_{\Phi,\mu}$ should be minimized for (θ, G). This means:

$$e^G_{\Phi,\theta}(f) = N_{\Phi,\theta}(H_f) \leq e^G_{\Phi,\theta}(\tilde{f}), \tag{7}$$

for any other measurable function $\tilde{f} : \Omega \to I\!R$. Such an f is called an *estimator* of $F(\mu)$, and it is termed *unbiased* if the following constraint is satisfied:

$$\int_\Omega f d\mu = F(\mu), \qquad \text{for all } \mu \in M. \tag{8}$$

The existence and uniqueness of such an f is a problem of considerable importance in the theory of statistical estimation. For a good account of the latter, see Lehmann (1983). Some conditions on the Young functions θ and Φ (and also on G) should be found. We can present the following solution.

Theorem 4. *Let (Φ, Ψ) be a complementary Young pair, θ another Young function and $G : M \to (0, \infty)$ be a weight function, where (Ω, Σ, M) is the basic family of probability spaces. If the unit ball $U^\Phi(\theta, G)$ of $L^\Phi(\theta, G)$ is compact for the Γ_Ψ-topology (cf. Definition 2.1) and $F : M \to I\!R$ satisfies $F \in \ell^\theta(M, \mathcal{P}(M), G)$, then there exists an $f_0 \in L^\Phi(\theta, G)$ for which (7) holds, i.e., the minimization problem has a solution. It is moreover unique if both θ and Ψ are strictly convex continuous Young functions, $\theta \in \Delta_2$ and Ψ an N-function. Further, if there exists at least one unbiased estimator (i.e., some f in $L^\Phi(\theta, G)$ satisfying (8)), then there is an optimal unbiased f_0 which then satisfies (7) among all unbiased estimators. It is also unique if θ, Φ again satisfy the preceding strict convexity and growth conditions.*

Proof: The statement is in three parts: (i) the existence of the solution without preconditions, (ii) uniqueness of the solution, and (iii) the preceding two questions when the estimator is desired to be unbiased. It should be noted that the estimator f should only be a function of ω and must not contain any other (unknown) parameters such as μ or any function of μ. We present the argument for (i)–(iii) in that order.

(i) Let $\alpha = \inf\{e^G_{\Phi,\theta}(f) : f \in L^\Phi(\theta, G)\}$. Since $\mu(\Omega) = 1$ for all $\mu \in M$, it is clear that $F(\mu) \in L^\Phi(\mu)$ and hence $H_f(\mu) = \|f - F(\mu)\|_{\Phi,\mu} < \infty$. The hypothesis that $F \in \ell^\theta(G)$ implies $N^G_\theta(f) < \infty$ where $N^G_\theta(\cdot)$ is the norm of $\ell^\theta(G)$, so $0 \leq \alpha \leq e^G_{\Phi,\theta}(0) < \infty$. Also $K = \{f \in L^\Phi(\theta, G) : e^G_{\Phi,\theta}(f) < \infty\}$ is convex and there exist $f_n \in K$ such that $e^G_{\Phi,\theta}(f_n) \downarrow \alpha$. Since $\|f_n\|_{\Phi,\mu} \leq H_{f_n}(\mu) + H_0(\mu) < \infty$, we get

$$N^G_\theta(\|f_n\|_{\Phi,\mu}) \leq e^G_{\Phi,\theta}(f_n) + \Psi^{-1}(1)N^G_\theta(F), \quad n \geq 1. \tag{9}$$

Hence there is an n_0 such that $n \geq n_0 \Rightarrow$

$$N^G_\theta(\|f_n\|_{\Phi,\mu}) \leq \alpha + 1 + \Psi^{-1}(1)N^G_\theta(F) < \infty.$$

Thus for $\varepsilon > 0$, there is $n_1(\varepsilon)$ such that $n \geq n_1(\varepsilon)$ implies $N^G_\theta(H_{f_n}(\cdot)) \leq \alpha + \varepsilon$, i.e.,

$$\sum_{\mu \in M} \theta\left(\frac{\|f_n - F(\mu)\|_{\Phi,\mu}}{\alpha + \varepsilon}\right) G(\mu) \leq 1. \tag{10}$$

This implies that the set $\{\|f_n - F(\mu)\|_{\Phi,\mu}, n \geq 1\}$ is bounded for each μ. If we let

$$M' = \bigcup_{m=1}^{\infty} \bigcup_{n=n_1(\frac{1}{m})}^{\infty} \left\{\mu \in M : \theta(\|f_n - F(\mu)\|_{\Phi,\mu}(\alpha + \frac{1}{m})^{-1}) > 0\right\}$$

then by (10) the set in braces { } being at most countable, since $G(\mu) > 0$ for all $\mu \in M$, it follows that M' is a countable set, say $\{\mu_1, \mu_2, \ldots\}$. We can use the standard diagonal procedure to extract a convergent subsequence of $\{\|f_n - F(\mu_m)\|_{\Phi,\mu_m}, n \geq 1, m \geq 1\}$, a bounded set of positive numbers. Thus

$$a_{kn} = \|f_{kn} - F(\mu_k)\|_{\Phi,\mu_k} \to \gamma_k \geq \alpha, \qquad \text{as } n \to \infty,$$

where each $\{a_{kn}, n \geq 1\} \subset \{a_{(k-1)n}, n \geq 1\}$. If $g_k = f_{kk}$, then $\|g_k - F(\mu_k)\|_{\Phi,\mu_k} \to \gamma_k$, $k \geq 1$, and since this is bounded, it is contained in a ball $\beta U^\Phi(\theta, G)$ for some $\beta > 0$. By hypothesis, this β-ball is Γ_Ψ-compact. Let f_0 be a cluster point of the g_k-sequence for the Γ_Ψ-topology. Hence for each bounded h with $N_{\Psi,\mu}(h) \leq 1$,

$$\int_\Omega f_0 h d\mu = \lim_{k \to \infty} \int_\Omega g_{n_k} h d\mu,$$

and then

$$\left|\int_{\Omega}(f_0-F(\mu))h\,d\mu\right| \leq \limsup_{k\to\infty} \|g_{n_k}-F(\mu)\|_{\Phi,\mu} \leq \limsup_{k\to\infty} \|g_k-F(\mu)\|_{\Phi,\mu}.$$

Taking the supremum over all h with $N_{\Psi,\mu}(h) \leq 1$, we get

$$\|f_0 - F(\mu)\|_{\Phi,\mu} \leq \limsup_{k\to\infty} \|g_k - F(\mu)\|_{\Phi,\mu}, \quad \mu\in M. \tag{11}$$

But for $\mu = \mu_i \in M'$, we get $\|g_k - F(\mu_i)\|_{\Phi,\mu_i} \to \gamma_i$, by the choice of g_k. Since $\theta(\cdot)$ is nondecreasing and continuous on a positive interval $[0, a]$, $a \geq 1$, we get by Fatou's lemma the following:

$$\sum_{i=1}^{\infty} \theta\left(\frac{\|f_0 - F(\mu_i)\|_{\Phi,\mu_i}}{\alpha + \frac{1}{m}}\right)G(\mu_i) \leq \sum_{i=1}^{\infty} \theta\left(\frac{\gamma_i}{\alpha + \frac{1}{m}}\right)G(\mu_i)$$

$$\leq \sum_{i=1}^{\infty} \liminf_{k\to\infty} \theta\left(\frac{\|g_k - F(\mu_i)\|_{\Phi,\mu_i}}{\alpha + \frac{1}{m}}\right)G(\mu_i)$$

$$\leq \liminf_{k\to\infty} \sum_{i=1}^{\infty} \theta\left(\frac{\|g_k - F(\mu_i)\|_{\Phi,\mu_i}}{\alpha + \frac{1}{m}}\right)G(\mu_i)$$

$$\leq 1, \quad \text{by (10)}. \tag{12}$$

But if $\mu \in M-M'$, then for $n \geq n_1(\frac{1}{m})$, $m \geq 1$, $\theta(\|f_n - F(\mu)\|_{\Phi,\mu}/(\alpha + \frac{1}{m})) = 0$. Hence the corresponding sum of (12) for $\mu \in M-M'$ vanishes because of (11). This and (12) imply that

$$\sum_{\mu\in M} \theta\left(\frac{\|f_0 - F(\mu)\|_{\Phi,\mu}}{\alpha + \frac{1}{m}}\right)G(\mu) \leq 1, \quad m\geq 1. \tag{13}$$

It then follows from (7) and (13) that $e^G_{\Phi,\theta}(f_0) \leq \alpha + \frac{1}{m}$. Since m is arbitrary,

$$e^G_{\Phi,\theta}(f_0) \leq \alpha \leq e^G_{\Phi,\theta}(f), \quad f\in L^{\Phi}(\theta, G). \tag{14}$$

Hence the existence assertion (i) follows.

(ii) By the conditions on θ and Φ, we conclude that both $(\ell^{\theta}(G), N_{\theta}(\cdot))$ and $(L^{\Phi}(\mu), \|\cdot\|_{\Phi})$ are strictly convex Orlicz spaces because of Theorem VII.1.3 and Corollary VII.1.7. The uniqueness is then an easy consequence. Indeed, if f_1, f_2 are two minimal elements, as guaranteed by (i), let $H_i(\mu) = \|f_i - F(\mu)\|_{\Phi,\mu}$, $i=1,2$ and $H_0(\mu) = \|\frac{f_1+f_2}{2} - F(\mu)\|_{\Phi,\mu}$. Then $f_1, f_2, \frac{f_1+f_2}{2}$ are in K (cf. (9) above) and

$$N_{\theta}(H_1) = N_{\theta}(H_2) \leq N_{\theta}(H_0). \tag{15}$$

However, $H_0(\mu) \leq \frac{1}{2}(\|f_1 - F(\mu)\|_{\Phi,\mu} + \|f_2 - F(\mu)\|_{\Phi,\mu}) = \frac{1}{2}(H_1(\mu) + H_2(\mu))$, so that if $H_1 \neq H_2$ we get

$$N_\theta(H_0) < \frac{1}{2}[N_\theta(H_1) + N_\theta(H_2)] = N_\theta(H_1). \qquad (16)$$

The contradiction between (15) and (16) implies that $H_1 = H_2 \ (\geq H_0)$. But, if $f_1 \neq f_2$ on a set of positive μ_0-measure for some μ_0, then $H_0(\mu_0) < H_1(\mu_0)$, so that $N_\theta(H_0) < N_\theta(H_1)$ and hence $e^G_{\Phi,\theta}(\frac{1}{2}(f_1 + f_2)) = N_\theta(H_0) < e^G_{\Phi,\theta}(f_1) = N_\theta(H_1)$, which contradicts the minimality of $e^G_{\Phi,\theta}(f_1)$. Thus $f_1 = f_2$ a.e.$[M]$, and the uniqueness assertion (ii) follows.

(iii) Finally let K_0 be the set of unbiased estimators of the problem. It then follows that $K_0 \ (\subset K)$ is also a convex subset which by hypothesis is nonempty in $L^\Phi(\theta, G)$. This set can be easily verified to be Γ_Ψ-closed. Hence the argument of parts (i) and (ii) applies verbatim, and shows that both of the assertions hold for this class. ∎

As the reader may have noted, the above result does not say anything about the (global) existence of unbiased estimators of $F : M \to \mathbb{R}$. The reason is that they need not exist in general. We now present conditions for the existence of unbiased estimators of $F(\cdot)$. Note that $\ell_{h,\mu} : L^\Phi(\theta, G) \to \mathbb{R}$ defined by

$$\ell_{h,\mu}(f) = \int_\Omega f h \, d\mu, \quad f \in L^\Phi(\theta, G), h \in L^\Psi(\mu), \mu \in M, \qquad (17)$$

is linear and it is also bounded. For the latter, we have

$$|\ell_{h,\mu}(f)| \leq \|f\|_{\Phi,\mu} N_{\Psi,\mu}(h), \text{ by Hölder's inequality,}$$

and since

$$N_{\Phi,\theta}(f) = \inf\left\{k > 0 : \sum_{\mu \in M} \theta\left(\frac{\|f\|_{\Phi,\mu}}{k}\right) G(\mu) \leq 1\right\} < \infty,$$

for each $\varepsilon > 0$ and $\mu_0 \in M$, one has

$$\theta\left(\frac{\|f\|_{\Phi,\mu_0}}{N_{\Phi,\theta}(f) + \varepsilon}\right) G(\mu_0) \leq \sum_{\mu \in M} \theta\left(\frac{\|f\|_{\Phi,\mu}}{N_{\Phi,\theta}(f) + \varepsilon}\right) G(\mu) \leq 1,$$

so that

$$\|f\|_{\Phi,\mu_0} \leq (N_{\Phi,\theta}(f) + \varepsilon)\theta^{-1}\left(\frac{1}{G(\mu_0)}\right). \qquad (18)$$

Since $\varepsilon > 0, \mu_0 \in M$ are arbitrary, by (17) and (18),

$$|\ell_{h,\mu}(f)| \leq N_{\Phi,\theta}(f) \cdot \theta^{-1}\left(\frac{1}{G(\mu)}\right) N_{\Psi,\mu}(h), \qquad (19)$$

so that $\|\ell\|$ denoting the norm of the linear functional ℓ,

$$\|\ell_{h,\mu}\| \leq \theta^{-1}\left(\frac{1}{G(\mu)}\right) N_{\Psi,\mu}(h) < \infty.$$

Using this notation we can present the following.

Theorem 5. *If there is an unbiased estimator f of $F : M \to \mathbb{R}$, then, with $C_0 = N_{\Phi,\theta}(f)$,*

$$\left|\sum_{i=1}^{n} a_i F(\mu_i)\right| \leq C_0 \left\|\sum_{i=1}^{n} a_i \ell_{1,\mu_i}\right\| \qquad (20)$$

holds for all $a_i \in \mathbb{R}, \mu_i \in M, n \geq 1$. On the other hand, if (Ω, Σ, M) is (θ, G)-compact, Ψ is continuous and there is a positive element in $U^\infty(\theta, G)$, then the inequality (20) [i.e., for each finite subset $(\mu_1,...,\mu_n)$, constants $(a_1,...,a_n)$, with some constant C_0 (20)] implies the existence of an unbiased estimator f such that $N_{\Phi,\theta}(f) \leq C_0 N_\theta(\theta^{-1}(\frac{1}{G(\cdot)}))$. This last norm is finite when $\theta^{-1}(\frac{1}{G(\cdot)}) \in \ell^\theta$.

Sketch of Proof: If there exists an unbiased estimator of F, then by (8) we get (20) at once with $C_0 = N_{\Phi,\theta}(f)$. The converse part uses all the hypothesis. Suppose then (20) holds. If $x^{**} : (L^\Phi(\theta, G))^* \to \mathbb{R}$ is defined by the equation

$$x_0^{**}\left(\sum_{i=1}^{n} a_i \ell_{1,\mu_i}\right) = \sum_{i=1}^{n} a_i F(\mu_i),$$

it is unambiguous and is a bounded linear functional because of (20), and has a norm preserving extension by the Hahn-Banach theorem. Thus it is an element of $(L^\Phi(\theta, G))^{**}$ with norm bounded by C_0. Now using the compactness and other hypothesis one finds that there is an $f : \Omega \to \mathbb{R}$ (measurable for Σ) such that

$$x_0^{**}(\ell_{h,\mu}) = \int_\Omega fh d\mu, \qquad h \in L^\Psi(\mu) \cap L^\infty(\mu), \mu \in M.$$

Thus $x_0^{**}(\ell_{1,\mu}) = \int_\Omega f d\mu = F(\mu)$. Now one has to show that $f \in L^\Phi(\theta, G)$ when the last part of the hypothesis also holds. Here we need to use several results of Section 2 above. This application also shows that our work, in the early sections is essential for such problems. We omit the details here. ∎

The inequality (20) is the best condition for the solution of an infinite set of equations, namely, (8). Such problems are abstractly discussed in linear analysis, but their specialization in concrete applications is nontrivial and the problem considered is just an illustration, although an important one. We thus conclude this analysis here and turn to some key function spaces related to Orlicz spaces.

Bibliographical Notes: Spaces based on sets of measures appear in different contexts in applications. One of the several natural areas is the statistical theory of estimation. A comprehensive treatment of this subject can be found in Lehmann (1983). The sufficiency concept plays a fundamental role in this subject. This and unbiasedness are treated in detail in the papers by Halmos and Savage (1949), and Kolmogorov (1950). When the family of measures is not dominated, as we treated in this chapter, the study cannot be simplified. Numerous difficulties are pointed out in Burkholder (1961). For a detailed account of these questions in the context of statistical theory one may consult Heyer (1982). A general attempt of the undominated case, using the condition of compactness, began with the pioneering study by Pitcher (1965) in the context of L^p-spaces. An extension of this work to the general convex functions and hence $L^\Phi(\theta, G)$-spaces was undertaken by Rosenberg (1968) in his thesis. Most of the work in this chapter follows his paper (1970). However, parts of the treatment in Sections 3–5 in the text also use some unpublished work of Rosenberg's thesis.

One of the central difficulties in the undominated case is that the set of measures, considered as a vector, does not constitute a vector measure in any of the Banach space topologies. Consequently the theory of Dunford-Schwartz type integration theory relative to such measures is not generally suitable in the present work. The detailed compactness study of Brooks and Dinculeanu (1976) uses the σ-additivity of a vector measure in one of the Banach space topologies and hence there is a control measure at every stage. So we could not use this since the set of measures for us is not necessarily dominated. The condition of "compactness of (Ω, Σ, M)," due to Pitcher, is shown to be more general than domination and its potential has not been fully exploited. Thus a further analysis of $L^{\Phi}(\theta, G)$ still awaits, but Rosenberg's work, which we essentially presented here, should give a good motivation for such an attempt. Perhaps Wright's ($1969_{a,b}$) integration and the Randon-Nikodým theorems should be further extended. On the latter problem Maynard (1973) has presented some results without assuming σ-additivity of a vector measure and that work may be employed in this context. Using the Brooks-Dinculeanu theory, Roy and Chakraborty (1986, 1990) have studied Orlicz spaces based on families of measures. Their work corresponds to the dominated sets of measures in our context, and thus seems to be a special case of Rosenberg's. But a generalization of this point of view for the undominated case should be of interest for some future investigations.

The existence of an unbiased estimator is essentially finding a solution of an infinite set of linear equations, and it has been considered in abstract form by many mathematicians. See, for instance, Hille and Phillips (1957), p. 31, and references given there. Theorem 5.5 in the case of Orlicz spaces $L^{\Phi}(\theta, G)$ shows how a direct attack can be made, although it is not as simple as one desires. The construction problem is important in applications. Under various differentiability hypotheses, an explicit construction with convex functions was presented in Rao (1965) and its extension to the $L^{\Phi}(\theta, G)$-case, at least under the uniform convexity (or smoothness) conditions, is possible. But these as well as martingale problems are some of the natural and nontrivial applications of the theory presented in this chapter. We hope that these will lead to further research in this somewhat less developed part of the subject.

IX
Some Related Function Spaces

This chapter is devoted to a treatment of a few classes of spaces, inspired by Orlicz spaces, which find important applications in potential theory and differential equations among others. These include the Hardy-Orlicz and the Orlicz-Sobolev spaces. In the former case we prove, after the basic theory, some key interpolation theorems which are companions to those of Chapter VI. Then we introduce and discuss the structural aspects of the second class for finite and then briefly for infinite orders. Finally we touch on another related class, namely, Besicovitch-Orlicz spaces of almost periodic functions. All these spaces will illustrate the many possibilities for extensions and applications of the work detailed in the preceding chapters.

9.1. Hardy-Orlicz spaces

If D is the open unit disc in the complex plane with C as its boundary so that $D = \{z : |z| < 1\}$ and $C = \{z : |z| = 1\}$, then recall that $f : D \to \mathbb{C}$ is *analytic* if it has a convergent power series expansion and f is *harmonic* if it satisfies the Laplacian: $\frac{\partial^2 f}{\partial x^2} + \frac{\partial^2 f}{\partial y^2} = 0$ where $z = x + iy$. A harmonic function, defined in D, may be real valued while satisfying the above Laplace equation. We then can consider a *harmonic conjugate* $g : D \to \mathbb{R}$, which by definition is any function such that $f + ig$ is analytic in D. Because of their fine structure, the harmonic functions play important roles in both function theory and classical harmonic analysis, and this utility is considerably

enhanced when certain abstract functional analysis is brought into their
study. The key role played by these functions can be understood from
their use in trigonometric series (cf. A. Zygmund (1959)), and their
place in many other important parts of (abstract) analysis may be seen
from K. Hoffman (1962) as well as P. L. Duren (1970). A subclass
of harmonic functions, initially isolated by G. H. Hardy has been of
particular interest. It is denoted $\mathcal{H}^p$, called the Hardy space, and a deep
analysis based on the L^p-space theory has emerged. This class naturally
leads to a corresponding $\mathcal{H}^\varphi$ family based on the Orlicz space theory.
Such a class was considered in the middle 1950s and is being studied
with the classical results of the $\mathcal{H}^p$-spaces, $p > 0$, as a motivation. We
now introduce the concept, following G. Weiss (1956). [Since φ is not
necessarily a Young function, in what follows φ, not Φ, will be employed
for the $\mathcal{H}^\varphi$-spaces.]

   Definition 1. Let φ be a nonnegative function on $I\!R$ such that
$\varphi(x) \to 0$, as $x \to -\infty$, and φ is nondecreasing but $\varphi(x) > 0$ for some
$x \neq 0$. Then $\widetilde{\mathcal{H}}^\varphi$ is the set of all $f : D \to \mathbb{C}$, analytic and such that
$\int_0^{2\pi} \varphi(\log |f(re^{i\theta})|)d\mu(\theta)$ is bounded for $0 \leq r < 1$. If φ is, moreover,
convex then $\widetilde{\mathcal{H}}^\varphi$ is called a *Hardy-Orlicz class*. A *Hardy-Orlicz space*,
denoted $\mathcal{H}^\varphi$, is the set of $f : D \to \mathbb{C}$ such that $\alpha f \in \widetilde{\mathcal{H}}^\varphi$ for some α
$(= \alpha_f) > 0$. Here $\mu(\cdot)$ is the normalized Lebesgue measure on $(0, 2\pi]$,
i.e., $d\mu(\theta) = \frac{d\theta}{2\pi}$.
   If $\varphi(x) = x\chi_{I\!R^+}$ then $\widetilde{\mathcal{H}}^\varphi$ becomes the Nevanlinna class $\mathcal{N}$ on D
(cf., Zygmund (1959) Chapter VII), i.e., $f \in \mathcal{N}$ iff $\int_0^{2\pi} \log^+ |f(re^{i\theta})|d\mu(\theta)$
is bounded for $0 \leq r < 1$. If $\varphi(x) = e^{p|x|}$, $p > 0$, then $\mathcal{H}^\varphi = \mathcal{H}^p$, and
the classical Hardy space for $p > 0$ is obtained, i.e., $\int_0^{2\pi} |f(re^{i\theta})|^p d\mu(\theta)$ is
bounded in $0 \leq r < 1$; and similarly $\mathcal{H}^\infty$ results if $\varphi(x) = 0$ for $|x| \leq 1$,
and $= +\infty$ for $|x| > 1$, so that $\max_{0 \leq \theta < 2\pi} |f(re^{i\theta})|$ is bounded in $0 \leq r < 1$.
We soon introduce a metric in $\mathcal{H}^\varphi$ and present a few results to study its
structure. For this it is necessary to state some properties of analytic
functions and their boundary behavior from the classical theory.
   A subclass of $\mathcal{N}$ is given in a convenient form as:

$$\mathcal{N}^+ = \{f \in \mathcal{N} : \lim_{0 \leq r \uparrow 1} \int_0^{2\pi} \log^+ |f(re^{i\theta})| d\mu(\theta) = \int_0^{2\pi} \log^+ |f(e^{i\theta})| d\mu(\theta)\}, \quad (1)$$

where $\mu(\cdot)$ is again the normalized Lebesgue measure of $(0, 2\pi]$. The fact that the integral $\int_0^{2\pi} \varphi(\log |f(re^{i\theta})|) d\mu(\theta)$ is nondecreasing as $r \uparrow 1$ for the φ, of type considered in Definition 1, is known from the classical theory. Since μ is a finite measure, it follows from definitions that $\tilde{\mathcal{H}}^\varphi \subset \mathcal{H}^\varphi \subset \mathcal{N}^+ \subset \mathcal{N}$. In fact, only the middle inclusion needs a proof. Thus, $\mu(\partial D) = \mu((0, 2\pi]) = 1$ so that by Jensen's inequality applied to the convex φ, and then using Fatou's lemma as well as a representation of the analytic function f on D, one gets the following: $f = BSg$ where B is a Blaschke product, S is a singular function and g is an outer function. This means,

$$B(z) = z^m \prod_n \frac{|a_n|}{a_n} \frac{a_n - z}{1 - \bar{a}_n z}, \quad \sum_n (1 - |a_n|) < \infty, \quad 0 \leq m < \infty,$$

$$S(z) = \exp \left\{ - \int_0^{2\pi} \frac{e^{it} + z}{e^{it} - z} d\nu(t) \right\}, \quad \nu \text{ is bounded and } \nu \perp \mu,$$

$$g(z) = e^{i\gamma} \exp \left\{ \int_0^{2\pi} \frac{e^{it} + z}{e^{it} - z} \log h(t) d\mu(t) \right\}, \quad \gamma \in \mathbb{R}, h(t) \geq 0 \text{ with}$$

$$\log h \in L^1(\mu). \qquad (2)$$

This is a classical canonical factorization, going back to F. Riesz, and will be used here. [See Duren (1970), p. 24 for more on this.] Also $|B(z)| \leq 1$ for $z \in D$ and $0 < |S(z)| \leq 1$. Hence φ being convex and increasing we have by Jensen's inequality

$$\varphi(\log |f(re^{i\theta})|) \leq \varphi(\log |g(re^{i\theta})|)$$

$$\leq \int_0^{2\pi} \varphi(\log |f(re^{it})|) d\zeta_\theta(t), \qquad (3)$$

where

$$\zeta_\theta(A) = \int_A \text{Re}\left(\frac{e^{it} + re^{i\theta}}{e^{it} - re^{i\theta}}\right) d\mu(t) = \int_A P(r, \theta - t) d\mu(t), \qquad (4)$$

with

$$P(r, \theta) = \frac{1 - r^2}{1 - 2r \cos \theta + r^2}$$

as the Poisson kernel, so that $\zeta_\theta(\cdot)$ is also a probability measure. Thus integrating (3) relative to $d\mu(\theta)$ and using Fubini's theorem, one gets

$$\int_0^{2\pi} \varphi(\log |f(re^{i\theta})|) d\mu(\theta) \leq \int_0^{2\pi} \varphi(\log |f(e^{it})|) d\mu(t), \qquad (5)$$

since $h(re^{i\theta}) = \int_0^{2\pi} P(r, \theta - t) h(e^{it}) d\mu(t)$. Hence with Fatou's lemma and the fact that $\lim_{r \uparrow 1} |f(re^{it})| = |f(e^{it})|$ a.e., one simplifies (5) to obtain

$$\int_0^{2\pi} \varphi(\log |f(e^{it})|) d\mu(t) \leq \liminf_{r \uparrow 1} \int_0^{2\pi} \varphi(\log |f(re^{i\theta})|) d\mu(\theta). \qquad (6)$$

From (5) and (6) we deduce that

$$\lim_{r \uparrow 1} \int_0^{2\pi} \varphi(\log |f(re^{i\theta})|) d\mu(\theta) = \int_0^{2\pi} \varphi(\log |f(e^{it})|) d\mu(t). \qquad (7)$$

This shows that $\mathcal{H}^\varphi \subset L^\varphi(\mu) \cap \mathcal{N}^+ \subset L^1(\mu) \cap \mathcal{N}^+ \subset \mathcal{N}^+$, as desired. This detailed computation also gives us a clue about the introduction of a norm in $\tilde{\mathcal{H}}^\varphi$ in order that we may embed it in the space $L^\varphi(\mu)$. [Note that φ is a Young function only on $\mathbb{R}^+$.]

For convenience we denote by $\varphi : x \mapsto \Phi(\log x)$, and call the mapping φ a log-convex φ-function for $x > 0$, $\varphi(x) = 0$, iff $x = 0$, where Φ is convex on $\mathbb{R}$, $\Phi(x) \to 0$ as $x \to -\infty$, and to $+\infty$ when $x \to +\infty$. The structure of such φ is given by:

Lemma 2. *A log-convex φ-function φ is representable as*

$$\varphi(x) = \int_0^x \frac{u(t)}{t} dt, \qquad x > 0 \qquad (8)$$

where $u : \mathbb{R}^+ \to \mathbb{R}^+$ is a nondecreasing function and $u(t) \to \infty$ if $\varphi(t) \to \infty$, as $t \to \infty$.

Proof: By definition $\varphi(e^x) = \Phi(x)$, $x \in \mathbb{R}$, with Φ convex. since $\Phi(x) \to 0$ as $x \to -\infty$, Theorem I.3.1 implies

$$\varphi(e^x) = \Phi(x) = \int_{-\infty}^{x} p_1(t)dt, \qquad x \in \mathbb{R} \tag{9}$$

with p_1 nondecreasing. A change of variables gives

$$\varphi(x) = \int_{-\infty}^{\log x} p_1(t)dt = \int_0^x t^{-1} p_1(\log t)dt,$$

which is (8) with $p_1(\log t) = u(t) \geq 0$. Note that $\Phi(x) - \Phi(0) \leq p_1(x)$ in (9) so that $\varphi(x) - \varphi(1) \leq p_1(\log x) = u(x) \to \infty$ as $x \to \infty$ whenever $\varphi(\cdot)$ does. ∎

We also want to consider φ-functions of the form $\varphi(x) = \Phi(x^s)$, $0 < s \leq 1$ with $\Phi(\cdot)$ as a Young function. This φ is a log-convex φ function since $\varphi(e^x) = \Phi(e^{sx})$ is convex increasing and tends to zero as $x \to -\infty$. Such a φ is also called an *s-convex* function. A norm functional will be introduced here to take both the above types of φ-functions into account. Since by the classical theory, for each $f \in \mathcal{H}^\varphi$, the radial limit of $f(re^{i\theta})$ is $f(e^{i\theta}) = \tilde{f}(\theta)$, for almost all $\theta[\mu]$ as $r \to 1+$, and since (7) holds for all such f, we can introduce the following:

Definition 3. For each $f \in \mathcal{H}^\varphi$ if $\tilde{f}$ is its radial limit, we consider

$$\|f\|_{\mathcal{H}^\varphi} = \|\tilde{f}\|_\varphi = \inf\left\{ k > 0 : \int_0^{2\pi} \varphi\left(\frac{\tilde{f}}{k}\right) d\mu \leq k \right\}, \tag{10}$$

and call this the *Fréchet-* or *F-norm* of f for $\mathcal{H}^\varphi$. We use $\|\cdot\|_{\mathcal{H}^\varphi}$ and $\|\cdot\|_\varphi$ interchangeably.

To say $\|\cdot\|_\varphi$ is an *F*-norm means that it has the following properties: (i) $\|\tilde{f} + \tilde{g}\|_\varphi \leq \|\tilde{f}\|_\varphi + \|\tilde{g}\|_\varphi$, (ii) $\|\tilde{f}\|_\varphi = 0$ iff $\tilde{f} = 0$ a.e.$[\mu]$, for all f, g in $\mathcal{H}^\varphi$. Although $\|a\tilde{f}\|_\varphi \to 0$ as $a \to 0$, is also a consequence of (10), it is $\neq |a|\|\tilde{f}\|_\varphi$.

For the *s*-convex functions one defines the *s-homogeneous F-norm* as follows:

$$\|\tilde{f}\|_{s\varphi} = \inf\left\{ k > 0 : \int_0^{2\pi} \varphi\left(\frac{|\tilde{f}|}{k^{1/s}}\right) d\mu \leq 1 \right\}, \quad f \in \mathcal{H}^\varphi, \varphi(x) = \Phi(x^s). \tag{11}$$

Thus $\|\cdot\|_{s\varphi}$ has the same properties as an F-norm, but also with

$$\|a\tilde{f}\|_{s\varphi} = |a|^s \|\tilde{f}\|_{s\varphi}, \qquad a \in \mathbb{C}, f \in \mathcal{H}^\varphi. \tag{12}$$

Then, when $s = 1$, one has $\|\tilde{f}\|_{1\varphi} = N_\Phi(\tilde{f})$, where $N_\Phi(\cdot)$ is the gauge norm. For s-convex functions the F-norms defined by (10) and (11) are equivalent in the sense that $\|\tilde{f}\|_\varphi < \infty$ iff $\|\tilde{f}\|_{s\varphi} < \infty$ and a sequence $\{f_n, n \geq 1\} \subset \mathcal{H}^\varphi$ is Cauchy in one F-norm iff it has the same property in the other. These results can be established by standard, although somewhat tedious, computations which will be left to the reader. (See, e.g., the solution given in a series of Exercises in Rao (1987), pp. 211–212 on similar statements.)

We include the details of the following result to indicate the flavor:

Theorem 4. *If φ is a convex φ-function then $(\mathcal{H}^\varphi, \|\cdot\|_\varphi)$ is a complete vector (i.e., a Fréchet) space, and is a Banach space when φ is 1-convex. In any case, if φ is Δ_2-regular, then trigonometric polynomials are dense in $(\mathcal{H}^\varphi, \|\cdot\|_\varphi)$ and hence $\mathcal{H}^\varphi$ is also separable. [If φ is not Δ_2-regular, then $\mathcal{H}^\varphi$ is nonseparable although $\mathcal{H}^\varphi \cap M^\varphi$ is separable as in the L^φ-theory.]*

Proof: The argument about completeness is essentially classical. Thus let $\{f_n, n \geq 1\} \subset \mathcal{H}^\varphi$ be a Cauchy sequence so that $\|f_n - f_m\|_\varphi \to 0$ as $n, m \to \infty$. This implies the uniform convergence of f_n to an element f in the interior of the unit disc D. Indeed, since $f_n - f_m$ is analytic in D, it admits a representation as

$$(f_n - f_m)(z) = \int_0^{2\pi} P(r, \theta - t)(f_n - f_m)(e^{it})d\mu(t), \tag{13}$$

by the classical theory (cf., e.g., Duren (1970), p. 34), where $P(\cdot, \cdot)$ is the Poisson kernel. Replacing $f_n - f_m$ by $\varphi(|f_n - f_m|)$ in (13) and replacing the unit disc by an r-disc where $0 < r < 1$ is arbitrarily fixed, we get for $0 \leq \rho < r < 1$,

$$\varphi(|(f_n - f_m)(\rho e^{i\theta})|) = \frac{1}{2\pi} \int_0^{2\pi} P(\rho e^{i\theta}, re^{it}) \varphi(|(f_n - f_m)(re^{it})|)dt. \tag{14}$$

But we also have

$$P(\rho e^{i\theta}, re^{it}) = \frac{r^2 - \rho^2}{r^2 - 2r\rho\cos(\theta - t) + \rho^2} \leq \frac{r + \rho}{r - \rho} \leq \frac{2}{r - \rho}. \tag{15}$$

Hence (14) and (15) yield

$$\varphi(|(f_n - f_m)(\rho e^{i\theta})|) \leq \frac{1}{\pi(r - \rho)} \int_0^{2\pi} \varphi(|(f_n - f_m)(re^{it})|)dt$$

$$\leq \frac{2}{(1 - \rho)} \int_0^{2\pi} \varphi(|\tilde{f}_n - \tilde{f}_m|)d\mu, \qquad (16)$$

where $\tilde{f}_n$ is the boundary value of f_n and where the nondecreasing property of the integral on the right side (cf. (6) above) is used. Replacing f_n by $\frac{f_n}{\varepsilon}$ in (16), and recalling the definition of norm in (10), we get

$$|(f_n - f_m)(z)| \leq \left[\varphi^{-1}\left(\frac{2\varepsilon}{(1 - \rho)}\right)\right]\varepsilon. \qquad (17)$$

But $\|f_n - f_m\|_\varphi \leq \varepsilon$ for large enough m, n and hence (17) implies $|f_n - f_m| \to 0$ uniformly in $z \in \{|z| \leq \rho < 1\}$, as desired. Thus $f_n \to f$ uniformly in this set, so that f is analytic there. Writing r for $\rho < 1$ here, one finds

$$\int_0^{2\pi} \varphi\left(\left|\frac{(f_n - f_m)(re^{it})}{\varepsilon}\right|\right)d\mu(t) \leq \varepsilon, \qquad \text{by (10)},$$

and letting $m \to \infty$ here we see that $f_n - f \in \mathcal{H}^\varphi$ and that $\|f_n - f\|_\varphi < \varepsilon$ for n large enough. Thus $\|f\|_\varphi \leq \|f - f_n\|_\varphi + \|f_n\|_\varphi < \infty$ implying $f \in \mathcal{H}^\varphi$, so that $\mathcal{H}^\varphi$ is complete.

Observing that an s-convex function is a special type of φ-function admitted above, and for the 1-convex function the F-norm is equivalent to the homogeneous- (i.e., the Banach-)norm, we deduce that $\mathcal{H}^\varphi$ in this case is a Banach-space (cf., (12)).

We now consider the case that φ is Δ_2-regular (but not necessarily s-convex), and establish the last part by extending the classical ideas for the $\mathcal{H}^p$-spaces (cf. Duren (1970), p. 36). Since the modifications are not obvious, we include the detail.

Thus if $f \in \mathcal{H}^\varphi$, let $\tilde{f}$ denote its boundary value (or its radial limit). Then by our work prior to the statement of the theorem $\|f\|_\varphi = \|\tilde{f}\|_\varphi$ where $\|\tilde{f}\|_\varphi = \inf\{k > 0 : \int_0^{2\pi} \varphi(\frac{|f|}{k})d\mu \leq k\}$. Let $L^\varphi(d\mu)$ be the (vector) space of all measurable functions on $(0, 2\pi)$ with this F-norm. [These

$L^\Phi(\mu)$ will be discussed in Section 10.1.] Then it is a complete (or F-)
space and when φ is Δ_2-regular then continuous functions are dense in
this norm in $L^\varphi(\mu)$. The proof is standard and similar to the $L^p(\mu)$-
case (cf., e.g., Rao (1987), p. 194 for the latter case). Thus given $\varepsilon > 0$,
there is a g_ε, continuous, such that $\|\tilde{f} - g_\varepsilon\|_\varphi < \varepsilon$. If now we take the
Poisson integral of g_ε, i.e.,

$$G_\varepsilon(re^{i\theta}) = \int\limits_0^{2\pi} P(r, \theta - t) g_\varepsilon(t) d\mu, \qquad 0 \le r < 1,$$

then G_ε is continuous in D and $\lim\limits_{r \to 1-} G_\varepsilon(re^{i\theta}) = g_\varepsilon(\theta)$ uniformly. Thus
$\|\tilde{f} - \tilde{G}_\varepsilon\|_\varphi < \varepsilon$. [This statement does not hold if φ is not Δ_2-regular, as
seen from our work on the Orlicz spaces in Chapter III.] Further $\| \cdot \|_\varphi$
is absolutely continuous. Although we proved it there only for Young
functions, the same argument extends to all Δ_2-regular φ-functions
(just as in the L^p-case for $0 < p < \infty$).

Now consider for any $f \in \mathcal{H}^\varphi$ its Taylor series expansion at the
origin. Thus $f(z) = \sum\limits_{n=0}^{\infty} a_n z^n$ and let $S_n(z) = \sum\limits_{k=0}^{n} a_k z^k$, so that
$S_n(z) \to f(z)$ uniformly in each disc $\{z : |z| \le \rho < 1\}$. Writing
$z = re^{i\theta}$, $S_n(\theta)$ is a trigonometric polynomial. Consider, with $f(r\cdot)$ for
$f(re^{i\cdot})$ and similarly for $S_n(\cdot)$,

$$\|\tilde{f} - S_n(r\cdot)\|_\varphi \le \|\tilde{f} - f(r\cdot)\|_\varphi + \|f(r\cdot) - S_n(r\cdot)\|_\varphi. \qquad (18)$$

Given $\varepsilon > 0$, choose $r_0 < 1$ such that $\|\tilde{f} - f(r\cdot)\|_\varphi < \frac{\varepsilon}{2}$. This can
be done by the absolute continuity of the F-norm and the fact that
$f(r\cdot) \to \tilde{f}$ a.e. as $r \uparrow 1$. Since on the disc $\{z : |z| \le r_0 < 1\}$, $S_n(\cdot) \to$
$f(\cdot)$ uniformly as $n \to \infty$, the second term on the right of (18) tends
to zero. So there is n_0 such that $n \ge n_0 \Rightarrow \|f(r_0\cdot) - S_n(r_0\cdot)\|_\varphi < \frac{\varepsilon}{2}$.
Hence for $r_0 \le r < 1$ and $n \ge n_0$, we get

$$\|f(r\cdot) - S_n(r\cdot)\|_\varphi \le \|\tilde{f} - S_n(r\cdot)\|_\varphi < \varepsilon. \qquad (19)$$

Since S_n is a polynomial of the desired kind, the result follows. But
because the coefficients of these polynomials can be taken rational, the
separability of $\mathcal{H}^\varphi$ is also a consequence. ∎

It is clear from the preceding discussion that the $\mathcal{H}^\varphi$ spaces are
of a special kind having a rich structure. The last part of the preceding theorem implies that each $\mathcal{H}^\varphi$ is infinite dimensional and may be

identified as [or embedded in] a subspace of $L^\varphi(\mu)$. In particular, if φ is 1-convex, so that it is a Young function, the Banach space $\mathcal{H}^\varphi$ has nicer properties than the corresponding $L^\varphi(\mu)$ space. For instance, the $L^1(\mu)$ is known *not* to be an adjoint space of any Banach space, in contrast to the sequence space ℓ^1 ($= (c_0)^*$, c_0 being the space of scalar sequences converging to zero and it is a Banach space under the supremum norm). However, the Hardy space $\mathcal{H}^1$ *is* an adjoint space just as all the $\mathcal{H}^p$, $p > 1$ are. We include an analog of this fact for the $\mathcal{H}^\varphi$-spaces as follows.

Theorem 5. *For each Young function φ, the space $\mathcal{H}^\varphi$ is a conjugate Banach space in the sense that it is topologically isomorphic to the conjugate or adjoint space of a Banach space.*

Proof: The argument will be facilitated if we have an alternative definition of $\mathcal{H}^1$. [Since the measure μ is finite, and φ is a Young function, $\mathcal{H}^\varphi \subset \mathcal{H}^1$ such a characterization of $\mathcal{H}^1$ will be sufficient.] Thus let $f \in \mathcal{H}^1$ so that it is analytic in D and hence it may be expanded as $f(z) = \sum\limits_{n=0}^{\infty} a_n z^n$, for $|z| < 1$. Then its boundary function $\tilde{f} : t \mapsto f(e^{it})$ is in $L^1(\mu)$, and let $b_n = \int\limits_0^{2\pi} \tilde{f}(t)e^{-int}d\mu(t)$ be its Fourier coefficient, $n \in \mathbb{Z}$. On the other hand, letting $z = re^{it}$, $0 < r < 1$, we have

$$r^n a_n = \int\limits_0^{2\pi} f(re^{it})e^{-int}d\mu(t), \qquad n \geq 0. \tag{20}$$

Hence $|r^n a_n - b_n| \leq \|f_r - \tilde{f}\|_1$ where $f_r(e^{it}) = f(re^{it})$. Since the norm of $L^1(\mu)$ is absolutely continuous, we get $\lim\limits_{r \uparrow 1} \|f_r - \tilde{f}\|_1 = 0$, and hence $a_n = b_n$ for $n \geq 0$. Also $\int\limits_0^{2\pi} f(re^{it})e^{-int}d\mu(t) = 0$ for $n < 0$, by the dominated convergence and the fact that $\lim\limits_{r \uparrow 1} f_r(e^{it}) = \tilde{f}(t)$ a.e., we get $b_n = 0$ for $n < 0$. Thus $f \in \mathcal{H}^1 \Rightarrow \tilde{f} \in L^1(\mu)$ whose Fourier coefficients vanish for $n < 0$.

This property characterizes $\mathcal{H}^1$ in the following sense: If $f \in L^1(\mu)$ with Fourier coefficients vanishing for $n < 0$, then its Poisson integral transform

$$F(re^{i\theta}) = \int\limits_0^{2\pi} P(r, \theta - t)f(t)d\mu,$$

defines $F : z \mapsto F(z)$, $z = re^{i\theta}$, and $F \in \mathcal{H}^1$, i.e., F has a power series expansion with integrability properties. Thus $\mathcal{H}^1$ can be obtained in this manner which constitutes an alternative definition. With this we can present the same argument as in the $\mathcal{H}^p$-case.

Consider φ to be such that $\frac{\varphi(x)}{x} \to \infty$, i.e., φ may be assumed to be an N-function. If its complementary Young function is denoted by ψ, then let $\mathcal{M}^\psi$ be the closed linear space spanned by all polynomials of $L^\psi(\mu)$. Then our work of Chapter IV implies that $(\mathcal{M}^\psi)^* \cong L^\varphi(\mu)$ (cf., Theorem IV.1.7). By the argument of the first part above, each f in $\mathcal{H}^\varphi$ satisfies $\int\limits_0^{2\pi} \tilde{f}(t)e^{int}d\mu(t) = 0$, $n \geq 1$. If $\mathcal{M}_1$ is the linear span of $\{e^{in(\cdot)}, n \geq 1\} \subset \mathcal{M}^\psi$, then $\mathcal{H}^\varphi$ annihilates $\mathcal{M}_1$ and hence also its completion $\overline{\mathcal{M}}_1$ since $\int\limits_0^{2\pi} fg d\mu = x_f^*(g)$ defines $x_f^* \in \mathcal{M}_1^*$ for $f \in \mathcal{H}^\varphi$. Thus $\mathcal{H}^\varphi$ may be identified with $(\mathcal{M}^\psi)^\perp$ which by the general results in linear analysis is isometrically isomorphic to the adjoint space of the quotient space $L^\varphi(\mu)/\overline{\mathcal{M}}_1$, i.e., $\mathcal{H}^\varphi$ is topologically equivalent to $(L^\varphi(\mu)/\overline{\mathcal{M}}_1)^*$. (See Dunford-Schwartz (1958), II.4.18(b) for this classical result.) This shows that $\mathcal{H}^\varphi$ is an adjoint space if φ is an N-function. The same result also holds if φ is discontinuous so that its complementary function is continuous, since then $L^\varphi(\mu) = L^\infty(\mu)$ [and $L^\psi(\mu) = L^1(\mu)$] which is the adjoint of $L^1(\mu)$, and since $\mathcal{H}^\infty \subset \mathcal{H}^1$, this computation shows that $\mathcal{H}^\infty$ is identifiable with $(L^\varphi(\mu)/\overline{\mathcal{M}}_1)^*$. Finally let φ be continuous, but $\frac{\varphi(x)}{x} \leq K_0 < \infty$. Then $L^\varphi(\mu) = L^1(\mu)$ which is not an adjoint space. This case was already treated in the $\mathcal{H}^p$-case (cf., e.g., Hoffman (1962), p. 137), and we include the detail for the sake of completeness.

Since $\nu : A \mapsto \int\limits_A f d\mu$, for $f \in L^1(\mu)$, is a signed measure on the Borel σ-algebra $\mathcal{B}$ of $[0, 2\pi]$, and $\|\nu\| = \|f\|_1$ where $\|\nu\|$ is the total variation norm of ν, we may embed (properly) $L^1(\mu) \subset rca([0, 2\pi])$, the space of σ-additive regular signed measures on $\mathcal{B}$. It is well known that $rca([0, 2\pi])$ is the adjoint space of $C([0, 2\pi])$, the Banach space of continuous real functions on $[0, 2\pi]$ under the uniform norm. We again have $\int\limits_0^{2\pi} f(t)e^{int}d\mu(t) = 0$, $n \geq 1$ and $f \in \mathcal{H}^\varphi$ ($= \mathcal{H}^1$). To apply the proof of the preceding paragraph, we must also know the behavior of $\nu \in rca([0, 2\pi])$ with $\int\limits_0^{2\pi} e^{int}d\nu(t) = 0$, $n \geq 1$. Fortunately in this case, by an important theorem of F. and M. Riesz, such measures ν

are absolutely continuous relative to μ and $d\nu(t) = f(t)d\mu(t)$ with f agreeing a.e. with an $h \in \mathcal{H}^1$ and $h(e^{it}) = f(e^{it})$, a.a.(t) (cf. Duren (1970), p. 42). Thus all these measures that annihilate $\widetilde{\mathcal{M}}_1 = \overline{\mathrm{sp}}\{e^{in(\cdot)},$ $n\geq1\} \subset C([0, 2\pi])$ come precisely from $\mathcal{H}^1$. By the preceding argument, therefore $\mathcal{H}^\varphi$ (= $\mathcal{H}^1$) is topologically equivalent to $(C([0,2\pi])/\widetilde{\mathcal{M}}_1)^*$, and hence is again an adjoint Banach space. Consequently the statement of the theorem is established for all Young functions. ∎

Remark. The result corresponding to Theorem IV.1.7 for the $\mathcal{H}^\varphi$-spaces would have simplified the earlier part of the above proof. We do not have such a representation in this book. A general study of linear functionals on Hardy-Orlicz spaces is conducted by R. Lésniewicz (1973). But one needs to consider essentially a parallel treatment of that given in Chapter IV. This is why we chose the preceding argument although we then had to use some general results from Linear Analysis. However, the result on quotient space is relatively elementary.

The alternative representations used in the preceding proof suggest that the $\mathcal{H}^\varphi$ spaces may have additional structural properties. We include a brief discussion to affirm this prospect. Regarding $[0, 2\pi)$ as a (compact) group under addition (modulo 2π), and using $\mu(\cdot)$ as its normalized translation invariant measure, one can define the convolution operation in $L^\varphi(\mu)$ ($\subset L^1(\mu)$):

$$(f * g)(x) = \int_0^{2\pi} f(x - t)g(t)d\mu(t), \quad f, g \in L^\varphi(\mu). \qquad (21)$$

Integration and Fubini-Torelli theorems show that $\|f * g\|_1 \leq \|f\|_1 \|g\|_1 < \infty$. If now $h \in L^\psi(\mu)$ [ψ is complementary to φ], then

$$\left| \int_0^{2\pi} (f*g)(x)h(x)d\mu(x) \right| = \left| \int_0^{2\pi} f(x) \left[\int_0^{2\pi} h(t)g(x-t)d\mu(t) \right] d\mu(x) \right|$$
$$\leq \|f\|_1 \|g\|_\varphi N_\psi(h), \text{ by Hölder's inequality. (22)}$$

Taking the supremum over h in the unit ball of $L^\psi(\mu)$, we get from (22) that $\|f * g\|_\varphi \leq \|f\|_1 \|g\|_\varphi \leq \alpha \|f\|_\varphi \|g\|_\varphi$, with $\alpha = (\psi^{-1}(1))^{-1}$. Thus $L^\varphi(\mu)$ is closed under convolutions, and if we "normalize" φ, ψ such that $\psi(1) = 1$, then $L^\varphi(\mu)$ becomes a Banach algebra. Using the alternative definition of $\mathcal{H}^\varphi$ employed in the preceding proof, one can verify that for each $f, g \in \mathcal{H}^\varphi$, $f * g$ is also analytic in the (open) disc D and conclude that $f * g \in \mathcal{H}^\varphi$. We thus have shown that the following result holds.

Proposition 6. *If φ is 1-convex [whence it is a Young function] such that its complementary function ψ satisfies $\psi(1) = 1$, then $(\mathcal{H}^\varphi, \| \cdot \|_\varphi)$ is a (commutative) Banach algebra under convolution as multiplication.*

A great deal of the Gel'fand transform of Banach algebras may be considered for the $\mathcal{H}^\varphi$-spaces and their special structure can be analyzed by extending the $\mathcal{H}^\infty$-theory discussed in Hoffman (1962). We shall omit that study here. From another point of view implied by the result of Theorem 5, the $\mathcal{H}^\varphi$-spaces are suitable for extremum problems. We recall that a point x of a convex set A of a vector space is *extreme* if $x = (1 - \alpha)y + \alpha z$ for y, z in A and $0 < \alpha < 1$ implies that $x = y = z$. The classical *Krein-Milman (and Šmulian) theorem* of Functional Analysis says that if $\mathcal{Y}$ is an adjoint Banach space and $A \subset \mathcal{Y}$ is a compact convex subset, in the weak-star topology of $\mathcal{Y}$, then it has extreme points, and in fact A is the weak-star closure of the convex hull of its extreme points. Since $(\mathcal{H}^\varphi, \|\cdot\|_\varphi)$ is an adjoint space, it follows that its closed unit ball [which is weak-star compact and convex] has sufficiently many extreme points in the sense that their closed hull has the stated property. [Regarding the above noted theorem, we refer again to Dunford-Schwartz (1958), p. 440 and p. 429.] Since $L^\varphi(\mu)$ here is rotund iff φ is Δ_2-regular and strictly convex (cf. Corollary VII.1.5) and $\mathcal{H}^\varphi$ can be identified as a closed subspace of $L^\varphi(\mu)$, it follows that each point on the boundary of the unit ball of $\mathcal{H}^\varphi$ is an extreme point if φ is Δ_2-regular and strictly convex. For $\mathcal{H}^1$ there is the following characterization of these points [since not every point on the unit ball is extreme] due to K. de Leeuw and W. Rudin (1958):

Theorem 7. *A function f on the boundary of the unit ball of $\mathcal{H}^1$ is extreme iff it admits a representation as:*

$$f(z) = \exp \left\{ \int_0^{2\pi} \frac{e^{it} + z}{e^{it} - z} \log k(t) d\mu(t) \right\}, \tag{23}$$

*where $0 \leq k(\cdot) \in L^1(\mu)$ and $\log k(\cdot) \in L^1(\mu)$. (Such an f is called an **outer** function.)*

We do not give a proof of this result (see, e.g., Hoffman (1962), p. 140). The $\mathcal{H}^p$ (and $\mathcal{H}^\varphi$) spaces play an important role in the extreme point [or variational] problems. A useful result in this connection is that there exists a bounded projection on $L^p(\mu)$ onto $\mathcal{H}^p$ if $1 < p < \infty$. This is a classical result due to M. Riesz. A similar result for the $\mathcal{H}^\varphi$ spaces appears to hold if $L^\varphi(\mu)$ is reflexive or perhaps even if φ is only an N-function. [For $\mathcal{H}^1$ and $\mathcal{H}^\infty$ the corresponding result is false.] In this connection the following result stated by V. M. Terpigoreva (1962) is revealing.

Proposition 8. *Let φ be an N-function and $\mathcal{H}^\varphi$ and $L^\varphi(\mu)$ be the spaces as before. Then for each $f \in L^\varphi(\mu)$ we have the following evaluations: there is an extremal function $f_0 \in \mathcal{H}^\varphi$ such that*

$$\|f - f_0\|_\varphi^{\bar{\ }} = \inf\{\|f - g\|_\varphi^{\bar{\ }} : g \in \mathcal{H}^\varphi\}$$

$$= \sup\{|\int_0^{2\pi} f(e^{it})g(e^{it})d\mu(t)| : g \in \mathcal{H}^\psi, \|g\|_\psi \le 1\}$$

$$= \sup\{|\int_0^{2\pi} f(e^{it})g(e^{it})d\mu(t)| : g \in \mathcal{H}^\psi \cap M^\psi, \|g\|_\psi \le 1\},$$

where $\|\cdot\|_\varphi^{\bar{\ }}$ is the (Orlicz) norm functional defined on $\mathcal{H}^\varphi$ as

$$\|f\|_\varphi^{\bar{\ }} = \sup\Big\{ \sup \Big[\int_0^{2\pi} |f(re^{it})g(e^{it})|d\mu(t) :$$

$$\int_0^{2\pi} \psi(|g(e^{it})|)d\mu(t) \le 1 \Big] : 0 < r < 1 \Big\}.$$

Here as in (10)–(12) the outer supremum can be replaced by limit since the inside quantity is an increasing function of r. One can also show that (11) with $s = 1$ and $\|\cdot\|_\varphi^{\bar{\ }}$ are equivalent norms. We omit further discussion of the general theory and present some analogs of the interpolation results of Chapter VI, since they are of interest in applications.

9.2. *Interpolation of sublinear operators*

In order to treat interpolation problems in $\mathcal{H}^\varphi$ spaces, we need to use a few properties of (sub) harmonic functions. Let us recall the concept. A real continuous function g defined on a bounded domain D in the complex plane (i.e., D is an open connected set) is termed *subharmonic* in D if for each $z_0 \in D$ there is an $r_0 > 0$ such that the disc $B = \{z : |z - z_0| < r_0\} \subset D$ and for all $z = z_0 + re^{it}$ in B we have

$$g(z_0) \le \int_0^{2\pi} g(z_0 + re^{it})d\mu(t), \qquad 0 < r < r_0. \tag{1}$$

This is not the original definition of a subharmonic function, as formulated by F. Riesz, but is equivalent to it. [See, e.g., T. Rado (1949), and, for the equivalence, P. L. Duren (1970), p. 7.] Similarly g is called *superharmonic* if $-g$ is subharmonic, and g is *harmonic* if it is both sub- and super-harmonic simultaneously so that there is equality in (1).

Using the fact that μ is a normed (= probability) measure in (1), this form of the definition admits immediate and far-reaching extension to functions on abstract probability spaces using the theory of (sub-) martingales. Here we shall include a few classical properties to be considered for interpolation of operators on $\mathcal{H}^\varphi$-spaces. Since the real and imaginary parts of an analytic function f are harmonic and $|f|^\alpha$, $\alpha \ge 0$, subharmonic, their appearance in the $\mathcal{H}^\varphi$ space theory is to be expected.

We now present an interpolation theorem. This and most of the work in the rest of this section largely follows W. T. Kraynek (1972). We only treat the s-convex functions which however include $\varphi(x) = x^p$, $p > 0$ so that the $\mathcal{H}^p$ spaces are covered in their entirety. For this work one has to give an extension of Theorem VI.3.5 for the more general φ-functions. These are contained in the following result.

Theorem 1. *Let φ_i, Q_i be r_i- and s_i-convex functions where $0 < r_i, s_i \le 1$, $i = 1, 2$. Consider the spaces $L^{\varphi_i}(\mu)$ and $L^{Q_i}(\nu)$, on (Ω, Σ, μ) and $(S, \mathcal{S}, \nu)$. If T is a sublinear operator defined on $L^{\varphi_i}(\mu)$ into $L^{Q_i}(\nu)$ satisfying*

$$\|Tf\|_{rQ_i} \le M_i \|f\|_{r\varphi_i}, \qquad i = 1, 2, f \in L^{\varphi_i}(\mu), \tag{2}$$

for some $0 < r \le \min(r_1, r_2, s_1, s_2)$, then we have

$$\|Tf\|_{rQ_t} \le K M_1^{1-t} M_2^t \|f\|_{r\varphi_t}, \qquad 0 < t < 1 \tag{3}$$

*for all simple functions f, where φ_t and Q_t are intermediate to (φ_1, φ_2)
and (Q_1, Q_2) as before. [I.e., φ_t is the inverse of $\varphi_t^{-1} = (\varphi_1^{-1})^{1-t}(\varphi_2^{-1})^t$
and similarly Q_t.] Here $0 < K \leq 4$ can be assumed.*

*If moreover, φ_1, φ_2 are Δ_2-regular, T is a sublinear operator on
$\mathcal{H}^{\varphi_i}$ $(\subset L^{\varphi_i}(\mu))$ into $L^{Q_i}(\nu)$ and (2) holds, then T is defined on $\mathcal{H}^{\varphi_t}$
into $L^{Q_t}(\nu)$ and (3) holds again when f is replaced by a trigonometric
polynomial.*

Remark. In many respects the argument is analogous to that of the
proof of Theorem IV.3.5, and so we detail the new points while outlining
and shortening the essentially repetitive parts. Recall that an s-convex
φ is given as $\varphi(x) = \Phi(x^s)$, $0 < s \leq 1$, and Φ is a Young function.
The idea is again to reduce the problem to the place where the result
is obtained from the Three-Line-Theorem for sublinear operators.

For convenience of reference we restate the desired form of the
"three-line-theorem."

Proposition 2. *Let $f : D \to \mathbb{R}^+$ be bounded and such that $\log f$
is continuous and subharmonic in the interior of $D = \{z \in \mathbb{C} : \alpha \leq
\mathrm{Re}\, z \leq \beta\}$. If $f(\alpha + iy) \leq M_1, f(\beta + iy) \leq M_2$ for all $y \in \mathbb{R}$, then
$f(z) \leq \max(M_1, M_2)$ for all $z \in D$. In particular, if $\alpha = 0, \beta = 1$ and
$0 < t < 1$, we have $f(t + iy) \leq M_1^{1-t} M_2^t$, $y \in \mathbb{R}$.*

Proof: Let $M = \max(M_1, M_2)$. By a classical characterization of the
subharmonicity of $\log f$ in a domain (cf. T. Rado (1949), p. 15), this will
be so iff $e^{h(z)} f(z)$ has the same property for each harmonic function
$h : D \to \mathbb{R}$. Hence in particular, since we are given that $\log f$ is
subharmonic, $f_n(z) = f(z) \cdot \exp(\frac{1}{n}(x^2 - y^2))$ is subharmonic for all $n \geq 1$.
Since $f(z) \leq M$, this implies that $\lim_{y \to \pm\infty} f_n(x + iy) = 0$ uniformly for
$x \in [\alpha, \beta] \subset \mathbb{R}$. Hence there is a $y_0 > 0$ such that for all $|y| \geq y_0$
we have $f_n(x + iy) \leq Me^{\gamma^2/n}$ where $\gamma = \max(|\alpha|, |\beta|)$. If $D_1 = \{z =
x + iy : \alpha < x < \beta, -y_0 \leq y \leq y_0\}$, then f_n is subharmonic on D_1
and continuous on its closure, so that by the maximum principle for
subharmonic functions its maximum is attained on the boundary of D_1
so that $f_n(x + iy_0) \leq Me^{\gamma^2/n}$. Consequently, using the bounds on f,
and the fact that on $D - D_1$, f_n is never larger than the preceding
bound, we get

$$f_n(\alpha + iy) \leq Me^{\gamma^2/n} \quad \text{and} \quad f_n(\beta + iy) \leq Me^{\gamma^2/n}. \qquad (4)$$

Hence $f_n(z) \le Me^{\gamma^2/n}$ for all $z \in D$. Since $\lim\limits_{n\to\infty} f_n(z) = f(z)$ and the bound in (4) tends to M, we get $f(z) \le M$, for all $z \in D$.

The last part is reduced to the preceding one by considering g defined as $g(z) = f(z)M_1^{z-1}M_2^{-z}$ where $z = x + iy$. Then $g : D \to \mathbb{R}^+$ is bounded by 1 for $x = 0$ ($= \alpha$) and $x = 1$ ($= \beta$), and $\log g$ is subharmonic in D. Hence $g(t + iy) \le 1$ for $0 \le t \le 1$, $y \in \mathbb{R}$. But this is simply the desired result in a different form. ∎

Proof of Theorem 1: Let $0 < r \le \min(r_i, s_i, i{=}1,2)$, and if $h_i : x \mapsto |x|^{r_i/r}$, $g_i : s \mapsto |x|^{s_i/r}$, let Φ_i, ξ_i be defined as $\Phi_i = \tilde{\varphi}_i \circ h_i$, $\xi_i = \tilde{Q}_i \circ g_i$, $i = 1, 2$, $\tilde{\varphi}_i, \tilde{Q}_i$ being Young functions. Then $\Phi_i(x^r) = \tilde{\varphi}_i(x^{r_i})$, $\xi_i(x^r) = \tilde{Q}_i(x^{s_i})$, so that all the new functions Φ_i, ξ_i, $i = 1, 2$ [which are the same as φ_i, Q_i] are r-convex. Let ζ_i be complementary to ξ_i, and Φ_t, ξ_t, ζ_t, $0 < t < 1$, be the intermediate Young functions to (Φ_1, Φ_2), (ξ_1, ξ_2), and (ζ_1, ζ_2) in the sense of Definition VI.3.1. Thus $\varphi_t(x) = \Phi_t(x^r)$ and $Q_t(u) = \xi_t(x^r)$. If $\alpha_i = \varphi_i^{-1}$ and $\beta_i = \zeta_i^{-1}$, $i = 1, 2$, $z = x + iy$, let $\alpha_z(u) = \alpha_1^{1-z}(u)\alpha_2^z(u)$, and $\beta_z(u) = \beta_1^{1-z}(u)\beta_2^z(u)$ which are analytic in the strip $0 \le x \le 1$, $y \in \mathbb{R}$, and are bounded there for each u, i.e.,

$$|\alpha_z(u)| = \alpha_1(u)^{1-x}\alpha_2(u)^x \le \max\{1, \alpha_1(u)\} \cdot \max\{1, \alpha_2(u)\}, \ 0 \le x \le 1,$$

and similarly $|\beta_z(u)|$. If $\mathcal{S}(\mu)$ denotes the set of μ-simple functions f in $L^{\varphi_i}(\mu)$, then for the sublinear operator T, $Tf \in L^{Q_1}(\nu) \cap L^{Q_2}(\nu)$, and

$$\|(Tf)^r\|_{\xi_t} = \sup \left\{ \int\limits_S |Tf|^r |g| d\nu : \rho_{\zeta_t}(g) \le 1, g \in \mathcal{S}(\nu) \right\}. \tag{5}$$

Since ζ_t and the complementary function of ξ_t define the same Orlicz space (cf. Lemma VI.3.3), the left side of (5) is an equivalent Orlicz norm in $L^{\xi_t}(\nu)$. *All our computations are given for the $\|\cdot\|_{\xi_t}, \|\cdot\|_{\Phi_t}$, $\|\cdot\|_{\zeta_t}$- norms, and at the end $\|\cdot\|_{\varphi_t}, \|\cdot\|_{Q_t}$-norms are obtained easily.* This is the basis of the following proof.

Let f and g be arbitrarily fixed such that $\|f\|_{r\Phi_t} = 1$ and $\rho_{\zeta_t}(g) \le 1$. We wish to estimate $I = \int_S |Tf|^r |g| d\nu$, following the procedure of Section 6.3, to use ultimately the three-line-theorem for subharmonic functions. For this purpose we represent the complex valued simple functions f, g as follows:

$$f = \sum_{j=1}^{m_1} a_j \chi_{F_j}, \ g = \sum_{\ell=1}^{m_2} b_\ell \chi_{G_\ell} \tag{6}$$

so that F_j's and G_ℓ's are disjoint. Next associate with them

$$f_z = \sum_{j=1}^{m_1} \alpha_z(\varphi_t(|a_j|))e^{i\theta_j}\chi_{F_j}, \ g_z = \sum_{\ell=1}^{m_2} \beta_z(\zeta_t(|b_\ell|))e^{i\theta'_\ell}\chi_{G_\ell}, \quad (7)$$

and consider

$$I(z) = \int_S |T(f_z)|^r|g_z|d\nu. \quad (8)$$

Then $I(t) = I$, and we obtain the bounds on $I(z)$ with the methods used before. The details of proof in the analysis of $I(z)$ differ somewhat from those of Theorem VI.3.5 since the power r in the integrand of (8) makes it nonlinear; in addition it is to be shown that $\log|Tf|$ is subharmonic. Thus the nontrivial differences from the previous ones will be detailed.

To simplify $I(z)$ of (8) and find its bounds, define the complex functions $\lambda_{j\ell}(\cdot)$, $\gamma_\ell(\cdot)$, and $\Gamma_\ell(\cdot)$ as

$$\lambda_{j\ell}(z) = \beta_z^{\frac{1}{r}}(\zeta_t(|b_\ell|))\alpha_z(\Phi_t(|a_j|)),$$

$$\gamma_\ell(z) = \sum_{j=1}^{m_1} \lambda_{j\ell}(z)\chi_{F_j}, \text{ and } \Gamma_\ell(z) = \int_{G_\ell} |T(\gamma_\ell(z))|^r d\nu. \quad (9)$$

Thus $I(z) = \sum_{\ell=1}^{m_2} \Gamma_\ell(z)$. It is clear that $\lambda_{j\ell}(z)$ is bounded, analytic in the strip $D = \{z \in \mathbb{C} : 0 < \text{Re}(z) < 1\}$, and continuous on the closure $\overline{D}$. Also using the definition of sublinearity of T, the fact that $u \mapsto u^r$, $0 < r \leq 1$, is concave on $\mathbb{R}^+$ so that it is subadditive, and the Hölder inequality for Orlicz spaces, one gets the continuity and boundedness of Γ_ℓ on $\overline{D}$. [Note that Γ_ℓ is a finite sum involving $\lambda_{j\ell}$ which have these properties, so $\gamma_\ell(\cdot)$ also has these properties.] We leave these straightforward computations to the reader, and verify the (not so simple) *claim:* $\log \Gamma_\ell(z)$ is subharmonic in D.

To verify this, it suffices to establish the property in a neighborhood of each point $z_0 \in D$. As seen in the proof of Proposition 2, this holds if for each harmonic function h, $e^{h(z)}\Gamma_\ell(z)$ is subharmonic at z_0. Let $z_1, \ldots, z_p$ be a set of equally spaced points on the boundary of the disc at z_0, of radius $\rho > 0$. Let H be an analytic function whose real part is h, and consider

$$\gamma_\ell^*(z) = \gamma_\ell(z)\exp(\tfrac{1}{r}H(z)), \text{ and } \lambda_{j\ell}^*(z) = \lambda_{j\ell}(z)\exp(\tfrac{1}{r}H(z)). \quad (10)$$

Then using (10) in (9), we get $\Gamma_\ell^*(z) = \int\limits_{G_\ell} |T\gamma_\ell^*(z)|^r d\nu \ (= e^{h(z)}\Gamma_\ell(z))$.

If now $\log|T\gamma_\ell^*(z)|$ is shown to be subharmonic in D, a.e.$[\nu]$, then so will $|T\gamma_\ell^*(z)|^r$ be for $r > 0$, since using (1) we have

$$|T\gamma_\ell^*(z)|^r = \exp\{r\log|T\gamma_\ell^*(z)|\} \le \exp\{\int\limits_0^{2\pi} \log|T\gamma_\ell^*(z + \rho e^{i\theta})|^r d\mu(\theta)\},$$

$$\le \int\limits_0^{2\pi} |T\gamma_\ell^*(z + \rho e^{i\theta})|^r d\mu(\theta), \tag{11}$$

by Jensen's inequality for concave functions. When this is shown, (11) implies that $\Gamma_\ell^*(z)$ is also subharmonic since

$$\Gamma_\ell^*(z) = \int\limits_{G_\ell} |T(\gamma_\ell^*(z))|^r d\nu$$

$$\le \int\limits_{G_\ell} \int\limits_0^{2\pi} |T(\gamma_\ell^*(z + \rho e^{i\theta}))|^r d\mu(\theta) d\nu$$

$$= \int\limits_0^{2\pi} \Gamma_\ell^*(z + \rho e^{i\theta}) d\mu(\theta), \text{ by Fubini's theorem.} \tag{12}$$

We need to show that $\log|T(\gamma_\ell^*)|$ is subharmonic, or that $e^{k(z)}|T(\gamma_\ell^*(z))|$ has the same property. Let $K(\cdot)$ be an analytic function having $k(\cdot)$ as its real part, and let $\gamma_\ell^{**}(z) = e^{K(z)}\gamma_\ell^*(z)$, $\lambda_{j\ell}^{**}(z) = e^{K(z)}\lambda_{j\ell}^*(z)$ and hence

$$e^{K(z)}|T\gamma_\ell^*(z)| = |T(\gamma_\ell^{**}(z))|. \tag{13}$$

Since $K(\cdot)$, $H(\cdot)$, and $\lambda_{j\ell}(\cdot)$ are analytic in D, $\gamma_{j\ell}^{**}(\cdot)$ is analytic. Let $z_n = \rho e^{i\Delta\theta_n}$, $1 \le n \le p$, be equidivision points of the circle at z_0, radius ρ. Then we get, on using a Riemann-type approximation to the integral,

$$\lambda_{j\ell}^{**}(z_0) = \int\limits_0^{2\pi} \lambda_{j\ell}^{**}(z_0 + \rho e^{i\theta}) d\mu(\theta)$$

$$= \lim_{p\to\infty} \frac{1}{p} \sum_{n=1}^p \lambda_{j\ell}^{**}(z_n). \tag{14}$$

On the other hand, applying the (reverse) triangle inequality to the sublinear T, one gets

$$\left| |T\gamma_\ell^{**}(z_0)| - \left| T\left(\frac{1}{p}\sum_{n=1}^{p}\gamma_\ell^{**}(z_n)\right)\right| \right| \leq \left| T\left(\gamma_\ell^{**}(z_0) - \frac{1}{p}\sum_{n=1}^{p}\gamma_\ell^{**}(z_n)\right)\right|$$

$$\leq \sum_{j=1}^{m_1}\left|\lambda_{j\ell}^{**}(z_0) - \frac{1}{p}\sum_{n=1}^{p}\lambda_{j\ell}^{**}(z_n)\right|\chi_{F_j}$$

$$\to 0, \text{ as } p \to \infty \text{ by (14).} \qquad (15)$$

From this we deduce that $|T(\gamma_\ell^{**}(z_0))| \leq \int_0^{2\pi}|T(\gamma_\ell^{**}(z_0 + \rho e^{i\theta}))|d\mu(\theta)$, and hence, by the earlier reduction $\log|T(\gamma_\ell^*(z))|$ is subharmonic for a.a.(ω). Since the sum is finite, this implies that $I(z) = \sum_{\ell=1}^{m_2}\Gamma_\ell(z)$, is bounded and continuous in $\overline{D}$, as desired.

We next calculate the bounds on $I(0 + iy)$ and $I(1 + iy)$. In fact,

$$I(iy) = \int_S |Tf_{iy}|^r|g_{iy}|d\nu$$

$$\leq 2\||Tf_{iy}|^r\|_{\xi_t}\|g_{iy}\|_{\zeta_t}$$

$$\leq 2\|(Tf_{iy})\|_{rQ_t}\|g_{iy}\|_{\zeta_t}, \text{ by the relation between } \xi_t \text{ and } Q_t.(16)$$

By definition $\rho_{\zeta_1}(|f_{iy}|) = \rho_{\zeta_t}(|g|) \leq 1$, and $\rho_{\varphi_1}(|f_{iy}|) = \rho_{\varphi_t}(|f|) \leq 1$ so that $I(iy) \leq 2M_1$. A similar computation shows $I(1 + iy) \leq 2M_2$. Hence by Proposition 2 we get

$$I(t) \leq 2M_1^{1-t}M_2^t. \qquad (17)$$

Then (5) simplifies with (17) to:

$$\|(Tf)\|_{rQ_t} = \|(Tf)^r\|_{\xi_t} \leq 4M_1^{1-t}M_2^t, \qquad f \in S(\mu), \|f\|_{r\Phi_t} = 1.$$

If the $0 \neq f \in S(\mu)$ is arbitrary, then set $f' = f/\|f\|_{r\varphi_t}^{\frac{1}{r}}$ to get

$$\|Tf\|_{rQ_t} \leq 4M_1^{1-t}M_2^t\|f\|_{r\varphi_t}. \qquad (18)$$

This shows that (3) is true as stated.

It remains to establish the last part, by reducing it to the above case. But by the classical representation of the functions of Hardy spaces (cf., e.g., Hoffman (1962), p. 32) by the boundary values on the unit disc, we have

$$(Pf)(re^{i\theta}) = \int\limits_{0}^{2\pi} P(r, \theta - t)f(t)d\mu(t), \qquad 0 \le r < 1, f \in L^{\varphi}(\mu),$$

and $Pf \in \mathcal{N}^{+}$ the Nevanlinna class, where $P(\cdot, \cdot)$ is the Poisson kernel and μ is the normalized Lebesgue measure. Since φ is an s-convex (hence log-convex) function, for f in $L^{\varphi}(\mu)$ we have (cf., Eq. (11) of Section 1) that $(Pf)(re^{i(\cdot)}) \in \mathcal{H}^{\varphi}$ and $\|Pf\|_{\varphi} = \|f\|_{\varphi}$. On the other hand, if $F \in \mathcal{H}^{\varphi}$, and $f(t) = F(e^{it})$, the boundary value, then the mapping $R : F \mapsto f$ is also norm preserving, i.e., $\|RF\|_{\varphi} = \|f\|_{\varphi}$. Thus if $\widetilde{T}$ is defined on $L^{\varphi}(\mu)$ into $L^{Q}(\mu)$ as

$$\widetilde{T}(f) = (RTP)(f), \qquad (19)$$

then for $\varphi(u) = \tilde{\varphi}(u^{r})$, $0 \le r < 1$,

$$\|\widetilde{T}f\|_{r\tilde{Q}} = \|RTPf\|_{r\tilde{Q}} \le \|TPf\|_{r\tilde{Q}} \le M\|Pf\|_{r\tilde{Q}} = M\|f\|_{r\tilde{Q}}. \qquad (20)$$

Let φ, Q in the above now be replaced by φ_{i}, Q_{i} as in the theorem on the same measure space $([0, 2\pi), \mathcal{B}, \mu)$. We can apply (20) in place of (2). Instead of simple functions, one can use polynomial functions F. If f denotes the boundary value of F, then (3) becomes

$$\|Tf\|_{rQ_{t}} \le KM_{1}^{1-t}M_{2}^{t}\|f\|_{r\varphi_{t}}, \qquad 0 < t < 1, 0 \le r < 1. \qquad (21)$$

If the φ_{i} are Δ_{2}-regular, then by Theorem 1.4, polynomial functions are dense in $\mathcal{H}^{\varphi_{i}}$. Hence for each $g \in \mathcal{H}^{\varphi}$, there exist polynomials $F_{n} \to g$ in the metric of $\mathcal{H}^{\varphi_{i}}$ and by the sublinearity of T, $\|\,|TF_{n}| - |Tg|\,\|_{rQ_{t}} \le \|T(F_{n} - g)\|_{rQ_{t}} \le M_{1}\|F_{n} - g\|_{r\varphi_{t}} \to 0$. Using the principle of extension by uniform continuity, we conclude that (21) holds on all of $\mathcal{H}^{\varphi_{t}}$ itself. This implies all parts of the result. ∎

Since now the necessary modifications are detailed in the above demonstration, we can give an extension of some of the results of Chapter VI. For instance, we may present the following analog of Theorem VI.3.12 (also using Proposition VII.5.7) on interpolation with change of measures. We note that the actual details of proof are lengthy. The

work proceeds to employ the three-line-theorem for subharmonic functions in contrast to the real variable method used in Theorem VI.6.12.
But the necessary computations are quite similar to the preceding result. We omit the demonstration, which can be found in Kraynek
(1972), and present the desired statement as follows.

Theorem 3. *Let φ_i, Q_i, $i = 1, 2$ be r_i, s_i-convex φ-functions, each
Δ_2-regular, and let $0 < r \leq \min(r_i, s_i)$, $i = 1, 2$. If $(\Omega_i, \Sigma_i, \mu_i)$, $i = 1, 2$,
are measure spaces, and a_i [b_i] are nonnegative measurable functions
on $(\Omega_1, \Sigma_1, \mu_1)$ [$(\Omega_2, \Sigma_2, \mu_2)$], suppose that $a_i \chi_E \in L^{\varphi_i}(\mu_1)$, $E \in
\Sigma_{10} = \{A \in \Sigma_1 : \mu_1(A) < \infty\}$, $i = 1, 2$. Let $T : L^{\varphi_i}(\mu_1) \to L^{Q_i}(\mu_2)$ be
a sublinear operator such that*

$$\|b_1 T(f)\|_{rQ_1} \leq M_1 \|a_1 f\|_{r\varphi_1}, \quad a_1 f \in L^{\varphi_1}(\mu_1), \tag{22}$$

and

$$\|b_2 T(f)\|_{rQ_2} \leq M_2 \|a_2 f\|_{r\varphi_2}, \quad a_2 f \in L^{\varphi_2}(\mu_1). \tag{23}$$

*If φ_t, Q_t are intermediate functions to (φ_1, φ_2) and (Q_1, Q_2), $0 < t < 1$,
and $a_t = a_1^{1-t} a_2^t, b_t = b_1^{1-t} b_2^t$, then*

$$\|b_t T(f)\|_{rQ_t} \leq 4 M_1^{1-t} M_2^t \|a_t f\|_{r\varphi_t}, \quad f \in \mathcal{S}(\mu_1) \tag{24}$$

where $\mathcal{S}(\mu_1)$ is the set of simple functions on $(\Omega_1, \Sigma_1, \mu_1)$.

It should be noted that this result actually holds for φ_i, Q_i which
are Δ_2-regular and not only for the Δ'-regular class as is the case with
the former result. However, it is useful to have both methods at our
disposal since they can be employed in other problems of interest. Note
also that the result is not stated for the $\mathcal{H}^\varphi$-spaces on different measure
spaces. This is because all $\mathcal{H}^\varphi$-spaces (contained in $\mathcal{N}^+$) are based on
the unit disc and using different discs to employ the change of measure
techniques seem artificial. So that point of view will not be pursued.

Let us include a new type of interpolation to complete this class of
ideas and results. It is related to considering a collection of operators
in place of just one, the family satisfying certain conditions.

Definition 4. (a) Let $(\Omega_i, \Sigma_i, \mu_i)$, $i = 1, 2$, be measure spaces and
$D = \{z \in \mathbb{C} : 0 < \text{Re}(z) < 1\}$ be the open strip in $\mathbb{C}$. If $\mathcal{T} = \{T_z, z \in D\}$ is a
family of (sub) linear operators on the set of simple scalar functions on
$(\Omega_1, \Sigma_1, \mu_1)$ into the set of measurable scalar functions on $(\Omega_2, \Sigma_2, \mu_2)$,

then $\mathcal{T}$ is termed an *analytic family* if $z \mapsto \int_{\Omega_2} (T_z f) g d\mu_2$ is well defined, analytic in D and continuous in its closure $\overline{D}$, for each simple f and g.

(b) An analytic family $\mathcal{T} = \{T_z, z \in D\}$ is, as above, of *admissible growth (in the strict sense)* if

$$I_r(z) = \int_{\Omega_2} |T_z(f)|^r |g| d\mu_2, \qquad (25)$$

then $\sup\{\sup[\log I_r(z) : 0 \le \mathrm{Re}(z) \le 1] : |\mathrm{Im}(z)| \le r\} \le Ae^{ar}$, for a pair of constants $0 < a < \pi, 0 < A < \infty$, for $r = 1$ [for all $0 < r \le 1$] and each pair of simple functions f and g.

We now present an interpolation theorem if T_z maps an $\mathcal{H}^\varphi$ into $L^Q(\mu)$ when $\{T_z, z \in D\}$ is an analytic family satisfying an admissible (strict sense) growth condition. Here now is the precise version of the desired result.

Theorem 5. *Let* φ_i, Q_i *be* r_i, s_i*-convex* φ*-functions,* $i = 1, 2$, *and* $\{T_z, z \in D\}$ *be an analytic family of sublinear operators of admissible growth in the strict sense from* $(\Omega_1, \Sigma_1, \mu_1)$ *into* $(\Omega_2, \Sigma_2, \mu_2)$ *as in Definition 4. Let* φ_t *and* Q_t *be intermediate functions for the pairs* (φ_1, φ_2) *and* (Q_1, Q_2). *If* $0 < r \le \min(r_i, s_i)$, $i = 1, 2$, *and*

$$\|T_{iy} f\|_{rQ_1} \le A_1(y) \|f\|_{r\varphi_1}, \|T_{(1+iy)} f\|_{rQ_2} \le A_2(y) \|f\|_{r\varphi_2} \qquad (26)$$

for each simple f *in* $(\Omega_1, \Sigma_1, \mu_1)$ *and for some* $0 < A_j(y) \le C_j e^{ar}$, $|y| \le r$, $j = 1, 2$, *and* $0 < a < \pi$, *then we have*

$$\|T_t f\|_{rQ_t} \le 4A_t \|f\|_{r\varphi_t}, \qquad 0 < t < 1, \qquad (27)$$

where A_t *is a constant independent of* f, *but may depend on* φ_i, Q_i, a, *and* C_i, $i = 1, 2$.

Remark. If the T_z are linear transformations but only of admissible growth and $\varphi_i(x) = |x|^{p_i}, Q_i(x) = |x|^{q_i}$, $p_i > 0, q_i > 0$, $i = 1, 2$, this result was proved by E. M. Stein and G. Weiss (1957); a similar theorem for $p_i \ge 1, q_i \ge 1$ was considered by Stein (1956) using analogous ideas. In this setup these authors were able to establish their result for $T_z : \mathcal{H}^{\varphi_i} \to L^{Q_i}(\mu)$ and then apply it to an important problem in trigonometric series. We shall comment further on this application later.

Sketch of Proof: As in the proof of Theorem 1, let $\Phi_i = \tilde{\varphi}_i \circ h_i$, $\xi_i = \tilde{Q}_i \circ g_i$, and ζ_i be complementary to ξ_i, $i = 1, 2$, so that Φ_i, ξ_i are r-convex. Here $h_i(x) = |x|^{r_i/r}$, $g_i(x) = |x|^{s_i/r}$. Let $\varphi_t, \Phi_t, \xi_t, \zeta_t$ be the corresponding intermediate functions. If α_z, β_z are $\alpha_1^{1-z} \alpha_2^z$ and $\beta_1^{1-z} \beta_2^z$ where $\alpha_i = \varphi_i^{-1}$ and $\beta_i = \zeta_i^{-1}$, consider for simple functions f, g on $(\Omega_1, \Sigma_1, \mu_1)$ and $(\Omega_2, \Sigma_2, \mu_2)$ respectively, with $\|f\|_{r\varphi_t} = 1$, $\|g\|_{r\zeta_t} \leq 1$,

$$I = \int_{\Omega_2} |T_t f|^r |g| d\mu_2. \tag{28}$$

If $f = \sum_{i=1}^{n_1} a_i \chi_{F_i}$, $g = \sum_{j=1}^{n_2} b_j \chi_{G_j}$ in the above, define f_z, g_z as:

$$f_z = \sum_{j=1}^{n_1} \alpha_z(\varphi_t(|a_j|)) e^{i\theta_j} \chi_{F_j}, \quad g_z = \sum_{j=1}^{n_2} \beta_z(\zeta_t(|b_j|)) e^{i\theta'_j} \chi_{G_j}, \tag{29}$$

and also

$$\gamma_k(z) = \sum_{j=1}^{n_1} \beta_z^{\frac{1}{r}}(\zeta_t(|b_k|)) \alpha_z(\varphi_t(|a_j|)) e^{i(\theta_j + \theta'_k)} \chi_{F_j}.$$

Then the $\gamma_k(\cdot)$ are analytic simple functions for $k = 1, \ldots, n_2$, and we may extend I of (28) as:

$$I(z) = \int_{\Omega_2} |T_z(f_z)|^r |g_z| d\mu_2 = \sum_{k=1}^{n_2} \int_{G_k} |T_z(\gamma_k(z))|^r d\mu_2. \tag{30}$$

Using a computation similar to that in the proof of Theorem 1, one finds the following properties of $I(z)$:

(a) $I(z) \geq 0$, $\log I(z)$ is subharmonic in $D = \{z : 0 < \mathrm{Re}(z) < 1\}$, and continuous on its closure $\overline{D}$;

(b) $I(z)$ is of admissible growth in the strict sense in $\overline{D}$, since $\mathcal{T} = \{T_z, z \in D\}$ is an analytic family satisfying the desired growth conditions.

Finally, for the bounds,

$$I(iy) = \int_{\Omega_2} |T_{iy}(f_{iy})|^r |g_{iy}| d\mu_2 \leq 2\|T_{iy}(f_{iy})\|_{rQ_1} \|g_{iy}\|_{\zeta_1}$$

$$\leq 2A_1(y) \|f_{iy}\|_{r\varphi_1} \|g_{iy}\|_{\zeta_1} \leq 2A_1(y).$$

Similarly $I(1 + iy) \leq 2A_2(y)$. Also $I(t) = I$. Hence by the three-line-theorem for subharmonic functions we get, as in (18), that

$$\|T_t f\|_{rQ_t} \leq 4A_t \|f\|_{r\varphi_t} \tag{31}$$

for each simple function, where A_t is a certain constant depending only on A_1, A_2, t, φ_1, φ_2, Q_1, and Q_2 but not on f. Using some work of I. I. Hirschman (1953), one can show that A_t satisfies the following equation:

$$\log A_t = \int_{\mathbb{R}} \omega(1 - t, y) \log(2A_1(y)) dy + \int_{\mathbb{R}} \omega(t, y) \log(2A_2(y)) dy$$

where $\omega(x, y) = \frac{1}{2} \tan \frac{\pi x}{2} \cdot \operatorname{sech}^2(\frac{\pi y}{2}) \cdot \left[\tan^2(\frac{\pi x}{2}) + \tanh^2(\frac{\pi y}{2}) \right]^{-1}$. This exact value does not play any role in our considerations except to deduce that $0 < A_t < \infty$. ∎

As an important application of this result, when $\varphi_i(x) = |x|^p = Q_i(x)$, with $0 < p \leq 1$, and $L^\varphi(\mu_1)$ and $L^Q(\mu_2)$ are replaced by $\mathcal{H}^p$ on the disc, Stein and Weiss (1957) have established the following.

Theorem 6. *Let $F \in \mathcal{H}^p$, with $0 < p \leq 1$ and $\sigma_n^\alpha(F, \cdot)$ as the Cesàro mean of order α ($= \frac{1}{p} - 1$), for the Fourier series expansion of the boundary value $F(e^{i\theta})$. Then*

$$\int_0^{2\pi} \sup_{n \geq 0} \left[\frac{|\sigma_n^\alpha(F, \theta)|^p}{\log(n + 2)} \right] d\mu(\theta) \leq A_p \|F\|_p^p, \tag{32}$$

where A_p depends on p but not on F.

Although the proof of this result uses the preceding theorem as a key tool, it is still necessary to establish several auxiliary facts, and hence the details need care. We refer the interested reader to the original paper. The corresponding applications for functions F in $\mathcal{H}^\varphi$ would be equally important. However, such problems have not yet been treated in the literature. The preceding discussion should motivate that work. Noting the fact that the functions F in $\mathcal{H}^\varphi$ are (infinitely) differentiable, it is possible to improve the integral inequalities of the type (31). We discuss this aspect briefly based on the work of S. L. Sobolev and some extensions in the following two sections.

9.3. Orlicz-Sobolev spaces I: Finite order

If (Ω, Σ, μ) is taken as $(I\!R^n, \mathcal{B}, \mu)$, the Lebesgue space, the differential structure of $I\!R^n$ can be utilized to obtain many integral inequalities for functions using their (weak) derivatives. Without demanding analyticity, one can consider functions that are continuously differentiable for a finite or infinite number of times. We treat the former aspect here, and discuss the infinite case later.

Since the work depends on the differential structure of the underlying space, let $\Omega \subset I\!R^n$ be an open nondegenerate set and (Ω, Σ, μ) be the Lebesgue measure space. We need to introduce some essential notations. Let $\alpha = (\alpha_1, \ldots, \alpha_n)$ be a vector of nonnegative integers and $|\alpha| = \alpha_1 + \cdots + \alpha_n$. If $x = (x_1, \ldots, x_n) \in I\!R^n$, let $D^\alpha = D_1^{\alpha_1} \cdots D_n^{\alpha_n}$ where the partial differential operator $D_i = \frac{\partial}{\partial x_i}$, $D_1^0 = $ identity. Similarly for $\alpha = (\alpha_1, \ldots, \alpha_n), \beta = (\beta_1, \ldots, \beta_n)$ one has $\alpha + \beta = (\alpha_1 + \beta_1, \ldots, \alpha_n + \beta_n)$, $\alpha! = \alpha_1! \alpha_2! \cdots \alpha_n!$, $\alpha \leq \beta$ means $\alpha_i \leq \beta_i$, and $\binom{\alpha}{\beta} = \binom{\alpha_1}{\beta_1} \cdots \binom{\alpha_n}{\beta_n}$ the latter being the binomial coefficients. Let $J : I\!R^n \to I\!R^+$ be an infinitely differentiable function, i.e., $D^\alpha J$ exists for each $\alpha \geq 0$. Then J is termed a *mollifier* if (i) $J(x) = 0$ for $\|x\|^2 = \left[\sum_{i=1}^{n} x_i^2 \right] \geq 1$, and (ii) $\int_{I\!R^n} J(x) d\mu(x) = 1$. Defining $J_\varepsilon(x) = \varepsilon^{-n} J(\frac{1}{\varepsilon} x)$, one has $J_\varepsilon(\cdot)$ to be a mollifier for each $\varepsilon > 0$. An example of a mollifier is given by $J(x) = c(\exp[(\|x\|^2 - 1)^{-1}]) \chi_{[\|x\| < 1]}$ for a suitable $c > 0$, to make $\int_{I\!R^n} J(x) d\mu = 1$. For any continuous function $f : I\!R^n \to I\!R$, and mollifier J_ε we find by a simple computation that $J_\varepsilon * f \to f$ uniformly on the closure $\bar{\Omega} \subset I\!R^n$, as $\varepsilon \downarrow 0$, where $J_\varepsilon * f$ is the convolution. This property of J_ε is called an *approximate identity* relation.

Let $C_c^\infty(\Omega)$ denote the space of scalar infinitely differentiable functions on Ω with compact supports. [The support of f is the set $\overline{\{x : f(x) \neq 0\}}$.] A sequence $\{f_n, n \geq 1\} \subset C_c^\infty(\Omega)$ is said to be convergent to an element f in the space $C_c^\infty(\Omega)$, if all f_n vanish outside a fixed compact set $K \subset \Omega$, and $D^\alpha f_n \to D^\alpha f$ uniformly on K, for each α. Note that $\operatorname{supp}(f) \subset K$ also. This notion of convergence defines a locally convex (linear) topology in the vector space $C_c^\infty(\Omega)$, denoted $\mathcal{D}(\Omega)$ and termed a *test space*. It can be verified that $\mathcal{D}(\Omega)$ is a complete space. It is not normable, and hence is *not* a Banach space. The adjoint space $\mathcal{D}'(\Omega)$ is called the *space of Schwartz distributions* of

$\mathcal{D}(\Omega)$, the latter is sometimes termed the space of *test functions*. Thus $T \in \mathcal{D}'(\Omega)$ iff T is linear and for each $f_n \to 0$ in $\mathcal{D}(\Omega)$, $T(f_n) \to 0$, as $n \to \infty$. This makes $\mathcal{D}'(\Omega)$ also a locally convex linear topological space. If $f : \Omega \to \mathbb{C}$ is locally integrable, i.e., $f\chi_A \in L^1(\mu)$ for each compact set $A \subset \Omega$, then f is said to have α^{th} *weak or distributional or generalized derivative*, denoted $D^\alpha f$ if there is a locally integrable function $g_\alpha : \Omega \to \mathbb{C}$ such that

$$\int_\Omega f D^\alpha \varphi \, d\mu = (-1)^{|\alpha|} \int_\Omega g_\alpha \varphi \, d\mu, \qquad \varphi \in \mathcal{D}(\Omega). \tag{1}$$

In case f has an αth order derivative, then an integration by parts $|\alpha|$-times shows that $D^\alpha f = g_\alpha$ and (1) is satisfied. In this case $D^\alpha f$ is the strong or ordinary derivative. Also if a weak derivative exists, then it is unique. This follows from (1) and the fact that $C_c^\infty(\Omega)$ is (norm) dense in $L^1(\mu)$. However, there are many functions which have weak but not necessarily strong derivatives. We can now introduce the concept of an Orlicz-Sobolev space.

Definition 1. If (Ω, Σ, μ) is the Lebesgue space with $\Omega \subset \mathbb{R}^n$ as an open set ($=$ domain) and (θ, Φ) are Young functions, then the *Orlicz-Sobolev space of order* $m \geq 0$, denoted $W_\theta^{m,\Phi}(\Omega)$, is the set of equivalence classes of functions from the Orlicz space $L^\Phi(\mu)$ given by

$$W_\theta^{m,\Phi}(\Omega) = \{f \in L^0(\mu) : D^\alpha f \in L^\Phi(\mu), \ |\alpha| \leq m, \ D^\alpha f \text{ being the}$$
$$\text{distributional derivative of } f\}. \tag{2}$$

For each $f \in W_\theta^{m,\Phi}(\Omega)$, let $N_{\theta,G}^{m,\Phi}(\cdot)$ be defined as:

$$N_{\theta,G}^{m,\Phi}(f) = \inf \left\{ k > 0 : \sum_{|\alpha| \leq m} \theta\left(\frac{\|D^\alpha f\|_\Phi}{k}\right) G(\alpha) \leq 1 \right\}, \tag{3}$$

where $\|D^\alpha f\|_\Phi$ is the Orlicz norm of $L^\Phi(\mu)$, and $0 \leq G(\alpha) < \infty$ are weights.

When $1 \leq m < \infty$, the space $W_\theta^{m,\Phi}(\Omega) = W^{m,\Phi}(\Omega)$ for any Young function θ since the sum in (3) is finite and hence does not depend on θ or the weights $G(\alpha)$. With $G(\alpha) \equiv 1$ and $\theta(x) = 0$ for $0 \leq |x| \leq 1$, $= +\infty$ for $|x| > 1$, and also when $\theta(x) = |x|$, the spaces $(W_\theta^{m,\Phi}(\Omega), N_{\theta,G}^{m,\Phi}(\cdot))$ were discussed by R. A. Adams (1975). These

are termed *Orlicz-Sobolev spaces of finite order.* If $m = +\infty$, and
$\theta(x) = |x|$, $G(\alpha) \geq 1$, such classes are considered by Ju. A. Dubinskij
(1986). These are called *Orlicz-Sobolev spaces of infinite order* and they
will be discussed in the next section.

It may be noted at the outset that if $\theta(x) = |x|$, $G(\alpha) \equiv 1$, and
$\Phi(x) = |x|^p$, $p \geq 1$, then $W^{m,\Phi}_{\theta,G}(\Omega) = W^{m,p}(\Omega)$, is the classical Sobolev
space and one has $C^\infty_c(\Omega) \subset W^{m,\Phi}_{\theta,G}(\Omega)$ if $m < \infty$. These spaces are
extremely important in the study of partial differential equations. In
the latter theory certain embedding results play an important role.
However, a Sobolev space of the type $W^{m,p}(\Omega)$ cannot always be embedded in a Lebesgue type space. But such embeddings are possible
if Orlicz spaces are admitted. We shall present briefly this material as
a motivation for the study of Orlicz-Sobolev spaces, and then include
some results on the latter topic to indicate the potential uses for future
analysis and applications.

Recall that if $\mathcal{X}$ and $\mathcal{Y}$ are topological vector spaces, then $\mathcal{X}$ is
embedded (or injected) into $\mathcal{Y}$, denoted $\mathcal{X} \hookrightarrow \mathcal{Y}$, if there is a mapping
$\tau : \mathcal{X} \to \mathcal{Y}$ which is one-one and continuous. The embedding is *compact*
if the mapping τ is moreover compact, i.e., it takes bounded sets into
relatively compact sets. The embedding problem often depends on the
geometry of the domain Ω in our case, and so we discuss two of the
desirable properties of Ω.

The domain Ω in $I\!\!R^n$ is said to have

(i) the *cone property* if at each point $x \in \Omega$, there is finite cone
$C_x \subset \Omega$, which is congruent to a fixed (right spherical) cone C [C_x
being obtained merely by a rigid motion from C];

(ii) the *segment property* if there is an open cover $\{\mathcal{U}_j, j \in J\}$ of the
boundary $\partial\Omega$ of Ω such that each compact subset of $\partial\Omega$ intersects at
most finitely many members of the cover, and a sequence $\{y_j, j \in J\} \subset$
$\partial\Omega$, $y_j \neq 0$, such that for any $x \in \bar{\Omega} \cap \mathcal{U}_j$, one has $x + t y_j \in \Omega$, $0 < t < 1$.

Other properties can be formulated (cf. Adams (1975), p. 66).
We now formulate a classical form of the Sobolev embedding, without
proof, as follows:

Theorem 2 (Sobolev). *Let $\Omega \subset I\!\!R^n$ have the cone property and
denote by Ω_k the intersection of Ω with a k-dimensional hyperplane in
$I\!\!R^n$, $1 \leq k \leq n$. Then we have for the integers $0 \leq m \leq n$, and real
$1 \leq p < \infty$,*

(a) $mp < n \Rightarrow W^{m,p}(\Omega) \hookrightarrow L^q(\Omega,\mu)$ for $p \leq q \leq np(n-mp)^{-1}$,
and $W^{m,p}(\Omega) \hookrightarrow L^q(\Omega_r,\mu)$ for $n - mp < r \leq n$, and $p \leq q \leq$
$rp(n-mp)^{-1}$ with $r = n - m$ when $p = 1$;

(b) $mp = n \Rightarrow W^{m,p}(\Omega) \hookrightarrow L^q(\Omega_r,\mu)$ for $1 \leq r \leq n, p \leq q < \infty$,
and when $p = 1$, $q = \infty$ is also possible.

Adding to or strengthening the hypothesis one has numerous other embedding statements that can be established. The following is an important reason for these embedding results and for considering Orlicz-Sobolev spaces: A form of the classical *Dirichlet problem* is to find a solution of the (not necessarily linear) "elliptic" operator equation $L(f) = g$ where

$$L(f) = \sum_{|\alpha| \geq 0} (-1)^{|\alpha|} D^\alpha (a_\alpha |D^\alpha f|^{p_\alpha} D^\alpha f), \qquad (4)$$

with $a_\alpha \geq 0$ and $p_\alpha \geq 1$ are real number sequences, and the boundary condition is of the form

$$D^\alpha f \big|_{\partial\Omega} = 0, \qquad 0 \leq |\alpha| \leq m \ (m \text{ an integer or } +\infty), \qquad (5)$$

Ω being a suitable domain in $\mathbb{R}^n$. Here f, g are to belong to a certain class of "admissible" functions on Ω. The operator $L(\cdot)$ could be of finite or infinite order. One considers the set of possible f's ($\theta(x) = |x|$) as

$$W^{m,\{p_\alpha\}}_{\theta,\{a_\alpha\}}(\Omega) = \{f \in C_c^\infty(\Omega) : D^\alpha f \in L^{p_\alpha}(\Omega,\mu), \sum_{|\alpha| \leq m} a_\alpha \|D^\alpha f\|^{p_\alpha}_{p_\alpha} < \infty\}. \quad (6)$$

Assuming m in (5) to be finite we see that the solution set (6) is a Sobolev space and hence its structural analysis is important for the problem. If one replaces $a_\alpha |D^\alpha f|^{p_\alpha}$ by $\Phi_\alpha(a_\alpha |D^\alpha f|)$ in (4), one naturally has the Orlicz-Sobolev spaces. Moreover, to simplify the analysis of these spaces one should be able to embed them in an L^p-type space for the smallest possible $p \geq 1$. However, if one restricts to the L^p-spaces, then it is not always possible to inject a $W^{m,p}(\Omega)$ space into an L^p-space. But it is possible to embed these spaces into an Orlicz space $L^\Phi(\mu)$ for a suitable Young function. In fact it is even possible to do this for the $W^{m,\Phi}_{\theta,G}(\Omega)$-spaces in most cases, due to the extensiveness of Orlicz spaces. Because of this circumstance we present a few results on $W^{m,\Phi}_{\theta,G}(\Omega)$-spaces for finite m in the following, and discuss the case

$m = +\infty$ in the next section. Let $\overset{\circ}{W}{}_{\theta,G}^{m,\Phi}(\Omega)$ be the closed linear span of $C_c^{\infty}(\Omega)$ in the norm $N_{\theta,G}^{m,\Phi}(\cdot)$ of (3), and $\mathcal{H}_{\theta,G}^{m,\Phi}(\Omega)$ be a similar span of $C^m(\Omega)$ in the same norm where $C^m(\Omega)$ is the space of functions f for which $D^{\alpha}f$ is continuous for $|\alpha| \leq m$.

We have the following analog of Theorem VIII.1.1 in a more complete form.

Theorem 3. *The set* $(W_{\theta,G}^{m,\Phi}(\Omega), N_{\theta,G}^{m,\Phi}(\cdot))$ *is a Banach space and* $\mathcal{H}_{\theta,G}^{m,\Phi}(\Omega) \subset W_{\theta,G}^{m,\Phi}(\Omega)$. *There is moreover equality in the latter if* Φ *is* Δ_2-*regular [and* $0 \leq m < \infty$], *and further* $\mathcal{H}_{\theta,G}^{m,\Phi}(\Omega) = \overset{\circ}{W}{}_{\theta,G}^{m,\Phi}(\Omega) = W_{\theta,G}^{m,\Phi}(\Omega)$ *in that case.*

Proof: Since the $W_{\theta,G}^{m,\Phi}(\Omega)$ is evidently a normed linear space only its completeness needs to be verified. Thus let $\{f_n, n \geq 1\}$ be a Cauchy sequence in it. Identifying $W_{\theta,G}^{m,\Phi}(\Omega)$ as a subspace of $L^{\theta,\Phi}(G)$ of Theorem VIII.1.1, we see that $f_n \to f$ in the norm (3) to $f = \{f_\alpha, |\alpha| \leq 1\}$ in $L^{\theta,\Phi}(G)$ where $f_\alpha \in L^{\Phi}(\Omega,\mu)$ for each α such that $|\alpha| \leq m$. So we only need to show that $D^{\alpha}f = f_\alpha$, the weak derivative, so that $f \in W_{\theta,G}^{m,\Phi}(\Omega)$ and the completeness follows.

Since by definition of norm in (3) [we could also use the Orlicz norm in the argument here since it and the gauge norms are equivalent in these spaces], $\|D^{\alpha}f_n - f_\alpha\|_{\Phi} \to 0$ as $n \to \infty$, for any $g \in C_c^{\infty}(\Omega)$. Consider $T_{f_n} \in \mathcal{D}'(\Omega)$ and T_f:

$$|T_{f_n}(g) - T_f(g)| = \left| \int_{\Omega} (f_n - f)g d\mu \right| \leq \|g\|_{\Psi}\|f_n - f\|_{\Phi} \to 0. \qquad (7)$$

Replacing f_n and f by $D^{\alpha}f_n$ and f_α, the same result (7) holds, so that T_{f_α} is an element of $\mathcal{D}'(\Omega)$, and we have

$$\int_{\Omega} f_\alpha g d\mu = T_{f_\alpha}(g) = \lim_{n\to\infty} T_{D^{\alpha}f_n}(g)$$

$$= \lim_{n\to\infty} (-1)^{|\alpha|} T_{f_n}(D^{\alpha}g), \text{ by definition of the}$$

$$\text{weak derivative,}$$

$$= (-1)^{|\alpha|} T_f(D^{\alpha}g) = (-1)^{|\alpha|} \int_{\Omega} f(D^{\alpha}g) d\mu.$$

Since $C_c^{\infty}(G)$ is norm determining in $L^{\Phi}(\mu)$, we conclude that $f_\alpha = D^{\alpha}g$ in the weak sense for all $|\alpha| \leq m$, so that $f \in W_{\theta,G}^{m,\Phi}(\Omega)$ and the latter is thus a complete space.

If $f \in C^m(\Omega)$ so that it has all its partial derivatives up to and including the mth order continuous, and $N_{\theta,G}^{m,\Phi}(f) < \infty$ implies that $f \in W_{\theta,G}^{m,\Phi}(\Omega)$, we infer, by the completeness of the latter space just established, that $\mathcal{H}_{\theta,G}^{m,\Phi}(\Omega)$ is a subspace of $W_{\theta,G}^{m,\Phi}(\Omega)$. Taking Φ to be discontinuous or growing exponentially fast, so that simple functions [hence $C_c^\infty(\Omega)$] are not dense, the inclusion will be proper. We now show that there is equality under the given growth conditions.

Thus let Φ be Δ_2-regular. Let $d(x, A)$ be the distance of a point $x \in \mathbb{R}^n$ to the set $A \subset \mathbb{R}^n$ so that $d(x, A) = \inf\{\|x - y\| : y \in A\}$ where $\|\cdot\|$ is the (Euclidean) norm in $\mathbb{R}^n$. Define $\Omega_{-1} = \Omega_0 = \emptyset$ and for $k \geq 1$ let Ω_k be

$$\Omega_k = \{x \in \Omega : \|x\| < k, d(x, \partial\Omega) > k^{-1}\} \quad (\partial\Omega = \text{boundary of } \Omega).$$

If $U_k = \Omega_{k+1} - \bar{\Omega}_{k-1}$, then $\{U_k, k \geq 1\}$ is a bounded open covering of the domain Ω with $U_k \cap \Omega \neq \emptyset$ for finitely many k. By a classical result, there exists a partition of unity subordinate to the above covering which is denoted $\{\psi_k, k \geq 1\}$ with the following properties: (i) $0 \leq \psi_k(x) \leq 1$, $x \in \mathbb{R}^n$, (ii) $\psi_k \in C_c^\infty(\mathbb{R}^n)$, (iii) $\text{supp}(\psi_k) \subset U_j$ for some $j \geq 1$, and (iv) $\sum_{k=1}^{\infty} \psi_k(x) = 1$, $x \in \Omega$. [A proof of this fact is in many books, cf., e.g., Yosida (1971), p. 61.] Then using some properties of mollifiers and the fact that Φ is Δ_2-regular, so that $C_c^\infty(\Omega)$ is norm dense in $L^\Phi(\Omega, \mu)$, we have the following estimates. Let $f \in W_{\theta,G}^{m,\Phi}(\Omega)$ and an $\varepsilon > 0$. By relabeling if necessary, we take $\psi_k \in C_c^\infty(U_k)$ so that we may consider $f\psi_k$ in lieu of f, and choose $0 < \varepsilon < [(k+1)(k+2)]^{-1}$ such that

$$\|J_{\varepsilon_k} * D^\alpha(\psi_k f) - D^\alpha(\psi_k f)\|_\Phi < 2^{-k}\varepsilon, \qquad |\alpha| \leq m, \qquad (8)$$

where '$*$' denotes convolution. Note that $\text{supp}(J_{\varepsilon_k} * D^\alpha(\psi_k f)) \subset U_k$, by the properties of mollifiers. The estimate (8) is possible since Φ is Δ_2-regular. If $h = \sum_{k=1}^{\infty} J_{\varepsilon_k} * D^\alpha(\psi_k f)$, then the series converges pointwise and $h \in C^\infty(\Omega)$. Hence for any integer $s \geq 0$, we have

$$\int_{\Omega_s} \Phi\left(\left|\frac{D^\alpha[h-f]}{\varepsilon}\right|\right) d\mu = \int_{\Omega_s} \Phi\left(\left|\sum_{k=1}^{s+1}[J_\varepsilon * D^\alpha(\psi_k f) - D^\alpha(\psi_k f)]\right|\frac{2^k}{2^k\varepsilon}\right) d\mu$$

$$\leq \sum_{k=1}^{s+1} \frac{1}{2^k} \int_{\Omega_s} \Phi\left(\left|J_\varepsilon * D^\alpha(\psi_k f) - D^\alpha(\psi_k f)\right|\frac{2^k}{\varepsilon}\right) d\mu,$$

$$\leq 1, \qquad \text{by (8) and the convexity of } \Phi.$$

Observing that $\Omega_s \nearrow \Omega$ and using the Fatou property of the function norm here, we get $\|D^\alpha f - D^\alpha h\|_\Phi \le \varepsilon$ so that, since $m < \infty$, $N_{\theta,G}^{m,\Phi}(f-h) \le \varepsilon K(m,G)$ where $K(m,G)$ is a constant depending only on m and $(G(\alpha), |\alpha| \le m)$. Here we also use the fact that when $m < \infty$, all the θ-norms are equivalent since ℓ^θ is a finite dimensional vector space. This and the arbitrariness of $\varepsilon > 0$ show that (i) $N_{\theta,G}^{m,\Phi}(h) < \infty$ implying $h \in \mathcal{H}_{\theta,G}^{m,\Phi}(\Omega)$, and (ii) f is in this space. This is the desired result. The final statement is obvious. ∎

The last part of the above result was proved by N. G. Meyers and J. Serrin (1964) when $\theta(x) = |x|$ and $\Phi(x) = |x|^p$. Their argument extends to the general case with simple modifications, as also noted by A. Kufner, O. John and S. Fucik (1977).

Letting $\theta(x) = |x|$, $G(\alpha) \equiv 1$ for $|\alpha| \le m$, and Ω again as a domain in the above, we denote the resulting spaces $\mathcal{H}^{m,\Phi}$, $W^{m,\Phi}$ and the norm $N^{m,\Phi}$. We then have the following specialization.

Corollary 4. $\mathcal{H}^{m,\Phi} = W^{m,\Phi}$ *and, when Ω is bounded, the same result holds if Φ is a Young function whose principal part is $|x|\log^+|x|$, or if $\Phi(x) = (1+|x|)\log(1+|x|)$, $x \in \mathbb{R}$. In any case $\overset{\circ}{W}_{\theta,G}^{m,\Phi}(\Omega) \hookrightarrow \mathcal{H}_{\theta,G}^{m,\Phi}(\Omega) \hookrightarrow W_{\theta,G}^{m,\Phi}(\Omega)$ (continuous embeddings).*

Remark. It should be noted (since $m < \infty$) that in the definition of $W_{\theta,G}^{m,\Phi}(\Omega)$, we may allow Φ's to depend on α's, i.e., we can consider $\tilde{\Phi} = \{\Phi_\alpha, |\alpha| \le m\}$ of Young functions and then

$$W_{\theta,G}^{m,\tilde{\Phi}}(\Omega) = \Big\{ f \in L^0(\mu) : D^\alpha f \in L^{\Phi_\alpha}(\mu),\ \sum_{|\alpha| \le m} \theta\Big(\frac{\|D^\alpha f\|_{\Phi_\alpha}}{k}\Big) G(\alpha) < \infty$$

$$\text{for a } k > 0 \Big\}. \qquad (9)$$

The completeness in the corresponding norm holds, with the same proof. If all the Φ_α are Δ_2-regular, then the corresponding statements also hold. While for $m < \infty$ this generalization is vacuous; it can be seen from (4) that for $m = \infty$ such an extension plays an essential role in the theory. This is discussed in the next section. For the present, we only consider a single Φ for simplicity.

In Theorem 2 on Sobolev embedding, we have seen that $W^{m,p}(\Omega) \hookrightarrow L^q(\Omega, \mu)$ where $q \le np(n-mp)^{-1}$. This upper bound with $m = 1$ is determined by p and n ($= \dim(\Omega)$), and is called the *Sobolev conjugate* of p since this number plays an important role in the theory. The corresponding concept for a Young function Φ is defined as follows. This is motivated by the fact that $\Phi(t) = |t|^p$ then the above conjugate Φ^* is given by $\Phi^*(t) = |t|^q$ where $q = np(n-p)^{-1}$ for $1 < p < n$. Rewriting this as an integral leads to:

Definition 5. Let Φ be a Young function. Then a first order *Sobolev conjugate* for $I\!\!R^n$, denoted Φ_1^*, is given by the equation

$$(\Phi_1^*)^{-1}(t) = \int\limits_0^t \frac{\Phi^{-1}(s)}{s^{1+\frac{1}{n}}} ds, \qquad t \geq 0$$

whenever this is finite and satisfies $\lim\limits_{t \to \infty} (\Phi_1^*)^{-1}(t) = \infty$. An mth *order Sobolev* conjugate Φ_m^* of Φ is then obtained as $\Phi_0^* = \Phi$, and $\Phi_j^* = (\Phi_{(j-1)}^*)^*$.

Since Φ is a Young function iff Φ^{-1} is concave increasing and $\Phi^{-1}(0) = 0$, it follows that Φ_j^* is also a convex function. However for $m \geq n$, $\Phi_m^*(t) \leq K_0 < \infty$ for all t, and this can happen even if $m < n$. We denote the smallest value by $m_0 \geq 1$. With such a concept, one has the following embedding theorem due to R. A. Adams (1977) and this refines and extends the earlier work of T. K. Donaldson and N. S. Trudinger (1971). We take $G(\alpha) \equiv 1$ and $\theta(x) = |x|$ without comment.

Theorem 6. *Let $\Omega \subset I\!\!R^n$ be a domain having the cone property (see the definition prior to Theorem 2). Let Φ be an N-function and $1 \leq m_0 \leq n$ be the smallest integer defining $\Phi_{m_0}^*$ after which it remains bounded. Then*

    (i) $1 \leq m \leq m_0 \Rightarrow W^{m,\Phi}(\Omega) \hookrightarrow L^{\Phi_{m_0}^*}(\mu)$, *and*

    (ii) $m > m_0 \Rightarrow W^{m,\Phi}(\Omega) \hookrightarrow C_b^{m-m_0}(\Omega)$

where $C_b^k(\Omega)$ is the Banach space of k-times continuously differentiable bounded functions on Ω with norm $\|f\| = \max\{\|D^\alpha f\|_u : |\alpha| \leq k\}$, $\|\cdot\|_u$ being the uniform norm.

The proof of this result is somewhat involved and we refer the reader to the original paper for details and other related results. But the point of this theorem is that by allowing the Orlicz spaces in the embedding theory not only can one embed $W^{m,p}$ for all p into Orlicz spaces, but even $W^{m,\Phi}$ can be admitted in this work, and this is a useful bonus for applications.

For compact embeddings the following result, also due to Adams (1977), can be stated:

Proposition 7. *Let Ω be a bounded domain with the cone property, and m_0, Φ_m^* be as in Theorem 6. Then $W^{m,\Phi}(\Omega) \hookrightarrow L^{\tilde{\Phi}}(\mu)$ compactly for $m \leq m_0$ where $\tilde{\Phi}$ is a Young function such that $\lim_{t\to\infty} [\tilde{\Phi}(rt)/\Phi_m^*(t)] = 0$ for each $r > 0$ (i.e., $\tilde{\Phi} \prec\prec \Phi_m^*$ in terms of* Theorem II.2.2).

Finally we note that by identifying $W_{\theta,G}^{m,\Phi}(\Omega)$ as a subspace of $L_{\mathcal{X}_\alpha}^\Phi(\mu)$ where $\mathcal{X}_\alpha$ is a finite dimensional space [and μ being Lebesgue measure] we can immediately read off several geometric properties of the Orlicz-Sobolev spaces. For instance, the space is separable if Φ is Δ_2-regular and is reflexive if Φ and its complementary function Ψ are both Δ_2-regular. Since m is finite we can have Φ replaced by $\tilde{\Phi}$ as in (9) with all Φ_α (and their complementary functions Ψ_α) being Δ_2-regular since then $W_{\theta,G}^{m,\tilde{\Phi}}(\Omega) \subset \underset{|\alpha|\leq m}{\times} L^{\Phi_\alpha}(\mu)$, the latter being normed, for instance, by the sum of the individual norms. The adjoint space of $W_{\theta,G}^{m,\Phi}(\Omega)$ can also be obtained. Omitting this discussion, we turn to the case that $m = \infty$ in the next section.

9.4. *Orlicz-Sobolev spaces II: Infinite order*

It was already seen in the last section that the Dirichlet problem leads to Sobolev spaces of infinite order with possibly different norms for different (weak) differentials. The corresponding Orlicz-Sobolev spaces should include those given by (3.6), and are generalizations of spaces introduced in Definition 3.1. We present the concept as follows:

Definition 1. Let $\Theta = \{\theta_\alpha, |\alpha| = 0, 1, 2, \ldots\}$ and $\Phi = \{\Phi_\alpha, |\alpha| = 0, 1, 2, \ldots\}$ be two sequences of Young functions and $\Omega \subset \mathbb{R}^n$ be a domain. Then the Orlicz-Sobolev spaces of infinite order, denoted $W_{\Theta,G}^{\infty,\Phi}(\Omega)$, is given by

$$W_{\Theta,G}^{\infty,\Phi}(\Omega) = \{f \in L^0(\mu) : D^\alpha f \in L^{\Phi_\alpha}(\mu), N_{\Theta,G}^\Phi(f) < \infty\} \tag{1}$$

where

$$N_{\Theta,G}^\Phi(f) = \inf\{k>0 : \sum_{|\alpha|\geq 0} \theta_\alpha \left(\frac{\|D^\alpha f\|_{\Phi_\alpha}}{k}\right) G(\alpha) \leq 1\} \tag{2}$$

with $0 \leq G(\alpha) < \infty$ as some weights. $\mathring{W}_{\Theta,G}^{\infty,\Phi}(\Omega)$ is obtained if in the above we only take $f \in C_c^\infty(\Omega)$ and $D^\alpha f|_{\partial\Omega} = 0, |\alpha| \geq 0$.

In contrast to the theory of finite order, the very first question in the infinite order case is the nontriviality of $W^{\infty,\Phi}_{\Theta,G}(\Omega)$, since it may not have anything other than the zero function. We thus present conditions for this space to have nonzero functions, and then indicate some results in the latter case. Actually the nontriviality aspect depends not only on Θ, Φ and G but also on the geometry of the domain Ω together with some properties of "quasi-analytic" functions.

First we note that after a comparison of Definition 3.1 and the above expression (1), $W^{m,\Phi}_{\Theta,G}(\Omega) \supset W^{m+1,\Phi}_{\Theta,G}(\Omega)$. As seen in the remark following Corollary 3.4, the θ and Φ here can be (finite) sequences. Consequently the corresponding norms (cf. (3.4) and (2)) satisfy $N^{m,\Phi}_{\theta,G}(f) \leq N^{m+1,\Phi}_{\theta,G}(f)$. From this it follows that $W^{\infty,\Phi}_{\Theta,G}(\Omega) = \bigcap_{m=1}^{\infty} W^{m,\Phi}_{\Theta,G}(\Omega)$. Moreover, if $\{f_n, n \geq 1\}$ is a Cauchy sequence in both the norms $N^{m,\Phi}_{\theta,G}(\cdot)$ and $N^{m+k,\Phi}_{\theta,G}(\cdot)$, for some $k \geq 1$, then the norm limit is seen to be the same. In other words, the norms have the consistency property. Since $(W^{m,\Phi}_{\theta,G}(\Omega), N^{m,\Phi}_{\theta,G}(\cdot))$ is a Banach space for each $m \geq 0$ by Theorem 3.3, we deduce that $(W^{\infty,\Phi}_{\Theta,G}(\Omega), N^{\Phi}_{\Theta,G}(\cdot))$ is also a Banach space. Hence we have

Proposition 2. *The space* $(W^{\infty,\Phi}_{\Theta,G}(\Omega), N^{\Phi}_{\Theta,G}(\cdot))$ *is a Banach space, and the same is true of* $\overset{\circ}{W}^{\infty,\Phi}_{\Theta,G}(\Omega)$ *for the same norm functional.*

As noted above, it is possible that $W^{\infty,\Phi}_{\Theta,G}(\Omega)$ is $\{0\}$ even though $W^{m,\Phi}_{\theta,G}(\Omega)$ contains $C^{\infty}_c(\Omega)$ for each $m \geq 1$. Thus to find conditions for the nontriviality of the limit space, one appeals to complex function theory to construct nontrivial elements. The following result is a simple extension of one given in Dubinskij [(1986), p. 19] when $\Omega = \mathbb{R}^n$.

Proposition 3. *The space* $(W^{\infty,\Phi}_{\Theta,G}(\mathbb{R}^n), N^{\Phi}_{\Theta,G}(\cdot))$ *contains nonzero functions (i.e., nontrivial) if there is a point* $\xi = (\xi_1, \ldots, \xi_n)$ *in* $\mathbb{R}^n$ *with* $\xi_0 = \min_{1 \leq i \leq n}(|\xi_i|) > 0$ *such that*

$$\sum_{|\alpha| \geq 0} \theta_\alpha(K_0 \xi_0^{|\alpha|} |\alpha|^{n+1}) G(\alpha) < \infty, \tag{3}$$

for any $K_0 > 0$; *in particular if* θ_α *is uniformly* Δ_2 *(i.e.,* $\theta_\alpha(2x) \leq C\theta_\alpha(x)$, $x > 0$, $|\alpha \geq 0$ *for a constant* $0 < C < \infty$ *) and* (3) *holds.*

Proof: By definition of $W^{\infty,\Phi}_{\Theta,G}(\mathbb{R}^n)$, we need to show the existence of a nonzero f in this space, so that $f \in L^0(\mu)$ and $N^{\Phi}_{\Theta,G}(f) < \infty$. We

find an $f \in C^\infty(I\!\!R^n)$ in this space under (3). Indeed, if $H = \{t \in I\!\!R^n :$
$\|t\| \le t_0 < \xi_0\}$, let $g \in C^\infty(H)$ such that $D^\alpha g|_{\partial H} = 0$ for all $|\alpha| \ge$
0. Define $f(x) = \int\limits_H g(t) e^{i\langle x,t\rangle} d\mu(t)$ where $\langle x, t\rangle$ is the inner (or dot)
product of $I\!\!R^n$. Clearly f is bounded and $f \in C^\infty(I\!\!R^n)$. We assert
that $N^\Phi_{\Theta,G}(f) < \infty$ under (3), by the following calculation.

Since $g \in C^\infty(H)$, we may integrate it by parts componentwise
repeatedly, using the fact that $(D^\beta g)|_{\partial H} = 0$, and then consider $D^\alpha f$
for the resulting integral. This gives, on using $|t_j| \le t_0$,

$$|(D^\alpha f)(x)| \le C_1 t_0^{|\alpha|} |\alpha|^m \prod_{j=1}^m \min\left(1, \frac{1}{|x_j|^{\frac{m}{n}}}\right). \qquad (4)$$

Here $|\beta| \le m$, and C_1 is a constant depending on $\max\{|(D^\beta g)(t)| : t \in$
$H, |\beta| \le m\}$ and hence on ξ_0 and m but not on α. Hence (4) gives, on
taking $m \ge n + 1$,

$$\|D^\alpha f\|_{\Phi_\alpha} \le C_1 t_0^{|\alpha|} |\alpha|^m \cdot C_2 \qquad (5)$$

where C_2 is the integral of the bounded function on the right side of
(4). This is clearly in each $L^{\Phi_\alpha}(\mu)$ because each of these spaces contains
$C_c(I\!\!R^n)$ and $\theta_\alpha(|x|^{-\frac{m}{n}}) = o(|x|^{-\frac{m}{n}})$ for large x. Taking $m = n + 1$, (3)
implies that for some $K > 0$

$$\sum_{|\alpha|\ge 0} \theta_\alpha\left(\frac{\|D^\alpha f\|_{\Phi_\alpha}}{K}\right) G(\alpha) \le \sum_{|\alpha|\ge 0} \theta_\alpha(K_0 \cdot t_0^{|\alpha|} |\alpha|^{n+1}) < \infty,$$

where $K_0 = \frac{C_1 C_2}{K} > 0$. Hence $f \in W^{\infty,\Phi}_{\Theta,G}(I\!\!R^n)$ as desired. ∎

The work in Dubinskij (1986) shows that in many cases the condition (3) is also necessary. Similar techniques can be used for Ω being
a torus or a strip. The geometry of these regions demands that we
use some results of quasi-analytic functions in the following sense: If
$M_n \ge 0, n \ge 0$ are given numbers and $C(\{M_n\}) = \{f \in C^\infty(a,b) :$
$|D^n f| \le K M_n, n \ge 0$, uniformly$\}$, then $C(\{M_n\})$ is *quasianalytic* if it
satisfies the Hadamard criterion; namely, $(D^n f)(x_0) = (D^n g)(x_0)$ for
some $x_0 \in (a,b) \Rightarrow f \equiv g$, $f,g \in C(\{M_n\})$. The space is nonquasi-
analytic if the Hadamard conditions fails. Numerous equivalent conditions to quasianalyticity have been given by S. Mandelbrojt (1952)
[see Chapitre IV in particular]. Using these, Dubinskij discusses the

nontriviality of the spaces in the case of Sobolev spaces of infinite order, and a few extensions to the Orlicz space set up have been given
by D. V. Chan (1980). However, it seems to us that all these results
are in a tentative form and a simple enough condition is not yet found.
We therefore omit further discussion on the conditions for nontriviality
and add a few remarks on other aspects of these spaces, *assuming* their
nontriviality.

The work of Chapter IV on the adjoint spaces of Orlicz spaces
shows that $(W_{\theta,G}^{m,\Phi}(\Omega))^*$ consists of additive set functions unless Φ is
Δ_2-regular. However, $(\mathring{W}_{\theta,G}^{m,\Phi}(\Omega))^*$ has only point functions. If Ψ and
ζ are the complementary Young functions of Φ and θ respectively, then
our previous theory gives $(\mathring{W}_{\theta,G}^{m,\Phi}(\Omega))^* = W_{\zeta,G}^{m,\Psi}(\Omega)$. We replace m by
$-m$ ($m > 0$) for the reason that, as m increases, the original spaces
decrease and hence the adjoint spaces increase. More precisely, we set

$$W_{\zeta,G}^{-m,\Psi}(\Omega) = \{g : g = \sum_{|\alpha| \leq m} D^{\alpha} g_{\alpha}, g_{\alpha} \in L^{\Psi}(\mu)\} \qquad (6)$$

and the norm is taken to be

$$\|g\|_{-m,\Psi} = \sup\{|\int_{\Omega} g h d\mu| : h \in \mathring{W}_{\theta,G}^{m,\Phi}(\Omega), N_{\theta,G}^{\Phi}(h) \leq 1\}. \qquad (7)$$

Using the general theory of Banach spaces, one can define the limit
space $W_{\zeta,G}^{-\infty}(\Omega)$ as the closure of $\bigcup_{m=1}^{\infty} W_{\zeta,G}^{-m,\Psi}(\Omega)$ relative to the norm
$\|g\|_{-\infty,\Psi} = \lim_{m\to\infty} \|g\|_{-m,\Psi}$ and can discuss the relations between the
spaces $W_{\theta,G}^{\infty,\Psi}(\Omega)$ and $W_{\zeta,G}^{-\infty,\Psi}(\Omega)$. If the former is nontrivial, then the
Hahn-Banach theorem shows that the latter is also nontrivial. The
converse is true only under some additional conditions. The structure
and geometry of these limit spaces, of some independent interest, can
be studied. However, the general theory of the infinite order space is
not well developed. Many problems needed a serious investigation, and
so we conclude their discussion at this point.

9.5. *Besicovitch-Orlicz spaces of almost periodic functions*

In this final section of the chapter we consider briefly Besicovitch-
Orlicz spaces of almost periodic functions, and related Stepanoff and
Weyl classes of a similar type in the context of Orlicz spaces. These

will show how some new directions are available in this treatment to
be used in future work.

Recall from Section VI.6 that a (necessarily) separable Banach
space $\mathcal{X}$ has a Schauder basis $\{x_1, x_2, \ldots\}$ if each x in $\mathcal{X}$ admits a unique
expansion $x = \sum_{i=1}^{\infty} a_i x_i$, the series converging in mean, the (unique) sequence $a_i \, (= a_i(x))$ are some scalars, called the *coefficient functionals*.
While each separable Banach space need not have such a basis, for
Orlicz spaces $L^{\Phi}[0,1]$, such a basis exists if Φ satisfies a Δ_2-condition
(cf., Corollary VI.5.2). If T is an invertible continuous linear operator
on $\mathcal{X}$, then $T x_n = y_n$, $n \geq 1$, is also a basis whenever $\{x_n, n \geq 1\}$ is, and
hence we may seek convenient bases when one is found. In the case of
special spaces such as $L^2[0,1]$, the trigonometric functions, with good
differentiability properties, are such a basis. This circumstance leads,
in other $L^p[0,1]$ spaces, to find similar bases and this search takes us
to the class of almost periodic functions invented by H. Bohr, and later
extended by Stepanoff, Weyl and Besicovitch. It turns out that in these
extensions, to be discussed below, the separability of the spaces (hence
the existence of a basis) is submerged but a considerably general class
of spaces emerges.

To introduce the desired spaces, let Φ be a Young function and
$L_{\text{loc}}^1(\mathbb{R})$ be the space of locally integrable scalar functions on the
Lebesgue line $(\mathbb{R}, \mathcal{B}, \mu)$. We define four (semi-) distances on this space:
If $f, g \in L_{\text{loc}}^1(\mathbb{R})$ define

$$D_u(f,g) = \sup_{x \in \mathbb{R}} |f(x) - g(x)|, \qquad D_{S_{\ell}^{\Phi}}(f,g) = \sup_{x \in \mathbb{R}} N_{\Phi_{x,\ell}}(f-g),$$
$$D_{W^{\Phi}}(f,g) = \lim_{\ell \to \infty} D_{S_{\ell}^{\Phi}}(f,g), \text{ and } D_{B^{\Phi}}(f,g) = \limsup_{T \to \infty} N_{\Phi_T}(f,g), \tag{1}$$

where the gauge norms $N_{\Phi_{x,\ell}}$ and N_{Φ_T} are given as

$$N_{\Phi_{x,\ell}}(f) = \inf \left\{ k > 0 : \frac{1}{\ell} \int_{x}^{x+\ell} \Phi\left(\frac{f}{k}\right) d\mu \leq 1 \right\},$$

$$N_{\Phi_T}(f) = \inf \left\{ k > 0 : \frac{1}{2T} \int_{-T}^{T} \Phi\left(\frac{f}{k}\right) d\mu \leq 1 \right\}.$$

These functionals are called uniform, Stepanoff, Weyl and Besicovitch
(semi-) norms. We write $D_{G^{\Phi}}(\cdot, \cdot)$ if G is to denote any of the four

functionals. It may be verified that all symbols in (1) are well defined, and one has, with a notation $D_{G^{\Phi}}(f) = D_{G^{\Phi}}(f,0)$, the following:

$$D_u(f) \geq k_1 D_{S_l^{\Phi}}(f) \geq k_2 D_{W^{\Phi}}(f) \geq k_3 D_{B^{\Phi}}(f) \geq k_4 D_B(f), \quad (2)$$

where $k_i > 0$, $i = 1, \ldots, 4$, are some constants and $D_B(\cdot)$ is $D_{B^{\Phi}}(\cdot)$ with $\Phi(x) = |x|$. Further one can verify that for discontinuous Φ (i.e., $\Phi(x) = 0$ for $0 \leq |x| \leq x_0$, $= +\infty$ for $|x| > x_0$) $D_{B^{\Phi}}(\cdot)$ and $D_u(\cdot)$ (and hence all) are equivalent. If functions agreeing outside μ-null sets are identified, then $D_{S_l^{\Phi}}(\cdot)$ is also a norm (as is always true of D_u). However $D_{W^{\Phi}}(f) = 0 \not\Rightarrow f = 0$ a.e.$[\mu]$. In fact $\mu\{|f| > 0\} > 0$ is possible and $D_{W^{\Phi}}(f) = 0$. Consequently $D_{B^{\Phi}}(\cdot)$ is also only a semi-norm in general.

Definition 1. Let $\mathcal{A}$ be the set of all finite expressions $\sum\limits_{j=1}^{n} a_j e^{i\lambda_j x}$, i.e., of the trigonometric polynomials on $\mathbb{R}$. Let $C_G(\mathcal{A})$ be the closure of $\mathcal{A}$ in the (semi-) metric $D_G(\cdot)$ of (1) where D_G denotes either uniform, Stepanoff, Weyl, or Besicovitch (semi-) norm. Then the elements of $C_G(\mathcal{A})$ are termed *G-almost periodic* functions on $\mathbb{R}$. In particular $C_u(\mathcal{A})$ is the space of H. Bohr, or *uniform almost periodic* (or u.a.p.) functions, $C_{S^{\Phi}}(\mathcal{A})$ [can be shown to be $= C_{S_l^{\Phi}}(\mathcal{A}), \ell \geq 1$] is the *Orlicz-Stepanoff class of almost periodic* (or S^{Φ}-a.p.) functions, $C_{W^{\Phi}}(\mathcal{A})$ is the *Orlicz-Weyl class of almost periodic* (W^{Φ}-a.p.) functions and $C_{B^{\Phi}}(\mathcal{A})$ is the *Besicovitch-Orlicz class of almost periodic* (B^{Φ}-a.p.) functions.

It must be noted that the original definition of H. Bohr's is as follows. A continuous scalar f on $\mathbb{R}$ is almost periodic if for each $\varepsilon > 0$, there is $\ell = \ell_\varepsilon > 0$ such that each interval $(x, x + \ell)$ contains a τ satisfying $|f(t + \tau) - f(t)| < \varepsilon$ uniformly in t. That this concept is equivalent to the one given above is a basic result in the original theory of H. Bohr's. Several other equivalent forms, including one which constitutes an aspect of the Peter-Weyl theory and group representations, make the u.a.p play such a vital role in modern analysis. Also since by (2) B^{Φ}-a.p class is the widest class known in this line of investigation we present some properties of B^{Φ}, the omitted details of which may be found in T. R. Hillmann (1986) [see also J. Albrycht (1962)]. The basic ideas are an extension of the classical treatment in A. S. Besicovitch (1932).

We first observe the following.

Proposition 2. *If $S \subset L^1_{loc}(\mathbb{R})$, then $C_{G^\Phi}(S) = C_{G^\Phi}(C_u(S))$, i.e., the G^Φ-closure [in $D_{G^\Phi}(\cdot)$ semi-norm] of S is the same as the G^Φ-closure of the uniform closure $C_u(S)$. In particular, $C_{G^\Phi}(A) = C_{G^\Phi}(\{u.a.p.\})$, where A is the set of trigonometric polynomials on $\mathbb{R}$.*

Proof: Since $S \subset C_u(S)$ we have $C_{G^\Phi}(S) \subset C_{G^\Phi}(C_u(S))$. For the opposite inclusion let f be in the right side. If it is in $C_u(S)$, then by (2) $f \in C_{G^\Phi}(S)$. In the general case, there exists $f_n \in C_u(S)$ such that $D_{G^\Phi}(f, f_n) \to 0$ as $n \to \infty$. But then there exist $g_n \in S$ such that $D_u(f_n, g_n) < \frac{1}{n}$. Hence by (2) $D_{G^\Phi}(f_n, g_n) < \frac{1}{kn}$ where $k = \max(k_i, i = 1, \ldots, 4)$. But then by the triangle inequality

$$D_{G^\Phi}(f, g_n) \leq D_{G^\Phi}(f, f_n) + D_{G^\Phi}(f_n, g_n) \to 0,$$

as $n \to \infty$. Hence $f \in C_{G^\Phi}(S)$. Thus equality holds. Since, by Definition 1, $C_u(A) = \{u.a.p.\}$, the last part is a consequence of the preceding result. ∎

One of the important and nontrivial properties of this class is that for each B^Φ-a.p function f the mean value $M(f) = \lim\limits_{T \to \infty} \frac{1}{2T} \int\limits_{-T}^{T} f(t)d\mu$ exists. This was established by Bohr himself for the u.a.p functions and was extended by Besicovitch for the B^p-a.p functions (i.e., $\Phi(x) = |x|^p$). Moreover, if f is B^p-a.p and g is either a trigonometric polynomial or a simple function ($g = \sum\limits_{i=1}^{n} a_i \chi_{[\alpha_i, \alpha_i + \delta_i)}, \delta_i > 0$), then fg and $|f|$ are B^Φ-a.p. If f is B^Φ-a.p and $e_\lambda(x) = e^{i\lambda x}$, then the Fourier transform of f is defined as $M(fe_\lambda) = a(\lambda)$ and this exists by the preceding statement. We thus have

Theorem 3. *The Fourier expansion of $f \in C_{G^\Phi}(A)$ exists and if $a(\lambda) = M(fe_\lambda)$, then $a(\lambda) = 0$, for all but at most a countable set of λ's in $\mathbb{R}$.*

Proof: By the relations (2) between the semi-norms, we note that

$$C_u(A) \subset C_{W^\Phi}(A) \subset C_{B^\Phi}(A) \subset C_B(A). \tag{3}$$

Hence f and fe_λ are in $C_B(A)$ so that $a(\lambda)$ exists, $\lambda \in \mathbb{R}$. Regarding the second statement, by Proposition 2, f is the strong (= norm) limit of trigonometric polynomials $P_n \in A$, where the norm is $D_B(\cdot)$ of (1). But e_λ being bounded, it also follows that $D_B[(f - P_n)e_\lambda] \to 0$, as

$n \to \infty$, for each $\lambda \in \mathbb{R}$. Then $M(P_n e_\lambda) \to M(f e_\lambda) = a(\lambda)$, as $n \to \infty$. However, $M(P_n e_\lambda) \neq 0$ only for a finite number of λ's so that the limit $= a(\lambda)$ can differ from zero for at most a countable number of λ's as asserted. ∎

One shows that if $a(\lambda) = 0$, $\lambda \in \mathbb{R}$, then $D_{G^*}(f) = 0$ so that if f, g have the same Fourier expansions, then $D_{G^*}(f, g) = 0$, and for B^2-a.p functions (i.e., $\Phi(x) = x^2$) the Parseval and the Riesz-Fischer type theorems hold. However, there are distinctions between the B^p-a.p and the B^Φ-a.p classes when Φ is simply a Young function. To exemplify these, we need to introduce:

Definition 4. A set E ($\subset \mathbb{R}$) is termed *relatively dense* in $\mathbb{R}$ if for some $0 < \ell < \infty$, every interval of length ℓ in $\mathbb{R}$ contains at least one element of E. A function f in $L^1_{\text{loc}}(\mathbb{R})$ is said to have the S^Φ_ℓ-translation property (S^Φ_ℓ-t.p.) if for each $\varepsilon > 0$ the set $E_{\varepsilon,f} = \{t \in \mathbb{R} : D_{S^*_\ell}(f^t, f) < \varepsilon\}$ is relatively dense for this $\ell > 0$ where $f^a(x) = f(x + a)$. The function f has W^Φ-*translation property* (W^Φ-t.p.) if for each $\varepsilon > 0$ there is an $\ell_\varepsilon > 0$ such that f has $S^\Phi_{\ell_\varepsilon}$-t.p.

As the definition indicates, the translation property of a function is not dependent on its almost periodicity. If $\{G^\Phi$-t.p.$\}$ denotes the set of all W^Φ-t.p [or S^Φ_ℓ-t.p.] functions from $L^1_{\text{loc}}(\mathbb{R})$, then using the fact that $D_{G^*}(f^t) = D_{G^*}(f)$ and (2), we deduce the following inclusion relation indicating the first difference with the Lebesgue theory applied to the a.p functions.

Proposition 5. *If G^Φ denotes either S^Φ_1 or W^Φ, then each G^Φ-a.p function has also G^Φ-t.p., i.e., $C_{G^*}(\mathcal{A}) \subset \{G^\Phi$-t.p.$\}$. These two classes coincide iff Φ satisfies a Δ_2-condition.*

The inclusion relation is deduced as noted above. Regarding the last part, if Φ is Δ_2 then one shows that for each f with G^Φ-t.p., $f_N = f\chi_{[|f| \leq N]} + N\frac{f}{|f|}\chi_{[|f| > N]}$ also has G^Φ-t.p., and that $D_{G^*}(f, f_N) \to 0$ as $n \to \infty$ iff Φ is Δ_2. [This involves nontrivial computation.] The converse part is shown by constructing a g having a G^Φ-t.p but $N_\Phi(g - g_N) \not\to 0$ so that $g \notin C_{G^*}(\mathcal{A})$.

A similar statement holds in the case of B^Φ spaces. The concept is more involved and is given by the following.

Definition 6. A sequence $\{t_i, -\infty < i < \infty\} \subset \mathbb{R}$ is *satisfactorily uniform* if there is an $0 < \ell < \infty$ such that $\nu^\circ(\ell) < 2\mu^\circ(\ell)$ where $\nu^\circ(\ell)$

$[\mu^\circ(\ell)\,]$ is the maximum [minimum] number of t_i in intervals of length ℓ in $\mathbb{R}$. Then an $f \in L^1_{\mathrm{loc}}(\mathbb{R})$ is said to have the B^Φ-*translation property* $[B^\Phi$-t.p.$]$ if for each $\varepsilon > 0$, there is a satisfactorily uniform sequence $\{t_i, -\infty < i < \infty, t_i < t_{i+1}\}$ such that for any $c > 0$, and some $\alpha \geq 1$ we have

$$\limsup_{T\to\infty} \limsup_{n\to\infty} \frac{1}{2T} \int_{-T}^{T} \frac{1}{2n+1} \sum_{i=-n}^{n} H^{f,\alpha}_{\Phi,c}(x,i)d\mu(x) < \varepsilon, \qquad (4)$$

where

$$H^{f,\alpha}_{\Phi,c}(x,i) = \frac{1}{c} \int_{x}^{x+c} \Phi\left(\frac{|f^{t_i} - f|}{\alpha}\right)(x)d\mu(x). \qquad (5)$$

Here $\alpha = 1$ can be taken if Φ satisfies a Δ_2-condition.

The analog of Proposition 5 can now be stated.

Proposition 7. *If* $\{B^\Phi$-*t.p.*$\}$ *denotes the set of* f *in* $L^1_{\mathrm{loc}}(\mathbb{R})$ *having the* B^Φ-*t.p., then* $C_{B^\Phi}(\mathcal{A}) \subset \{B^\Phi$-*t.p.*$\}$, *and there is equality here iff* Φ *satisfies a* Δ_2-*condition.*

The proof involves, as may be expected, several detailed computations.

In addition to this, the proofs of the next two results are definitely nontrivial, and the interested reader will find all details in Hillmann (1986).

Theorem 8. *The spaces* B^Φ-*a.p.,*S^Φ-*a.p are complete (i.e., are complete seminormed vector spaces) while* W^Φ-*a.p is an incomplete (seminormed) vector space. Moreover, in contrast to the* $\{u.a.p.\}$, *none of these spaces is separable.*

Since the D_{G^Φ} is not a true norm in B^Φ- [and W^Φ-]spaces, we now consider the quotient spaces $\widetilde{B}^\Phi$ and discuss their duality to round out the picture. Let $\mathcal{N}^\Phi = \{f \in \{B^\Phi$-a.p.$\} : D_{B^\Phi}(f) = 0\}$. Similarly let $\mathcal{N}^\Phi_1 = \{f \in \{B^\Phi$-t.p.$\} : D_{B^\Phi}(f) = 0\}$. Clearly both these are closed linear spaces. Denote by $\widetilde{B}^\Phi$-a.p and $\widetilde{B}^\Phi$-t.p the spaces $B^\Phi/\mathcal{N}^\Phi$ and $\{B^\Phi$-t.p.$\}/\mathcal{N}^\Phi_1$ with their quotient norms. [Recall that if $\mathcal{X}$ is a Banach space and $\mathcal{Y} \subset \mathcal{X}$ is a closed subspace, $\mathcal{X}/\mathcal{Y}$ denoting their quotient, we have the norm in the latter as:

$$\|[x]\| = \inf\{\|x - y\| : y \in \mathcal{Y}\}, \qquad [x] = x + \mathcal{Y}, \ x \in \mathcal{X}, \qquad (6)$$

relative to which $\mathcal{X}/\mathcal{Y}$ becomes a Banach space.] The space $\tilde{B}^{\Phi}$-a.p corresponds to our M^{Φ} space of Chapter IV, while $\tilde{B}^{\Phi}$-t.p does not have that property. In fact we have the following characterization of their dual spaces, due to Hillmann. Recall that, if the real line $I\!R$ is denoted $I\!R_d$ with discrete topology, and $I\!R'$ is its dual group (thus compact), then $I\!R'$ is the *Bohr compactification* of $I\!R$. [$I\!R'$ has many more elements than those of $I\!R$ ($\subset I\!R'$).]

Theorem 9. *Let* $(\Phi.\Psi)$ *be a pair of Young complementary functions and* Φ *be continuous* $\Phi(x) = 0$ *iff* $x = 0$. *Let* $I\!R'$ *denote the Bohr compactification of* $I\!R$, *which thus is a compact group with* μ' *as its Haar measure. Then the dual space of* $\{\tilde{B}^{\Phi}$-*a.p.*$\}$ *is isometrically isomorphic to* $L^{\Psi}(I\!R', \mu')$ *the latter having the Orlicz norm. Thus* $\{\tilde{B}^{\Phi}$-*a.p.*$\}^* \cong L^{\Psi}(I\!R', \mu')$, *and for each* x^* *there is a unique* g' *in* $L^{\Psi}(I\!R', \mu')$ *such that*

$$x^*(f) = M(fg'), \qquad f \in \{\tilde{B}^{\Phi}\text{-a.p.}\}. \tag{7}$$

where $M(\cdot)$ *is the mean value on* B^1-*a.p functions.*

On the other hand $\{\tilde{B}^{\Phi}$-t.p.$\}$ is also a Banach space under the norm derived from $D_{B^{\Phi}}(\cdot)$, and is isometrically isomorphic to $(L^{\Psi}(I\!R', \mu'),$ $N_{\Phi}(\cdot))$. Hence (by Theorem IV.2.11), $\{\tilde{B}^{\Phi}$-t.p.$\}^*$ is isometrically isomorphic to the space $\mathcal{A}_{\Psi}(\mu')$ of finitely additive set functions, on the Borel algebra of $I\!R'$, vanishing on μ'-null sets and having Ψ-variation finite.

The corresponding statements for $\{S^{\Phi}$-t.p.$\}$ and $\{W^{\Phi}$-t.p.$\}$ are not considered, although they may be obtained by similar work.

Thus far we discussed the spaces of almost periodic functions on $I\!R$. This work shows how the group structure and the metric of $I\!R$ are used in G^{Φ}-a.p functions. Can we define the corresponding functions on a general locally compact (always Hausdorff) group? If $\Phi(x) = |x|^p$, $p \geq 1$, such a study was undertaken by H. Davis (1967) who also extended the classical u.a.p theory by J. von Neumann (1934). Some of Davis' work has been discussed in the context of Orlicz spaces by Hillmann (1977). We shall indicate merely the form of these concepts for B^{Φ}-a.p spaces.

If Ω is a topological group and $B(\Omega)$ is the Banach space of bounded Borel functions on Ω, under the uniform norm, then a function f is *uniformly almost periodic* (u.a.p.) if the set $\{f^a, a \in \Omega\}$ of translates is relatively compact in $B(\Omega)$ where $f^a(x) = f(xa)$, $x \in \Omega$, is the right translate. It is known that there exists a *mean* $M(\cdot)$ on the class of all u.a.p functions $\mathcal{U}(\Omega)$ which is a positive linear functional such that $M(1) = 1$ and $M(^b f^a) = M(f)$, $a, b \in \Omega$. $[^b f(x) = f(bx)$, is the left translate.] Let $N = \{(U_d, \nu_d), d \in I\}$ be a directed family, called a *Bohr net*, where U_d is a Borel set and ν_d is a finite Borel measure on U_d, if for each $f \in \mathcal{U}(\Omega)$ we have

$$\lim_{d \in I} \frac{1}{\nu_d(U_d)} \int_{U_d} f d\nu_d = M(f). \tag{8}$$

If Ω is a locally compact group with μ as a left Haar measure and $\nu_d(\cdot) = \mu(\cdot)/\mu(U_d)$ in (8), then N is called a *homogeneous Bohr net*. It is also known that in each locally compact group there *exists* a homogeneous Bohr net.

It is now possible to introduce the desired concept as:

Definition 10.   Let Ω be a locally compact group and $N = \{(U_d, \nu_d), d \in I\}$ be a Bohr net in Ω. If Φ is a continuous Young function and $f \in L^1_{\mathrm{loc}}(\Omega)$, let

$$\|f\|_{B^\Phi, N} = \limsup_{d \in I} \inf\{k > 0 : \frac{1}{\nu_d(U_d)} \int_{U_d} \Phi\left(\frac{|f|}{k}\right) d\nu_d \leq 1\}. \tag{9}$$

Then f is called B^Φ-*almost periodic* if f is in $C_{B^\Phi, N}(\{\text{u.a.p.}\}) =$ the $\|\cdot\|_{B^\Phi, N}$-closure of the u.a.p functions. $[S^\Phi$-a.p and W^Φ-a.p classes can also be defined similarly.]

The u.a.p theory is well developed. [See, e.g., Hewitt and Ross (1963), Sec. 18, and the references for earlier work on the subject.] If $\Phi(x) = |x|^p$, $p \geq 1$, some properties of S^Φ-a.p and W^Φ-a.p functions were studied by H. Davis (1967), and the corresponding theory of S^Φ a.p for general continuous Young functions Φ was sketched by Hillmann, but the full treatment, analogous to the B^Φ a.p work above, is not yet available. There is also the possibility of studying the above problems if the functions are Banach space valued. If $\Phi(x) = |x|^p$, $p \geq 1$, such an account, mostly for the u.a.p case, was given in the volume by L. Amerio and G. Prouse (1971) motivated by some interesting applications. This work also admits an extension to the Orlicz

space context. These studies have not been available and we shall omit
further discussion on them.

Bibliographical Notes: Several of the function spaces discussed in this
chapter, from the Lebesgue to the Orlicz space context, are good can-
didates for a deep analysis and interesting applications. The classical
Hardy spaces with their beautiful structure theory is available in many
books. We note especially the classical treatise of A. Zygmund (1959),
and for somewhat abstract extensions, one should refer to K. Hoffman
(1962) and P. L. Duren (1970). Theorem 1.5 is an extension of the
L^p-case that is discussed in Hoffman's book. The point of interest here
is that the Young function φ is not restricted. We have utilized some
work of R. Lésniewicz (1971). Further study, especially the duality
analysis of the Hardy-Orlicz spaces on the lines presented in Section
1 will be of interest. The interpolation theory of sublinear operators
in the $\mathcal{H}^p$-spaces was given by E. M. Stein and G. Weiss (1957), and
our treatment of $\mathcal{H}^\varphi$-spaces with s-convex functions, follows essentially
that of W. T. Kraynek (1972), and utilizes some work in Chapter VI.
The presentation of Section 2 combined with that of Chapters VI and
VII includes a fairly comprehensive account of interpolation theory that
is useful in numerous applications. Using this, it is possible to study
further extensions of an important thesis by L. Lumer-Naim (1967)
which we did not include due to space limitations. The work on $\mathcal{H}^\varphi$
spaces initiated there, admits interesting extensions, and these should
be considered as a next item of research.

Section 3 and 4 contain a certain amount of Sobolev space theory,
and it is largely influenced by the treatment in R. A. Adam's book
(1975), and the infinite order case in Ju. A. Dubinskij's book (1985).
Here we could merely give the flavor of the subject and only wanted
to show the great many avenues available for future research. We also
refer to V. L. Shapiro (1977) where it is shown how Orlicz space theory
is very effective in partial differential equations. The recent monograph
due to V. G. Maz'ja (1985) on Sobolev spaces (of finite order) contains
the most up-to-date treatment of the subject in the L^p-context. In the
work we discussed, especially for the finite order case, the number of
derivatives is always an integer. However, fractional differentiation, as
discussed in Zygmund's book (1959), and illustrated in Section VI.4 in
the Orlicz space context, shows that one should discuss this aspect also.

Using the interpolation methods that can be done, and this extension brings in the Besov spaces (in the L^p-context). A detailed discussion of the latter spaces is in H. Triebel (1978). See also P. Brenner, V. Thomée and L. B. Wahlbin (1975). It is clear that a detailed account of Besov-Orlicz spaces can be developed since especially the necessary interpolation theory is now available in the preceding chapters. One should also note that in place of the Lebesgue measure, it is possible to develop the subject with Lebesgue-Stieltjes measures perhaps dominated by the Lebesgue measure. For an account in the classical L^p-case, see, e.g., Kufner (1985).

In discussing such function spaces of considerable concrete structure, we could not overlook the beautiful theory of Besicovitch's (1932) almost periodic function spaces B^p. Their extension B^Φ, in the Orlicz space context, is available. See, e.g., the thesis of T. R. Hillmann (1977) and his paper (1986). We indicated the highlights of his work in Section 5, mostly sketching or omitting the proofs. There is also a related extension, under the name Marcinkiewicz-Orlicz spaces, by J. Albrycht (1962). This gives the point of view of "modular" extensions in contrast to Hillmann's Banach space approach, but both studies complement each other and should be of interest for the reader. Because of its relation to Fourier transform theory and harmonic analysis in general, one could study certain subalgebras, such as Beurling algebras, of $L^\Phi(I\!R)$, closed under convolution. A good treatment of Beurling and related algebras is in H. Reiter (1968,1971) in the L^p-space context.

X

Generalized Orlicz Spaces

This final chapter is devoted to some generalizations of Orlicz spaces. These are the $L^\varphi(\mu)$ spaces in which φ is only monotone increasing, $\varphi(0) = 0$, called a φ-function, including the s-convex class. These are given indications of a few applications to stochastic analysis. The details appear in Section 1. The following section is devoted to Orlicz spaces of additive set functions and of operator algebras, instead of point functions. The last section gives a comparison of Lorentz and Orlicz spaces and includes a common generalization of both.

10.1. *Nonlocally convex spaces and applications*

In most of the preceding work the $L^\varphi(\mu)$ spaces are Banach spaces, so that they are locally convex. We also considered $\mathcal{H}^\varphi$ spaces when φ is just a φ-function which need not be convex. But most of these spaces are locally convex, i.e., they contain a convex neighborhood of the origin. However the $L^p(\mu)$ spaces for $0 < p < 1$ need not have the latter property. They are complete metric spaces [or Fréchet spaces] and we discuss their extension in the Orlicz space context. Let us observe immediately that an important motivation for considering this generalized class is their use in the theory of stochastic integration, among others. It will not be detailed here. [See, e.g., Urbanik (1967) who shows the essential use of the generalized Orlicz spaces, with Fréchet norms, in such an application.] However, some results on the sample paths of stochastic processes in these spaces are indicated later in this section.

Just as in Section 9.1, but slightly more generally, we consider a φ-*function*, denoted again by φ, which is nondecreasing, left continuous, and $\varphi(0) = 0$ on $\mathbb{R}^+$. This may be extended to all of $\mathbb{R}$ by setting $\varphi(-x) = \varphi(x)$. We now introduce the corresponding L^φ-spaces.

Definition 1. If (Ω, Σ, μ) is a measure space and φ is a φ-*function*, let the *generalized Orlicz space* $\{L^\varphi(\mu), \|\cdot\|_\varphi\}$ be the set

$$L^\varphi(\mu) = \{f : \text{scalar and measurable for } \Sigma \text{ and } \rho_\varphi(\alpha f) =$$

$$\int_\Omega \varphi(\alpha|f|)d\mu < \infty, \text{ for some } \alpha = \alpha_f > 0\}, \quad (1)$$

and

$$\|f\|_\varphi = \inf\left\{k > 0 : \rho_\varphi\left(\frac{|f|}{k}\right) \leq k\right\}. \quad (2)$$

We justify the name by proving the following.

Theorem 2. $\{L^\varphi(\mu), \|\cdot\|_\varphi\}$ *is a complete linear metric space if we identify f_1 and f_2 whenever $\|f_1 - f_2\|_\varphi = 0$. Moreover, if φ is also convex [so that it is a Young function], then $\|\cdot\|_\varphi$ and the gauge (hence Orlicz) norm are equivalent in the sense that a sequence $\{f_n, n \geq 1\}$ is convergent [or bounded] in $\|\cdot\|_\varphi$ iff it has the same property in the $N_\varphi(\cdot)$-norm.*

Proof: Most of the details are similar to the convex case, and so we just outline the argument, adding the necessary steps when a new idea is needed. By definition of $\|\cdot\|_\varphi$ in (2), it follows that $\|\alpha f\|_\varphi \leq \max(1, \alpha)\|f\|_\varphi$ for all $\alpha > 0$, and that $|f_1| \leq |f_2| \Rightarrow \|f_1\|_\varphi \leq \|f_2\|_\varphi$. Also if $0 \leq f_i \in L^\varphi(\mu)$, then there exist $\alpha_i > 0$ such that $\int_\Omega \varphi\left(\frac{|f_i|}{\alpha_i}\right)d\mu \leq \alpha_i$, and hence

$$\int_\Omega \varphi\left(\frac{f_1 + f_2}{\alpha_1 + \alpha_2}\right)d\mu = \int_{\Omega_1} \varphi\left(\frac{a(f_1 + f_2)}{\alpha_1}\right)d\mu + \int_{\Omega_2} \varphi\left(\frac{b(f_1 + f_2)}{\alpha_2}\right)d\mu,$$

where $a = \alpha_1(\alpha_1 + \alpha_2)^{-1}, b = 1 - a$, and

$$\Omega_1 = \Omega_2^c = \{\omega : \alpha_2 f_1(\omega) \geq \alpha_1 f_2(\omega)\},$$

$$\leq \int_\Omega \varphi\left(\frac{f_1}{\alpha_1}\right)d\mu + \int_\Omega \varphi\left(\frac{f_2}{\alpha_2}\right)d\mu \leq \alpha_1 + \alpha_2.$$

Here we also used the fact that $\varphi(\cdot)$ is increasing. This implies that $L^\varphi(\mu)$ is linear and that $\|\cdot\|_\varphi$ satisfies the triangle inequality. It is easily seen, by the Dominated Convergence theorem, that $\|\alpha f_n\|_\varphi \to 0$ as $\alpha \searrow 0$ or $\|f_n\|_\varphi \searrow 0$. These facts imply that $\{L^\varphi(\mu), \|\cdot\|_\varphi\}$ is a linear metric space if we take the distance, as usual, as $d(f,g) = \|f - g\|_\varphi$ and f, g are identified when $d(f,g) = 0$. The completeness follows exactly as in the standard case (cf. the proof of Theorem III.3.10). Thus $\{L^\varphi(\mu), \|\cdot\|_\varphi\}$ is a Fréchet space. We now show that it is a Banach space if φ is also convex. This assertion is due to S. Mazur and W. Orlicz (1958).

Thus let $\{f_n, n \geq 1\}$ be a convergent sequence with limit f in the $N_\varphi(\cdot)$-norm topology. Replacing f_n, by $f_n - f$ if necessary, we take $f = 0$ for simplicity. Then for each $\varepsilon > 0$, there is an n_0 ($= n_0(\varepsilon)$) such that $n \geq n_0 \Rightarrow N_\varphi(f_n) < \varepsilon^2$. But by the convexity of φ, with $N_\varphi(g) \leq 1$ we have $\rho_\varphi(g) = \rho_\varphi\left(\frac{N_\varphi(g)\cdot g}{N_\varphi(g)}\right) \leq N_\varphi(g)\cdot 1$ using the definition of gauge norm. Hence for $n \geq n_0$ we get on taking ε^2 ($< \varepsilon$) < 1 so that

$$\rho_\varphi\left(\frac{f_n}{\varepsilon}\right) \leq N_\varphi\left(\frac{f_n}{\varepsilon}\right) < \frac{1}{\varepsilon}\cdot \varepsilon^2 = \varepsilon.$$

Hence by (2) $\|f_n\|_\varphi \leq \varepsilon$ and so $f_n \to 0$ in $\|\cdot\|_\varphi$-metric also.

Conversely let $\|f_n\|_\varphi \to 0$ as $n \to \infty$. Then for $1 \geq \varepsilon > 0$ there is n_1 ($= n_1(\varepsilon)$) such that $n \geq n_1 \Rightarrow \|f_n\|_\varphi < \varepsilon$ and hence $\rho_\varphi\left(\frac{f_n}{\varepsilon}\right) < \varepsilon \leq 1$. By definition of the gauge norm, this implies $N_\varphi(f_n) \leq \varepsilon$, and $N_\varphi(f_n) \to 0$ as $n \to \infty$.

A similar argument applies to bounded sets. By the linearity of $L^\varphi(\mu)$, it suffices to verify that the unit balls are the same in both topologies. Thus if $N_\varphi(f_n) \leq 1$, then $\rho_\varphi(f) \leq N_\varphi(f) \leq 1$, as noted above. So $\|f\|_\varphi \leq 1$. On the other hand, since $\rho_\varphi(\alpha f) \leq \max(\alpha, 1)\|f\|_\varphi$, $\|f\|_\varphi \leq 1 \Rightarrow N_\varphi(f) \leq 1$ on taking $\alpha = 1$. Consequently the unit balls are the same in both topologies. ∎

Remark. If $\varphi(x) = |x|^p$, $0 < p < 1$, then the Lebesgue spaces $L^p(\mu)$ are complete metric (but *not* locally convex) spaces. So from the classical work, we can conclude that in general $L^\varphi(\mu)$ (i.e., φ is not also convex) need not be locally convex so that, for instance, its unit ball (or any neighborhood of the origin) need not contain a convex set. Indeed, if $\mu(\Omega) < \infty$ and $\varphi(x) = |x|(1 + |x|)^{-1}$, then $L^\varphi(\mu) = L^0(\mu)$ is included here. Thus, unlike the Banach space case, in $\{L^\varphi(\mu), \|\cdot\|_\varphi\}$ for linear functionals boundedness and continuity are different. In view of this, most of the structural results of Chapter III have two versions and $(L^\varphi(\mu))^*$ can be trivial, i.e., $\{0\}$.

To gain some facility with analysis on these spaces, also called F-normed Orlicz spaces, we include a few results. As before, let $M^\varphi = \{f \in L^0(\mu) : \rho_\varphi(\alpha f) < \infty,$ for all $\alpha > 0\}$. It is clear that $M^\varphi \subset L^\varphi(\mu)$ and it is also easy to see that M^φ is a closed linear subset. Using simple modifications of the methods and ideas of Sections III.4 and III.5, the following result can be established. We omit the details, which are tedius but similar to the earlier work.

Proposition 3. *Let* φ *be continuous and* $\varphi(x) > 0$ *for* $x > 0$. *Then*

(i) *simple functions are dense in* M^φ *and the latter is separable if* (Ω, Σ, μ) *is also separable;*

(ii) *for each* $f \in M^\varphi - \{0\}$, $\rho_\varphi\left(\frac{f}{\|f\|_\varphi}\right) = \|f\|_\varphi$, *and the converse holds if also* $\lim\limits_{x \to \infty} \varphi(x) = \infty$, μ *is diffuse on a set of positive measure, and* φ *is* Δ_2-*regular (i.e.,* $\varphi(2x) \leq k_0 \varphi(x)$, $x \geq x_0 \geq 0$); *and*

(iii) *supposing* φ *to be as in (ii) [i.e., continuous,* $\varphi(x) \neq 0$ *for* $x > 0$ *and* $\varphi(x) \nearrow \infty$ *as* $x \nearrow \infty$ *], we have* $\lim\limits_{\mu(E) \to 0} \|f\chi_E\|_\varphi = 0$ *for each* $f \in L^\varphi(\mu)$ *iff* $M^\varphi = L^\varphi(\mu)$. *[This may again be called absolute continuity of the* F-*norm* $\|\cdot\|_\varphi$, *which reduces to the earlier concept when* φ *is a Young function.]*

We now present some typical analogs of certain comparison results of Chapter V for convenience and applications. Following the relations introduced in Section II.2 with Young functions, we may restate them for the φ-functions as follows:

Definition 4. Let φ_1, φ_2 be a pair of φ-functions. Then $\varphi_1 \succ \varphi_2$ (φ_1 is stronger than φ_2) if for some constants $0 < c, k < \infty$ and $x_0 \geq 0$,

$$\varphi_2(x) \leq c\varphi_1(kx), \qquad x \geq x_0; \tag{3}$$

and $\varphi_1 \succ\!\!\!\succ \varphi_2$ (φ_1 is completely stronger than φ_2) if for each $\varepsilon > 0$,

$$\varphi_2\left(\tfrac{x}{\varepsilon}\right) \leq K\varphi_1(x), \quad x \geq x_0(\varepsilon) \text{ and } K = K_\varepsilon > 0. \tag{4}$$

Thus $\varphi_1 \sim \varphi_2$ if $\varphi_1 \succ \varphi_2$ and $\varphi_2 \succ \varphi_1$, so that for some $0 < c_i, k_i < \infty$, $i = 1, 2$, we have

$$c_1\varphi_1(k_1 x) \leq \varphi_2(x) \leq c_2\varphi_1(k_2 x), \qquad x \geq x_0. \tag{5}$$

Note that $\varphi \in \Delta_2$ iff $\varphi \rightarrowtail \varphi$ and $\varphi_1 \rightarrowtail \varphi_2 \Rightarrow \varphi_1 \succ \varphi_2$ from the definition itself.

Thus following the earlier work in Chapters II and V we can establish the assertions of the next result.

Proposition 5. *Let φ_i, $i = 1, 2$, be φ-functions and (Ω, Σ, μ) a finite measure space. If $\{L^{\varphi_i}(\mu), \|\cdot\|_\varphi\}$, $i = 1, 2$, are the corresponding spaces, consider the following statements:*

(i) $\varphi_1 \succ \varphi_2$,   (i)' $\varphi_1 \rightarrowtail \varphi_2$,

(ii) $L^{\varphi_1}(\mu) \subset L^{\varphi_2}(\mu)$,   (ii)' $L^{\varphi_1}(\mu) \subset M^{\varphi_2}$,

(iii) $\lim_{\|f\|_{\varphi_1} \to 0} \|f\|_{\varphi_2} = 0$,   (iii)' $\lim_{\rho_{\varphi_1}(f) \to 0} \|f\|_{\varphi_2} = 0$,

(iv) *each $\|\cdot\|_{\varphi_1}$-bounded set is $\|\cdot\|_{\varphi_2}$-bounded,* (iv)' *each $\|\cdot\|_{\varphi_1}$-bounded set is ρ_{φ_2}-bounded,*

(v) $M^{\varphi_1} \subset M^{\varphi_2}$,   (v)' $\lim_{\rho_{\varphi_2}(f) \to \infty} \|f\|_{\varphi_1} = +\infty$.

Then (i) $\Rightarrow \{(ii)-(v)\}$, and (i)' $\Rightarrow \{(ii)'-(v)'\}$. If moreover, φ_i are continuous, $\lim_{x \to \infty} \varphi_i(x) = +\infty$, and μ is diffuse on a set of positive measure, all the statements (i)-(v) and (i)'-(v)' are respectively mutually equivalent in their collections.

The results (i)–(v) for the relation '$\succ$' are due to Matuszewska and Orlicz (1961), and the corresponding statements (i)'–(v)' for '$\rightarrowtail$' are verified by Ren (1986$_c$). The details are completely analogous to those of Section V.3 which will not be repeated here. It should be noted that $\varphi_1 \succ \varphi_2 \not\Rightarrow \|f\|_{\varphi_2} \leq C\|f\|_{\varphi_1}$, $f \in L^{\varphi_1}(\mu)$, for some $0 < C < \infty$ even if $\mu(\Omega) < \infty$ which is in contrast to the case that φ_i are Young functions [and $\mu(\Omega) < \infty$].

As a consequence of the above result we have the following.

Corollary 6. *If (Ω, Σ, μ) is a finite diffuse measure space, φ a continuous increasing φ-function with $\lim_{x \to \infty} \varphi(x) = +\infty$, then the following are equivalent statements:*

(i) $\varphi \in \Delta_2$,

(ii) $L^\varphi(\mu) = M^\varphi$,

(iii) $\lim_{\rho_\varphi(f) \to 0} \|f\|_\varphi = 0$,

(iv) $\lim_{\rho_\varphi(f) \to \infty} \|f\|_\varphi = +\infty$, *and*

(v) $\sup\{\|f\|_\varphi : f \in S\} < \infty$ *iff* $\sup\{\rho_\varphi(f) : f \in S\} < \infty$.

Note. In many computations, when $\varphi(x) \nearrow \infty$, as $x \nearrow \infty$, φ may be replaced by a strictly increasing φ-function φ_1 such that $\varphi \sim \varphi_1$ (hence $L^\varphi(\mu) = L^{\varphi_1}(\mu)$). This may be seen as follows: Let $x_0 = 0$ and $x_1 = \sup \varphi^{-1}(\{1\})$ where $\varphi^{-1}(\{1\})$ is an interval when φ is not strictly increasing. Similarly let $x_n = \sup \varphi^{-1}(\{n\})$. Consider the polygonal line through $\{(x_n, n), n \geq 0\}$. Let φ_1 be the function which coincides with the above polygonal line for $x \geq 0$, and set $\varphi_1(x) = \varphi(-x)$ for $x < 0$. Then φ_1 is a φ-function and $|\varphi_1(x) - \varphi(x)| \leq 1$, so $\varphi \sim \varphi_1$.

Recall that in Theorem II.2.2(b) we have seen for the N-functions, a relation between φ_1, φ_2 with '$\succ\!\!\succ$,' called *essentially stronger*. We state it for φ-functions as: $\varphi_1 \succ\!\!\succ \varphi_2$ if for each $\varepsilon > 0$

$$\lim_{x \to \infty} \frac{\varphi_1(\varepsilon x)}{\varphi_2(x)} = +\infty. \tag{6}$$

It is clear that $\varphi_1 \succ\!\!\succ \varphi_2 \Rightarrow \varphi_1 \succ\!\!\!-\!\!\prec \varphi_2$ and hence $\varphi_1 \succ \varphi_2$. With (6) we can establish the following relations between $L^{\varphi_i}(\mu)$, $i = 1, 2$.

Proposition 7. *Let (Ω, Σ, μ) be a finite measure space and φ_1, φ_2 be a pair of φ-functions. Consider the statements:*

    (i) *$\varphi_1 \succ\!\!\succ \varphi_2$,*

    (ii) *for each set S in a ball of $L^{\varphi_1}(\mu)$, we have $\lim\limits_{\mu(A) \to 0} \sup\limits_{f \in S} \|f\chi_A\|_{\varphi_2}$*
    $= 0$ where the A's are measurable sets, and

    (iii) *for each sequence $\{f_n, n \geq 1\}$ in a ball of $L^{\varphi_1}(\mu)$ such that $f_n \to$*
    0 in measure, we have $\lim\limits_{n \to \infty} \|f_n\|_{\varphi_2} = 0$.

Then (i) $\Rightarrow \{$(ii) and (iii)$\}$, and if μ is also diffuse and $\varphi_i(x) \nearrow \infty$ as $x \nearrow \infty$, φ_i being continuous, all the three statements are equivalent.

That (i) $\Leftrightarrow$ (iii) under the given conditions was shown by S. V. Lapin (1977) and the part that (i) $\Leftrightarrow$ (ii) was verified by Z. D. Ren (1985$_b$). For the proofs one may assume that the φ_i are strictly increasing on using some equivalent φ-functions as shown in the Note above, if necessary. The converses are shown by an example. The details proceed along familiar lines, although there are several tedious computations. We again omit them here.

To illuminate the distinctions among the generalized, s-convex and the original Orlicz spaces, we include a discussion on their structure. Recall that a set A in a Fréchet space $\mathcal{X}$ is *(metrically) bounded* if it can be enclosed in some ball, and *topologically bounded* if $A \subset aU$ for some $a \, (= a_U \,) > 0$ where U is *any* neighborhood of zero, given by the

linear (metric) topology of $\mathcal{X}$. (These two concepts need not coincide!)
The space is *locally bounded* if it contains at least one *bounded neigh-
borhood* of zero. An F-norm $\|\cdot\|$ is *s-homogeneous* if $\|ax\| = a^s\|x\|$,
for $a \geq 0$ and $x \in \mathcal{X}$. A set $B \subset \mathcal{X}$ is *solid* if $x \in B \Rightarrow \alpha x \in B$ for
$0 \leq \alpha \leq 1$. A solid set B is *pseudoconvex* if there is a number $\alpha_B \geq 0$
such that $B + B \subset \alpha_B B$ where we used the vector addition and scalar
multiplication notation. The smallest such α_B is called the *modulus
of pseudoconvexity*. For the $(L^\varphi, \|\cdot\|_\varphi)$-spaces, we have the following
result.

Proposition 8. *If a Fréchet space* $(\mathcal{X}, \|\cdot\|)$ *is locally bounded,
then there is an s-homogeneous norm* $\|\cdot\|'$ *on* $\mathcal{X}$, $0 < s \leq 1$, *equivalent
to* $\|\cdot\|$. *In particular if* $(L^\varphi(\mu), \|\cdot\|_\varphi)$ *is locally bounded, then there
exists an s-convex* φ*-function* $\tilde{\varphi}$ *equivalent to* φ *so that* $L^\varphi(\mu) = L^{\tilde{\varphi}}(\mu)$,
on (Ω, Σ, μ).

A proof of the main part of this result may be found in S. Rolewicz
(1972), p. 61, and the specialization follows from the last part of Theorem 2. Since there exist $L^\varphi(\mu)$-spaces where φ is not necessarily s-
convex, the significance of local boundedness can be understood. If
$\alpha_0 = \inf\{\alpha_U : U \subset \mathcal{X}$ is a neighborhood of zero$\}$, then one can verify
that $0 < s_0 = \frac{\log \alpha_0}{\log 2}$ works for the last statement of the above proposition, α_U being the modulus of U defined before.

The following statement elaborates the local boundedness.

Proposition 9. *Let* φ *be a* Δ_2*-regular* φ*-function with* $\varphi(x) \nearrow \infty$,
as $x \nearrow \infty$. *Then the generalized function space* $(L^\varphi(\mu), \|\cdot\|_\varphi)$ *is locally
bounded provided* φ *is restricted further to:*

$$r = \inf\{\inf[a > 0 : \varphi(at) \geq \tfrac{1}{2}\varphi(t)] : 0 < t < \infty\} > 0. \tag{7}$$

Proof: We need to show that $L^\varphi(\mu)$ has a bounded neighborhood of
zero. For this it is enough to verify that each set bounded in the
metric topology is absorbed by each neighborhood of zero. Thus if
$U_k = \{f \in L^\varphi(\mu) : \|f\|_\varphi \leq k\}$, a ball, is absorbed in the above sense,
i.e., if $f_n \in U_k$ is any sequence and $\delta_n \searrow 0$, then $\|\delta_n f_n\|_\varphi \to 0$. Because
of the hypothesis, $\varphi \in \Delta_2$ [cf. Corollary 6] is equivalent to showing that
$\rho_\varphi(\delta_n f_n) \to 0$, as $n \to \infty$. We claim that this holds if (7) is true.

Indeed by the same corollary, the metric boundedness is equivalent
to $\{f \in U_k : \rho_\varphi(f) \leq k_0\}$ for some k_0. Now by (7), we have for all $f \in U_k$,

$$\rho_\varphi(rf) = \int\limits_\Omega \varphi(rf)d\mu \le \frac{1}{2}\int\limits_\Omega \varphi(f)d\mu. \tag{8}$$

Hence replacing f by rf and iterating, we get $\rho_\varphi(r^n f) \le 2^{-n}\rho_\varphi(f) \le k_0 2^{-n}$. Given $\varepsilon > 0$, choose n_0 such that $n \ge n_0 \Rightarrow k_0 2^{-n} < \varepsilon$ and $\delta_n \le r^{n_0}$. Then

$$\rho_\varphi(\delta_n f_n) \le \rho_\varphi(r^n f_n) \le k_0 2^{-n} < \varepsilon.$$

This implies $\rho_\varphi(\delta_n f_n) \to 0$ as $n \to \infty$, for any sequence $\{f_n, n\ge 1\} \subset U_k$. Since U_k is a subset of a metric space, sequences suffice, and we are done. ∎

The condition (7) is essentially the best for the conclusion. Several related results for Fréchet spaces are discussed in the book by Rolewicz (1972).

We now discuss briefly another class of $L^\varphi(\mu)$ spaces where φ is a *concave* φ-function. By introducing a suitable norm in this set of measurable functions, we show that it becomes a Banach space, thereby establishing that there exist such spaces with an honest norm even though φ is only concave. For such a class, it is again possible to develop many results analogous to those of Chapters I–VII. For this, a new concept is needed. It is the nonincreasing rearrangements of scalar functions which we now introduce. These are closely related to Lorentz spaces, to be discussed in Section 3 below.

Recall that we can attach a measure function λ_f with each measurable $f : \Omega \to \mathbb{R}$ (or $\mathbb{C}$) as $\lambda_f(y) = \mu[\omega : |f(\omega)| > y]$ which is thus nonincreasing. Its inverse, denoted f^*, is called the *decreasing rearrangement* of f. Thus

$$f^*(t) = \inf\{y : \lambda_f(y) \le t\}, \qquad t \ge 0, \tag{9}$$

with $\inf(\emptyset) = +\infty$, as usual. If $f_i : \Omega \to \mathbb{R}$ (or $\mathbb{C}$), $i = 1, 2$, measurable, $|f_1| \le |f_2|$, then $f_1^* \le f_2^*$.

Using the classical image measure results which are popular in probability theory (under the name Fundamental Law of Probability, c.f., e.g., Rao (1984), p. 19) we get in the present context

$$\int\limits_\Omega \varphi(f)d\mu = \int\limits_{\mathbb{R}^+} \varphi(t)d(\mu\circ f^{-1}) = \int\limits_{\mathbb{R}^+} \mu(f^{-1}(t,\infty))d\varphi(t)$$

$$= \int\limits_{\mathbb{R}^+} \lambda_f(t)d\varphi(t) = \int\limits_0^{\mu(\Omega)} \varphi(f^*)(t)dt. \tag{10}$$

Here we also used the integration by parts for the Lebesgue-Stieltjes integrals and (9). From this one can verify that for all $A \in \Sigma$ and f, g measurable,

$$\int_A |f| d\mu \le \int_0^{\mu(A)} f^*(t) dt, \qquad \int_A |fg| d\mu \le \int_0^{\mu(A)} (f^* g^*)(t) dt. \qquad (11)$$

If now φ is also concave, $\varphi(0) = \varphi(0^+) = 0$, then $(-\varphi$ is a Young function) by Theorem I.3.1 we get, φ' denoting the (left) derivative of φ,

$$\varphi(x) = \int_0^x \varphi'(t) dt, \qquad (12)$$

where $\varphi' \ge 0$ is nonincreasing and left continuous. For such a concave φ we define a functional $\| \cdot \|'_\varphi : f \to \overline{I\!R}^+$ by

$$\|f\|'_\varphi = \int_{I\!R+} \varphi(\lambda_f(y)) dy, \qquad (13)$$

for any measurable scalar f. It is clear that $\|\alpha f\|'_\varphi = |\alpha| \|f\|'_\varphi \ge 0$, a consequence of change of variables. Now if $\varphi(x) > 0$ for $x > 0$, then $\|f\|'_\varphi = 0 \Leftrightarrow f = 0$ a.e. To see that (13) also satisfies the triangle inequality, we use the fact that the difference quotient is nonincreasing in (12). More particularly if $\alpha < \beta \le \delta$, and $\alpha \le \gamma < \delta$, then since $\varphi'(\cdot)$ is decreasing, we get

$$\frac{\varphi(\beta) - \varphi(\alpha)}{\beta - \alpha} \ge \frac{\varphi(\delta) - \varphi(\gamma)}{\delta - \gamma}. \qquad (14)$$

Hence if $A_1 A_2 \in \Sigma$, and $\mu(A_i) < \infty$, then one has, for $\nu(A_i) = \varphi(\mu(A_i))$, the following useful inequality [called *strong subadditivity* of $\nu(\cdot)$] from (14). Thus let $\alpha = \mu(A_1 \cap A_2)$, $\beta = \mu(A_2)$, $\gamma = \mu(A_1)$ and $\delta = \mu(A_1 \cup A_2)$ with $\mu(A_2 \triangle A_1) > 0$. Then

$$\frac{\nu(A_2) - \nu(A_1 \cap A_2)}{\mu(A_2) - \mu(A_1 \cap A_2)} \ge \frac{\nu(A_1 \cup A_2) - \nu(A_1)}{\mu(A_1 \cup A_2) - \mu(A_1)}.$$

Since the denominators are equal, the above inequality becomes

$$\nu(A_1 \cup A_2) + \nu(A_1 \cap A_2) \le \nu(A_1) + \nu(A_2). \qquad (15)$$

But this always holds if $\nu(A_i) = +\infty$ or $\mu(A_1 \triangle A_2) = 0$, and so (15) holds for *all* $A_i \in \Sigma$, $i = 1, 2$. Since clearly $0 \le f_n \nearrow f$ a.e. $\Rightarrow \|f_n\|'_\varphi \uparrow \|f\|'_\varphi$ [this is the Fatou property of $\|\cdot\|'_\varphi$] by the monotone convergence theorem, it suffices to verify the triangle inequality for simple functions. It is enough to consider the following type of simple functions:

$$f = \sum_{i=1}^{n} i\chi_{A_i}, \qquad g = \sum_{j=1}^{m} j\chi_{B_j}, \tag{16}$$

where $\{A_i\}_1^n$ and $\{B_j\}_1^m$ are disjoint measurable sets. This is because if $\tilde{f}, \tilde{g}$ are simple functions with rational coefficients, taking some $A_i[B_i]$ to be empty, $f = k_1\tilde{f}$, $g = k_2\tilde{g}$ for some integers k_1, k_2. But the triangle inequality with such f, g in (16) implies the same for $\tilde{f}$ and $\tilde{g}$ since

$$\|\tilde{f} + \tilde{g}\|'_\varphi = \left\| \frac{f}{k_1} + \frac{g}{k_2} \right\|'_\varphi = (k_1 k_2)^{-1} \|k_2 f + k_1 g\|'_\varphi$$

$$\le (k_1 k_2)^{-1} [k_2 \|f\|'_\varphi + k_1 \|g\|'_\varphi] = \|\tilde{f}\|'_\varphi + \|\tilde{g}\|'_\varphi.$$

Then for a general pair f, g there exist f_n, g_n simple with rational coefficients such that $f_n \to f$, $g_n \to g$ and the inequality for f, g follows by the Fatou property. Thus consider f_0, g_0 of the form (16). Let $A_0 = \left(\bigcup_{i=1}^{n} A_i \right)^c$ and $B_0 = \left(\bigcup_{j=1}^{m} B_j \right)^c$. Then letting $C_i = \bigcup_{\ell=i}^{n} A_\ell$, $D_j = \bigcup_{\ell=j}^{m} B_\ell$ and $E_k = \bigcup \{A_i \cap B_j : i + j \ge k\}$, the desired triangle inequality for the f_0, g_0 is equivalent to proving:

$$\sum_{k=1}^{m+n} \nu(E_k) \le \sum_{i=1}^{n} \nu(C_i) + \sum_{j=1}^{m} \nu(D_j). \tag{17}$$

If $m = 1 = n$, then (17) reduces, since then $C_1 = A, D_1 = B$ and $E_1 = A \cup B, E_2 = A \cap B$, to (15) which was shown to be true. Next we assume that (17) is true for m and n, and using (double) induction it is shown that (17) always holds. We leave this (somewhat tedius) computation to the reader. Thus (13) gives a (semi-)norm and a norm if $\varphi(x) > 0$ for $x > 0$. Let $(L^\varphi(\mu), \|\cdot\|'_\varphi)$ be the class of all scalar measurable functions f on (Ω, Σ, μ), for which $\|f\|'_\varphi < \infty$. Then we have shown that it is a normed linear space. In fact, more is true. Let $\mathcal{A}_\varphi(\mu)$ be the class of σ-additive scalar set functions ζ on Σ vanishing on μ-null sets such that $\|\zeta\| = \|\zeta\|(\Omega)$, where

$$\|\zeta\|(\Omega) = \sup\{|\zeta(A)|(\nu(A))^{-1} : 0 < \mu(A) < \infty, A \in \Sigma\} < \infty. \tag{18}$$

Here $\nu(A) = \varphi(\mu(A))$ as in (15). We have the following representation, due to M. S. Steigerwalt and A. J. White (1971).

Theorem 10. *For each concave increasing φ-function φ, the set $(L^\varphi(\mu), \|\cdot\|'_\varphi)$ is a Banach space. Moreover, if $\varphi(t) \nearrow \infty$, as $t \nearrow \infty$, or μ is σ-finite, then $(L^\varphi(\mu))^*$ is isometrically isomorphic to $(\mathcal{A}_\varphi(\mu), \|\cdot\|)$ of (18). Explicitly, for each $x^* \in (L^\varphi(\mu))^*$ there is a unique $\zeta \in \mathcal{A}_\varphi(\mu)$ such that*

$$x^*(f) = \int_\Omega f d\zeta, \qquad f \in L^\varphi(\mu), \tag{19}$$

$$\|x^*\| = \sup\{|x^*(f)| : \|f\|'_\varphi \le 1\} = \|\zeta\|, \ of\,(18). \tag{20}$$

If moreover $\lim_{u \to \infty} \frac{\varphi(u)}{u} = 0$ *(and $\varphi(u) \nearrow \infty$ as $u \nearrow \infty$) then (19) becomes*

$$x^*(f) = \int_\Omega f g d\mu, \qquad f \in L^\varphi(\mu), \tag{19}'$$

for a unique measurable g such that

$$\|x^*\| = \|g\|''_\varphi = \sup \left\{ \frac{1}{\nu(A)} \int_A |g| d\mu : 0 < \mu(A) < \infty, A \in \Sigma \right\}. \tag{21}$$

Sketch of Proof: Only completeness of $(L^\varphi(\mu), \|\cdot\|'_\varphi)$ has to be verified. Let $\{f_n, n \ge 1\} \subset L^\varphi(\mu)$ be a Cauchy sequence. Then using (12) and the image measure property (10), we have for any $x > 0$,

$$x\varphi(\lambda_{f_n-f_m}(x)) \le \int_0^x \varphi(\lambda_{f_n-f_m}(t)) dt \le \|f_n - f_m\|'_\varphi, \text{ by (13).}$$

Since $x > 0$ is arbitrary, this shows that $\{f_n, n \ge 1\}$ is Cauchy in measure and hence $f_n \to f$ in measure. So $\lambda_{f_n-f}(t) \to 0$, as $n \to \infty$, for each $t > 0$. Consider

$\|f_n - f\|'_\varphi$

$$= \int_0^\infty \varphi(\lambda_{f_n-f}(t)) dt \le 2 \varlimsup_{N \nearrow \infty} \varlimsup_{\delta \searrow 0} \left[\varlimsup_{m \nearrow \infty} \int_\delta^N \varphi(\lambda_{f-f_m}(t)) dt \right.$$

$$\left. + \varlimsup_{m \nearrow \infty} \int_\delta^N \varphi(\lambda_{f_m-f_n}(t)) dt \right], \text{ since } \lambda_{f+g}(t) \le 2\left(\lambda_f\left(\frac{f}{2}\right) + \lambda_g\left(\frac{g}{2}\right)\right),$$

$$\le 2 \varlimsup_{N \nearrow \infty} \varlimsup_{\delta \searrow 0} \left[(N-\delta) \varlimsup_{m \nearrow \infty} \varphi(\lambda_{f-f_m}(\delta)) dt + \varlimsup_{m \nearrow \infty} \|f_n - f_m\|'_\varphi \right]$$

$$= 2 \varlimsup_{m \nearrow \infty} \|f_n - f_m\|'_\varphi \to 0, \text{ as } n \to \infty.$$

Hence $\|f\|'_\varphi \le \|f - f_n\|'_\varphi + \|f_n\|'_\varphi < \infty \Rightarrow f \in L^\varphi(\mu)$ and that $\|f_n - f\|'_\varphi$
$\to 0$. So $(L^\varphi(\mu), \|\cdot\|'_\varphi)$ is complete.

If x^* is defined as in (19), then

$$|x^*(f)| \le \int_\Omega |f| d|\zeta| = \int_0^\infty |\zeta|(\omega : |f(\omega)| > t) dt$$

$$\le \|\zeta\| \int_0^\infty \varphi(\mu(\omega : |f(\omega)| > t)) dt, \text{ by (18)},$$

$$= \|\zeta\| \|f\|'_\varphi, \text{ by (13)}.$$

Hence $x^* \in (L^\varphi(\mu))^*$, and $\|x^*\| \le \|\zeta\|$. The opposite inequality is also
simple:

$$\|x^*\| = \sup\{|x^*(f)| : \|f\|'_\varphi = 1\}$$
$$\ge \sup\{|x^*(\chi_A)|/\nu(A) : 0 < \mu(A) < \infty, A \in \Sigma\}$$
$$= \sup\{|\zeta(A)/\nu(A)| : 0 < \mu(A) < \infty, A \in \Sigma\} = \|\zeta\|.$$

Thus (20) holds.

To get the representation, let $\zeta(A) = x^*(\chi_A)$ for $A \in \Sigma$, $\mu(A) < \infty$.
Then one verifies that $\zeta(\cdot)$ is additive, and $|\zeta(A)| \le \|x^*\| \cdot \|\chi_A\|'_\varphi$.
From this one gets easily that ζ is σ-additive and $\zeta \in \mathcal{A}_\varphi(\mu)$. Then
the representation (19) holds for simple functions. The first part also
implies easily, under the present hypothesis, that simple functions are
dense so that (19) holds. The Radon-Nikodým theorem further shows
that $g = \frac{d\zeta}{d\mu}$ is well defined (μ being σ-finite) and one can check without
difficulty that (21) holds and (19) becomes (19)′, as asserted. ∎

The following relation between the norm (13) and the decreasing
rearrangement function f^* of f is useful in some computations.

Proposition 11. *If $\varphi(t) \nearrow \infty$ as $t \nearrow \infty$, then $\|f\|'_\varphi = \int_0^\infty \varphi'(t) f^*(t) dt$,
$f \in L^\varphi(\mu)$, where φ' is the left derivative of φ as usual.*

Proof: It suffices to establish this formula for simple functions. Thus
let $f \ge 0$, and $0 = a_0 < a_1 < \cdots < a_n < \infty$,

$$f = \sum_{i=1}^n a_i \chi_{A_i} = \sum_{i=1}^n (a_i - a_{i-1}) \chi_{B_i}, \qquad B_i = \bigcup_{j=i}^n A_j, \, 1 \le i \le n.$$

Then $\lambda_f(t) = \sum\limits_{i=1}^{n} \mu(B_i)\chi_{[a_{i-1},a_i)}(t)$, and (13) becomes:

$$\|f\|'_\varphi = \sum_{i=1}^{n}(a_i - a_{i-1})\varphi(\mu(B_i))$$

$$= \sum_{i=1}^{n} a_i(\nu(B_i) - \nu(B_{i+1})), \text{ with } B_{n+1} = \emptyset,$$

$$= \sum_{i=1}^{n} a_i \int_{\mu(B_{i+1})}^{\mu(B_i)} \varphi'(t)dt = \int_{0}^{\infty} \varphi'(t)f^*(t)dt,$$

since $f^* = \sum a_i \chi_{[\mu(B_{i+1}),\mu(B_i))}$. This gives the result. ∎

Several other properties of this space were considered by the authors of the above referenced paper. This indicates certain other aspects of the spaces with such φ-functions.

We shall now present an application when φ is a Young function but does not satisfy a Δ_2-type condition.

If (Ω, Σ, P) is a probability space (i.e., $P(\Omega) = 1$) and T is a topological set, then a curve $X : T \to L^0(\Omega, \Sigma, P) = L^0(P)$, is called a *stochastic* or *random function* (*process* if $T \subset \mathbb{R}$ and *field* if $T \subset \mathbb{R}^n$, $n > 1$). Alternately, this may be stated as $X : T \times \Omega \to \mathbb{R}$ [or $\mathbb{C}$] such that $X(t,\cdot) \in L^0(P)$, a *random variable* for each t in T. If $\omega \in \Omega$ is fixed, $X(\cdot,\omega)$ is called a *sample function*, as it describes the position of the "particle" ω at the instant t. The problem of interest here is (using the topology of T) to study the continuity properties of the path $X(\cdot,\omega)$ for each ω. This is the deterministic problem and can be considered only for a relatively small collection of ω's. So one enlarges the question (by slightly weakening the requirement) to find the continuity of $X(\cdot,\omega)$ for almost all $\omega \in \Omega$, which uses the existence of the probability measure on (Ω, Σ). In this form, a classical result due to A. N. Kolmogorov gives "good" sufficient conditions for the continuity of almost all sample paths as follows: Suppose there are constants $\alpha > 0$, $\beta > 0$, and $K > 0$ such that for any pair of points t_1, t_2, of $T \subset \mathbb{R}$, and $\varepsilon > 0$

$$P(\omega : |X(t_1,\omega) - X(t_2,\omega)| > \varepsilon) \leq K\varepsilon^{-\beta}|t_1 - t_2|^{1+\alpha}, \qquad (22)$$

then for almost all ω ($\in \Omega$), $X(\cdot,\omega)$ is a continuous function on T. [For a proof of the result in even a slightly more general form, see, e.g. Rao

(1979), p. 189.] This type of theorem has motivated the function theory aspect of stochastic analysis and numerous other regularity properties. Let T be a metric space and $(T, \mathcal{T})$ be the Borelian space and $\mathcal{P}(T)$ be the set of all probability measures on $(T, \mathcal{T})$. Then when is it true that $\int_T \exp(|X(t, \omega)|/\alpha)^q d\mu(t) < \infty$ for almost all $\omega \in \Omega$ and $\mu \in \mathcal{P}(T)$? For Gaussian as well as certain other processes, this problem has been investigated and important results were obtained by M. B. Marcus and G. Pisier (1985), M. Talagrand (1990) and others (see the references to other works of these authors and related results in these papers). Noting that $\Phi_p(x) = (\exp|x|^p) - 1$ is a φ-function for $p > 0$ (and a Young function without Δ_2-property for $p \geq 1$ and not convex for $p < 1$), the Orlicz space theory played a key part in proving various results. We shall formulate a useful result to give the flavor. Let $d : T \times T \to I\!\!R^+$ be a metric function on T. Then: *Problem.* Suppose that $X : T \times \Omega \to \mathbb{C}$ is jointly $\mu \otimes P$-measurable and for each t, u in T, $N_{\Phi_p}(X(t, \cdot) - X(u, \cdot)) \leq d(t, u)$ where $N_{\Phi_p}(\cdot)$ is the gauge norm of $L^{\Phi_p}(\mu)$, $\mu \in \mathcal{P}(T)$. Then under what (additional) conditions can we conclude that $X(\cdot, \omega) \in L^{\Phi_p}(\mu)$ for a.a.(ω) and all $\mu \in \mathcal{P}(T)$? I.e., $X(\cdot, \omega) \in \cap\{L^{\Phi_p}(\mu) : \mu \in \mathcal{P}(T)\}$? (Compare also this with the work of Chapter VIII.) The same question can be asked if Φ is any N-function replacing Φ_p here.

Since T is taken as a metric space, we consider $N(T, d, r)$, the *covering number* of T, i.e., the smallest number of open balls of radius $r > 0$ in the metric d, with centers in T, that cover T. Then we have the solutions to the above discussed problems as follows.

Theorem 12. *Let Φ be an N-function and $X : T \to L^{\Phi}(P)$ be a stochastic function such that for each t, u in T one has*

$$\int_{\Omega} \Phi\left(\frac{X(t, \cdot) - X(u, \cdot)}{d(t, u)}\right) dP \leq 1. \tag{23}$$

Then $X(\cdot, \omega)$ can be taken to be continuous for a.a.(ω) whenever

$$\int_{\Omega} \text{ess.} \sup_{t, u \in T} (X(t, \cdot) - X(u, \cdot)) dP \leq K \int_0^D \Phi^{-1}(N(T, d, x)) dx < \infty, \tag{24}$$

where $K > 0$ is a constant and D is the diameter of T, i.e., $D = \sup\{d(x, y) : x, y \text{ in } T\}$.

Here the essential supremum can be taken to be measurable since P is a finite (hence localizable) measure. The result extends Kolmogorov's theorem in many ways. A complete solution to the next problem can be given in the following case.

Theorem 13. *Let (T, d) be a compact metric space. Then the problem has a solution iff the covering numbers $N(T, d, x)$ satisfy for $0 < p < p' < \infty$,*

$$\sup_{x>0} x(\log N(T, d, x))^{\frac{1}{p} - \frac{1}{p'}} < \infty. \tag{25}$$

The proofs of both these results involve intricate estimates of the sizes of sets in the Orlicz space under study. Complete details are to be found in Talagrand (1990), and we shall omit them here since they additionally involve several probabilistic notions.

It should also be noted that in the same way, the Orlicz space theory plays a key role in differentiation theory and martingale convergence with directed index sets, using certain generalizations of the known "Vitali conditions." For a substantial contribution in this direction, we refer to M. Talagrand (1986). This work also fits in here, but for space reasons, we cannot include it. Instead, we indicate in the next section, another direction of the theory of Orlicz spaces.

10.2. *Spaces of set and operator functions*

In all the preceding work we have considered point functions, i.e., those mappings that are defined on a point set Ω, which may be real or complex, and on occasion vector valued. But there are at least two closely related functions which are not defined on point sets. The first one is the space of (additive) set functions for which the measurability questions can be reduced to a minimum. The second one, of great interest in modern analysis, is to consider spaces formed of algebras of (linear but perhaps unbounded) operators especially on a Hilbert space. We now indicate these in the context of Orlicz spaces. We first consider convex Φ (i.e., Young) functions.

If Σ is an algebra of subsets of Ω, $\mu : \Sigma \to \overline{I\!R}^+$ is additive, and $\nu : \Sigma \to \overline{I\!R}$ is another additive function, then ν is termed μ-continuous if $\lim_{\mu(A) \to 0} |\nu(A)| = 0$, for $A \in \Sigma$. A partition π of a set $E \in \Sigma$ is a finite disjoint collection of sets $\{E_i, 1 \leq i \leq n\} \subset \Sigma$ such that $\bigcup_{i=1}^{n} E_i \subseteq E$.

The collection of all partitions is partially ordered by refinement, i.e., $\pi_1 \prec \pi_2$ if each element of π_1 is the union of some elements of π_2. if Φ is a Young function, then a $\nu : \Sigma \to \overline{I\!R}$ is said to be of Φ-*bounded variation* on E, if $I_\Phi(\nu, E) < \infty$, where, for $E \in \Sigma$,

$$I_\Phi(\nu, E) = \sup_\pi \left\{ \sum_{i=1}^n \Phi\left(\frac{|\nu(E_i)|}{\mu(E_i)}\right)\mu(E_i) : \{E_i\}_1^n = \pi \subset \Sigma, E_i \subset E \right\}, \quad (1)$$

the supremum being taken over all partitions of E. It can be verified that the supremum in (1) can be replaced by the (Moore-Smith) limit as partitions are refined. [If $\Phi(x) = |x|$, (1) is the *ordinary (= Vitali) variation* of ν and μ does not come into the picture. This point has some interest since generally, the Φ-variation also depends on μ.] If $E = \Omega$, then we denote $I_\Phi(\nu, \Omega)$ simply by $I_\Phi(\nu)$.

Definition 1. For each Young function Φ, the space $V^\Phi(\mu)$ is the class of additive set functions ν such that

$$V^\Phi(\mu) = \left\{ \nu : \Sigma \to I\!R \mid I_\Phi\left(\frac{\nu}{k}\right) < \infty \text{ for some } k = k_\nu < \infty \right\}; \quad (2)$$

and let the functional $N_\Phi(\cdot) : V^\Phi(\mu) \to \overline{I\!R}^+$ be given by

$$N_\Phi(\nu) = \inf\left\{ k > 0 : I_\Phi\left(\frac{\nu}{k}\right) \leq 1 \right\}. \quad (3)$$

Clearly it is also possible to consider $L^\Phi(\mu)$, on (Ω, Σ, μ), of point functions $f : \Omega \to \overline{I\!R}$, such that $\int_\Omega \Phi\left(\frac{f}{k}\right)d\mu < \infty$ for some $k < \infty$ where the integral for a finitely additive μ is as in Chapter IV. Here we define the integral first for simple functions f_n and then for all f which are limits in μ-measure of simple functions such that $\int_\Omega |f_n - f_m|d\mu \to 0$. [We follow Dunford-Schwartz (1958), Section III.2.] If $N_\Phi(f) = \inf\{k > 0 : \int_\Omega \Phi\left(\frac{f}{k}\right)d\mu \leq 1\}$, then $(L^\Phi(\mu), N_\Phi(\cdot))$ is easily seen to be a normed linear space, but it is incomplete if μ is merely finitely additive. It can be seen that $N_\Phi(\cdot)$ of (3) is the gauge norm. One may also define the Orlicz norm:

$$\|\nu\|_\Phi = \sup_\zeta \left\{ \sup_\pi \sum_{i=1}^n \frac{|\nu(E_i)| \, |\zeta(E_i)|}{\mu(E_i)} : N_\Psi(\zeta) \leq 1 \right\}, \quad (4)$$

where Ψ is the complementary Young function to Φ. A systematic study of the $V^\Phi(\mu)$ spaces, even when ν takes values in a Banach space, was made by J. J. Uhl (1967). We follow his treatment, and present a few key results.

It can be verified that the functionals (3) and (4) are related by

$$N_\Phi(\nu) \le \|\nu\|_\Phi \le 2N_\Phi(\nu). \tag{5}$$

The spaces $V^\Phi(\mu)$ and $L^\Phi(\mu)$ are connected by the following fundamental:

Theorem 2. *The set $(V^\Phi(\mu), N_\Phi(\cdot))$ is always a Banach space while $(L^\Phi(\mu), N_\Phi(\cdot))$ is a normed linear space if in (Ω, Σ, μ), μ is only finitely additive (but need not be complete). Moreover, if $\lambda : L^\Phi(\mu) \to V^\Phi(\mu)$ is defined as*

$$(\lambda f)(A) = \int_A f d\mu, \qquad A \in \Sigma, \ \mu(A) < \infty, \ f \in L^\Phi(\mu), \tag{6}$$

then (λf) is additive and is an element of $V^\Phi(\mu)$. In fact

$$N_\Phi(\lambda f) = N_\Phi(f), \qquad f \in L^\Phi(\mu), \tag{7}$$

so that $f \mapsto \lambda f$ is an isometric injunction into $V^\Phi(\mu)$. The mapping λ is onto if (Ω, Σ, μ) is a localizable (or σ-finite) measure space and Φ is an N-function.

It is clear that (λf) is an additive set function on $\Sigma_0 = \{A \in \Sigma : \mu(A) < \infty\}$. If μ is σ-additive, then one gets the Φ-variation of λf to be bounded by $\int_\Omega \Phi(f) d\mu$ for simple functions using Jensen's inequality. However, in the present case (μ being finitely additive), an alternative argument is needed. With the Orlicz norm (4), using Young's inequality, one shows that $\|\lambda f\|_\Phi \le 2N_\Phi(f)$, and then establishes that $I_\Phi(\frac{\lambda f}{k}, A) = \int_A \Phi(\frac{|f|}{k}) d\mu$. A further argument shows that (7) holds. Here the extension is obtained with the definition of the integral relative to (the additive) μ. The details can be found in Uhl (1967). Taking $\Phi(x) = |x|$, so that $V^\Phi(\mu) = ba(\Omega, \Sigma)$, which does not depend on μ, it is clear that $\lambda(L^1(\mu)) \underset{\ne}{\subseteq} V^1(\mu)$ properly even if μ is σ-additive. The classical theory shows that $\lambda(L^1(\mu)) \underset{\ne}{\subseteq} ca(\Omega, \Sigma)$, and hence the injection is generally proper. That $V^\Phi(\mu)$ is a Banach space is also not very hard.

To analyze the structure of the adjoint space $(V^\Phi(\mu))^*$, it will be useful to consider the subspace of $V^\Phi(\mu)$ determined by *simple set functions*, i.e., those of the form $\sum_{i=1}^{n} a_i \mu(E_i \cap \cdot)$. If $\nu \in V^\Phi(\mu)$, and $\pi = \{E_i, 1\leq i\leq n\}$ is a partition, then $\nu_\pi = \sum_{i=1}^{n}(\nu(E_i)/\mu(E_i))\mu(E_i \cap \cdot) \in V^\Phi(\mu)$, and is called the *projection* of ν on the partition π. If $\mathcal{S}^\Phi$ is the closed subspace of step functions of $(V^\Phi, N_\Phi(\cdot))$ and $\widetilde{V}^\Phi = \{\nu \in V^\Phi(\mu) : I_\Phi(\nu) < \infty\}$, then, as in the point function case, we have $\mathcal{S}^\Phi \subset \widetilde{V}^\Phi \subset V^\Phi$ with equality when $\Phi \in \Delta_2$. Here $\mathcal{S}^\Phi$ plays the same role as $\mathcal{M}^\Phi$ of Chapters III and IV. For each $\nu \in \mathcal{S}^\Phi$ one has $\lim_\pi N_\Phi(\nu - \nu_\pi) = 0$ where ν_π is the projection of ν on π. Considering only the real valued set functions, and ordering $\nu_1 \leq \nu_2$ iff $\nu_1(A) \leq \nu_2(A)$, $A \in \Sigma$, we find that $V^\Phi(\mu)$ is a Banach lattice since $\nu_1 \leq \nu_2$ is a partial ordering with the lattice operations '$\vee$' and '$\wedge$' given by

$$\begin{aligned}\nu_1 \vee \nu_2 &: A \mapsto (\nu_1 \vee \nu_2)(A) = \sup\{\nu_1(E \cap A) + \nu_2(E^c \cap A) : E \in \Sigma\},\\ \nu_1 \wedge \nu_2 &: A \mapsto (\nu_1 \wedge \nu_2)(A) = \inf\{\nu_1(E \cap A) + \nu_2(E^c \cap A) : E \in \Sigma\}.\end{aligned} \tag{8}$$

The abstract Banach lattice theory implies that $V^\Phi(\mu)$, $\mathcal{S}^\Phi$ as well as $(V^\Phi)^*$ are (AB)-lattices.

Extending the methods of Chapter IV, one can establish:

Theorem 3. *If Φ is continuous ($\Phi(x) > 0$ for $x > 0$) then $(\mathcal{S}^\Phi)^* = V^\Psi(\mu)$ where Ψ is complementary to Φ. More explicitly, for each $\nu \in \mathcal{S}^\Phi$ and $F \in V^\Psi(\mu)$, the functional x^* defined by*

$$x^*(\nu) = \lim_\pi \sum_{E_i \in \pi} \frac{\nu(E_i)F(E_i)}{\mu(E_i)}, \tag{9}$$

is in $(\mathcal{S}^\Phi)^*$, and $\|x^*\| = \sup\{|x^*(\nu)| : \|\nu\|_\Phi \leq 1\} = N_\Psi(F)$. The norms $\|\cdot\|_\Phi$ and $N_\Psi(\cdot)$ can be interchanged to $N_\Phi(\cdot)$ and $\|\cdot\|_\Psi$ respectively in these spaces. On the other hand, every x^* in $(\mathcal{S}^\Phi)^*$ is uniquely represented by (9) for an $F \in V^\Psi(\mu)$. [The limit in (9) is always taken in the Moore-Smith sense.]

This result, when the set functions take values in a Banach space, was proved by Uhl (1967). From this representation it follows that $V^\Phi(\mu)$ is reflexive whenever the complementary pair (Φ, Ψ) both obey the Δ_2-condition.

However, using ther lattice structure, a general determination of $(V^{\Phi}(\mu))^*$ itself can be made. The computations in the present case become considerably more complicated. If an x^* in $(V^{\Phi}(\mu))^*$ admitting the representation (9) for some (hence unique) $F \in V^{\Psi}(\mu)$ is called *absolutely continuous*, and one which annihilates $\mathcal{S}^{\Phi}$ *singular*, then Uhl (1966) has obtained:

Theorem 4. *Let Φ be continuous ($\Phi(x) > 0$ for $x > 0$) and $\mu(\Omega) < \infty$. Then each x^* in $(V^{\Phi}(\mu))^*$ admits a unique decomposition as*

$$x^* = x_a^* + x_s^* \tag{10}$$

where x_a^* is absolutely continuous and x_s^* is singular. Moreover, $x^* \in (V^{\Phi}(\mu), N_{\Phi}(\cdot))^*$ implies, with (10), that

$$\|x^*\| = \|x_a^*\| + \|x_s^*\|. \tag{11}$$

While x_a^* is representable as (9), the singular functional x_s^* of $(V^{\Phi}(\mu))^*$ has not been given a reasonable representation. Further work needs to be done, and we have to leave it for interested readers. Uhl's methods are generalizations of those presented in Chapter IV, and some additional ideas and simplification of his representation seems desirable.

We observe that the preceding point of view has important dividends in that one can formulate and prove strong convergence results for nets of martingales of set (and even point) functions. These considerations hold also for Banach space valued set martingales, as shown in Uhl (1969). Such applications are a strong motivation for the preceding study.

Let us now briefly discuss spaces of operators on a Hilbert space. They have both algebraic and vector space structures, playing important roles in modern analysis. We start with the basic treatment of R. Schatten (1960) and include a few results, in the Orlicz space context, to indicate a still underdeveloped area which has rich theoretical and applicational potential.

We start with a complex infinite dimensional separable Hilbert space $\mathcal{H}$ and denote by $B(\mathcal{H})$ the algebra of all bounded linear operators on $\mathcal{H}$ into itself. Of interest in many studies are several subalgebras of $B(\mathcal{H})$. Some of these, useful in applications, have been discussed in I. C. Gokhberg and M. G. Kreĭn (1969). We include an aspect related to Orlicz spaces here. Thus if Φ is a continuous Young function with $\Phi(x) > 0$ for $x > 0$, and $A \in B(\mathcal{H})$, let us consider the canonical polar decomposition $A = U_A[A]$ where $[A]$ is the square root of the positive operator A^*A, and U_A is unitary. Then by the spectral theorem, $\Phi([A])$ is well defined. [In fact, this is defined even if Φ is only a φ-function on $\mathbb{R} \to \mathbb{R}^+$, and some of the following discussion can be extended to this case; for instance, if $\varphi(x) = |x|^p$, $0 < p < \infty$.] Let k denote the trace of $\Phi([A])$ so that $0 \leq k \leq \infty$, $k = t(\Phi([A]))$. Note that for any $\alpha > 0$, $\Phi(\frac{1}{\alpha}[A])$ is also a positive operator. We may now introduce the following concepts.

Definition 5. (i) For each Φ, let $L^\Phi(\mathcal{H})$ be given by

$$L^\Phi(\mathcal{H}) = \{A \in B(\mathcal{H}) : N_\Phi(A) = \inf\{k > 0 : t(\Phi(\tfrac{1}{k}[A])) \leq 1\} < \infty\}. \quad (12)$$

[If φ is a φ-function, then we consider instead

$$L^\varphi(\mathcal{H}) = \{A \in B(\mathcal{H}) : \|A\|_\varphi = \inf\{k > 0 : t(\varphi(\tfrac{1}{k}[A])) \leq k\} < \infty\}, \quad (13)$$

in analogy with the earlier treatment.]

(ii) If $\mathcal{A} \subset B(\mathcal{H})$ is a closed subalgebra ($B(\mathcal{H})$ with the uniform operator norm) having a topology which is at least as fine as that induced by $B(\mathcal{H})$, then $\mathcal{A}$ is called *approximately tame* whenever the following conditions hold: Let $\{e_1, e_2, \ldots\}$ be an orthonormal basis of $\mathcal{H}$, $\mathcal{H}_n = \mathrm{sp}\{e_1, \ldots, e_n\}$, and $P_n : \mathcal{H} \mapsto \mathcal{H}_n$ be the orthogonal projection. If $Q_n(A) = P_n A P_n$, for $A \in B(\mathcal{H})$, so that $Q_n : B(\mathcal{H}) \to B(\mathcal{H}_n)$ is a projection, then $\cup_n B(\mathcal{H}_n) \subset B(\mathcal{H})$ and $Q_n(A) \to A$, for all $A \in \mathcal{A}$ this limit existing in the topology of $\mathcal{A}$.

The concept of approximate tameness is useful in studying certain homotopy types of infinite dimensional manifolds by inductive limits of finite dimensional submanifolds. This is a topological property (cf. R. S. Palais (1965)) of interest in global analysis. The spaces $L^\Phi(\mathcal{H})$ have importance in Fourier analysis on (nonabelian) groups. In this study the trace functional 't' has to be replaced by a more general "gauge" or "state" and one has to use noncommutative integration. When once this is done the interpolation theory of Chapter VI has a well-defined counterpart. Also the Hausdorff-Young theorem has an extension. This was given in the L^p-context, by R. A. Kunze (1958), in a very general form for harmonic analysis on unimodular groups. A simple extension for compact (or locally compact abelian) groups in the context of Orlicz spaces was considered in Rao (1968$_b$). The general study needs an integration theory based on unbounded ("measurable") operators, as found, for instance, in W. F. Steinspring (1959). We can indicate some interesting results already in the special case of the spaces given by (12), to motivate future studies of the full-fledged theory in the context of Orlicz spaces.

Theorem 6. *The space $(L^\Phi(\mathcal{H}), N_\Phi(\cdot))$ is a Banach algebra where Φ is a continuous Young function satisfying $\Phi(1) = 1, \Phi(x) = 0$ iff $x = 0$ and Φ is continuous. Also $N_\Phi(A) = N_\Phi(A^*)$, for $A \in L^\Phi(\mathcal{X})$ (A^* is the adjoint of A), and $\|A\|_\infty \leq N_\Phi(A)$ [$\|A\|_\infty = \sup\{\|Ax\|_\mathcal{H} : \|x\|_\mathcal{H} = 1, x \in \mathcal{H}\}$],*

$$N_\Phi(AB) \leq \|A\|_\infty N_\Phi(B) \leq N_\Phi(A)N_\Phi(B). \tag{14}$$

Ideas of Proof: Using the spectral theorem for self-adjoint operators $[A]$, one notes that $\Phi(t[A])$ is a well-defined positive operator and by hypothesis $t(\Phi(t[A])) < \infty$ so that the above operator is compact (i.e., it maps bounded sets of $\mathcal{H}$ into relatively compact sets) and hence so is A. If λ_n are the eigenvalues arranged in decreasing order $\lambda_n \downarrow 0$, of $[A]$, then

$$t\left(\Phi\left(\frac{1}{k}[A]\right)\right) = \sum_{n=1}^\infty \Phi\left(\frac{\lambda_n}{k}\right). \tag{15}$$

The λ_n (denoted $s_n(A)$) are also the Schmidt numbers, and $N_\Phi(A)$ defined in (12) is then verified to be the "symmetric norming function" in the sense of Gokhberg and Kreĭn [(1969), p. 71]. In the terminology of the latter reference, $N_\Phi(A) = \Phi(s(A))$ where $s(A) = (s_n(A), n \geq 1)$. Then the result is deduced from Theorem III.4.1 of the above reference where it is shown that $L^\Phi(\mathcal{H}) \subset B(\mathcal{H})$ is an ideal generated by the symmetric norming function. This gives all the conclusions of the theorem. [The work of the first three chapters of Gokhberg and Kreĭn (1969) will be essentially used in this work.] The space has the properties of $(\ell^\Phi, N_\Phi(\cdot))$ and so we may think of a "linear injection" into the latter. However it should be remembered that this is not a true isometry since ℓ^Φ is a commutative (algebra) while $L^\Phi(\mathcal{H})$ is a noncommutative one.

It must be noted that a direct computation needed for the complete proof of the above theorem is also possible but quite long. The case

for $\Phi(x) = |x|^p$, $0 < p < \infty$, has been considered by C. A. McCarthy
(1967), by elementary arguments. For $1 \le p < \infty$ the result is also
established by R. S. Palais (1965) which is shorter than the preceding
paper. R. Schatten (1960) has shown that the adjoint space of $L^1(\mathcal{H})$
is $B(\mathcal{H})$ and that $\mathcal{C}$, the class of all compact operators on $\mathcal{H}$, has its
adjoint space as $L^1(\mathcal{H})$. It is known that $L^2(\mathcal{H})$, the space of Hilbert-
Schmidt operators, is a Hilbert space.

Also C. A. McCarthy (1967) and (later on independently) P. T. Lai
(1973) have shown that $(L^p(\mathcal{H}))^* = L^q(\mathcal{H})$, for $1 < p < \infty$, $q = \frac{p}{p-1}$.
The case $p = 1$ (and $p = 2$) was already treated by R. Schatten. See
also L. Klotz (1988) on a closely related class of spaces of interest in
applications. Actually McCarthy also established that for $1 < p < \infty$,
$L^p(\mathcal{H})$ is uniformly convex. If $x^* \in (L^p(\mathcal{H}))^*$, then

$$x^*(A) = t(AB), \qquad A \in L^p(\mathcal{H}), \tag{16}$$

for a unique B in $L^q(\mathcal{H})$. It should be possible to extend this result
to $L^\Phi(\mathcal{H})$ spaces also when Φ satisfies suitable growth restrictions as
in Chapter VII. In the general case the theory is more involved since
the trace functional is not appropriate and A, B are not necessarily
bounded. However, in the present context we can state the following.

Theorem 7. *Let Φ be as in* Theorem 6 *which further satisfies*
$\Phi(ax) \le b\Phi(x)$, $0 \le x \le x_0$ *for some positive a, b and x_0. Then the algebra*
$L^\Phi(\mathcal{H})$ *of* (12) *is approximately tame in the sense of* Definition 5(ii).

Using the method of proof of the $L^p(\mathcal{H})$-space given in R. S. Palais
(1965), one can establish the above result which employs a few properties of tensor products. We omit the details. Both of the above results
are formulated by Rao (1971) who included outlines of the arguments.

The classical Riesz-Thorin interpolation theorem for the $L^p(\mu)$-
spaces has been simply extended to the $L^p(\mathcal{H})$ spaces of operators,
$1 \le p \le \infty$, of the type given in Definition 5 by P. T. Lai (1973).
This gives an inducement of extending the corresponding results of
Chapter VI. These should be relatively straightforward, unlike the case
of the "measurable" [and unbounded] operators considered by R. Kunze
(1958). Several of these possibilities are available for future reseach.

10.3. *Lorentz and Orlicz spaces*

In this final paragraph we make a brief comparison of Lorentz
spaces with Orlicz spaces having parametric weights. First we need to

recall these classes so that they can be viewed together.

It was noted in Section 1 above (cf., Eq. (1.10)) that by the image measure relations of Real Analysis, for any measurable $f : \Omega \to I\!R$ we have

$$\int_{\Omega} \varphi(f)d\mu = \int_{0}^{\mu(\Omega)} \varphi(f^*)(t)dt, \tag{1}$$

where f^* is the decreasing rearrangement of $|f|$ and φ is a φ-function. Taking $\varphi(x) = |x|^p$, $p \geq 1$, (1) may be written as

$$\|f\|_p = \left[\int_{\Omega} |f|^p d\mu \right]^{\frac{1}{p}} = \left[\frac{p}{p} \int_{0}^{\mu(\Omega)} (t^{\frac{1}{p}} f^*(t))^p \frac{dt}{t} \right]^{\frac{1}{p}}. \tag{2}$$

This motivates the definition of a general *Lorentz space* $L^{p,q}(\mu)$ as the set of measurable real functions $f : \Omega \to I\!R$ such that $\|f\|_{p,q} < \infty$ where

$$\|f\|_{p,q} = \begin{cases} \left[\frac{q}{p} \int\limits_{0}^{\infty} (t^{\frac{1}{p}} f^*(t))^q \frac{dt}{t} \right]^{\frac{1}{q}}, & 0<p<\infty, 0<q<\infty, \\ \sup\limits_{t>0} t^{\frac{1}{p}} f^*(t), & 0<p\leq\infty, q=\infty. \end{cases} \tag{3}$$

In this notation $\|f\|_p = \|f\|_{p,p}$ of (2) and $L^p(\mu) = L^{p,p}(\mu)$. Slightly more specialized versions of $L^{p,q}(\mu)$ were introduced by G. G. Lorentz (1950), and studied by him and others extensively because of the usefulness of these spaces in Fourier analysis, multiplicity and interpolation theories among others. We only refer to the papers by R. O'Neil (1963) and especially ((1968), Sec. 6), R. A. Hunt (1966) and the general treatment of A. P. Calderón (1964). Here we discuss a few results of $L^{p,q}(\mu)$ spaces, compare them with Orlicz spaces and indicate a common generalization of both which may be useful for future work on the subject.

Regarding the Lorentz spaces, we have the following basic information.

Theorem 1. *The classes* $(L^{p,q}(\mu), \|\cdot\|_{p,q})$, $1 \leq q \leq p < \infty$, *are Banach spaces and if* $1 \leq p < \infty$ *but* $0 < q < 1$ *or* $0 < q \leq p$, *then* $(\|\cdot\|_{p,q})^q$ *is an F-norm under which the* $L^{p,q}(\mu)$ *is a complete linear metric space (i.e., a Fréchet space). In case* $0 < p \leq 1$ *and* $p < q \leq \infty$, *then also* $(L^{p,q}(\mu), (\|\cdot\|_{p,q})^{\alpha})$ *is a Fréchet space for each* $0 < \alpha < p$.

In particular, for $1 \leq p < \infty$, the adjoint space $(L^{p,1}(\mu))^*$ of $L^{p,1}(\mu)$ is $L^{p',\infty}(\mu)$ whenever μ is localizable, and for $1 < p < \infty$, and $1 < q < \infty$, the $L^{p,q}(\mu)$ is a reflexive Banach space. In this case the $(L^{p,q}(\mu))^*$ is identifiable with $L^{p',q'}(\mu)$ where $p' = \frac{p}{p-1}$ and $q' = \frac{q}{q-1}$.

The proof of this theorem involves some auxiliary results on non-negative rearrangements [the f^*'s of f's]. Especially the following facts which are consequences of the definitions (as in Section 1) are needed: If $f : \Omega \to \mathbb{R}$ is measurable and f^* is its rearranged function, let f^{**} be defined for each $x > 0$ as

$$f^{**}(x) = \frac{1}{x}\int_0^x f^*(t)dt \quad [= \frac{1}{x}\sup\{\int_E |f|d\mu : \mu(E) \leq x, E \in \Sigma\}]. \quad (4)$$

Then f^{**} is also monotone nonincreasing, and if $f = f_1 + f_2$ (f_i measurable), we have

$$f^{**}(t) \leq f_1^{**}(t) + f_2^{**}(t), \quad (5)$$

and (using integration by parts and image law formulas) one also has

$$xf^{**}(x) = xf^*(x) + \int_{f^*(x)}^{\infty} \lambda_f(t)dt. \quad (6)$$

Since $\|\cdot\|_{p,q}$ is monotone on positive monotone functions, (5) shows that $\|\cdot\|_{p,q}$ is a norm if $1 \leq p \leq q$, and an F-norm in the other cases of the statement. Since $L^{p,q}(\mu) \subset L^0(\mu)$, the latter being a Fréchet space under the metric of convergence in measure, one shows, under the usual (but tedius) computations that the space is complete. The last part on the adjoint space of $L^{p,q}(\mu)$, with μ localizable, proceeds along familiar methods using the Radon-Nikodým theorem and the fact that $\|f\|_{p,q_1} \geq \|f\|_{p,q_2}$ for $0 < q_1 \leq q_2$. The details will be omitted. They may be obtained from Hunt (1966) or O'Neil [(1968), Sec. 6]. An interesting direct comparison of the Lorentz and Orlicz spaces is included in Theorem 3 below.

We now indicate how a common extension of Orlicz and Lorentz spaces can be formulated and studied. Again let (Ω, Σ, μ) be a measure space and φ be a φ-function. Let $h = h_\varphi : \mathbb{R}^+ \to \mathbb{R}^+$ be a monotone increasing (or a *weight*) function. Consider the space $L^{\varphi,h}(\mu)$ defined as

$$L^{\varphi,h}(\mu) = \{f \in L^0(\mu) : \|f\|'_{\varphi,h} < \infty\} \quad (7)$$

where

$$\|f\|'_{\varphi,h} = \inf\{k > 0 : \int_0^\infty \varphi\Big(\frac{f^{**}}{k}\Big) dh(t) \le k\}. \tag{8}$$

Here f^{**} is given by (4), the averaged f^*, the nonincreasing rearrangement of f. For different φ-functions, h_φ may change [analogous to the change of measures in Chapter VI]. Since clearly $f^*(x) \le f^{**}(x)$ (cf. (4)), one gets an equivalent (F-) norm if f^* is replaced by f^{**} in (3) (cf. O'Neil (1963), pp. 136–137), and this is the motivation for (8). (See also Bennett and Sharpley (1988), Lemma IV.4.5.) In fact if $\varphi(x) = |x|^q$, $q > 0$, and $h(x) = |x|^{\frac{q}{p}}$, $p > 0$, then (8) becomes

$$\int_0^\infty [x^{\frac{1}{p}} f^{**}(x)]^q \frac{dx}{x} \le k^{q+1}, \tag{9}$$

and the infimum of k here is precisely $\|f\|'_{p,q}$ of (3) with f^{**} in place of f^*. But as noted above, the norms are equivalent and thus the $L^{p,q}(\mu)$ spaces are recovered from $L^{\varphi,h}(\mu)$ spaces. That the latter are linear and $\|f\|'_{\varphi,h}$ satisfies the triangle inequality follow from (5), with the computations of the proof of Theorem 1.2. Similarly replacing f by $\frac{f}{k}$ in (1) and taking $h(x) = x$, we see that the norm $\|\cdot\|'_{\varphi,h}$ is equivalent to $\|\cdot\|'_\varphi$ and it reduces to the (generalized) Orlicz norm. Thus the spaces $L^{\varphi,h}(\mu)$ are a common generalization of both these classes. We thus state the following result on this $L^{\varphi,h}(\mu)$.

Proposition 2. *The space $(L^{\varphi,h}(\mu), \|\cdot\|'_{\varphi,h})$ introduced by (7) and (8) is a Fréchet space which, with suitable specialization, reduces to the Lorentz and Orlicz spaces.*

Sketch of Proof: We have already noted how the spaces $L^{\varphi,h}(\mu)$ reduce to the Orlicz and Lorentz classes. Only completeness needs to be verified, since the linearity is clear. For simplicity we discuss the case that φ is also convex, using a nice trick due to Calderón (1964).

Let $\mathcal{X} = L^\varphi(\widetilde{I\!\!R}^+, \mathcal{B}, dh)$ be the real Orlicz space on the open half line $\widetilde{I\!\!R}^+ = (0, \infty)$ with the Lebesgue-Stieltjes measure determined by h [using the same symbol again], and consider $\mathcal{Y} = \{f \in L^0(\mu) : f^{**} \in \mathcal{X}\}$, defining the norm in $\mathcal{Y}$ by the formula $\|\cdot\|_\mathcal{Y} : f \mapsto \|f^{**}\|_\mathcal{X}$. We assert that $\mathcal{Y}$ is complete. It is clear that $\mathcal{Y} = L^{\varphi,h}(\mu)$ with equivalent norms. This will finish the argument.

Being a real Orlicz space, $\mathcal{X}$ is a Banach lattice. To see that $\mathcal{Y}$ is complete, let $\{f_n, n \geq 1\} \subset \mathcal{Y}$ be a Cauchy sequence. Choose a subsequence $\{f_{n_k}, k \geq 1\}$ such that $n_1 < n_2 < \cdots$, and $\|f_n - f_{n_k}\|_{\mathcal{Y}} < 2^{-k}$ for $n > n_k$, which is possible since $\|f_n - f_m\|_{\mathcal{Y}} \to 0$, as $n, m \to \infty$. If $g_k = f_{n_k} - f_{n_{k-1}}$ ($f_{n_0} = 0$), then $\sum_{k=1}^{\infty} \|g_k\|_{\mathcal{Y}} < \|g_1\|_{\mathcal{Y}} + 1 < \infty$. Hence $\sum_{k=1}^{m} g_k = f_{n_m} \to f$ if we show that the preceding condition implies that $\sum_{k=1}^{\infty} g_k$ converges (necessarily to f) and then conclude that $f \in \mathcal{Y}$. Thus consider $\{g_k, k \geq 1\} \subset \mathcal{Y}$ with $\sum_{k=1}^{\infty} \|g_k\|_{\mathcal{Y}} < \infty$. By definition of the $\mathcal{Y}$-norm, we have $\sum_{k=1}^{\infty} \|g_k^{**}\|_{\mathcal{X}} < \infty$. But this implies that $g_n^{**} \to 0$ in measure since $\|\cdot\|_{\mathcal{X}}$-norm convergence is stronger than that of $L^0(dt)$. This easily allows us to conclude that $\sum_{k=1}^{\infty} g_k^{**}(t)$ converges for a.a.(t). Now if t_0 is a point of convergence of this series, then, for each $n \geq 1$,

$$\left(\sum_{k=1}^{n} |g_k| \right)^{**}(t_0) \leq \sum_{k=1}^{n} g_k^{**}(t_0) \leq \sum_{k=1}^{\infty} g_k^{**}(t_0) < \infty, \text{ by (5).} \qquad (10)$$

Hence the finite subset property of μ, (10), and (4) yield

$$\int_A \sum_{k=1}^{n} |g_k| d\mu \leq \int_0^x t \sum_{k=1}^{n} g_k^{**} dt \leq \int_0^x t \sum_{k=1}^{\infty} g_k^{**}(t) dt < \infty, \qquad (11)$$

for each $A \in \Sigma$ with $\mu(A) \leq x < \infty$. From this we deduce that $\sum_{k=1}^{\infty} |g_k| < \infty$ a.e. Thus $g = \sum_{k=1}^{\infty} g_k$ converges a.e. and $g \in \mathcal{X}$. Note that by (11) and monotone convergence, we have $\left(\sum_{k=1}^{\infty} |g_k| \right)^{**} \leq \sum_{k=1}^{\infty} g_k^{**}$, a.e., and $g^{**} \leq \sum_{k=1}^{\infty} |g_k|^{**} = \sum_{k=1}^{\infty} g_k^{**} < \infty$ a.e. We claim that $g \in \mathcal{Y}$ and is the limit of f_n in $\mathcal{Y}$.

We already have $g = \lim_{m \to \infty} \sum_{k=1}^{m} g_k = \lim_{m \to \infty} f_{n_k}$ a.e. Since $\|f_n - f_m\|_{\mathcal{X}} \to 0$ and $f_{n_k} \to g$, a.e., the Fatou inequality implies that $\|f_{n_k} - g\|_{\mathcal{X}}$ can be made arbitrarily small, and that $\|g\|_{\mathcal{Y}} \leq \|g - f_{n_k}\|_{\mathcal{Y}} + \|f_{n_k}\|_{\mathcal{Y}} < \infty$, so $g \in \mathcal{Y}$. Also $\|f_n - g\|_{\mathcal{Y}} \leq \|f_n - f_{n_k}\|_{\mathcal{Y}} + \|f_{n_k} - g\|_{\mathcal{Y}} \to 0$ as $k \to \infty$ and then $n \to \infty$. Thus the space $\mathcal{Y}$ is complete, as asserted. Note that we only used the completeness of $\mathcal{X}$ and the metric properties used above are also available for the F-norm. Thus the argument holds for the general statement as well. ∎

One can restrict φ to be a Young function or a concave function [see Theorem 1.10] and consider the corresponding duality results. These and the ensuing interpolation theory for the $L^{\varphi,h}(\mu)$-spaces are interesting topics for research. The Lorentz spaces have been especially used for various extensions of the classical Marcinkiewicz L^p-interpolation theory which is later employed in harmonic analysis. Since the Orlicz space version of this theorem is given in Chapter VI, the corresponding results may be considered now.

As a final item, we present a result, due to G. G. Lorentz (1961), which gives precise conditions for a Lorentz and an Orlicz space to coincide. If (Ω, Σ) is a measurable space, let $\nu : \Sigma \to \mathbb{R}^+$ be an increasing, strongly subadditive continuous function. This means (i) $\nu(A) \leq \nu(B)$ for $A \subset B$, (ii) $\nu(A \cup B) + \nu(A \cap B) \leq \nu(A) + \nu(B)$, and (iii) $A_n \in \Sigma$, $A_n \downarrow \varphi \Rightarrow \nu(A_n) \to 0$. (See relation (15) of Section 1, and the the discussion that precedes it.) Let

$$\Lambda(\nu) = \{f : \Omega \to \mathbb{R}, \text{ measurable for } (\Sigma),$$

$$\|f\| = \int_0^\infty \nu(\omega : |f(\omega)| > y) dy < \infty\}.$$

Then $\Lambda(\nu)$ is a Lorentz space that is based on the set $\nu(A) = \Phi_0(\mu(A))$ where Φ_0 is a concave increasing function as in Theorem 1.10, (Ω, Σ, μ) is a finite measure space. This example is canonical in what we consider below.

The following result establishes the preceding statements.

Theorem 3. *Let (Ω, Σ, μ) be a finite diffuse measure space and Φ be an N-function with Ψ as its complementary function.*

If $\Phi_0(x) = x\Psi^{-1}(\frac{1}{x}) x > 0$, $\Phi_0(0) = 0$, and $\nu(A) = \Phi_0(\mu(A))$, $A \in \Sigma$, then $\Lambda(\nu) \subset L^\Phi(\mu)$ is the largest Lorentz space contained in the Orlicz space $L^\Phi(\mu)$, and there is equality between these spaces when and only when

$$\int_{\Psi^{-1}(\alpha^{-1})}^\infty \frac{\Psi(\delta t)}{\Psi(t)^2} d\Psi(t) < \infty \qquad \text{for some } \delta > 0, \qquad (12)$$

where $\alpha = \mu(\Omega)$.

The condition (12) implies that ($\delta \neq 1$), and Ψ must grow exponentially fast. In fact one can show that Φ must be Δ_2-regular. Examples of Ψ are $\Psi(x) = \exp(\log^p x)$, $p > 1$, and $\Psi(x) = x$, $\log \log x$ for large x. Thus $L^{\Phi}(\mu)$ must be "fairly close" to the $L^1(\mu)$ for this equality of the spaces.

An interesting aspect of this result is its first part. Namely, each such $L^{\Phi}(\mu)$ contains a largest $\Lambda(\nu)$ space when $\mu(\Omega) < \infty$. Hence subspaces of $L^{\Phi}(\mu)$ spaces, which themselves are not Orlicz spaces, can have finer topological structures than the original ones, and are thus subjects for a separate study. A proof of this theorem is given in Lorentz (1961), in detail, and will be omitted here. This also concludes our account of Orlicz spaces and several of their extensions.

Bibliographical Notes: We included here some results on not necessarily locally convex Orlicz spaces because of their appearance in some important applications. A detail analysis and general theory of this class is given in P. Turpin's thesis (1976). Also a key area of applications of these $L^{\Phi}(\mu)$ is in probability theory. They appear when studying stochastic measures with independent values in $L^0(P)$. The set of real functions integrable relative to such (vector) measures can be shown to be an L^{φ}-space where φ is a φ-function. An aspect of this may be seen, e.g., from Urbanik (1967). On the other hand "exponential Orlicz spaces" are of interest in the studies of sample path behavior of stochastic processes. The results given in Theorems 1.12 and 1.13 are only a few of important classes of such studies. For a background material on this, one should refer to the recent monograph by Marcus and Pisier (1981).

The spaces of additive set functions in the context of L^p-spaces was originally studied by S. Leader (1953), and the general case for $L^{\Phi}(\mu)$, and even with values in a Banach space, was given in Uhl (1967). To consider their adjoint spaces, when Φ is not Δ_2-regular, the work is more involved. The initial study of Uhl (1966) on this problem, given in his thesis should be simplified. We presented just an aspect of this, and a systematic study similar to that for point functions of Chapter IV is desirable. This will help a better understanding of both the point and set function cases. On the other hand the nonpoint function studies are also of interest in many areas of modern as well as global analysis.

The indication given in Theorems 1.6 and 1.7 should be studied in greater detail. A few other related results were also included in the announcement of Rao (1971). The monographs of Schatten (1960) and especially Gokhberg-Kreĭn (1969) will be indispensable. The general case depends on I. E. Segal's (1953) theory of noncommutative integration. (See also E. Nelson (1974) for a simplified version of the latter.) Many of the results of Chapters IV–VII admit nontrivial extensions to this class with such a theory of integration.

Another extension of the Lebesgue spaces, motivated by Fourier analysis problems and fractional integration, Lorentz (1950) introduced his class of functions. It was noted that these are effective for interpolation with weak types, such as those appearing in the Marcinkiewicz' theory. The latter was extended in Chapter VI. Now that all these interpolation results are availabe in a unified manner in this work, it would be interesting to consider them for the $L^{\varphi,h}(\mu)$-spaces. These were raised by one of us (Rao) in his lectures in 1975 and Proposition 3.2 was included there. Nothing more was done, since the relevant supporting Orlicz space account was not availabe at one place. The $L^{p,q}(\mu)$-spaces and their interpolation theory were treated recently in C. Bennett and R. Sharpley (1988), and this should give a necessary background for further development. The final result, Theorem 3.3, has interest in putting both the Lorentz and Orlicz spaces in perspective.

There is also an extension of the theory in which $\Phi : I\!\!R \times \Omega \to I\!\!R^+$ is such that $\varphi(\cdot, \omega)$ is a φ-function for each $\omega \in \Omega$ and $\varphi(t, \cdot)$ is measurable (Σ), for $t \in I\!\!R$. Let $L^\varphi(\mu)$ be the set $\{f : \Omega \to I\!\!R$, measurable, and $\|f\|_\varphi^* < \infty\}$, where $\|f\|_\varphi^* = \inf\{k > 0 : \int_\Omega \varphi(\frac{|f(\omega)|}{k}, \omega) d\mu \le k\}$. Then one can verify that $(L^\varphi(\mu), \|\cdot\|_\varphi^*)$ is a Fréchet space and the work of the preceding chapters can be considered for this class. A summary of the work on this general class with numerous references is given by J. Musielak (1983). A simple instance where such spaces arise is as follows. Let $\mathbf{a} = \{a_n, n \ge 1\}$ be a vector such that $\sum_{n=1}^{\infty} |a_n|^{p_n} < \infty$. The behavior of such vectors is of interest in summability theory. If we take $\Omega = I\!\!N$ (the natural numbers), $\varphi(x, \omega) = |x|^{p_\omega}$, $\omega \in I\!\!N$, and μ as counting measure, then the last sum becomes $\int_{I\!\!N} \varphi(|a_n|, n) d\mu(n) < \infty$. Many important contributions for this generalized case were made by Orlicz himself along with his students. These go under the name of modulared spaces. These were also considered by H. Nakano (1968) and his associates. (See, e.g., W. Orlicz (1966) and H. Nakano (1968) and references there for their contributions.) It is clearly possible to extend the concrete Orlicz space theory that we presented to these classes. In another direction B. Turett (1980) has studied Fenchel-Orlicz spaces and gave a detailed structural analysis. However, we have indicated a sufficient number of different directions where the Orlicz space methods and ideas play a key role in the current studies; and we conclude this account at this point, due also to the space and time constraints. But one notes that the interest in this subject and its applications remains high.

References

Adams, R. A. (1975). *Sobolev Spaces*, Academic Press, New York.
(1977). "On the Orlicz-Sobolev imbedding theorem," *J. Functional Anal.* **24**, 241–257.

Akimovič, B. A. (1972). "On uniformly convex and uniformly smooth Orlicz spaces," *Teor. Funckii Funkcional Anal. & Prilozen* **15**, 114–120 (Russian).

Albrycht, J. (1962). *The Theory of Marcinkiewicz-Orlicz Spaces*, Dissertationes Math., No. 27.

Amerio, L. and G. Prouse (1971). *Almost Periodic Functions and Functional Equations*, Van Nostrand Reinhold, New York.

Andô, T. (1960ₐ). "On some properties of convex function," *Bull. Acad. Polon. Sci. Sér. Math.* **8**, 413–418.
(1960_b). "On products of Orlicz spaces," *Math. Ann.* **140**, 174–186.
(1960_c). "Linear functionals on Orlicz spaces," *Nieuw Arch. Wisk.* **8**, 1–16.
(1961). "Certain classes of convex functions," *Soviet Math.* **2**, 139–142.
(1962). "Weakly compact sets in Orlicz spaces," *Canad. J. Math.* **14**, 170–176.

Aronszajn, N. and E. Gagliardo (1965). "Interpolation spaces and interpolation methods," *Ann. Mat. Pura ed Appl.* **68**, 51–117.

Bennett, C. and R. Sharpley (1988). *Interpolation of Operators*, Academic Press, Inc., New York.

Besicovitch, A. S. (1932). *Almost Periodic Functions*, Cambridge Univ. Press, London.

429

Billik, M. (1957). *Orlicz Spaces*, M.A. Thesis, MIT, Cambridge, MA.

Birnbaum, Z. W. and W. Orlicz (1931). "Über die Verallgemeinerung des Begriffes der zueinander konjugierten Potenzen," *Studia Math.* **3**, 1–67.

Brenner, P., V. Thomée and L. B. Wahlbin (1975). *Besov Spaces and Applications to Difference Methods for Initial Value Problems*, Lecture Notes in Math., No. 434, Springer-Verlag, New York.

Brooks, J. K. and N. Dinculeanu (1976). "Lebesgue-type spaces for vector integration, linear operators, weak completeness and weak compactness," *J. Math. Anal. Appl.* **54**, 348–389.

Burkholder, D. L. (1961). "Sufficiency in the undominated case," *Ann. Math. Stat.* **32**, 1191–1200.

Calderón, A. P. (1964). "Intermediate spaces and interpolation, the complex method," *Studia Math.* **24**, 113–190.

Chan, D.-V. (1980). "Nontriviality of Sobolev-Orlicz spaces of infinite order in a bounded domain of Euclidean space," *Soviet Math.* **21**, 335–338.

Clarkson, J. A. (1936). "Uniformly convex spaces," *Trans. Amer. Math. Soc.* **40**, 396–414.

Cleaver, C. E. (1972). "On the extension of Lipschitz-Hölder maps on Orlicz spaces," *Studia Math.* **42**, 195–204.
(1976). "Packing spheres in Orlicz spaces," *Pacific J. Math.* **65**, 325–335.

Cudia, D. F. (1963). "Rotundity," *Proc. of Symp. Pure Math. Amer. Math. Soc.* **7**, 73–97.
(1964). "The geometry of Banach spaces; smoothness," *Trans. Amer. Math. Soc.* **110**, 284–314.

Cui, Y. A. and T. F. Wang (1987). "Strong extreme points of Orlicz spaces," *J. Math Wuhan* **7**, 335–340 (Chinese).

Davis, H. (1967). "Stepanoff and Weyl a.p. functions on locally compact groups," *Duke Math. J.* **34**, 535–548.

Day, M. M. (1944). "Uniform convexity in factor and conjugate spaces," *Ann. of Math.* **45**, 375–385.
(1962). *Normed Linear Spaces*, Springer-Verlag, Berlin (3d ed., 1973).

de Jonge, E. (1973). *Singular Functionals on Köthe Spaces*, Ph.D. Dissertation, Univ. of Leiden, Leiden, The Netherlands.

de la Vallée Poussin, Ch. J. (1915). "Sur l'integrale de Lebesgue," *Trans. Amer. Math. Soc.* **16**, 435–501.

de Leeuw, K. and W. Rudin (1958). "Extreme points and extremum problems in H_1," *Pacific J. Math.* **8**, 467–485.

Diestel, J. and J. J. Uhl, Jr. (1977). *Vector Measures*, Amer. Math. Soc., Providence, RI.

Dinculeanu, N. (1967). *Vector Measures*, Pergamon Press, London.
(1974). *Integration on Locally Compact Spaces*, Noordhoff Int'l Publishers, Leiden, The Netherlands.

Donaldson, T. K. and N. S. Trudinger (1971). "Orlicz-Sobolev spaces and imbedding theorems," *J. Functional Anal.* **8**, 52–75.

Dubinskij, Ju. A. (1986). *Sobolev Spaces of Infinite Order and Differential Equations*, D. Reidel Publ. Co., Dordrecht.

Dunford, N. and J. T. Schwartz (1958). *Linear Operators, Part I: General Theory*, Wiley-Interscience, New York.

Duren, P. L. (1970). *Theory of H^p Spaces*, Academic Press, New York.

Edwards, R. E. and G. I. Gaudry (1977). *Littlewood-Paley and Multiplier Theory*, Springer-Verlag, New York.

Enflo, P. (1973). "A counterexample to the approximation problem in Banach spaces," *Acta Math.* **130**, 309–317.

Figiel, T. (1976). "On the moduli of convexity and smoothness," *Studia Math.* **56**, 121–155.

Gaposhkin, V. F. (1968). "Absolute bases in Orlicz spaces," *Siberian Math. J.* **9**, 211–217.

Gokhberg, I. C. and M. G. Kreĭn (1969). *Introduction to the Theory of Linear Nonselfadjoint Operators* (translation), Amer. Math. Soc., Providence, RI.

Gretsky, N. E. and J. J. Uhl, Jr. (1981). "Carleman and Korotkov operators on Banach spaces," *Acta Sci. Math.* **43**, 207–218.

Grothendick, A. (1955). *Produits Tensoriels Topologiques et Espaces Nucléaires*, Mem. Amer. Math. Soc., No. 16, Providence, RI.

Gustavsson, J. and J. Peetre (1977). "Interpolation of Orlicz spaces," *Studia Math.* **60**, 33–59.

Halmos, P. R. and L. J. Savage (1949). "Application of the Radon-Nikodým theorem to the theory of sufficient statistics," *Ann. Math. Stat.* **20**, 225–241.

———— and V. S. Sunder (1978). *Bounded Integral Operators on L^2 Spaces*, Springer-Verlag, New York.

Halperin, I. (1954). "Reflexivity in the L^λ function spaces," *Duke Math. J.* **21**, 205–208.

Hayden, T. L. and J. H. Wells (1971). "On the extension of Lipschitz-Hölder maps of order β," *J. Math. Anal. Appl.* **33**, 627–640.

Hewitt, E. and K. A. Ross (1963). *Abstract Harmonic Analysis*, Vol. I, Springer-Verlag, New York.

Heyer, H. (1982). *Theory of Statistical Experiments*, Springer-Verlag, New York.

Hille, E. and R. S. Phillips (1957). *Functional Analysis and Semigroups*, Amer. Math. Soc., New York.

Hillmann, T. R. (1977). *Besicovitch-Orlicz Spaces of Almost Periodic Functions*, Ph.D. Dissertation, Univ. of Calif., Riverside, CA. (1986). "Besicovitch-Orlicz spaces of almost periodic functions," *in: Real and Stochastic Analysis*, Wiley, New York, 119–167.

Hirschman, Jr., I. I. (1953). "A convexity theorem for certain groups of transformations," *J. d'Analyse Math.* **2**, 209–218.

Hoffman, K. (1962). *Banach Spaces of Analytic Functions*, Prentice-Hall, Englewood Cliffs, NJ.

Hudzik, H. (1987). "An estimation of the modulus of convexity in a class of Orlicz spaces," *Math. Japon.* **32**, 227–237.

Hunt, R. A. (1966). "On $L(p,q)$ spaces," *L'Enseign. Math.* **12**, 249–276.

James, R. C. (1957). "Reflexivity and the supremum of linear functionals," *Ann. of Math.* **66**, 159–169.

Jensen, J. L. W. V. (1906). "Sur les functions convexes et les inégalités entre les valeurs moyennes," *Acta Math.* **30**, 175–193.

Kakutani, S. (1941$_a$). "Concrete representation of abstract (L)-spaces and the mean ergodic theorem," *Ann. of Math.* **42**, 523–537. (1941$_b$). "Concrete representation of abstract (M)-spaces," *Ann. of Math.* **42**, 994–1024.

Kaminska, A. and W. Kurc (1986). "Weak uniform rotundity in Orlicz spaces," *Comment. Math. Univ. Carolinae* **27**, 651–664.

Kantorovich, L. V. and G. P. Akilov (1982). *Functional Analysis* (translation), Pergamon Press, Inc., New York.

Karlin, S. (1948). "Bases in Banach spaces," *Duke Math J.* **15**, 971–985.

Klotz, L. (1988). "Some Banach spaces of measurable operator-valued functions," *in: Naturwiss.-Theor. Zentrum (Sekt. Math.)*, Preprint No. 15, Karl-Marx-Univ., Leipzig, 17pp.

Kolmogorov, A. N. (1931). "Über Kompaktheit der Funktionenmengen bei der Konvergenz im Mittel," *Nachr. Ges. Wiss. Göttg.* **1**, 60–63. (1950). "Unbiased estimates," *Izvestiya Akad. Nauk SSSR, Ser. Mat.* **14**, 303–324.

Köthe, G. (1937). "Die Teilräume eines linearen Koordinatenraumes," *Math. Ann.* **114**, 99–125.

———— and O. Toeplitz (1934). "Lineare Räume mit unendlich vielen Koordinaten und Ringe unendlicher Matrizen," *J. Reine Angew. Math.* **171**, 193–226.

Kottman, C. A. (1970). "Packing and reflexivity in Banach spaces," *Trans. Amer. Math. Soc.* **150**, 565–576.

Krasnosel'skii, M. A. and Ya. B. Rutickii (1961). *Convex Functions and Orlicz Spaces* (translation), P. Noordhoff Ltd., Groningen.

Kraynek, W. T. (1970). "Interpolation of multilinear functionals on $L^{\Phi\Psi}$ spaces," *J. Math. Anal. Appl.* **31**, 414–430. (1972). "Interpolation of sublinear operators on generalized Orlicz and Hardy-Orlicz spaces," *Studia Math.* **43**, 93–123.

Kufner, A. (1985). *Weighted Sobolev Spaces*, Wiley, New York.

————, O. John, and S. Fučik (1977). *Function Spaces*, Academia, Prague.

Kunze, R. A. (1958). "L_p Fourier transforms on locally compact unimodular groups," *Trans. Amer. Math. Soc.* **89**, 519–540.

Lai, P. T. (1973). "L'analogue dans C^p des theorems de convexité de M. Riesz et G. O. Thorin," *Studia Math.* **46**, 111–124.

Lapin, S. V. (1977). "A theorem on a convergence condition in the space L^φ," *Math. Notes* **21**, 346–352.

Leader, S. (1953). "The theory of L^p spaces for finitely additive set functions," *Ann. of Math.* **58**, 528–543.

Lehman, E. L. (1983). *Theory of Point Estimation*, Wiley, New York.

Lésniewicz, R. (1971). "On Hardy-Orlicz spaces I," *Commentationes Math.* **15**, 3–56.
(1973). "On linear functionals in Hardy-Orlicz spaces I–III," *Studia Math.* **46**, 53–77; **46**, 259–295; **47**, 261–284.

Lions, J. L. (1961). "Sur les espaces d'interpolation; dualité," *Math. Skand.* **9**, 147–177.

Lorentz, G. G. (1950). "Some new functional spaces," *Ann. of Math.* **51**, 37–55.
(1961). "Relations between function spaces," *Proc. Amer. Math. Soc.* **12**, 127–132.

――――― and D. G. Wertheim (1953). "Representation of linear functionals of Köthe spaces," *Canad. J. Math.* **5**, 568–575.

Lumer-Naim, L. (1967). "H^p-spaces of harmonic functions," *Ann. L'Inst. Fourier, Grenoble* **17**, 425–469.

Luxemburg, W. A. J. (1955). *Banach Function Spaces*, Ph.D. Dissertation, Delft.

――――― and A. C. Zaanen (1963). "Compactness of integral operators in Banach function spaces," *Math. Ann.* **149**, 150–180.

Maleev, R. P. and S. L. Troyanski (1975). "On the moduli of convexity and smoothness in Orlicz spaces," *Studia Math.* **54**, 131–141.

Mandelbrojt, S. (1952). *Séries Adhérentes. Régularisation de Suites. Applications*, Gauthier-Villars, Paris.

Marcinkiewicz, J. (1939). "Sur l'interpolation d'operations," *C. R. Acad. Sci.* **208**, 1272–1273.

Marcus, M. B. and G. Pisier (1981). *Random Fourier Series with Applications to Harmonic Analysis*, Ann. Math. Studies, No. 101, Princeton Univ. Press, Princeton, NJ.

(1985). "Stochastic processes with sample paths in exponential Orlicz spaces," *in: Probability in Banach Spaces V*, Lecture Notes in Math., No. 1153, Springer-Verlag, New York, 325–358.

Matuszewska, W. and W. Orlicz (1961). "A note on the theory of s-normed spaces of φ-integrable functions," *Studia Math.* **21**, 107–115.

Maynard, H. B. (1973). "A general Radon-Nikodým theorem," *in: Vector and Operator Valued Measures and Applications*, Academic Press, Inc., New York, 233–246.

Maz'ja, V. G. (1985). *Sobolev Spaces* (translation), Springer-Verlag, New York.

Mazur, S. and W. Orlicz (1958). "On some classes of linear spaces," *Studia Math.* **17**, 97–119.

McCarthy, C. A. (1967). "c_p," *Israel J. Math.* **5**, 249–271.

McShane, E. J. (1950). "Linear functionals on certain Banach spaces," *Proc. Amer. Math. Soc.* **1**, 402–408.

Meyers, N. G. and J. Serrin (1964). "$H = W$," *Proc. Nat. Acad. Sci. USA* **51**, 1055–1056.

Mikusinski, J. (1978). *The Bochner Integral*, Academic Press, New York.

Milnes, H. W. (1957). "Convexity of Orlicz spaces," *Pacific J. Math.* **7**, 1451–1486.

Morse, M. and W. Transue (1950). "Functionals F bilinear over the product $A \times B$ of two pseudo-normed vector spaces II, Admissible spaces A," *Ann. of Math.* **51**, 576–614.

Musielak, J. (1983). *Orlicz Spaces and Modular Spaces*, Lecture Notes in Math., No. 1034, Springer-Verlag, New York.

Nakano, H. (1950). *Modulared Semi-ordered Linear Spaces*, Maruzen, Tokyo.
(1968). "Generalized modular spaces," *Studia Math.* **31**, 439–449.

Nelson, E. (1974). "Notes on noncommutative integration," *J. Functional Anal.* **15**, 103–116.

Nowak, M. (1986). "A characterization of the Mackey topology $\tau(L^\varphi,$ $L^{\varphi^*})$ on Orlicz spaces," *Bull. Polish Acad. Sci. Math.* **34**, 577–583.

Okikiolu, G. O. (1966). "A convexity theorem for multilinear functions II," *J. Math. Anal. Appl.* **16**, 394–404.

O'Neil, R. (1963). "Convolution operators and $L(p,q)$ spaces," *Duke Math. J.* **30**, 129–142.
    (1965). "Fractional integration in Orlicz spaces I," *Trans. Amer. Math. Soc.* **115**, 300–328.
    (1968). "Integral transforms and tensor products on Orlicz spaces and $L(p,q)$ spaces," *J. d'Anal. Math.* **21**, 1–276.

Orlicz, W. (1932) "Über eine gewisse Klasse von Räumen vom Typus B," *Bull. int'l. de l'Acad. Pol. série A* **8**, 207–220.
    (1936). "Über Räume (L^M)," *Bull. int'l. de l'Acad. Pol. série A* **10**, 93–107.
    (1966). "Some classes of modular spaces," *Studia Math.* **26**, 165–192.

Palais, R. S. (1965). "On the homotopy type of certain groups of operators," *Topology* **3**, 271–279.

Pettis, B. J. (1939). "A proof that every uniformly convex space is reflexive," *Duke Math. J.* **5**, 249–253.

Phillips, R. S. (1940). "On linear transformations," *Trans. Amer. Math. Soc.* **48**, 516–541.

Pitcher, T. S. (1965). "A more general property than domination for sets of probability measures," *Pacific J. Math.* **15**, 597–611.

Rabinovich, L. B. (1968). "On some imbedding theorems in Orlicz spaces," *Izv. Vysš. Učebn. Zaved. Mat.* **76**, 78–85 (Russian).

Radó, T. (1949). *Subharmonic Function*, Chelsea, New York.

Rankin, R. A. (1955). "On packings of spheres in Hilbert space," *Proc. Glasgow Math. Assoc.* **2**, 145–146.

Rao, M. M. (1964). "Linear functionals on Orlicz spaces," *Nieuw Arch. Wisk.* **12**, 77–98.
    (1965). "Smoothness of Orlicz spaces," *Indag. Math.* **27**, 670–690.
    (1966). "Interpolation, ergodicity and martingales," *J. Math. Mech.* **16**, 543–567.

(1967_a). "Characterizing Hilbert space by smoothness," *Indag. Math.* **29**, 132–135.

(1967_b). "Convolutions of vector fields and interpolation," *Proc. Nat. Acad. Sci. USA* **57**, 222–226.

(1968_a). "Linear functionals on Orlicz spaces: general theory," *Pacific J. Math.* **25**, 553–585.

(1968_b). "Extensions of the Hausdorff-Young theorem," *Israel J. Math.* **6**, 133–149.

(1968_c). "Almost every Orlicz space is isomorphic to a strictly convex Orlicz space," *Proc. Amer. Math. Soc.* **19**, 377–379.

(1971). "Approximately tame algebras of operators," *Bull. Acad. Polon. Sci. Sér. Math.* **19**, 43–47.

(1975). "Compact operators and tensor products," *Bull. Acad. Polon. Sci. Sér. Math.* **23**, 1175–1179.

(1979). *Stochastic Processes and Integration*, Stijthoff and Noordhoff, Alphen aan den Rijn, The Netherlands.

(1980). "Convolutions of vector fields - I," *Math. Z.* **174**, 63–79.

(1981_a). "Structure and convexity of Orlicz spaces of vector fields," *in: General Topology and Modern Analysis*, Academic Press, New York, 457–473.

(1981_b). *Foundations of Stochastic Analysis*, Academic Press, New York.

(1984). *Probability Theory with Applications*, Academic Press, New York.

(1987). *Measure Theory and Integration*, Wiley-Interscience, New York.

Reiter, H. (1968). *Classical Harmonic Analysis and Locally Compact Groups*, Oxford Univ. Press, London, New York.

(1971). *L^1-Algebras and Segal Algebras*, Lecture Notes in Math., No. 231, Springer-Verlag, New York.

Ren, Z. D. (1963). "A theorem for the ∇_2-condition," *Natur. Sci. J. Harbin Normal Univ.*, No. 1, 6–9 (Chinese).

(1981). "On the M_Δ-condition of N-function and some applications," *Natur. Sci. J. Xiangtan Univ.*, No. 1, 60–62 (Chinese).

(1983). "On linear integral operators in Orlicz spaces," *Nature Journal (Shanghai)* **6**, 238–239 (Chinese).

(1985_a). "Packing spheres in Orlicz function spaces with Luxemburg norm," *Natur. Sci. J. Xiangtan Univ.*, No. 1, 51–60 (Chinese).

(1985_b). "Some convergence theorems for Orlicz spaces," *Hunan Ann. Math.* **5**(2), 1–7 (Chinese).

Ren, Z. D. (1986$_a$). "On moduli and norm in Orlicz spaces," *Adv. in Math. (Beijing)* **15**, 315–320 (Chinese).
(1986$_b$). "Weakly compact imbedding theorem for Orlicz spaces," *Nature Journal (Shanghai)* **9**, 313–314 (Chinese).
(1986$_c$). "Some theorems on comparison of Orlicz spaces," *Natur. Sci. J. Xiangtan Univ.*, No. 2, 22–31 (Chinese).
(1987). "On reflexive Orlicz spaces and J. L. Lions lemma," *Sci. Bull., Acad. Sinica, Beijing* **32**, 354–355.

Riordan, W. J. (1956). *On the Interpolation of Operations*, Ph.D. Dissertation, Univ. Chicago (see also *Notices Amer. Math. Soc.* **5**(1958), 590).

Rohlin, V. A. (1949). "On the fundamental ideas of measure theory," *Amer. Math. Soc. Translations Ser. 1* **10**, 1–54.

Rolewicz, S. (1972). *Metric Linear Spaces*, Warzawa.

Rosenberg, R. L. (1968). *Compactness in Orlicz Spaces Based on Sets of Probability Measures*, Ph.D. Dissertation, Carnegie-Mellon Univ., Pittsburgh, PA.
(1970). "Orlicz spaces based on families of measures," *Studia Math.* **35**, 15–49.

Rosenthal, H. P. (1978). "Some recent discoveries in the isomorphic theory of Banach spaces," *Bull. Amer. Math. Soc.* **84**, 803–831.

Roy, S. K. and N. D. Chakraborty (1986). "Orlicz spaces for a family of measures," *Anal. Math. (Budapest)* **12**, 229–235.
(1990). "On Orlicz spaces for a family of measures," *J. Math. Anal. Appl.* **145**, 485–503.

Royden, H. L. (1968). *Real Analysis* (2d ed.), Macmillan, New York.

Rutickii, Ja. B. (1962). "Some theorems from the theory of Orlicz spaces," *Dopovidi Acad. Nauk Ukrain RSR*, 1278–1282.
(1963). "Scales of Orlicz spaces and interpolation theorems," *Soviet Math.* **4**, 305–308.

Salehov, D. V. (1963). "An example of a completely continuous integral operator from L_p to L_p with positive kernel not belonging to $L_r(r > 1)$," *Uspekhi Mat. Nauk* **18**(4), 179–182 (Russian).
(1968$_a$). "Some properties of functions in Orlicz spaces," *Math. Notes* **3**, 92–99.
(1968$_b$). "On a property of N-functions," *Math. Notes* **4**, 662–667.

REFERENCES   439

Schatten, R. (1960). *Norm Ideals of Completely Continuous Operators*, Springer-Verlag, Berlin.

Schauder, J. (1927). "Zur Theorie stetiger Abbildungen in Funktional-räumen," *Math. Z.* **26**, 47–65.

Schep, A. R. (1979). "Kernel operators," *Indag. Math.* **41**, 39–53.

Segal, I. E. (1954). "Equivalence of measure spaces," *Amer. J. Math.* **73**, 275–313.

Shapiro, V. L. (1977). "Partial differential equations, Orlicz spaces and measure functions," *Indiana Univ. Math. J.* **26**, 875–883.

Šilov, G. E. (1951). "Homogeneous rings of functions," *Uspekhi Mat. Nauk* **6**(1), 91–137 (Russian).

Singer, I. (1970). *Bases in Banach Spaces I*, Springer-Verlag, New York.

Šmulian, V. L. (1939). "On some geometrical properties of the unit sphere in the space of the type (*B*)," *Mat. Sbornik N. S.* **6**, 77–94.

Steigerwalt, M. S. and A. J. White (1971). "Some function spaces related to L_p," *Proc. London Math. Soc.* **22**, 137–163.

Stein, E. M. (1956). "Interpolation of linear operators," *Trans. Amer. Math. Soc.* **83**, 482–492.

————— and G. Weiss (1957). "On the interpolation of analytic families of operators acting on H^p-spaces," *Tôhoku Math. J.* **9**, 318–339.
(1958). "Interpolation of operators with change of measures," *Trans. Amer. Math. Soc.* **87**, 159–172.

Stinespring, W. F. (1959). "Integration theorems for gages and duality for unimodular groups," *Trans. Amer. Math. Soc.* **90**. 15–56.

Takahashi, T. (1935). "On the compactness of the function-set by the convergence in mean of general type," *Studia Math.* **5**, 141–150.

Talagrand, M. (1986). "Derivation, L^Ψ-bounded martingales and covering conditions," *Trans. Amer. Math. Soc.* **293**, 257–291.
(1990). "Sample boundedness of stochastic processes under increment conditions," *Ann. Prob.* **18**, 1–50.

Terpigoreva, V. M. (1962). "Extremum problems for Orlicz classes of functions analytic in the unit circle," *Soviet Math.* **3**, 30–34.

Triebel, H. (1978). *Spaces of Besov-Hardy-Sobolev Type*, Teubner-Texte zur Math., Leipzig.

Turett, B. (1976). "Rotundity of Orlicz spaces," *Indag. Math.* **38**, 462–469.

(1980). *Fenchel-Orlicz Spaces*, Dissertationes Math., No. 181.

Turpin, P. (1976). *Convexités dans les Espaces Vectoriels Topologiques Généraux*, Dissertationes Math., No. 131.

Uhl, Jr., J. J. (1966). *Orlicz Spaces of Additive Set Functions and Set-Martingales*, Ph.D. Dissertation, Carnegie-Mellon Univ., Pittsburgh, PA.

(1967). "Orlicz spaces of finitely additive set functions," *Studia Math.* **29**, 19–58.

(1969). "Martingales of vector valued set functions," *Pacific J. Math.* **30**, 533–548.

(1970). "Vector integral operators," *Indag. Math.* **32**, 463–478.

(1971). "On a class of operators on Orlicz spaces," *Studia Math.* **40**, 17–22.

Urbanik, K. (1967). "Some prediction problems for strictly stationary processes," *in: Proc. 5th Berkeley Symp. Math. Stat. & Prob.*, Vol II, Part I, Univ. Calif. Press, Berkeley, CA, 235–258.

Vala, K. (1964). "On compact sets of compact operators," *Ann. Acad. Sci. Fennicae, Ser. A(I), Math.* **351**, 1–9.

Von Neumann, J. (1934). "Almost periodic functions in a group I," *Trans. Amer. Math. Soc.* **36**, 445–492.

Wang, S. W. (1963). "Problem of products of Orlicz spaces," *Bull. Acad. Polon. Sci. Sér. Math.* **11**, 19–22.

Wang, T. F. and S. T. Chen (1987). "Smoothness and differentiability of Orlicz Spaces," *Chinese J. Eng. Math.* **14**(3), 113–115 (Chinese).

_____, Y. W. Wang, and Y. H. Li (1986). "Weakly uniform convexity in Orlicz spaces," *J. Math (Wuhan)* **6**, 209–214 (Chinese).

Wang, Y. W. (1981). "Inclusion relations among Orlicz sets," *J. Harbin Univ. Sci. & Tech.*, No. 1, 18–26 (Chinese).

(1985). "Weak sequential completeness of Orlicz spaces," *Donbei Shuxue* **1**(2), 241–246 (Chinese).

Weil, A. (1938). *L'Intégration dans les Groupes Topologique et ses Applications* (2d ed., 1953), Hermann, Paris.

Weiss, G. (1956). "A note on Orlicz spaces," *Portugaliae Math.* **15**, 35–47.

Wells, J. H. and L. R. Williams (1975). *Embeddings and Extensions in Analysis*, Springer-Verlag, New York.

Wright, J. D. M. (1969$_a$). "Stone algebra valued measures and integrals," *Proc. London Math. Soc.* **19**, 107–122.
(1969$_b$). "A Radon-Nikodým theorem for Stone algebra valued measures," *Trans. Amer. Math. Soc.* **139**, 75–94.

Wu, C. X. (1962). "A note on decomposition of linear operators," *Natur. Sci. J. Jilin Univ.* **1**, 67–68 (Chinese).

————, S. Z. Zhao and J. A. Chen (1978). "Formulae of Orlicz norm and conditions of rotundity of Orlicz spaces," *J. Harbin Inst. Tech.*, No. 3, 1–12 (Chinese).

Wu, Y. P. (1982). "A criterion for compactness in Orlicz spaces," *Nature Journal (Shanghai)* **5**, 234 (Chinese).

Ye, H. A. (1982). "On some problems of decomposition of operators in Orlicz spaces," *J. China Univ. Sci. Tech.* **2**, 21–24 (Chinese).

Yosida, K. (1971). *Functional Analysis*, 3d ed., Springer-Verlag, New York.

Young, W. H. (1912). "On classes of summable functions and their Fourier series," *Proc. Roy. Soc.* **87**, 225–229.

Zaanen, A. C. (1953). *Linear Analysis*, North-Holland Publ. Co., Amsterdam.
(1967). *Integration* (2d ed.), North-Holland Publ. Co., Amsterdam.
(1981). "Some remarks about the definition of an Orlicz space," *in: Measure Theory* (Oberwolfach), Lecture Notes in Math., No. 945, Springer-Verlag, Berlin, 1982, 261–268.

Zhou, A. H. (1986). "Zygmund's problem and some applications of Orlicz spaces in integration theory," *Natur. Sci. J. Xiangtan Univ.*, No. 4, 8–15 (Chinese).

Zygmund, A. (1959). *Trigonometric Series* (2d ed.), Cambridge Univ. Press, London.

Symbols and Notation

Chapter VIII

M: family of probability measures on (Ω, Σ), 319

$L^\Phi(\theta, G)$: Orlicz space defined by θ and $G(\cdot)$, 319

$N_{\Phi, \theta}(\cdot)$: gauge norm on $L^\Phi(\theta, G)$, 320

$U^\Phi(\theta, G)$: unit ball of $L^\Phi(\theta, G)$, 325

$L^\Phi(\mu, \ell^\theta(G))$: substitution space, 340

$(L^\Phi_{\mathcal{X}}(\theta, G), N_{\Phi, \theta}(\cdot))$: Orlicz space of $\mathcal{X}$-valued functions with (θ, G), 344

$ca(\Omega, \Sigma)$: σ-additive scalar set functions on Σ, 344

Chapter IX

$\tilde{\mathcal{H}}^\varphi$: Hardy-Orlicz class, 355

$\mathcal{H}^\varphi$: Hardy-Orlicz space, 355

$\mathcal{N}$: Nevanlinna class, 355

$\mathcal{N}^+$: subclass of $\mathcal{N}$, 356

$\|\tilde{f}\|_{s\varphi}$: s-homogeneous F-norm, 358

$C[0, 2\pi]$: space of continuous functions on $[0, 2\pi]$, 363

$rca[0, 2\pi]$ space of σ-additive regular signed measures, 363

$\sigma^\alpha_n(F, \cdot)$: Cesáro mean of order α for F, 377

$D^\alpha = D^{\alpha_1}_1 D^{\alpha_1}_1 \cdots D^{\alpha_n}_n$: partial differential operators, 378

$C^\infty_c(\Omega)$: space of infintely differentiable functions on Ω with compact
    supports, 378

$\mathcal{D}(\Omega)$: test function space, 378

$\mathcal{D}'(\Omega)$: space of Schwartz distributions, 378

$N^{m, \Phi}_{\theta, G}(f)$: gauge norm of f, 379

$W^{m, \Phi}_{\theta, G}(\Omega)$: Orlicz-Sobolev space of order m, 380

$\overset{\circ}{W}{}^{m, \Phi}_{\theta, G}(\Omega)$: closed linear span of $C^\infty_c(\Omega)$ in $W^{m, \Phi}_{\theta, G}(\Omega)$, 382

$\mathcal{H}^{m, \Phi}_{\theta, G}(\Omega)$: closed linear span of $C^m(\Omega)$ in $W^{m, \Phi}_{\theta, G}(\Omega)$, 382

$W^{m, \Phi}(\Omega)$: Orlicz-Sobolev space when $G(\alpha) = 1$ and $\theta(x) = |x|$, 384

$N^{m, \Phi}(\cdot)$: gauge norm on $W^{m, \Phi}(\Omega)$, 384

$W^{m, \tilde{\Phi}}_{\theta, G}(\Omega)$: Orlicz-Sobolev space with $\tilde{\Phi} = \{\Phi_\alpha, |\alpha| \leq m\}$, 384

Φ^*_m: m^{th} order Sobolev conjugate of Φ for $I\!R^n$, 385

Θ: $\{\theta_\alpha, |\alpha| = 1, 2, \ldots\}$, 386

Φ: $\{\Phi_\alpha, |\alpha| = 1, 2, \ldots\}$, 386

$W^{\infty, \Phi}_{\Theta, G}(\Omega)$: Orlicz-Sobolev space of infinite order, 386

$N^\Phi_{\Theta, G}(\cdot)$: gauge norm on $W^{\infty, \Phi}_{\Theta, G}(\Omega)$, 386

$L^1_{\text{loc}}(I\!R)$: space of locally integrable functions, 390

$D_u(f, g)$: uniform distance, 390

Index